S0-BMG-606

HEAT TRANSFER IN FLAMES

ADVANCES IN THERMAL ENGINEERING

Editors:
JAMES P. HARTNETT
THOMAS F. IRVINE, JR.

I	Blackshear	•	**Heat Transfer in Fires:** **thermophysics** **social aspects** **economic impact**
II	Afgan and Beer	•	**Heat Transfer in Flames**
III	deVries	•	**Heat and Mass Transfer in the Biosphere:** **plant growth and productivity** **soil effects, ecology and pollution**

A publication of the International Centre for Heat and Mass Transfer

Founding Members:

American Institute of Chemical Engineers
American Society of Mechanical Engineers
Canadian Society for Chemical Engineering
Canadian Society for Mechanical Engineering
Institution of Chemical Engineers, London
Institution of Mechanical Engineers, London
National Committee for Heat and Mass Transfer of the
Academy of Sciences of the U.S.S.R.
Societé Française des Thermiciens
Verein Deutscher Ingenieure
Yugoslav Society of Heat Engineers

Sponsoring Members:

Associazione Termotecnica Italiana
Egyptian Society of Engineers
Indian National Committee for Heat and Mass Transfer
Institution of Engineers of Australia
Israel Institute of Chemical Engineers
Koninklijk Instituut van Ingenieurs, Netherlands
Society of Chemical Engineers of Japan

HEAT TRANSFER IN FLAMES

N. H. AFGAN and J. M. BEER, Editors

SCRIPTA BOOK COMPANY
a division of hemisphere publishing corporation
1974 Washington, D.C.

A HALSTED PRESS BOOK

JOHN WILEY & SONS
New York Toronto London Sydney

Library of Congress Cataloging in Publication Data

Main entry under title:

Heat transfer in flames.

(Advances in thermal engineering, 2)
Selected chapters from the 1973 seminar of the International Centre for Heat and Mass Transfer.
1. Heat–Transmission–Congresses. 2. Mass transfer–Congresses. 3. Flame–Congresses.
I. Afgan, Naim, ed. II. Beér, János Miklós, ed.
III. International Centre for Heat and Mass Transfer.
QC319.8.H43 536'.2 74-8747
ISBN 0-470-00931-4

HEAT TRANSFER IN FLAMES

Scripta Book Company
a division of hemisphere publishing corporation
1025 Vermont Avenue, N.W.
Washington, D.C. 20005

Printed in the United States of America

CONTENTS

FOREWORD

The main theme of this volume is heat and mass transfer from flames and its application to engineering systems.

The practical application of the fundamentals of heat transfer in flames covers a wide range of important engineering systems, which include industrial furnaces, boilers, gas turbine combustors, internal combustion engines and open flames such as flares and fires. In recent years there has been a growing interest in the study of thermal radiation for industrial application. This is due to increasing combustion intensities in industrial applications and the need for prediction methods for detailed distribution of radiant heat flux to bounding surfaces.

A definite need exists for more detailed information on engineering design for heat transfer from flames to heat sinks and also on heat transfer within the flame itself, as the latter affects combustion intensity and the rate at which certain types of flames and fires propagate.

The material presented in this book contains three different aspects of heat transfer from the flame. The first part includes papers which deal with problems of heat transfer in open flames. Special attention is given to the mechanism of fire spread in forests and calculation methods for predicting radiative heat transfer in a diffusion flame. This section also includes a probability analysis of fabric ignition and the burn injury hazard.

The largest part of this volume is devoted to problems of heat transfer in steady confined flames. This group of papers is divided according to subject-matter. It covers radiative properties, measurement techniques, prediction methods (zone and flux methods) and the effect of turbulent fluctuation on radiant heat transfer. The analytical and experimental method for determination of radiative properties of the flame are presented. Special consideration is given to the effect of pressure on the mean absorption coefficient of luminous flames of liquid fuel spray combustion. The theoretical and experimental results of radiative properties in a dispersed system are also offered.

Although there are not many papers dealing with measurement techniques, which are mainly covered under particular topics, the papers presented in this group treat new techniques for radiative heat transfer and their employment in the measurement of scaling deposits in large thermal power stations.

Development of the prediction method for calculation of heat transfer from confined flames is of utmost importance. We are privileged to have Professor H. C. Hottel, the pioneer in radiative heat transfer, to lead this volume with his chapter "First Estimate of Industrial Furnace Performance–One Zone Model Reexamined." Recent advances in methods for predicting radiative heat flux distribution in furnaces and combustors have focused attention on comparison of the zone method and the flux method of analysis. Most of the chapters presented in this part of the volume are directly or indirectly connected with analyzing the advantages of particular methods of predicting radiant heat flux distribution. Those methods have been applied in different industrial systems, and the results of the heat transfer and flow distribution are given for a gas turbine combustion chamber, an industrial water-tube boiler, pulverized fuel furnaces and other types of furnaces.

The third part is devoted to heat transfer in unsteady confined flames. Most of the chapters in this group deal with heat transfer from flames in internal combustion engines. This section reviews the available information of heat transfer from the working fluid to the wall of internal combusion engines. Some chapters deal with the experimental and theoretical method for determination and prediction of radiant heat transfer in internal combustion engines and its correlation with other heat transfer mechanisms.

The scientific organization of the program of the 1973 International Seminar was the responsibility of the Seminar Committee, as follows: W. J. D. Annand, University of Manchester, U.K.; A. Blokh, Central Boiler and Turbine Institute, U.S.S.R.; J. M. Riviera, I.R.S.I.D., France; L. Kreuh, University of Zagreb, Yugoslavia; A. F. Sarofim, Massachusetts Institute of Technology, U.S.A.; P. H. Thomas, Fire Research Station, U.K.; and Chairman J. M. Beer, University of Sheffield, U.K. The Seminar was attended by 112 participants from 20 countries.

Organization of the lectures was financially supported by the following organizations: UNESCO, Paris; Council for Scientific Coordination of the Socialist Republic of Croatia, Zagreb, Yugoslavia; Djuro Djaković enterprise for design and construction of power and industrial plants, Slavonski Brod, Yugoslavia; National Science Foundation, Washington, D.C., U.S.A.; Academy of Sciences of the U.S.S.R., Moscow, U.S.S.R.; Boris Kidric Institute, Belgrade, Yugoslavia.

Acknowledgement should be given to UNESCO for special grants to participants from developing countries. The editors would also like to express their appreciation to the authors for their most helpful cooperation in preparing this volume.

N. H. Afgan

J. M. Beer

PART I: HEAT TRANSFER IN STEADY CONFINED FLAMES

SECTION I:

METHOD OF CALCULATION

Chapter 1

FIRST ESTIMATES OF INDUSTRIAL FURNACE PERFORMANCE – THE ONE-GAS-ZONE MODEL REEXAMINED

Hoyt C. Hottel

Massachusetts Institute of Technology, Cambridge, Mass., U.S.A.

Abstract

The one-gas-zone model of industrial furnace performance is set up, with allowance for wall losses, sink temperature variation, and departure from perfect stirring. Judicious choice of the mode of allowance for the last of these keeps the final relation simple enough to visualize. It is shown that although the model equations are in a form for prediction of furnace performance, they may instead be used in a graphical technique for correlating furnace data and for determining two constants in a semi-empirical relation. A solution of the problems of determining the effective sink area A_S, the mean beam length for gas radiation, and the effective sink emissivity is presented, for tubular heaters with wall-mounted tubes as well as for tubes internally mounted in planes with radiating gas on either side.

NOMENCLATURE

A_O area of furnace openings losing radiation

$A_r(A_p)$ area of refractory (of tube plane)

A_S effective sink area (old $(\overline{GS})_{R,c}$)

A_1 area of stock or sink in furnace chamber. If tubes on wall, plane of tubes.

$A_{1,e}$ that part of A_1 exclusive of any tube curtain across the gas-exit passage.

a_g energy fraction of black-body spectrum occupied by gray gas in a gray plus clear mixture

B ratio of center-to-center tube spacing to diameter

$\underline{C}$ "cold" fraction of furnace enclosure area, $A_1/(A_1+A_r)$

$c_{p,g}$ specific heat of combustion gases, mean value from gas-chamber exit to base temperature

D' reduced firing density, $\dot{H}_F/\sigma A_S T_F{}^4(1-T_O{}')$

d dimensionless constant in relation $\Delta' = (1-1/d)Q'$

$\mathcal{E}_3$ third exponential integral, $\int_1^\infty (e^{-xt}/t^3)dt$

$\underline{E}$ black emissive power, σT^4

$\overline{F}$ factor for radiation loss through walls, based on inside and outside temperatures and area of opening

$\overline{F}_{iso(gas)}$ fraction of isotropic (gas) radiation intercepted by tube row, directly plus by interception of returning beam from background

$\underline{F_{xy}}$ fraction of radiation from surface x which is intercepted by surface y

$\overline{g_x s_y}$ direct-interchange area between gas x and surface y

$\overline{G_x S_y}$ total-interchange area, ratio of net radiative flux between gas zone x and surface zone y, allowing for reflections at all surfaces, to the difference in black emissive powers (σT^4) of x and y. Sometimes, allowance made for refractory aid without appending sub-R.

$(\overline{GS})_R$ total-interchange area between gas and surface zones, with aid given by equilibrium refractory surfaces included.

$H(T_1)$ enthalpy of stream 1 (sink stream) dependent on temperature

$H_{1,i(o)}$ enthalpy of stream 1 at inlet (outlet)

$\dot{H}_F$ enthalpy rate of any entering streams affecting firing rate, including fuel, air, and recirculated flue gas if any, above dead state of completely burned gaseous products at T_O.

h_i convection heat-transfer coefficient on inside surface of refractory walls
$h_{c+r,o}$ heat transfer coefficient by convection plus radiation, on outside walls of refractory surfaces
K absorption coefficient, l^{-1}
k thermal conductivity of refractory walls
L dimension of parallelepiped
L_m mean beam length for gas radiation
L_0' dimensionless loss coefficient for radiative flux through furnace openings, $\overline{F}A_0/A_S$
L_r' dimensionless loss coefficient for heat loss through refractory walls, $U_rA_r/\sigma A_ST_F^3$
m_g mass flow rate of combustion gases
n number of tubes in a tube row
P pitch of tubes in row, center-to-center distance
p partial pressure of gas-radiating components, atm.
$\dot{Q}_G$ rate of heat transfer from combustion gases
Q' reduced rate of heat transfer from combustion gases, $\dot{Q}_G/\dot{H}_F)(1-T_0')$
r ratio of temperature rise of stock surface to the sum of inlet and outlet temperatures, $(T_{1,o} - T_{1,i})/(T_{1,o} + T_{1,i})$.
S speckledness; 1 for surface with A_1 and A_r intimately mixed
$\overline{s_xs_y}$ direct-interchange area for radiative exchange between surfaces x and y no gas absorption included
$\overline{s_xs_y}'$ same as above, except that gas absorption <u>is</u> included
$\overline{xy}(\overline{xy}')$ shorthand for $\overline{s_xs_y}$ $(\overline{s_x/s_y}')$
T_F adiabatic pseudo-flame temperature, based on Eq. following (2)
T_g mean radiating temperature of combustion gases
T_1 mean temperature of stock or sink surface
T_0 base temperature (also ambient)
U_r overall coefficient of heat transfer from combustion gases through refractory wall to ambient
W refractory wall thickness. Also thickness of gas slab, Figs. 7-9
$\overline{1r}$ shorthand for $\overline{s_1s_r}$, $\equiv A_1F_{1r} \equiv A_rF_{r1}$

α gas absorptivity
Δ gas-radiating temperature minus gas temperature leaving combustion chamber
$\varepsilon_g(\varepsilon_1)$ gas emissivity (effective emissivity (emittance, absorptance) of surface A_1)
ε_1' true emissivity of tube surface
η furnace efficiency, (flux to stock or sink)/$\dot{H}_F$
σ Stefan-Boltzmann constant
τ gas transmissivity ($\equiv$1-absorptivity, $=1-\varepsilon_g$ if gas gray)
ψ angle. See Fig. 7

<u>Subscripts</u>

i,o inlet, outlet
r refractory
1 sink surface, or stock surface
T total, applied to area

<u>Primes</u>

On T_g, T_1, T_0 designate the ratio of those temperatures to T_F.

Introduction

The design of a furnace, more specifically the prediction of the performance of a chosen design, can be carried out at several levels of sophistication. Although a determination of the distribution of heat-flux density over the surface of the stock is desirable — sometimes, in high flux-density systems, necessary — the attainment of the simpler objective of determining the total heat transfer rate as a function of firing rate and excess air is a proper orienting first step; and often it suffices. Even if that is the sole objective, knowledge of the detailed interaction of radiation and convection with mass transfer and combustion is in principle necessary. But integral formulations are tolerant of casual treatment of detail, especially in the presence of the leveling effect of radiation, responsive to a high power of temperature; and a surprisingly accurate overall performance is predictable from a relatively simple model. Even though the knowledge of flux distribution over the stock may be the ultimate objective, it is still good engineering practice to start with an almost-quantitative understanding of the overall process. In fact, it may be asserted that prospects of success with the zone method are poor if the simpler and less ambitious approach, which is after all a one-gas-zone example of the zone method, is not thoroughly understood.

It is the object here to set up a simple overall furnace performance model, in form as general as is consistent with the assumption of a single gas-radiating temperature, a single equilibrium refractory temperature and a single term characterizing the exchange area between the combustion gases and the sink, which allows for

1. The effect of the adiabatic flame temperature T_F, dependent on the entering fuel and air enthalpies and the gas heat capacity
2. The effect of stock or sink temperature T_1, measured by the ratio of its mean value to T_F
3. The effect of stock temperature variation, measured by the ratio r of stock temperature rise to its arithmetic mean temperature.
4. The value of the characteristic or effective sink area A_S, that area which, multiplied by the difference of black-body emissive powers of the gas and stock temperatures, gives the flux from gas to stock. The major problem of making the model describe realistically the effects of gas composition, furnace shape, and disposition of heat sinks in the furnace comes in the evaluation of A_S, the total gas-sink exchange area

5. The use of a single gas-radiating temperature T_g, but
6. a difference Δ between the gas-radiating and leaving-gas-enthalpy temperature, which varies with firing rate
7. The loss of heat through the refractory walls
8. The loss of heat by radiation through openings
9. Other factors, contributing to the evaluation of A_S.

The One-Gas-Zone Furnace Model

Although parts of the following development have appeared before[1,2a], parts are new; and for completeness the full derivation will be presented. The well stirred furnace gases are at temperature T_g in consequence of loss of heat (a) by radiative exchange with the stock or sink at T_1, (b) by convection to that part of the sink $A_{1,e}$ which does not include any curtain tubes across the gas exit from the chamber, (c) by convection to and through the refractory walls, and (d) by radiation through furnace openings of area A_o.

a) The net radiative flux to the sink — direct as well as with the aid of refractory surfaces which reflect diffusely or absorb and reradiate — must, if the refractory is radiatively adiabatic and the gas is gray, be proportional to the difference in black-body emissive power of the gas and sink. The proportionality constant, having the dimensions of area, is called the gas-surface total-exchange area $(\overline{GS})_R$ the formulation of which will be discussed later. (The subscript indicates that allowance has been made for the aid given by refractory surfaces). The flux is then $(\overline{GS})_R\ \sigma(T_g^4-T_1^4)$.

b) Convective flux to those surfaces A_{1e} which affect the stirred-gas enthalpy is $hA_{1,e}(T_g-T_1)$. Because this term is quite small compared to (a), it is convenient to combine the two by forcing the convection into a fourth-power form: $hA_{1,e}(T_g-T_1) \sim hA_{1,e}\sigma(T_g{}^4-T_1{}^4)/4\sigma T_{g1}^3$, where T_{g1} is the arithmetic mean of T_g and T_1. Then

$$(\overline{GS})_R\ \sigma(T_g{}^4-T_1{}^4)+hA_{1,e}(T_g-T_1) = [(\overline{GS})_R + hA_{1,e}/4\sigma T_{g1}^3]\ \sigma(T_g{}^4-T_1{}^4)$$

The bracket has in other contributions been called $(\overline{GS})_{R,c}$ to indicate that convection has been included. The simpler term A_S, the effective area of the sink, will be used here.

c) Convection to and through refractory walls. If the walls are in radiative equilibrium, convection — gas to wall — equals conduction through the wall. The flux is $U_rA_r(T_g-T_o)$, where T_o is the ambient temperature and U_R is given, conventionally, by $U_r = \dfrac{1}{\dfrac{1}{h_i} + \dfrac{W}{k} + \dfrac{1}{h_{c+r,o}}}$

The inside flux is not in fact equal to the flux through the wall, but the difference is so small compared to the radiative flux as hardly to negate the assumption of radiative equilibrium. Without that assumption one would need to introduce an additional unknown and an additional equation, and the slight improvement in final accuracy does not justify the complication.

d) Radiation through peep holes or other openings, of area A_O. Rigorous allowance for this usually small effect would introduce such complexities as to prevent obtaining a solution capable of easy engineering use. Although the view from the outside through furnace openings is a view of sink and refractory surfaces seen dimly through partly diathermanous gas, the assumption will be made that the effective furnace temperature (the inside plane of the openings) is T_g. With $\overline{F}$ representing the exchange factor to allow for wall thickness ([3]), the loss through the openings becomes $A_O\overline{F}\sigma(T_g{}^4-T_o^4)$. Furnaces with openings large enough to make this casual treatment inadequate are rare.

The equation of transfer from the gas is, from the above,

$$\dot{Q}_G = A_S\sigma(T_g{}^4-T_1{}^4) + U_rA_r(T_g-T_o) + \overline{F}A_o\sigma(T_g{}^4-T_o{}^4) \tag{1}$$

An energy balance on the gas is needed. Although a single gas radiating temperature has been postulated, it can be a space-mean value rather than the uniform gas temperature of a perfectly stirred chamber; and the gas temperature measuring the gas enthalpy leaving the chamber is usually lower. Let $T_g-\Delta$ represent the leaving-gas temperature, between which and the base temperature T_o the mean heat capacity is $\overline{C}_{p,g}$. Then the energy balance is

$$\dot{Q}_G = \dot{H}_F - (T_g-\Delta-T_o)\dot{m}_g\overline{C}_{p,g} \tag{2}$$

where $\dot{H}_F$ is the hourly entering enthalpy, chemical plus sensible, in the fuel and air and recirculated flue gas, if any, T_o is the enthalpy-base temperature, and $\dot{m}_g$ is the mass flow rate of gas/hour. Let the same mean heat capacity be used to define an adiabatic pseudo-flame temperature T_F

$$(T_F-T_o) = \dot{H}_F/\dot{m}_g\overline{C}_{p,g}$$

(T_F will in general be much higher than the true adiabatic flame temperature which allows for a temperature-varying C_p and for dissociation). The energy balance may then be written in the form

$$\frac{\dot{H}_F-\dot{Q}_G}{\dot{H}_F} = \frac{T_g-\Delta-T_o}{T_F-T_o} \tag{2a}$$

The additional relation needed is that giving furnace efficiency η.

$$\eta = \frac{\dot{Q}_G - \text{wall losses}}{\dot{H}_F} = \frac{\dot{Q}_G - [U_rA_r(T_g-T_o) + \overline{F}A_o\sigma(T_g{}^4-T_o^4)]}{\dot{H}_F} \tag{3}$$

Equations (1) and (2a) contain as unknowns T_g and $\dot{Q}_G$ (assuming rules available for finding $(\overline{GS})_R$ and choosing Δ). Solution for these and insertion into (3) gives the furnace efficiency. But much more understanding of the nature of furnaces can be obtained by further manipulation. Let (1) be made dimensionless by division through by $A_S\sigma T_F^4$ and let the dimensionless ratios of various temperatures to T_F be denoted by their primes.

$$\frac{\dot{Q}_G}{A_S\sigma T_F{}^4} = T_g'^4 - T_1'^4 + \frac{U_rA_r}{\sigma A_S T_F^3}(T_G' - T_o') + \frac{\overline{F}A_o}{A_s}(T_g'^4 - T_o'^4) \tag{4}$$

Equation (2a) may be written

$$\frac{\dot{Q}_G}{\dot{H}_F}(1-T_o') = 1 - T_g' + \Delta' \tag{5}$$

where $\Delta' = \Delta/T_F$. To complete the normalization, let

$\frac{\dot{Q}_G}{\dot{H}_F}(1-T_o') = Q'$, the reduced gas efficiency

$\frac{\dot{H}_F}{\sigma A_S T_F^4(1-T_o')} = D'$, the reduced firing density

$\frac{U_rA_r}{\sigma A_S T_F{}^3} = L_r'$ the refractory loss factor

$\frac{\overline{F}A_O}{A_S} = L_O'$ the furnace-opening loss factor

Equations (4) and (5) then become

$$Q'D' = T_g'^4 - T_1'^4 + L_R'(T_g' - T_o') + L_O'(T_g'^4 - T_o'^4) \tag{6}$$

and

$$Q' = 1-T_g' + \Delta' \tag{7}$$

When loss factors L_r' and L_O' are small enough to be neglected and Δ' is assumed zero, Q' becomes the reduced furnace efficiency — the furnace efficiency times $(1-T_O')$ — given by solution of the equation

$$Q'D' = (1-Q')^4 - T_1'^4 \tag{8}$$

This extraordinarily simple relation, giving furnace efficiency as a function of two dimensionless parameters, D', proportional to the firing rate per unit of effective sink area, and T_1', the ratio of sink temperature to adiabatic flame temperature, is the basic relationship governing the efficiency of furnaces of almost any class. Performance data on furnaces of a wide variety of types, from gas turbine combustors to openhearth furnaces, can be put on a diagram of efficiency versus firing rate,with sink temperature as a parameter (1,2b)(efficiency here refers to transfer of heat from the combustion gases rather than to the stock.)

Better agreement between prediction and experiment, however, may be expected if allowance is made for some difference between mean-radiating temperature and leaving enthalpy temperature. Experience with furnaces of various types as well as with computations based on the multi-zone model indicate that Δ varies inversely with H_F, and the assumption that Δ' is proportional to Q' is much more realistic than treating Δ as constant. For a reason which will emerge later let the proportionality constant be (1-1/d). (There is some evidence that d is about 4/3, putting Δ in the range of 150 to 240C for heavily fired cracking coils or marine boilers). The energy balance Equation (7) then becomes

$$T_g' = 1 - \frac{Q'}{d} \tag{9}$$

An additional modification is desirable. In many modern high-output furnaces — e.g., reformers and ethylene furnaces — the stock is often heated through a significant temperature interval; and the stock temperature leaving the combustion chamber may come to within a few hundred degrees of the leaving-gas temperature. It may readily be shown that, if the stock has a constant specific heat and T_1 rises from $T_{1,i}$ to $T_{1,o}$ within the combustion chamber, the term $(T_g'^4 - T_1'^4)$ in Eq. (6) should be replaced by

$$\frac{8T_g'^3 T_1' r}{\ln \dfrac{T_g' - T_1'(1-r)}{T_g' + T_1'(1-r)} \dfrac{T_g' + T_1'(1+r)}{T_g' - T_1'(1+r)} + 2\left(\tan^{-1}\dfrac{T_1'(1+r)}{T_g'} - \tan^{-1}\dfrac{T_1'(1-r)}{T_g'}\right)} \tag{10}$$

where T_1 is now $(T_{1,i} + T_{1,o})/2$ and $r = (T_{1,o}-T_{1,i})/(T_{1,o}+ T_{1,i})$. As before, $T_1' = T_1/T_F$. In the limit as $r \to 0$, (10) approaches $(T_g'^4 - T_1'^4)$.*

---

*Although chemical or phase change within the stock invalidates the assumption of constant specific heat, it is always possible to find an equivalent r to make (10) numerically correct. With $H(T_1)$ representing the stock enthalpy as a function of T_1 varying from $H_{1,i}$ at entry to $H_{1,o}$ at exit, the true value of $\dot{Q}_{stock}/\sigma A_S$ is

$$(H_{1,o} - H_{1,i}) \Big/ \int_{H_{1,i}}^{H_{1,o}} \frac{dH(T_1)}{T_g^4 - T_1^4} \quad ,$$

readily evaluated graphically. The result must equal T_F^4 times (10), or (10) with the primes missing. For the T_g and T_1 of interest, r can be found by trial and error. Since the ratio of r defined as $(T_o-T_i)/(T_o+T_i)$ to r from the above equality depends on the H-T_1 relation and on T_G/T_1 and the latter is substantially constant for a particular furnace type , the ratio of the two r's requires but infrequent determination. It is to be remembered that if the stock is a fluid in a tube, T_1 in the H-T_1 relation is the outer tube surface temperature, not the bulk temperature of the fluid. Enough is generally known in advance to construct an H-T_1 diagram allowing for temperature drop through the fluid film and tube wall.

With (10) replacing $(T_g'^4 - T_1'^4)$ in (6) and with T_g' replaced by its value from (8), Equation (6) becomes

$$\left(\frac{Q'}{d}\right)(dD') = \frac{8\left(1-\frac{Q'}{d}\right)^3 T_1' r}{\ln \dfrac{\left(1-\frac{Q'}{d}-T_1'(1-r)\right)}{1-\frac{Q'}{d}+T_1'(1-r)} \cdot \dfrac{1-\frac{Q'}{d}+T_1'(1+r)}{1-\frac{Q'}{d}-T_1'(1+r)} + 2\left(\tan^{-1}\dfrac{T_1'(1+r)}{1-Q'/d} - \tan^{-1}\dfrac{T_1'(1-r)}{1-Q'/d}\right)}$$

$$+ L_r'\left(1-\frac{Q'}{d}-T_o'\right) + L_o'\left[\left(1-\frac{Q'}{d}\right)^4 - T_o'^4\right] \qquad (11a)$$

with $r = 0$, this reduces to

$$\left(\frac{Q'}{d}\right)(dD') = \left(1-\frac{Q'}{d}\right)^4 - T_1'^4 + L_r'\left(1-\frac{Q'}{d}-T_o'\right) + L_o'\left[\left(1-\frac{Q'}{d}\right)^4 - T_o'^4\right] \qquad (11b)$$

The error in use of (11b) rather than (11a) is less than one percent when $T_1/T_g = 0.8$ and $r < 0.05$ or $T_1/T_g = 0.9$ and $r < 0.02$. r is often many times these values and the error mounts rapidly.

If loss coefficients L_r' and L_o' are both zero, (11b) has the same structure as (8), with (Q'/d) replacing Q' and $(D'd)$ replacing D'. Thus the form of relation chosen for Δ has permitted inclusion of allowance for it without increasing the number of dimensionless groups.

The furnace efficiency η, when there are wall losses, may be shown from (3) to take the form

$$\frac{\eta}{d} = \frac{Q'/d}{1-T_o'} - \frac{L_R'}{(dD')}\left(1-\frac{Q'/d}{1-T_o'}\right) - \frac{L_o'}{(dD')(1-T_o')}\left[\left(1-\frac{Q'}{d}\right)^4 - T_o'^4\right] \qquad (12)$$

The desired relation between efficiency η and reduced firing density D' — or rather between η/d and $D'd$ — is obtainable from (11a or b) and (12) considered as parametric equations in Q'/d. They express the relation

$$\eta/d = f(dD',\ T_1'(\text{or } T_1' \text{ and } r),\ L_r' \text{ and } L_o') \qquad (13)$$

which is in a form involving no commitment as to what value will be assigned to d, the measure of difference between radiating and enthalpy temperature of the combustion gases.

Overall Furnace Performance - Graphical Presentation

To indicate the character and use of relation (13), three graphs have been presented. Fig. 1 shows, on logarithmic coordinates, the consequences of allowing for wall losses, for the case of L_o' and r both zero, and for $T_o'=1/8$, $T_1'=0.5$, and $L_r'=0.02$ (a realistic wall-loss coefficient). The lower curve is the furnace efficiency, going to 0 when the normalized firing density drops to 0.015,* passing

* $(dD')_{\eta=0} = [L_r'(T_1'-T_o') + L_o'(T_1'^4 - T_o'^4)]/(1-T_1')$; $(Q'/d)_{\eta=0} = 1 - T_1'$

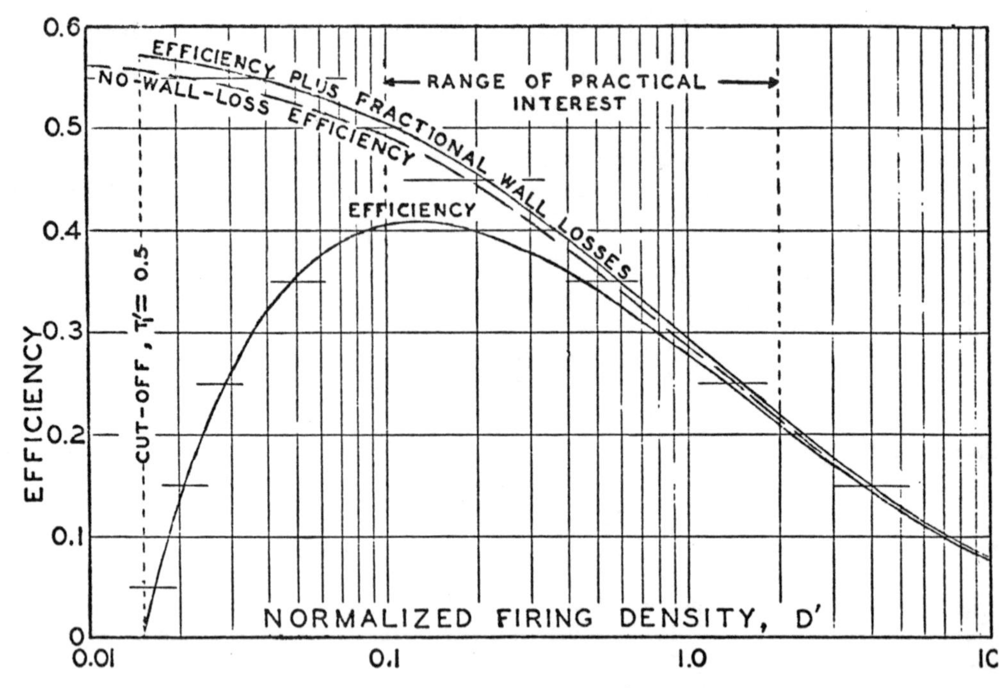

Fig. 1. Effect of Wall-Loss Factor L_r' on Combustion Chamber Performance. L_r'=0.02; T_1'=0.5; T_0'=1/8

through a maximum, and droping at high firing rate. The top curve is the sum of efficiency and fractional wall loss. The middle curve is the no-wall-loss efficiency, which increases continuously as the firing rate drops, and at $D' \to 0$ is asymptotic to $(1-T_1')/(1-T_0')$.

Figure 2 shows, on arithmetic coordinates, the (η/d) versus D'd relation, and gives the effect of sink temperature (T_1' = 0.2 to 0.7) and L_r(0.02 and 0.04).

Much more useful is the same picture on logarithmic coordinates, Figure 3. Its importance is that in this form the relationship is usable for correlating and extrapolating furnace data without concern over the sometimes difficult technique of evaluating accurately the quantity A_S. Suppose that furnace performance data are available in the form of efficiency η versus firing rate H_F. The quantitaties T_0, T_1, T_F, U_rA_r, A_0, $\overline{F}$ are known or readily calculable, and a value of A_S sufficiently good for determining the not too important quantities L_r' and L_0 should also be calculable. Let the efficiency η be plotted vs $H_F/\sigma T_F^4(1-T_0')$ on translucent logarithmic paper which matches Fig. 3, and let the plot be superimposed on that equivalent of Fig. 3 which has been constructed for the values of r and L_0' which characterize the furnace. Let the plot be displaced vertically and horizontally until the data fit the proper T_1' and L_r' curves. The relative vertical displacement of the two plots yields the value of d; their relative horizontal displacement yields the value $D'd/[H_F/\sigma T_F^4(1-T_0)]$, which is d/A_S. The two quantities d and A_S are a characterization of the furnace; one is an empirical constant taking account of Δ, the other a measure of the many factors affecting radiative exchange. They permit a computation of what will happen under other furnace operating conditions (provided those don't change the general character of the flow pattern or the system's radiation geometry); and when determined for each of several furnaces in a particular class they lead to an empirical determination of the effects of design variables on performance.

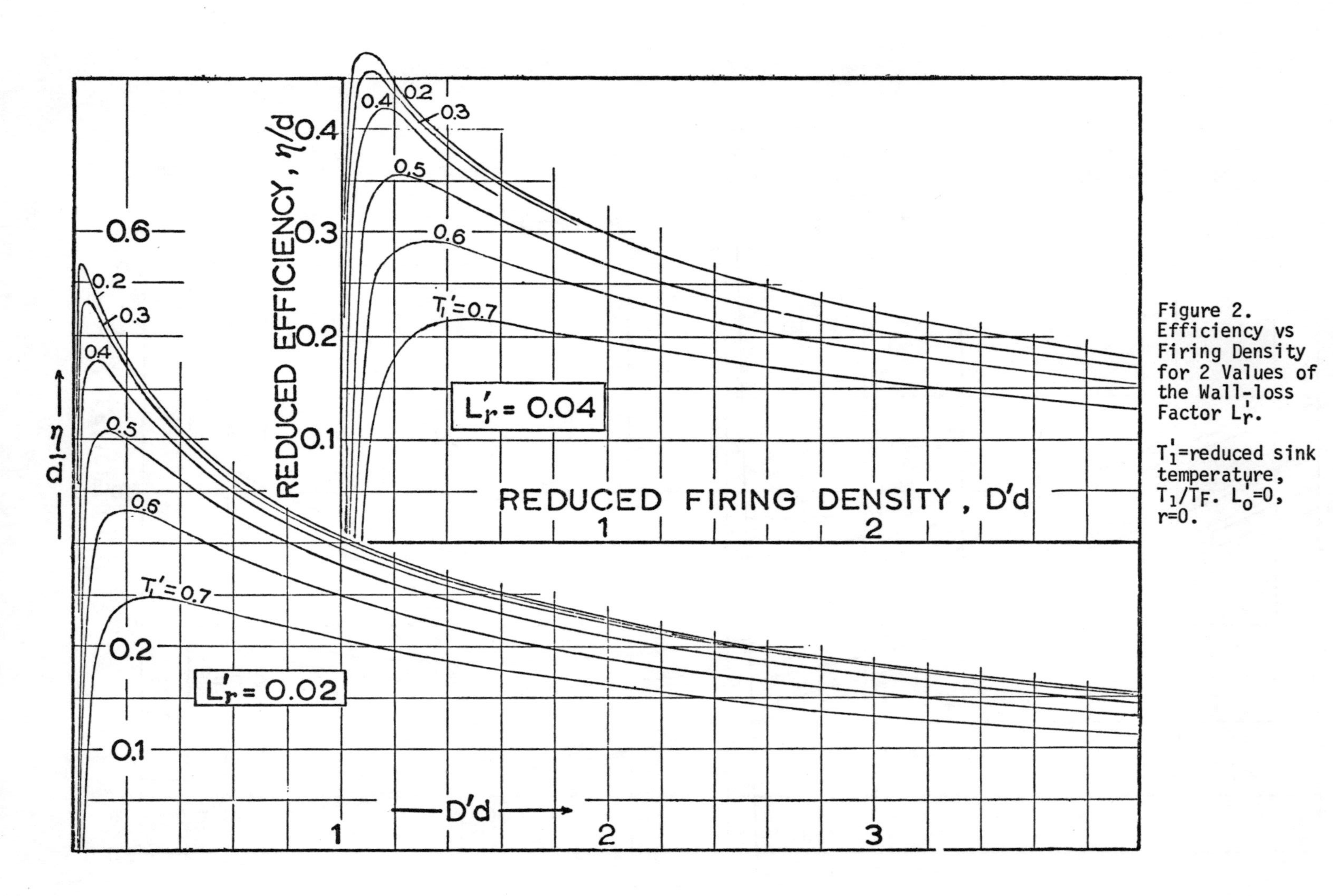

Figure 2. Efficiency vs Firing Density for 2 Values of the Wall-loss Factor L'_r.

T'_1=reduced sink temperature, T_1/T_F. L'_o=0, r=0.

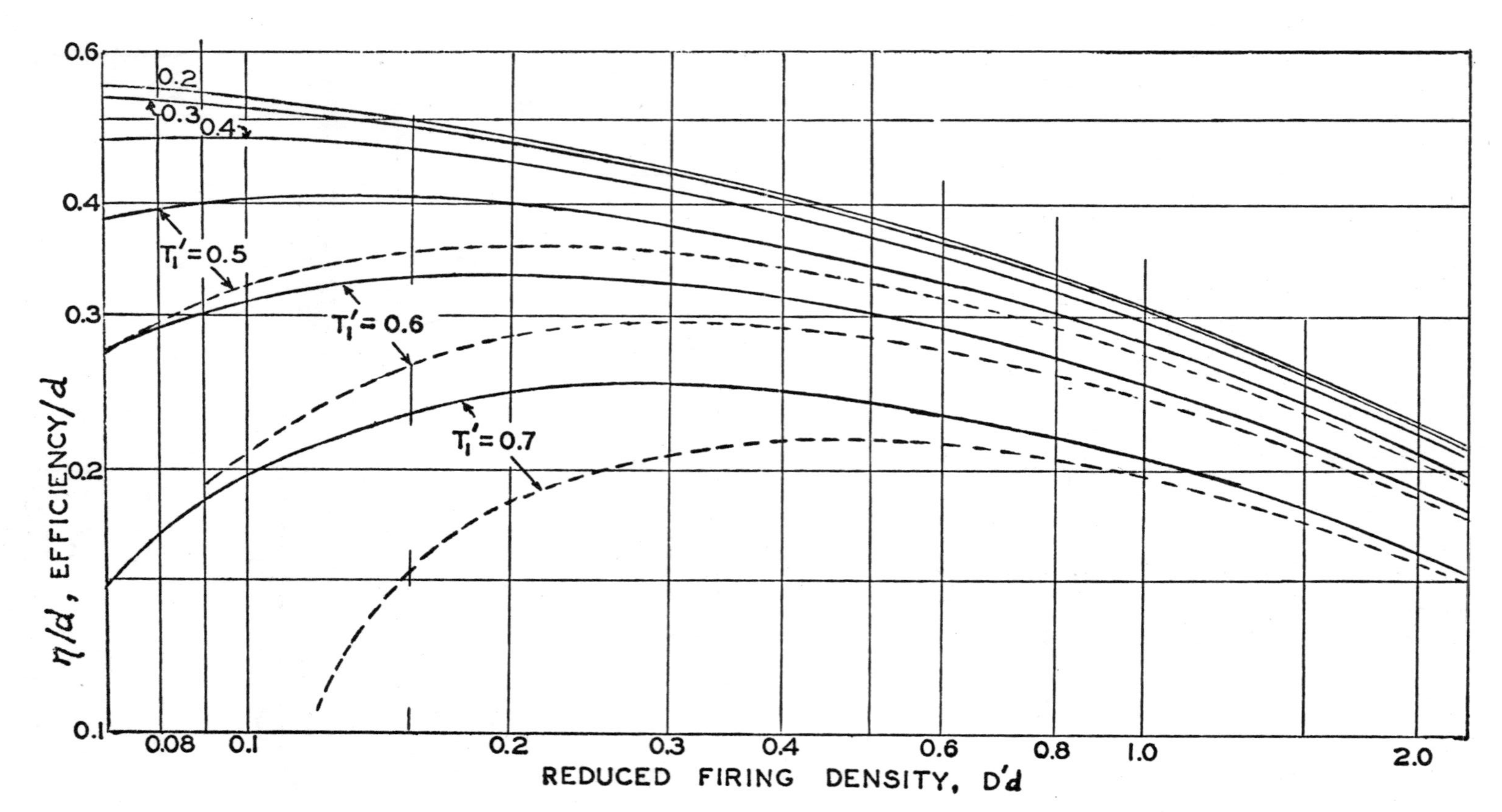

Figure 3. (η/d) vs (D'd) for furnace chambers. For comparison with data on efficiency vs. firing rate, to determine furnace performance characteristics d and A_e by curve fitting. Conditions: $T_0'=1/8$, $L_0'=0$, $r=0$, $L_r'=0.02$ (solid lines) and 0.04 (dotted lines).

For predictions requiring no prior data, however, it is necessary to be able to calculate A_S from first principles, and this depends primarily on the total-exchange area the evaluation of which will now be considered.

The Total-exchange Area

Given: a gray isothermal gas volume enclosed by an isothermal sink surface A_1 and by a refractory surface A_r in radiative equilibrium. The net flux from gas to sink is $\overline{GS}_1\sigma(T_g^4 - T_1^4)$ where $\overline{GS}_1$ is the total-exchange area. The restriction must be imposed that the gas and two surfaces are each treated as a single zone, implying not that all of surface A_1, for example, is segregated into a single area but that a single mean view that A_1 has of A_r can be used in evaluating all of the radiation emitted or reflected from A_1 toward A_r. The total-exchange area $\overline{GS}$ allows for multiple reflections at all surfaces and for assistance given by the refractory in absorbing gas radiation and reradiating a part of it to the sink. In the model under discussion it is clear that $\overline{GS}$ carries a major burden of making the model agree with reality. Many degrees of complexity exist in evaluating $\overline{GS}$, and the engineer has the choice of advancing as far along the path as the importance of his problem warrants.

It may be shown that the total-exchange area depends on sink emissivity and on the inter-zone direct-exchange areas (lower case letter pairs)

$$\overline{GS}_1 = \cfrac{1}{\cfrac{1-\varepsilon_1}{A_1\varepsilon_1} + \cfrac{1}{\overline{gs}_1 + \cfrac{1}{\cfrac{1}{\overline{gs}_r} + \cfrac{1}{\overline{s_r s}_1}}}} \tag{14}$$

with $\overline{gs}_1 \equiv A_1[\varepsilon_g(L_{m,1})]$ and $\overline{gs}_r \equiv A_r[\varepsilon_g(L_{m,r})]$ and $\overline{s_r s}_1 \equiv A_r F_{r1}\{1-[\varepsilon_g(L_{m,r1})]\}$. Here the gas emissivity ε_g is written to indicate its dependence on mean beam length L_m, and the latter in turn to indicate that its value depends on the source and sink of the radiation. If the three ε_g's are assumed representable by a single gas emissivity applicable to the whole enclosure, (14) becomes

$$\overline{GS}_1 = \cfrac{A_1}{\cfrac{1}{\varepsilon_1} - 1 + \cfrac{1}{\varepsilon_g\left(1 + \cfrac{A_r/A_1}{1 + \varepsilon_g/(1-\varepsilon_g)F_{r1}}\right)}} \tag{15}$$

This ancient relation* was replaced early by a simplified version (5) based on what later became known as the speckled-furnace approximation. If A_1 and A_r are intimately mixed over the surface of the enclosure rather than more or less segregated, the view factor either surface has of A_1 is $A_1/(A_1 + A_r)$. Substitution of this value for F_{r1} in (15) together with replacement of A_1 by CA_T and A_r by $(1-C)A_T$, where C is the fraction of the total envelope area A_T which is "cold" (i.e., which is A_1), yields the very simple relation

$$\overline{GS}_1 = \frac{A_T}{\frac{1}{C\varepsilon_1} + \left(\frac{1}{\varepsilon_g} - 1\right)} \tag{16}$$

*Its black-sink equivalent appeared in 1928(4), perhaps earlier.

If instead of A_1 and A_r being intimately mixed A_1 lies in a single plane, F_{1r} becomes 1 and, since $A_1F_{1r} \equiv A_r/F_{r1}$, F_{r1} is A_1/A_r or $C/(1-C)$. Substitution of this into (15) yields a result which may be shown to have the structure of (16) except that the parenthesis in the denominator is now multiplied by

$$\left(\frac{1-\varepsilon_g}{1-C\varepsilon_g}\right) \tag{16a}$$

The object of converting (15) to (16) or to (16) modified has of course been to dodge the often tedious formulation of F_{r1} (or F_{1r}). From the two results which appear to be limit cases it is tempting to seek an interpolation procedure, i.e., to find a function by which to multiply the parenthesis in (16) which varies from 1 for speckled furnaces to (16a) for the case of A_1 in a single plane.

It may be shown that Eq. (15), with A_1 and A_r replaced by CA_T and $(1-C)A_T$, becomes

$$\frac{GS_1}{A_1} \equiv \frac{GS_1}{CA_T} = \frac{1}{\dfrac{1}{\varepsilon_1} + \left(\dfrac{1}{\varepsilon_g} - 1\right)C\left[\dfrac{\varepsilon_g + F_{r1}(1-\varepsilon_g/C)}{C\varepsilon_g + F_{r1}(1-\varepsilon_g)}\right]} \tag{17}$$

The bracketed term can be written in a form more nearly symmetrical with respect to C and (1-C) by substitution of the exchange area $\overline{1r}$ for $A_1F_{1r}(\equiv A_rF_{r1})$ to give

$$\left[\frac{\varepsilon_g/C + (1-\varepsilon_g/C)\overline{1r}/A_TC(1-C)}{\varepsilon_g + (1-\varepsilon_g)\overline{1r}/A_TC(1-C)}\right] \tag{18}$$

The term takes the three limiting (?) forms

(a)	(b)	(c)
If A_1 lies in a single plane, $(C\leq 1/2)$, substitution of $F_{1r} = 1$ yields	If the walls are speckled, $(0<C<1)$ substitution of $F_{r1} = C$ or $F_{1r} = 1-C$ yields	If A_r lies in a single plane, $(C>1/2)$ substitution of $F_{r1} = 1$ yields
$[\ ] = \dfrac{1-\varepsilon_g}{1-C\varepsilon_g}$	$[\ ] = 1$	$[\ ] = \dfrac{1-\varepsilon_g(1-C)/C}{1-\varepsilon_g(1-C)}$

The first two of these three results have already been given. Since F_{1r} changes from (1-C) for case (b) above — the speckled system — to 1 for A_1 segregated in a plane, let us make the heuristic assumption that a term S — called the speckledness — measures the shift from complete speckledness to A_1 in a plane, according to the relation

$$\left.\begin{aligned} &F_{1r} = 1 - SC \\ &\text{or} \\ &\overline{1r}/A_T = C(1 - SC) \end{aligned}\right\} \quad C \leq 1/2 \tag{19}$$

with S = 1 for a speckled enclosure and 0 for A_1-segregation. Substitution of this into the bracket of (17) gives

$$[\ ] \equiv \left[\frac{1-\varepsilon_g - S(C-\varepsilon_g)}{1-C\varepsilon_g - SC(1-\varepsilon_g)}\right] \tag{20}$$

Values of S=1 and 0 substituted into (20) yield the results of cases (b) and (a) above.

Similarly, since F_{r1} changes from C for the speckled system to 1 for A_r in a single plane, we make the parallel assumption that S' measures the shift from speckledness to A_r-segregation according to the relation

$$\left.\begin{aligned} F_{r1} &= 1 - S'(1-C) \\ &\text{or} \\ \overline{1r}/A_T &= (1-C)\big(1-S'(1-C)\big) \end{aligned}\right\} \quad C \geq 1/2 \tag{21}$$

with S' = 1 for a speckled enclosure and 0 for A_r in a plane. Substitution of this into the bracket of (17) give

$$[\ ] \equiv \left[\frac{\varepsilon_g + (1-\varepsilon_g/C)\big(1-S'(1-C)\big)}{\varepsilon_g C + (1-\varepsilon_g)\big(1-S'(1-C)\big)}\right] \tag{22}$$

Values of S'=1 and 0 substituted into (22) yield the results of cases (b) and (c) above.

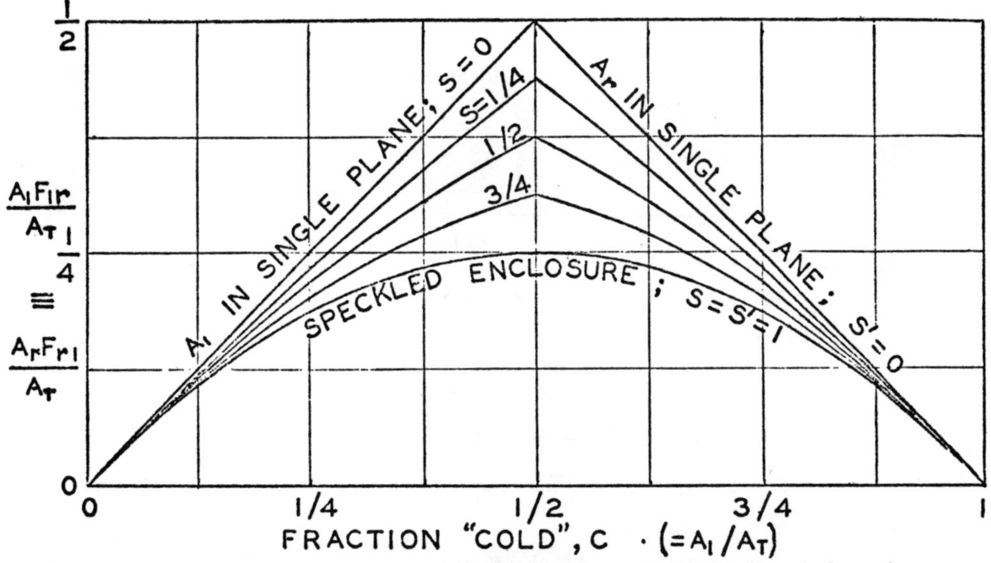

Figure 4. Evaluation of Ratio of Sink-refractory Direct-exchange Area to Total Furnace Envelope Area, in terms of Fraction Cold C, and Speckledness S.

On Figure 4 the value $\overline{Ir}/A_T$ appears as a function of C for the three limit values:

speckled-wall enclosure, lower parabola (S=S'=1)
A_1 in a plane, upper left straight line (S=0)
A_r in a plane, upper right straight line (S'=0)

From Equations (19) and (21) values of $\overline{Ir}/A_T$ appear on Fig. 4 at S and S' of 1/4, 1/2, 3/4. The question arises as to the meaning of these curves intermediate between S(or S') = 1 and 0 where the meanings are quantitatively identifiable. Asked another way, does the S-concept have utility in quick identification of $\overline{Ir}$ for use in (17) to evaluate $\overline{GS_1}$? How significant is an error in $\overline{Ir}/A_T$?

An examination of a few geometrical shapes is illuminating:

1. Spheres. The view that A_1 has of A_r depends on areas only, not on location. ($A_1F_{1r} = A_1A_r/A_T$). No matter what the disposition of surfaces or their degree of segregation, S=1=S', and all spherical enclosures are in the "speckled" category.

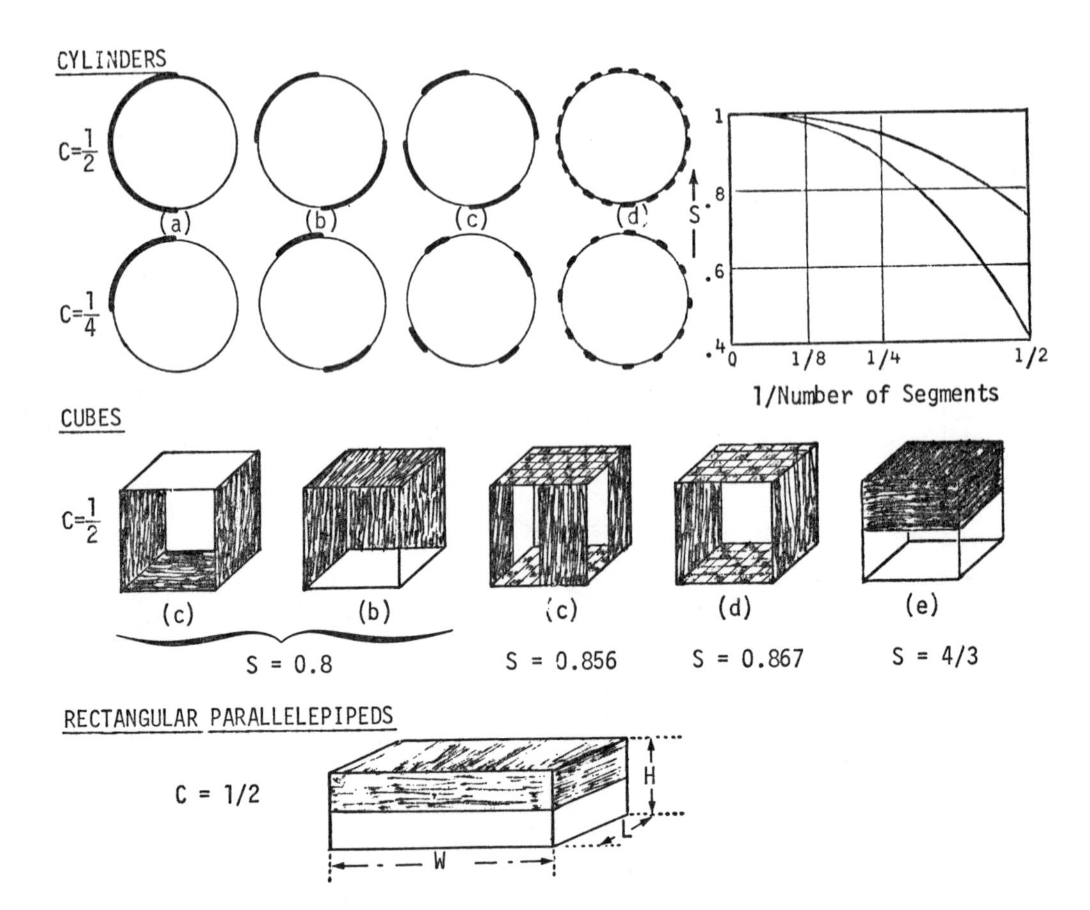

Figure 5. Effect of Division of the Surface of Various Enclosures into A_1 (sink) and A_r (refractory) on the Speckledness Necessary to Predict the Exchange Area $A_1F_{1r}(\equiv A_rF_{r1})$.

2. Long cylinders. Fig. 5 shows two sets of cylinders, C=1/2 and 1/4, with A_1 successively divided into smaller segments symmetrically disposed. The value of S necessary to give the known value of $\overline{Tr}/A_T$ is plotted versus reciprocal of the number of segments into which the surface is divided. Note that division of A_1 and A_r into only two segments each is almost sufficient to put S above 0.9.

3. Cubes, C=1/2. When three of the six faces are A_1, the only two possible arrangements will yield an S of 0.8, cases (a) and (b). Case (c) shows the four sidewalls symmetrically banded, with A_1 occupying one-half the area, and the roof and floor speckled. The overall S is 0.856. When this case is modified to (d), with A_1 occupying all of two opposite sides and a speckled half of the roof and floor, S=0.867. But when a value of C=1/2 is achieved (e) by a plane parallel to one of the faces, with all A_1 on one side of it, S=4/3. Values of S of 0 and 1 are seen not to be bounding values (see next case).

4. Rectangular parallelepipeds, C=1/2, A_1 and A_r on either side of a plane parallel to one face. It may be shown that for this case $S=2-2/[1+H(1/W + 1/L)]$. The larger S the poorer the performance. When H/(W+L) is small, S is small, zero in the limit as H→0(the case of infinite parallel planes). Under these conditions A_r makes a maximum contribution to transfer of heat to A_1. As H increases to 1/2W, with L=W, S becomes 1. With H great compared to W or L, S is increasingly greater than 1. Actual furnaces of such geometry as to make S>1 are rare, and should be.

A study of the cases presented justifies the conclusion that for most furnace chambers S, which is near 1 unless C is small(when A_1 may be in a single plane), may generally be estimated within 0.2 on the basis of a qualitative examination of the chamber. To examine the accuracy to which S need be established, $\overline{GS_1}/A_1$ was evaluated for the case of ε_1=0.9 and ε_g=0.3, realistic values for cracking coils and reformers. Figure 6 shows curves for S=S'=1 and for S=0(C≤1/2) and S'=0(C≥1/2). Since the maximum difference between the two curves is only about 10% at C = 0.4, it is clear that estimation of S to within 0.2 should be adequate for most design purposes, and that in consequence it should rarely be necessary to evaluate F_{r1} rigorously.

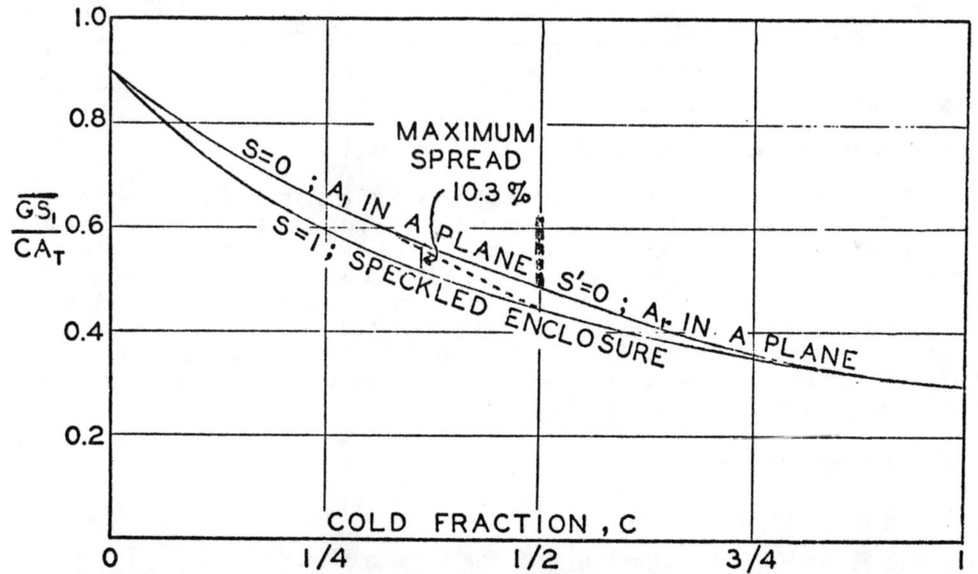

Figure 6. Example of Effect of Variation of Speckledness from 0 to 1 on the Relation Between Total Gas-Sink Exchange Area and "Cold" Fraction of a Furnace Chamber. (ε_1=0.9; ε_g=0.3).

It is interesting to note that Lobo and Evans [6], in an early application of the approximate equivalent of Eq. (15) to cracking-coil furnaces, recommended a procedure which can be shown to be the equivalent, in Fig.6, of assuming that $\overline{GS}_1/C\,A_T$ is uniquely determined by C. The relation recommended was the equivalent of assuming that S = 0 for C up to 1/3, that S undergoes transition from 0 to 1 as C varies from 1/3 to 1/2, and that S = 1 from C = 1/2 to 1. The transition is shown as a dotted line on Fig. 6. On Fig. 4, the Lobo and Evans recommendation is equivalent to following the top line from the left at C = 0 to C = 1/3, dropping from there to the S=1 curve at C = 1/2.

Effect of Non-gray Gas on $\overline{GS}$

Although refractory surfaces are in overall radiative equilibrium, they are capable of absorbing radiation from a gas and then reradiating it with a different spectral distribution, some of the radiation passing through the spectral windows of the gas directly to the sink A_1. The mixed-gray-gas assumption[7,2c] is consequently more realistic than the gray-gas assumption on which Eqs. (14) and (15) are based.

The simplest mixed-gray gas is a gray-plus-clear system, with a_g and $(1-a_g)$ representing the energy-fractions of the black-body spectrum in which the gas is gray (or the absorption coefficient is constant) and clear, respectively. The quantity a_g is obtainable from a determination of gas emissivity ε_{g,L_m} at the mean beam length L_m which characterizes the enclosure, and of $\varepsilon_{g,2L_m}$ at twice that length.

$$a_g = \frac{(\varepsilon_{g,L_m})^2}{2\varepsilon_{g,L_m} - \varepsilon_{g,2L_m}} \qquad (23)$$

The rather involved derivation [2d] leads to a reasonably simple expression for $\overline{GS}_1$.

$$\overline{GS}_1 = \frac{A_T}{\frac{1}{C\varepsilon_1} + \frac{1}{\varepsilon_{g,e}} - \frac{1}{a_g} + \left(\frac{1}{a_g} - 1\right)\frac{1}{C\varepsilon_1 + (1-C)\varepsilon_r}} \qquad (24)$$

Note that for a gray gas $a_g=1$, and (24) reduces to (16).

The new subscript e on gas emissivity requires explanation. For a gray gas, emissivity and absorptivity are the same; and the net gas-sink flux is the black-body emissive-power difference (E_g-E_1) multiplied by a $\overline{GS}_1$ dependent on ε_g. For a non-gray gas E_g and E_1 are separately multiplied by different values of $\overline{GS}_1$, based on gas emissivity and gas absorptivity, respectively. This complication may be avoided, however, by using in (24) a modified gas emissivity equal to the value at the arithmetic mean of T_g and T_1, then multiplied by the factor $[1+(a'+b-c)/4]$ which allows primarily for the way emissivity varies with absorption strength and temperature[6,2c].

$a' = \partial \ln \varepsilon_g / \partial \ln pL$

$b = \partial \ln \varepsilon_g / \partial \ln T_g$

c = 0.65 for CO_2, 0.45 for H_2O, 0.5 for average flue gas.

When $T_1 < T_g/2$, the simpler ε_g evaluated at gas temperature replaces $\varepsilon_{g,e}$ in (24). (Note that $\overline{GS}_1$ now depends on refractory emissivity, about 0.5).

Special Problems Associated with Tubular Heaters

The heat sink of many furnaces is in the form of a row or rows of tubes, often mounted in a plane parallel to and near a refractory backing wall. When so mounted each tube row - backwall combination acts, with respect to radiative interchange with the remainder of the chamber – gas and walls – like a plane gray surface of area equal to the continuous tube plane A_p and of effective emissivity given (7,2f) by

$$\varepsilon_1 = \frac{1}{\frac{1}{F+(1-F)F} + \frac{B}{\pi}\left(\frac{1}{\varepsilon_1'} - 1\right)} \tag{25}$$

Here ε_1' is the true emissivity of the tube metal (often taken as 0.9, lower for high-quality alloys), B is the ratio of tube center-to-center distance to diameter, and F is the fraction of radiation incident on the tube plane through 2π steradians which is intercepted by the tubes. The fraction (1-F) impinges on the backwall and is reradiated or reflected.

The fraction F, the view-factor for radiation incident on a row of tubes, is conventionally evaluated for incident radiation which is isotropic, of which black-body radiation is an example. For that case

$$F_{iso} = 1 - \frac{1}{B}\left[(B^2-1)^{1/2} - \cos^{-1}\frac{1}{B}\right] \tag{26}$$

(The numerical equivalent of (26), clumsily obtained, first appeared in 1930(8)). The question properly arises as to how much error is involved in using F from (26) when the incident radiation on a tube row is non-isotropic, as from a gray gas (the result for a real gas is then readily evaluated by use of the mixed-gray gas concept).

Consider a two-dimensionally infinite tube row, Fig. 7, irradiated by a slab of

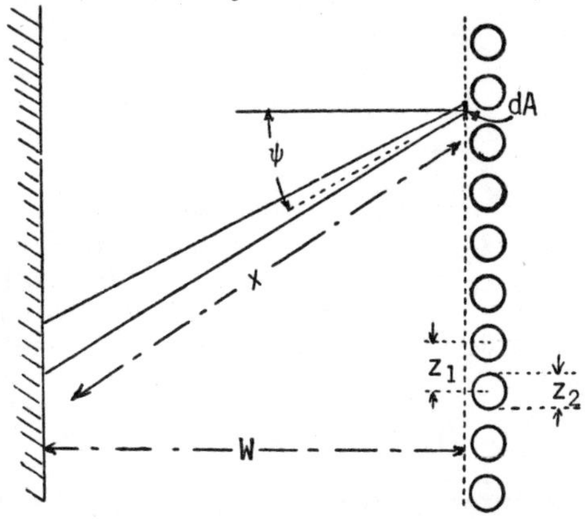

Figure 7. Section through a Gas Slab adjacent to a Row of Tubes. $z_1/z_2 \equiv B$

gas of width W and absorption coefficient K. The interchange-area $\overline{ss}$ between surface elements bounding the ends of the wedge-shaped space (2g) is

$$\frac{d^2\overline{ss}}{dA\cos\psi\ d\psi} = \frac{2}{\pi}\int_0^{\pi/2} e^{-Kx/\cos\theta}\cos^2\theta\, d\theta \tag{27}$$

Call the definite integral $f_3(Kx)$; its numerical value is given in (2g). The value of $\overline{ss}$ if the gas is non-absorbing is, from (27) with K=0, $(2/\pi)f_3(0)=1/2$. Since

$$(\overline{ss})_{clear} - (\overline{ss})_{absorbing\ gas\ present} = \overline{gs}$$

one may express $\overline{gs}$, with x replaced by $W/\cos\psi$, as

$$\frac{d^2\overline{gs}}{dA\ d(\sin\psi)} = \frac{1}{2} - \frac{2}{\pi} f_3(KW/\cos\psi) \qquad (28)$$

As dA is moved along the plane of the tube row without changing ψ, the interception of the beam varies intermittently from 0 to 1. Let the mean fractional interception for ψ-oriented radiation be F_ψ, given by

$$F_\psi = \frac{\sec\psi}{B} \quad \text{when } \sec\psi \leq B$$

$$1 \quad \text{when } \sec\psi \geq B$$

The interchange area ratio $\overline{gs}/A$ — the flux from gas to tubes per unit area of tube plane and per unit black emissive power difference of gas and tube surface -- is then given by multiplying the r.h.s. of (28) by F_ψ and integrating.

$$\frac{\overline{gs}}{A} = 2\int_0^1 \left[\frac{1}{2} - \frac{2}{\pi}f_3(KW/\cos\psi)\right]F_\psi d\sin\psi \qquad (29)$$

The same interchange-area ratio for a continuous plane receiver would be (29) evaluated with $F_\psi=1$; this may be shown to be $1-2\mathcal{E}_3(KW)$, where $\mathcal{E}_3$ is the third exponential integral. The ratio of these two values of $\overline{GS}$ is the desired mean fractional interception F_{gas} of radiation from a gas slab to a bounding tube row.

$$F_{gas} = \frac{\int_0^1\left[\frac{1}{2} - \frac{2}{\pi} f_3(KW/\cos\psi)\right][F_\psi(B)]d\sin\psi}{0.5 - \mathcal{E}_3(KW)} \qquad (30)$$

The ratio $F_{gas}/F_{isotropic}$, from (26)/(30), is the correction factor to the conventionally used graphs giving F_{iso} as a function of B. This ratio appears in Fig. 8. Since, for the gray-plus-clear gas model, tube furnaces have a KW of the order of 1, it is seen that the correction for non-isotropic incidence on the tube row is generally small.

If the tube-row is mounted on a backwall, F is required for use in (25) to obtain the effective plane emissivity ε_1. The term [F+(1-F)F], representing interception of the incoming beam plus interception of the beam returning from the refractory, is called $\overline{F}$. Because the returning beam is either almost-isotropic emission or almost-isotropic diffuse reflection,

$$\overline{F}_{gas} = F_{gas} + (1-F_{gas})F_{iso} \qquad (31)$$

Clearly, the ratio $\overline{F}_{gas}/\overline{F}_{iso}$ is even nearer 1 than the direct-radiation ratio F_{gas}/F_{iso}; and the correction can almost always be ignored.

From the above discussion it is clear that in using Equation (15)-(22) and (24) to predict performance of a furnace with tubes mounted on walls, A_r refers only to bare refractory, A_1 is the total area of the tube planes which, with their refractory backing act like a surface at T_1 and of emissivity ε_1 given by Eq. (25). There remains the evaluation of gas emissivity, which depends on the mean beam length. For rectangular parallepipeds varying from cubes to the space between infinite parallel planes an average mean beam length of 0.83x4 times the system mean hydraulic radius is an excellent approximation for absorption strengths in the range of industrial furnaces ($KL_m \sim 1$–3).

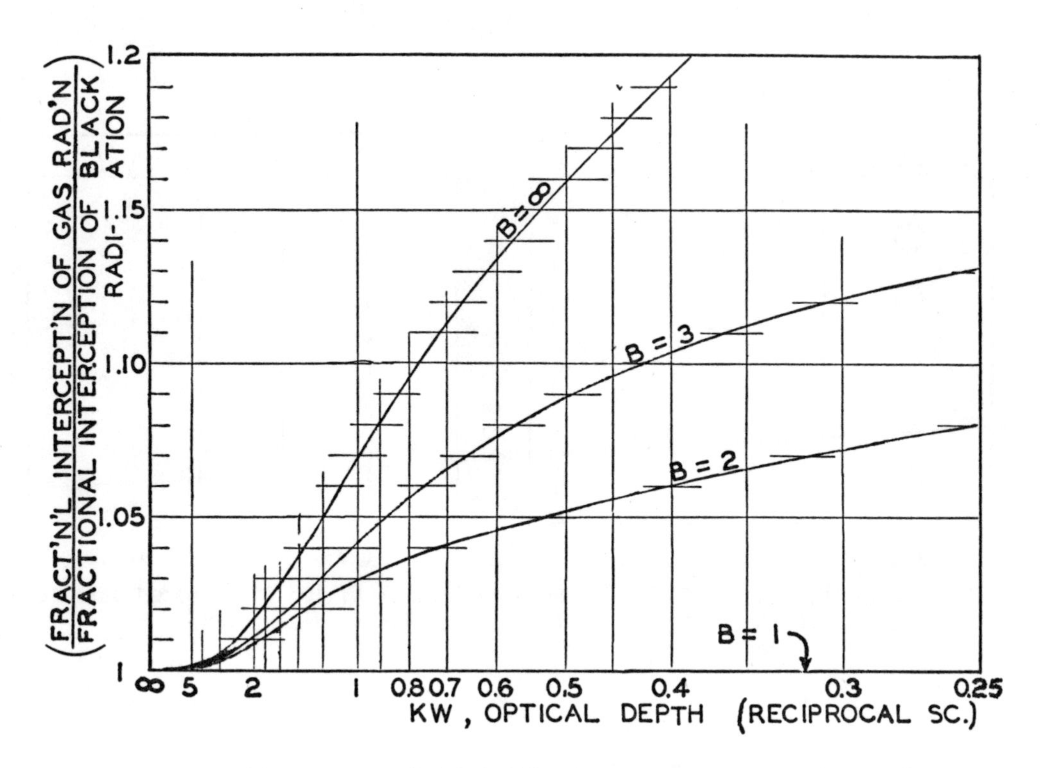

Figure 8. Ratio F_{gas}/F_{iso}, (fractional interception of gas radiation)/ (fractional interception of isotropic radiation) by a tube row. Tube center-to-center distance / diameter = B; optical thickness of gas slab = KW.

The use of the 1-gas-zone model on furnace chambers in which the tubes are enveloped in combustion gases rather than mounted in planes near refractory backwalls raises some difficult questions, such as how to define the sink area A_1 and what mean beam length L_m to use in evaluating ε_g. No longer is the system representable as an equivalent box, with the sink A_1 represented by continuous plane surfaces. A rigorous treatment of this problem has not to my knowledge been made. The multizone method could be used to determine the best recommendations for a one-zone model. In the interim an approximation will be suggested.

Consider a furnace in which equally spaced parallel tube screens are mounted, with firing between each pair, sufficient in number to justify the assumption that a repeating pattern is typical. Figs. 9(a) and (b) show the two choices available for picturing the repeating pattern. If there are several zones on either side of the one pictured, the dotted lines are planes of no net radiative flux. Specular mirrors replacing the dotted lines would make both sketches truly representative, and therefore completely equivalent. It is tempting to visualize perfectly diffuse mirrors instead, because for a gray-gas system with refractory surfaces in radiative equilibrium such surfaces are the equivalent of perfectly diffuse mirrors. A little consideration shows, however, that diffuse and specular mirrors produce different distributions of flux, particularly when the top and bottom refractory surfaces completing the enclosure are significant. Because the

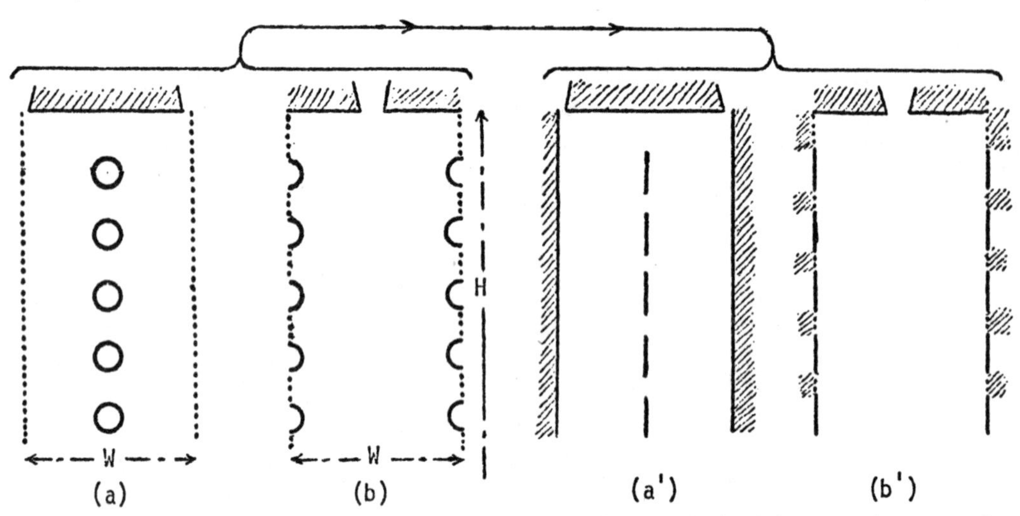

Fig. 9. Section through one Cell of a Furnace Chamber filled with spaced parallel vertical Tube Screens, and its two alternative Visualizations

scale of variation in image detail is small compared to W and particularly because H/W is usually large, the assumption that the dotted lines are replaceable by refractory surfaces, thereby defining a chamber independent of the rest of the system, is probably a good one. But it provides two alternative proceedures between which to choose.

Step one is to replace the tubes by plane interrupted surfaces — vertical strips — which intercept the same radiation that the tubes would. Since most of the radiation is from the gas, the height of a replacing strip should be the tube pitch P (the center-to-center spacing) multiplied by the F_{gas} of Eq. (30) __ the F_{iso} of Eq. (26), multiplied by a correction factor from Fig. 8 if warranted. (This will hereafter be called F, without subscript). Figs. 9(a) and (b) then become (a') and (b'). With the number of tubes in a screen equal to n, the total screen height is nP, generally little less than H; and the area A_1 is 2nPF (all areas are per unit dimension normal to the plane of the sketch). The tabulation below gives, for each of the two cases, values of A_1, A_r, A_T, $\overline{1r}/A_T (\equiv F_{r1}A_r/A_T)$ for use in Eq. (15) or (18), and the transverse dimension involved in determining mean beam length L_m.

	Case (a')	Case (b')
A_1 A_r $A_T \equiv A_1 + A_r$	$2nPF (\approx 2HF)$ $2(H + W)$ $2(H + W) + nPF)$	$2nPF (\approx 2HF)$ $2[H - nPF + W]$ $2(H + W)$
$\frac{\overline{1r}}{A_T}$, clear-gas value for use in Eq. (15)	$\frac{A_1 F_{1r}}{A_T} = \frac{nPF}{H+W+nPF} \equiv C \ (\Leftrightarrow S=0)$	$\frac{A_1(1-F_{11})}{2(H+W)} = \frac{F\{nP-[\sqrt{(nP)^2+W^2}-W]F\}}{H+W}$
Basis for L_m, related to transmittances associated with;		
$\overline{s_r s_1}$	Based on W/2	Based on W
$\overline{s_1 s_1}$	Inapplicable. $\overline{s_1 s_1}=0$	Based on W
$\overline{s_r s_r}$	Based on W	Based on W

Some of the terms need explanation. For case (b'), the term $[\sqrt{(nP)^2+W^2}-W]F$ in the expression for $\overline{1r}/A_T$ comes from setting $\overline{1r}(\equiv A_1 F_{1r})$ equal to $A_1(1-F_{11})$ and from visualizing $A_1 F_{11}$ to be formulated from the exchange-area between two continuous parallel planes of height nP and separated by $W(= \sqrt{(nP)^2+W^2}-W)$, then multiplied once by F to allow for emitter area and once more to allow for receiver area.

The gas emissivity affects $\overline{GS_1}$ through its appearance in the three direct-exchange areas appearing in Eq. (14). These may be written in terms of gas transmittance τ as follows, with primes indicating that gas absorption in present.

$$\overline{gs_1} = A_1 - \overline{11}' - \overline{1r}' = A_1 - \overline{s_1 s_1}\tau_{11} - \overline{s_1 s_r}\tau_{1r}$$

$$\overline{gs_r} = A_r - \overline{r1}' - \overline{rr}' = A_r - \overline{s_r s_1}\tau_{1r} - \overline{s_r s_r}\tau_{rr} \qquad (32)$$

$$\overline{s_r s_1}' = \overline{r1}' = \overline{s_r s_1}\tau_{1r}$$

All exchange areas on the r.h.s. are clear-gas values; those in Eq. (14) include gas emission or absorption. It is seen that $\overline{GS_1}$ involves three different values of gas transmittance $\tau(\equiv 1-\alpha_g, = 1-\varepsilon_g$ if the gas is gray), differentiated by the subscripts identifying the surfaces at the two ends of a beam.

Case (b'). For this case the three τ's are identical, and Eq. (15) may be used directly, with ε_g based on a rectangular parallelepiped H high and W wide. Substitution of values from the table into (15) yields

$$\frac{1}{\overline{GS_1}} = \frac{1}{2nPF}\left[\left(\frac{1}{\varepsilon_1} - 1\right) + \frac{1}{\varepsilon_g\left(1 + \cfrac{1}{\cfrac{nPF}{H+W-nPF} + \cfrac{\varepsilon_g}{1-\varepsilon_g}\cdot\cfrac{nP}{nP - F(\sqrt{(nP)^2+W^2} - W)}}\right)}\right] \qquad (33)$$

As a numerical example, if nP=H, W/H = 2/3, ε_g=0.3, B=2, and F=0.6576 x 1.02=0.67 (use of Fig. 8), the above yields

$$\frac{1}{\overline{GS}_1} = \frac{1}{2nPF}\left[\left(\frac{1}{\varepsilon_1} - 1\right) + 1.91\right]$$

If $\varepsilon_1 = 0.9$, $\overline{GS}_1 = 2nPF/2.021$

Case (a'). Although two different τ's (and ε_g's) are involved, assume temporarily that a single value of ε_g may be used. Substitution from the table into Eq. (15) yields

$$\frac{1}{\overline{GS}_1} = \frac{1}{2nPF}\left[\left(\frac{1}{\varepsilon_1} - 1\right) + \frac{1}{\varepsilon_g\left(1 + \frac{1}{\frac{nPF}{H+W} + \frac{\varepsilon_g}{1-\varepsilon_g}}\right)}\right] \qquad (34)$$

The same numerical example gives

$$\frac{1}{\overline{GS}_1} = \frac{1}{2nPF}\left[\left(\frac{1}{\varepsilon_1} - 1\right) + 1.511\right]$$

If $\varepsilon_1 = 0.9$, $\overline{GS}_1 = 2nPF/1.622$, which is 24.5% higher than Case (b'). Consider the other ε_g involved. τ_{r1} is based on the slab width W/2, τ_{rr} on W (the screen effect is allowed for in the F's). If the gas is gray and A_r is dominantly on the walls,

$\tau_{rr} = \tau_{r1}\,\tau_{1r}$, or $\left[1 - \varepsilon_g(W)\right] = [1 - \varepsilon_g(W/2)]^2$. Then

$$\varepsilon_g(W/2) = 1 - \sqrt{1 - \varepsilon_g(W)} \qquad (35)$$

Putting $\varepsilon_g(W/2)$ into (33) instead of $\varepsilon_g(W)$ gives

$$\frac{1}{\overline{GS}_1} = \frac{1}{2nPF}\left[\left(\frac{1}{\varepsilon_1} - 1\right) + 2.286\right]$$

If $\varepsilon_1 = 0.9$, $\overline{GS}_1 = 2nPF/2.397$, which is 15.7 percent lower than Case (b'). Clearly, the two bounds set by use of the limiting values of ε_g are too far apart to be useful, and it is necessary to go back to the more basic Eq. (14). Either the substitution of exchange areas from (32) into (14) or the use of a different approach(7) yields

$$\frac{1}{\overline{GS}_1} = \frac{1}{A_1}\left[\left(\frac{1}{\varepsilon_1} - 1\right) + \frac{1}{1 - \frac{\overline{11}'}{A_1} - \frac{(\overline{1r}')^2/A_1}{A_r - \overline{rr}'}}\right] \qquad (36)$$

From first principles and definitions

$$\overline{11}' = A_1(1-F_{1r})\,\tau_{11}$$

$$(\overline{1r}')^2/A_1 = A_1(F_{1r}\tau_{1r})^2$$

$$A_r - \overline{rr}' = A_r[1 - (1-F_{r1})\tau_{rr}] = A_r(1-\tau_{rr}) + A_1F_{1r}\tau_{rr}$$

Insertion of these into (36) yields

$$\frac{1}{\overline{GS}_1} = \frac{1}{A_1}\left[\left(\frac{1}{\varepsilon_1} - 1\right) + \frac{1}{1 - (1-F_{1r})\tau_{11} - \dfrac{(F_{1r}\tau_{1r})^2}{(A_r/A_1)(1-\tau_{rr}) + F_{1r}\tau_{rr}}}\right] \quad (37)$$

This is still general. For Case (a'), from the relation before (35), $\tau_{1r}^2 = \tau_{rr} = 1 - \varepsilon_g(W)$. In addition, $F_{1r} = 1$ and $A_r/A_1 = (H + W)/nPF$; and substitution into (37) gives

$$\frac{1}{\overline{GS}_1} = \frac{1}{A_1}\left[\left(\frac{1}{\varepsilon_1} - 1\right) + \frac{1}{1 - \dfrac{1}{1 + \dfrac{H+W}{nPF}\dfrac{\varepsilon_g}{1-\varepsilon_g}}}\right] = \frac{1}{A}\left[\frac{1}{\varepsilon_1} + \frac{1-\varepsilon_g}{\varepsilon_g}\frac{nPF}{H+W}\right] \quad (38)$$

Numerical substitution gives

$$\frac{1}{\overline{GS}_1} = \frac{1}{2nPF}\left(\frac{1}{\varepsilon_1} - 1\right) + 1.938$$

If $\varepsilon_1 = 0.9$, $\overline{GS}_1 = 2nPF/2.049$, which is but 1.3% lower than Case (b'). The excellent agreement between Eqs.(33) and (38) is no measure of the error introduced by the diffuse-mirror assumption (As $W/H \to 0$, both (a') and (b') reduce to: $A_1/\overline{GS}_1 = (1/\varepsilon_1) + (FnP/H)(1-\varepsilon_g)/\varepsilon_g$). Because Case (a'), Eq.(38), is simpler, its use is recommended. It is to be remembered that both derivations apply rigorously to two-dimensional systems; many reformer furnaces approximate that condition. Approximate allowance for the third dimension comes from adding the end enclosure areas to A_r and adjusting $\overline{11}$ in the conventional manner of allowing for opposed rectangles rather than opposed strips.

There remains the term ε_1, the effective emissivity of the strips which have replaces the tubes. The recesses formed by the tubes will give them an effective emissivity greater than that of plane metal. In the absence of gas, the interchange area between a tube row of true area A_2 and true emissivity of ε' and two parallel black plates on either side of it, of total area A_B and coextensive with the plane of the tubes, is given by

$$\frac{1}{\dfrac{1}{A_2}\left(\dfrac{1}{\varepsilon'} - 1\right) + \dfrac{1}{A_BF}}$$

For black plates surrounding gray strips of emissivity ε_1 and area A_BF replacing the tubes, the interchange area is $A_B F \varepsilon_1$. The two interchange areas are then equated to find ε_1. Replacement of A_B/A_2 by $2B/\pi$, where B is the ratio of center-to-center distance to diameter, gives

$$\varepsilon_1 = \frac{}{1 + \dfrac{2BF}{\pi}\left(\dfrac{1}{\varepsilon_1'} - 1\right)} \quad (39)$$

This is the effective emissivity of A_1. When B = 2 and $\varepsilon_1' = 0.8$, F = 0.66 and $\varepsilon_1 = 0.84$.

<u>Other tube arrangements</u>. Since furnaces with one central vertical screen of horizontal tubes between refractory walls are the complete equivalent of Case (a'), they present no new problem. If the central screen consists of two rows of tubes the only change is the increase in value of the interception factor F defining the fraction of the central plane occupied by equivalent strips. If one cell consists of a fired section bounded by a refractory wall on one side and a tube screen on the other, with another cell on the other side of the tube screen, there is a lack

of symmetry which makes the replacement of the tube row by mixed strips of refractory and sink not quite right; but such a model, modified, is recommended. If, when calculations of the performance of that cell and the one next to it have been completed, the flux densities onto the two sides of the tube row are different, there is then a basis for repeating the calculations allowing for net flux between the two cells.

The statement has been made that the gray-plus-clear gas model is more realistic and by implication better than the gray-gas model; but the only solution given for the former was restricted to the speckled box-type furnace, Eq.(24). It may be shown[2,h] that, if the constraint is put on the gray-plus-clear gas model that its refractory surfaces diffusely reflect all radiation incident on them (thereby failing to take advantage of shifting the incident gas radiation to the spectral windows of the gas on re-emission through gas to sink), the flux is obtained quite simply from the relation

$$\overline{GS}\text{ (gray-plus-clear-gas model, with refractory surfaces perfect diffuse reflectors)} = a_g \times (\overline{GS}\text{ based on a gray gas of emissivity } \varepsilon_g/a_g) \quad (39)$$

Furthermore, the result for the real-gas real-refractory model--when it is available--, with $\varepsilon_r = 0.5$, always lies roughly half-way between (39) and that obtained for the simple gray-gas model. The arithmetic mean of (39) and the gray-gas model is therefore suggested for those cases, such as the ones last discussed above, which are too complex to formulate rigorously. How important it is to make such a correction can only be established by correlation of a considerable body of furnace data.

Models (a') and (b) presented here have not been tested with furnace data. It will be especially necessary to allow for the temperature difference Δ in tube-screen furnaces with fuel fired between screens because of the absence of strong back-mixing with a reach comparable to the total gas-flow path. The fitting by Lobo and Evans (6) of a model similar to that based on Eq. (15) was done without the use of a Δ. Whether the striking success of that model—when fitted to and applied to box furnaces—indicated that Δ was in fact zero in such furnaces or whether there were compensating errors, such as the gray-gas assumption, was never established. It is reported that that model has in a number of cases been less than satisfactory in application to modern furnaces with interior tubes. The reason may lie in the incorrect evaluation of GS_1.

A correlation of performance data on modern furnaces, using a model such as that herein described, is highly desirable.

BIBLIOGRAPHY

1. Hottel, H. C., Melchett Lecture for 1960. Jl. of Inst of Fuel, 220-234, June, 1961
2. Hottel, H. C., and Sarofim, A. F.,Radiative Transfer, McGraw-Hill, New York, 1967. Page references as follows: (a), 311, 459; (b), 462; (c), 247;(d), 316; (e), 300; (f), 113; (g), 271
3. Hottel, H. C., and Keller, J. D. , Tr.Am.S .Mech.Eng.,IS 55, 39-49 (1933)
4. Haslam, R. T., and Hottel, H. C., Tr A.S.M.E., FSP 50 (1928)
5. Hottel, H. C., M.I.T. Notes on Heat Transfer in the Com ustion Chamber of a Furnace, Cambridge, Mass., April, 1940
6. Lobo, W. E., and Evans, J. E., Tr.A.I.Ch.E., 35, 743 (1939)
7. McAdams, W. H., Heat Transmission, 3rd Ed., Chapter 4, McGraw-Hill, New York, 1954
8. Hottel, H. C., Mech.Eng., 52, 699-704 (1930); Tr. A.S.M.E., FSP 53, 265-73, (1931)

Chapter 2

METHODS FOR CALCULATING RADIATIVE HEAT TRANSFER FROM FLAMES IN COMBUSTORS AND FURNACES

János M. Beér

Professor, Department of Chemical Engineering,
University of Sheffield, England.
Superintendent of Research, International Flame Research Foundation.
Ijmuiden, Holland

Abstract

Recent advances in methods for predicting radiative heat flux distribution in furnaces and combustors are reviewed with special reference to the zone method of analysis and the flux methods. Recent experimental studies specially designed to test these prediction procedures under sufficiently severe conditions are discussed.

NOMENCLATURE

$a_{g,n}$ $a_{s,n}$	:	weighting factor for emissivity equations.
c	:	mass concentration of soot (ML^{-3}).
d	:	particle diameter (L).
I	:	radiation intensity $(QL^{-2}\ T^{-1}\ \Omega^{-1})$.
$I_{b,w}$	:	monochromatic black body radiation intensity $(QL^{-2}\ T^{-1}\ \Omega^{-1})$.
k	:	gas or soot specific absorption coefficient (T^2M^{-1}) or $(L^2\ M^{-1})$ absorption index of material (L^{-1}).
k_w, k	:	monochromatic specific absorption coefficient.
K	:	gas absorption coefficient (L^{-1})
L	:	radiation path length (L).
m	:	complex refractive index.
n	:	refractive index; subscript referring to weighted grey gas.
p	:	gas partial pressure (MLT^{-2}).
$T(T_g)$	:	temperature (gas temperature) (θ).
X	:	soot optical parameter $(=\pi d/\lambda)$.
$\varepsilon_g(\varepsilon_s)$	:	gas (soot) emissivity.
ε_λ	:	monochromatic emissivity.
λ	:	radiation wavelength (L).
ρ	:	soot particle density (ML^{-3}).
σ	:	Stefan-Boltzmann radiation constant $(QL^{-2}T^{-1}\theta^{-4})$
ω	:	radiation wave number (L^{-1}).

1. Introduction

A rigorous calculation of radiative heat transfer from a flame to heat sinks in an enclosure requires the simultaneous solution of partial differential equations of fluid flow with chemical reaction and of energy transfer. The interdependence of these processes makes the problem highly complex and at present complete solution is not possible for cases of practical combustors and furnaces. The general approach, therefore, has been to develop simplified mathematical models of steady confined flame processes.

Simplified processes for calculating radiative heat transfer from flames require information on the flow and heat release patterns in the flame. Such information can be based either upon similarity considerations of confined turbulent jet flows without or with physical modelling or it can be based upon the numerical solution of the equations of conservation of mass and momentum in turbulent flow with chemical reaction. It is generally assumed that the overall rate of combustion in turbulent diffusion flames is determined by the rate of mixing of the reactants and that the effect of the flow upon the combustion reaction is therefore important while the effect of the combustion on the flow pattern is slight. There has been a major advance in the development of the numerical analysis of partial differential equations and also in the mathematical modelling of turbulent flows. There remain, however, difficulties when modelling complex flows such as exist in many practical combustion systems. Table 1. shows combinations of methods for predicting flow patterns with those for the calculation of radiative heat flux distribution in a combustion chamber. While there is no theoretical limitation to the combination of the more exact zone method with mathematical modelling techniques for the prediction of flow patterns, the flux methods which replace the exact integrodifferential equations of radiative transfer by approximate differential equations are more suitable for this combination as they can be easily cast in finite difference form, and can thus be solved conveniently together with the equations of convective transport. The choice of the appropriate combination of prediction-modelling methods has to be made in the light of available computer storage capacity and the accuracy and detail of the solution required.

The following discussion is largely based on the thesis work of T.R. Johnson (3) and on a recent review by Beér and Siddall (17).

A	B	C
Physical modelling	Physical modelling	Computer modelling
Determine velocity and local "air-fuel" ratio distributions by the use of isothermal modelling.	Determine residence time distribution in prototype or in isothermal model.	Use numerical solution procedures of the differential transport equations of convective flows to yield patterns of flow and of concentration in turbulent flames.
↓	↓	↓
Calculate heat release pattern and species concentration and temperature (adiabatic) distribution.	Calculate the size of the "well-stirred" and "plug flow" parts and the proportions of the mean residence time spent in these parts respectively.	Use the Flux methods for radiative heat flux distribution calculations by replacing the exact integrodifferential equations of radiative transfer by approximate differential ones which are cast in finite difference form.
↓	↓	
Use Zone Method for detailed radiative heat flux distribution calculation.	Use the Zone Method to calculate the radiative heat transfer in the well stirred part and the heat flux distribution along the flow in the plug flow part.	

Table 1. Combinations of methods for determining flow patterns with those for determining radiative heat flux distributions.

2. The zone method of analysis

2.1 Principles

One of the more complete mathematical models of radiation heat transfer is the zone method of analysis, which was developed by Hottel and Cohen (1). The starting point for the zone method of analysis is an assumed knowledge of the patterns of fluid flow, chemical heat release and radiating gas concentration within a furnace. From this point equations describing the conservation of energy within the furnace are drawn up. This is carried out by dividing the furnace into a large number of zones, both surface zones and gas zones, which are small enough for each zone to be assumed to be isothermal and to have constant properties. The radiative interchange in the enclosure is first obtained by determining radiative exchange factors for each zone pair combination in the furnace, taking account of attenuation by intervening gas, and including the effects of reflection within the enclosure. These exchange factors, the 'total exchange areas', are the constants of proportionality in equations relating the exchange of radiative energy to the difference of the fourth power of temperature of each respective zone pair, and give a solution for the radiation exchange within the furnace. Total energy balance equations for each zone can then be drawn up, taking account of all forms of energy transfer in the furnace, such as convection of heat due to gas mass flow, chemical heat release and net radiative transfer with other zones. This leads to a set of simultaneous non-linear, algebraic equations in terms of the temperature of each zone, which on solution yield the temperature field within the furnace, and this then enables the heat fluxes to be evaluated.

The evaluation of the total exchange areas forms an essential part of the zone method of analysis. They are constructed according to the laws of radiation geometry, taking account of the physical-chemical properties (emissivity and absorptivity) of the radiation media, and describe the way in which radiation is exchanged between the various combinations of emitting-absorbing volumes and emitting-absorbing-reflecting surfaces.

Mathematically tractable application of the zone method depends largely on the separation of the calculation of total exchange areas from the solution of the total energy balance equations for each zone, which is only possible when the total exchange areas are independent of the temperatures in the furnace. This situation is only realized when the gas is grey, i.e. the gas absorption coefficient, K_g, is independent of the gas temperature and the wavelength of the incident radiation.

3. Evaluation and representation of radiative properties.

3.1 Radiation from gases

Real gases, such as carbon dioxide and water vapour, are not grey but emit and absorb radiation only in discrete bands in the spectrum and, therefore, have mean absorption coefficients which vary with gas temperature and the temperature of the radiation source.

The emissivity-path length relationship for a grey gas is given by:

$$\varepsilon_g = 1 - e^{-kpL} \qquad (1)$$

where k is the specific absorption coefficient,
p is the partial pressure of the absorbing gas, and
L is the path length for radiation.
(the absorption coefficient K is given by K = k.p).

Hottel (2) has shown that emissivity of a real gas may be visualized as the weighted sum of grey gases, so that the ε_g - pL relationship may be represented by an exponential series as:

$$\varepsilon_g = \sum_n a_{g,n} (1 - e^{-k_n pL}) \qquad (2)$$

$$\text{where} \quad \sum_n a_{g,n} = 1 \qquad (3)$$

where $a_{g,n}$ is the fractional amount of energy in the spectral regions where grey gas of absorption coefficient k_n exists.

With this representation, the absorption coefficients, k_n, can be made independent of temperature and the entire temperature dependence of gas emissivity taken by the weighting factors $a_{g,n}$. Under these conditions the total exchange areas can be evaluated for each of the grey gases of absorption coefficient k_n, and the radiative exchange obtained as the $a_{g,n}$ weighted sum of the contributions for each grey gas. Thus, the total exchange areas can be evaluated without knowledge of the temperatures in the furnace.

The monochromatic emissivity of a gas at wavelength λ is related to the number of molecules of gas and the specific spectral absorption coefficient. The number of molecules can be represented by the product of partial pressure, p, and length, L, i.e. related to pL. Thus, the monochromatic emissivity ε_λ of a column of gas is given by:

$$\varepsilon_\lambda = 1 - e^{-k_\lambda pL} \qquad (4)$$

or, in terms of the wave number, ω (where $\omega = 1/\lambda$).

$$\varepsilon_\lambda = 1 - e^{-k_\lambda pL} \qquad (5)$$

where k_ω may be a function of gas temperature and pressure.

For non-continuous emission, such as that from CO_2 and H_2O, the absorption coefficient k_ω varies considerably over the spectrum, and the total emissivity can be found as the energy weighted mean emissivity, i.e. for a gas of temperature T_g:

$$\varepsilon_g = \frac{\int_0^\infty I_{b,\omega}(T_g)\left(1 - e^{-k_\omega (T_g) pL}\right) d\omega}{\int_0^\infty I_{b,\omega}(T_g)\, d\omega} \qquad (6)$$

where $I_{b,\omega}(T_g)$ is the black body spectral radiation intensity.

Thus, from a knowledge of the spectral absorption coefficient, the gas emissivity can be determined.

The problem of completely defining the spectral characteristics of the radiation from gases is very complex, and the usual procedure has been to rely on measurement of total (as distinct from spectral) emissivity for engineering calculations of heat transfer. The major compilations of gas emissivities are those given by

Hottel and his co-workers (2), which are based mainly on measurements of total emissivity. More recent investigations of spectral emissivities of CO_2 and H_2O (4)(5) have shown differences in detail, but in general have confirmed the validity of Hottel's original correlation.

3.2 Representation of a real gas for engineering calculations.

The mathematical formulation of the zone method is much simplified if the absorbing medium can be treated as a grey gas. Real gases, such as water vapour and carbon dioxide, do not behave as grey gases, because of the discrete nature of their absorption bands. However, it is possible to retain the simplicity of the mathematical formulation of the zone method by representing the emissivity of the real gas as the weighted sum of the emissivities of a number of grey gases, as given by equation (2).

If the spectral variation of k_ω can be imagined to be approximated by a number of groups of constant absorption coefficient bands, each having an absorption coefficient k_n, as shown in Fig.1, then the weighting factor $a_{g,n}$ for each term in the exponential series (equation (2)) represents the fraction of the black body energy in the wave number regions where the absorption coefficient is k_n. This is illustrated in Fig.1 where the representation is shown for three terms in equation (2).

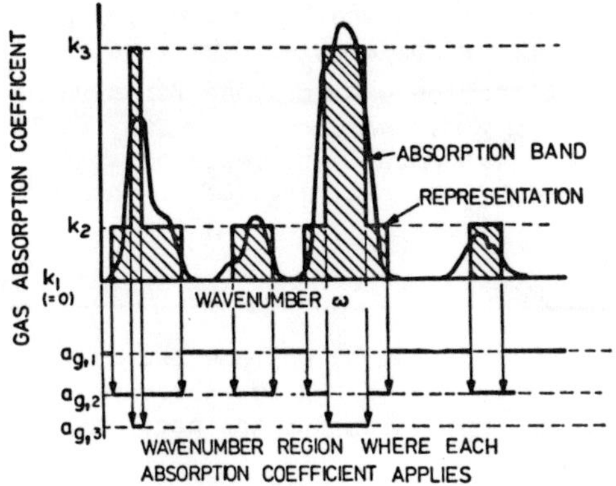

Fig.1. Representation of real gas absorption bands (after Johnson and Beér (15)).

Since the distribution of black body energy in the spectrum varies with temperature, according to Planck's law, and the size of the gas absorption bands may also depend on temperature, it follows that the weighting factors $a_{g,n}$ will vary with temperature. Thus, by suitable selection of the values of k_n and $a_{g,n}$, it is possible to use a fixed set of absorption coefficients, and let the full temperature dependence of emissivity be carried by the weighting factors $a_{g,n}$.

Although in theory the emissivity of a gas approaches 1 at sufficiently high values of pL, at all practically attainable values of pL the emissivity is considerably less than 1. Therefore, the fitted emissivity equation will usually have one term where $k_n = 0$, corresponding to the non-absorbing part of the spectrum between the strong bands. This is often referred to as the clear gas component.

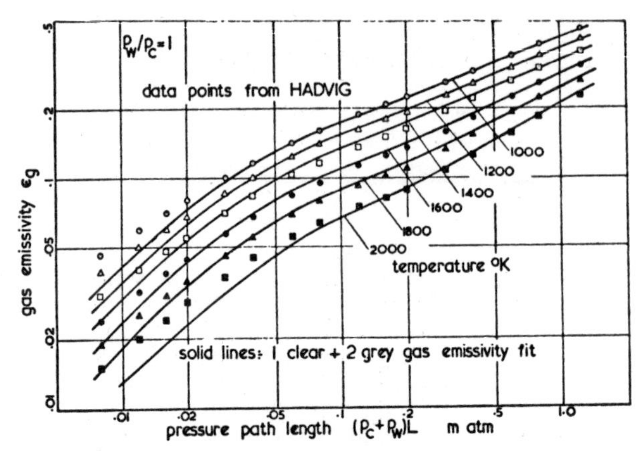

Fig. 2. Emissivity of a CO_2–H_2O gas mixture with $p_w/p_c = 1$. (after Johnson and Beér (15)).

The number of terms used in equation (3) depends on the range of pressure path lengths to be considered, and the desired accuracy of fit. The greater the number of terms the better will be the representation. In many cases a three term or even a two term equation will be satisfactory. Fig. 2 shows the fit of a three term equation (that is, a one clear–two grey gas representation) to the data from Hadvig's (7) charts for a CO_2–H_2O mixture with $p_w/p_c = 1$. This shows that a reasonable representation was obtained for a range of $(p_c + p_w)L$ from 0.02 metre atmospheres to 1.2 m atm. Improved accuracy could be obtained by using a larger number of terms in the emissivity representation: Hottel and Sarofim (6) obtained a good fit over a 3000 fold range of pL when using a four term representation.

3.3 Representation of soot emissivity for engineering calculations (8–15).

The emissivity of a soot cloud can be represented as the weighted sum of grey gases, in the same manner as a real gas is represented. Johnson (3,15) recommended the following expression for the soot emissivity:

$$\varepsilon_s = \sum_n a_{s,n}(T)\,(1 - e^{-k_{s,n}cL}) \tag{7}$$

where

$$\sum_n a_{s,n}(T) = 1 \tag{8}$$

where again the weighting factors $a_{s,n}$ can be thought of as the black body energy in the regions of the spectrum where the soot absorption coefficient is equal to $k_{s,n}$. This is illustrated in Fig. 3, where the soot spectral absorption coefficient is represented approximately as a number of bands at three different levels of absorption coefficient (note that the spectral parameter used is the wave number ω). Since the soot radiation is continuous, there are no 'windows' in the spectrum and hence all the absorption coefficients $k_{s,n}$ have non-zero values. With this approximation a good representation of the soot emissivity can be obtained for an extended range of concentration-path lengths cL.

Fig. 4. shows the fit of a three-term exponential series to soot emissivity data

3.4 The emissivity of a mixture of gas and soot.

In order to include the effect of soot radiation in the zone method of analysis, the emissivity of the combined gas-soot mixture should also be represented as the

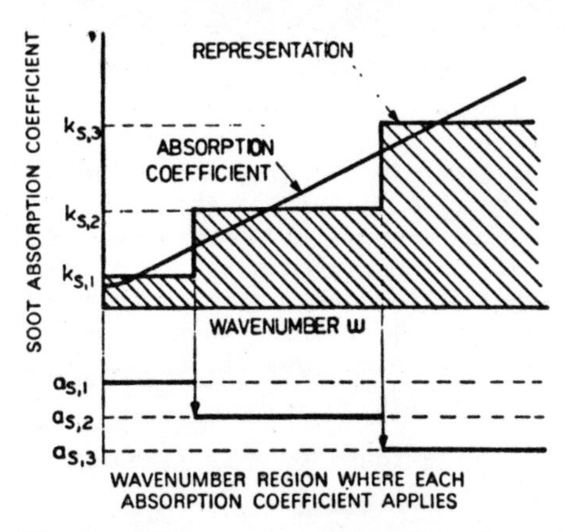

Fig.3. Representation of soot as a number of absorption bands. (after Johnson & Beér (15)).

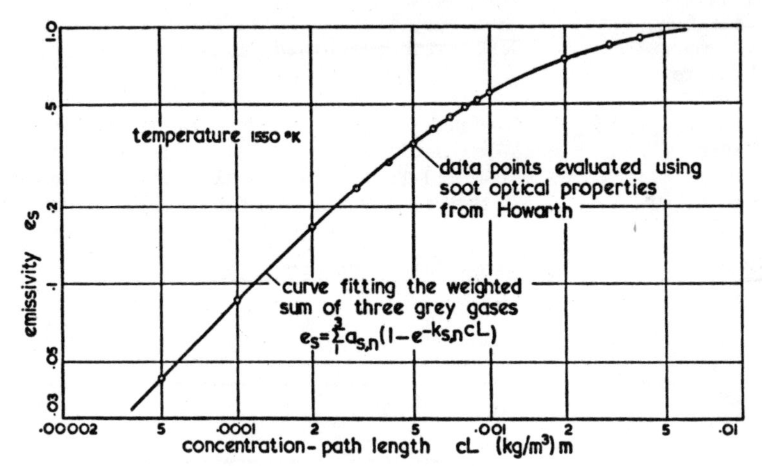

Fig.4. Fit of exponential series (weighted sum of grey gas) to soot emissivity data. (after Johnson & Beér (15)).

weighted sum of a number of grey gases. This is possible if the weighting factors $a_{s,n}(T)$ in the soot equations can be made equal to the weighting factors $a_{g,n}(T)$ in the gas equations. Fig.5 shows the way in which the soot spectral absorption coefficient can be represented, in an approximate manner, by bands of constant absorption coefficient $k_{s,n}$ which occupy the same spectral positions as the approximate absorption bands in the gas spectrum. With this approximation the groups of bands occupy the same fractions of the energy spectrum for both gas and soot, so that the condition

$$a_{s,n}(T) = a_{g,n}(T)$$

is satisfied.

The emissivity of the mixture of gases can then be found by adding the absorption coefficients for gas and soot in each spectral region. Thus, the emissivity of the gas-soot mixture is given by:

$$\varepsilon_m = \sum_n a_{g,n}(T).(1 - e^{K_n L}) \tag{9}$$

$$\text{where } K_n = k_{s,n}C + k_{g,n}(p_c + p_w) \tag{10}$$

The weighting factors and absorption coefficients for a three-term equation have been evaluated for the case of a mixture of the products of oil combustion ($p_w/p_c = 1$) and a soot having a spectral absorption coefficient given by Hammond (12). Because of the approximate nature of the soot representation, a degree of trial and error was necessary in order to obtain a suitable fit. The procedure adopted was to obtain a good representation of the gas emissivity, and to determine a set of soot absorption coefficients $k_{s,n}$ which gave an acceptable representation of soot emissivity.

The weighting factors $a_{g,n}$ were found to vary linearly with temperature, such that:

$$a_{g,n}(T) = b_{1,n} + b_{2,n}\,T \tag{11}$$

With this temperature dependence of the gas weighting factors imposed on the soot equations, it was not possible to obtain an exact fit for the soot emissivity, except at one temperature. However, the essential features of the soot emissivity temperature dependence were preserved, i.e. the increase in emissivity with temperature.

The values of $b_{1,n}$, $b_{2,n}$, $k_{s,n}$ and $k_{g,n}$ obtained for the mixture emissivity are given in Table 2. With these values equations (9), (10) and (11) define the gas and flame emissivity in an oil-fired furnace where the radiation path length is in the range of 0.2 metre to 6 metre, and the temperature is in the range of 1100 K to 1800 K.

Table 2. Constants in gas/soot emissivity equation.

n	Weighting factors $a_{g,n} = b_{1,n} + b_{2,n}\,T$		Gas absorption coefficients $k_{g,n}$	Soot absorption coefficients $k_{s,n}$
	$b_{1,n}$	$b_{2,n}$		
1	0.130	0.000265	0	3460
2	0.595	-0.000150	0.835	960
3	0.275	-0.00115	26.25	960

Units of gas absorption coefficients $(m.atm.)^{-1}$
Units of soot absorption coefficients $(m.kg/m^3)^{-1}$
Temperature in K.

3.4.1 The experimental testing of the emissivity model.

The emissivity model, equation (9) was tested by comparing calculated and measured traverses of unidirectional radiation intensity across a luminous flame. Beer and Claus (14) developed this method and showed that the calculated radiation intensity was a sensitive function of emissivity variations along the optical beam. In this experimental method a narrow angle radiometer probe fitted with a long water cooled extension is passed through the flame and sighted on a cold target at the other side of the flame. Radiation intensity is then measured as the probe is progressively withdrawn through the flame. The intensity distribution can be predicted from measurements of temperature and gas and soot concentration across the flame. The flame is divided into slices

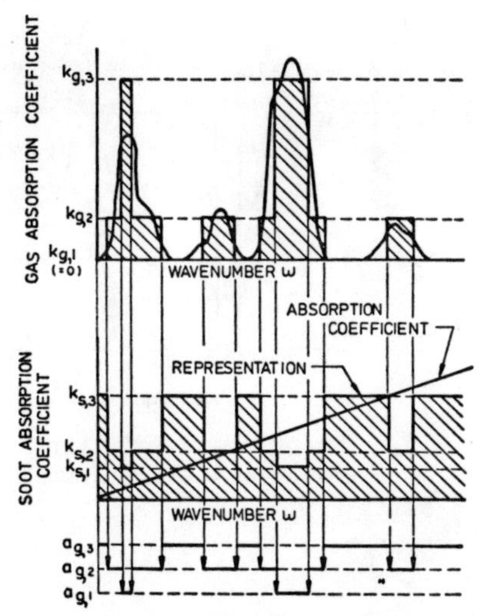

Fig.5. Representation of gas and soot absorption coefficients with bands in the same spectral ranges.
(after Johnson & Beér (15)).

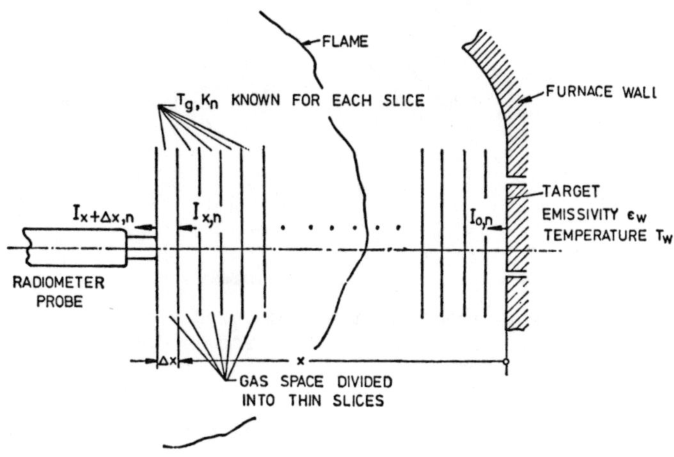

Fig.6. Radiation intensity change across a thin slice of flame.
(after Beér & Claus (14)).

of thickness Δx and the radiation leaving a slice $I_{x+\Delta x}$ is related to the "grey gas" radiations $I_{x,n}$ entering the slice by

$$I_{(x+\Delta x)} = \sum_n \left\{ I_{x,n} e^{-K_n \Delta x} + a_{g,n}(T) \cdot (1 - e^{-K_n \Delta x}) \frac{\sigma}{\pi} T^4 \right\} \qquad (12)$$

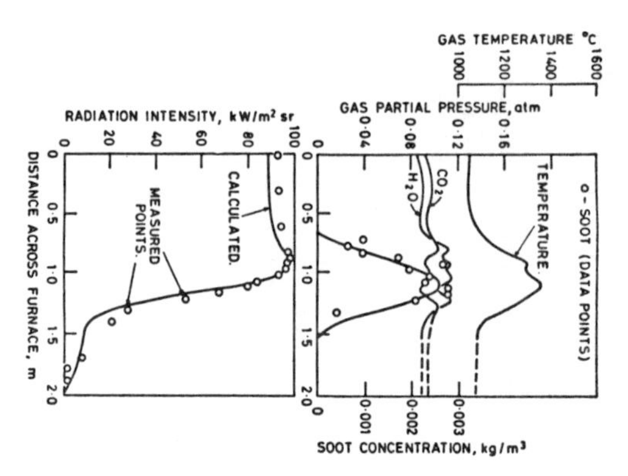

Fig.7. Narrow angle radiation traverse - oil flame AD 150. (after Johnson & Beér (15)).

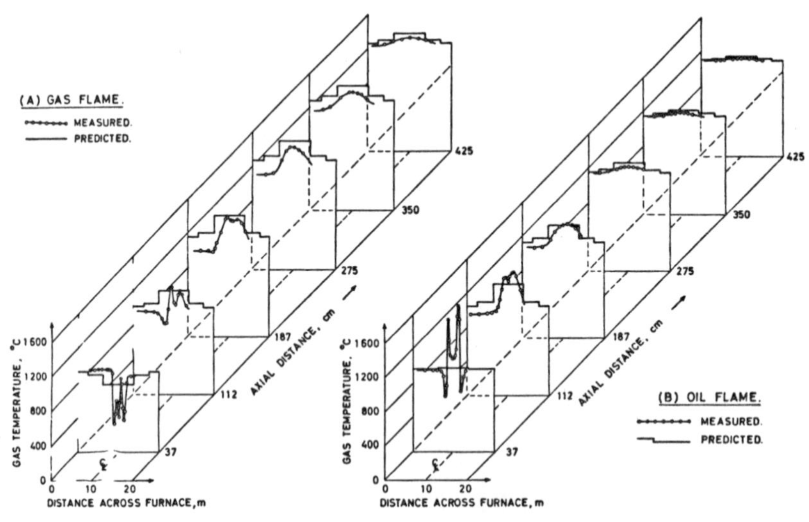

Fig.8. Comparison of predicted and measured gas temperature profiles at several axial distances in the Ijmuiden furnace (zone method). (after Johnson & Beér (16)).

where $a_{g,n}(T)$ and K_n are weighting factors and absorption coefficients defined in equations (9), (10) and (11). The boundary conditions at the target can be defined in terms of emissivity and temperature as:

$$I_{0,n} = a_{g,n} \cdot \varepsilon_W \cdot \frac{\sigma}{\pi} T_W^{\,4} \tag{13}$$

Thus the distribution of radiation intensity across the flame can be predicted using the target leaving radiation as the initial condition (Fig.6.).

The comparison between calculated and measured radiation traverses for an axial station in the flame is shown in Fig.7 together with measured distributions of temperature and of CO_2, H_2O and soot. The value of the soot absorption coefficient which gave this good agreement between calculated and measured data was $k\ \lambda = 0.056^{\,-1.086}$ which is the same as that given by Hammond (12).

4. The experimental testing of the Zone Method

A zone method analysis incorporating the new emissivity model for luminous radiation, allowance for the spatial variation of the absorption coefficient and for mixed surface zones (15) has been tested against measurements carried out in one of the experimental furnaces of the International Flame Research Foundation using both a non luminous gas flame and a highly luminous oil flame under such conditions that the flow patterns and the heat input rates were the same for both flames (16). Initial flow and heat release patterns and the composition of the gas and concentration of soot were obtained from measurements within the experimental furnace. Any major discrepancies between measured and predicted data should then be due to deficiencies in the calculation method rather than inaccurate or unreliable initial data. For calculation purposes the furnace was treated as cylindrical and was subdivided into annular coaxial zones (51 gas zones and 23 surface zones). Gas temperature and heat flux distributions were predicted and compared with measured values. The maximum discrepancies (200° K in 1200° K) were found with the gas flame. Maximum errors with the oil flame were less than 100° K. Fig.8 illustrates the comparison of measured and predicted data for the oil and gas flames at several axial distances from the burner. For both types of flame the predicted furnace exit temperature differed from the measured value by less than 50° K. It was suggested that the errors in the predicted gas temperatures in the neighbourhood of the burner are probably due to the inability of the mathematical model to predict large temperature gradients accurately because of the coarse zoning.

Comparison of the heat absorption along the furnace showed a maximum error of 10%. The curves of cumulative heat absorption (Fig.9.) show a very rapid rise in heat absorption along the first two metres of the furnace (about one furnace width) coupled with a very slow increase over the remaining four metres. Predictions for the oil flame somewhat overestimated the measured values while those for the gas flame were under estimated.

The distribution of heat absorption in the furnace is shown in Fig.10. for the oil flame. This is given as a histogram with each bar representing one surface zone. The furnace had three types of heat sink; refractory wall, water cooled surfaces such as access doors and cooling tube loops. The three fractions have been plotted cumulatively so that the height of the bars give the amount of heat absorbed by all heat sinks in each surface zone. Here again the measured and predicted data show very good agreement for all zones. The largest differences, 10%, were found in the heat absorption by the cooling tubes. These tests have

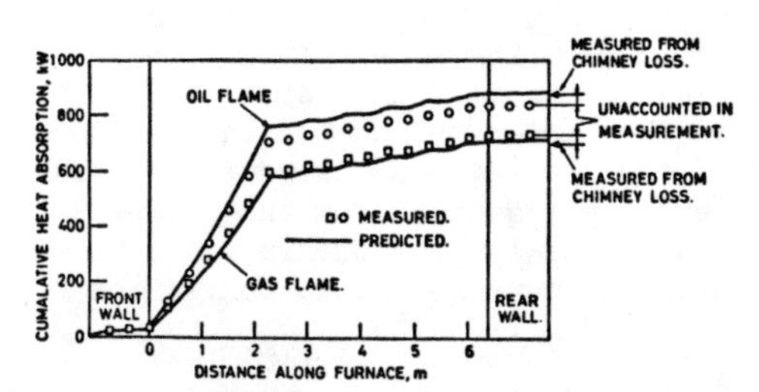

Fig.9. Comparison of predicted and measured cumulative heat absorptions (zone method). (after Johnson & Beér (16)).

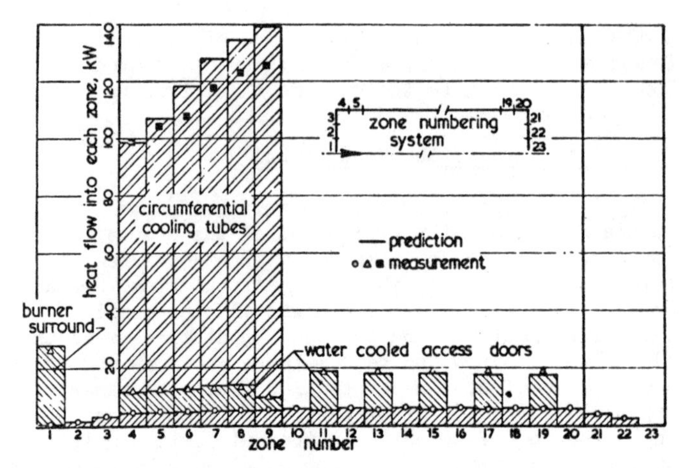

Fig.10. Histogram showing distribution of heat absorption in furnace-oil flame.

shown that the zone method is capable of predicting heat flux distribution in combustors or furnaces with great detail and satisfactory accuracy and that it can be used with sufficient confidence for luminous flames.

5. Flux methods*

Over the last few years methods have been developed for the detailed prediction of two dimensional (axi-symmetrical) flow and reaction in enclosures (48)(49) and the successful extension of these methods to three dimensional asymmetrical flow situations is expected in the near future. Complete evaluation of the three dimensional behaviour of an enclosure from a knowledge of the initial and boundary conditions will then be possible for the first time by combining these methods with the simultaneous treatment of radiative transfer. Unfortunately the zone method of analysis is likely to be too demanding in both computer time and storage capacity to be incorporated economically into the complete prediction procedure. An alternative approach to the calculation of radiant heat transfer is therefore necessary, which involves less computer storage capacity and arithmetical effort, without too great a loss of accuracy. It is hoped that such an approach will be provided by flux methods, which have the additional advantage of permitting relatively straight forward incorporation of the effects of scattering and of rapidly varying absorption coefficients.

5.1 Basis of flux methods

The basis of most flux methods of analysis is a balance for the flux of radiant energy in a specified direction through an elementary volume, derived on the assumption that the material within the element is optically grey (greater accuracy can be achieved by relaxing this restriction). The balance equates the difference in outward and inward fluxes of radiant energy in the specified direction to the difference between the sum of the emitted and inscattered radiation and the sum of the absorbed and outscattered radiation. It takes the form of an integro-differential equation for the intensity of radiation in the specified direction at the general point within the enclosure. For the purposes of approximate solution of the equation it is convenient to assume that the solid

*For the detailed discussion of flux methods the reader is referred to a recently published paper by R.G. Siddall (18).

angle surrounding the point can be subdivided into a small number (2 N) of equal solid angles, in each of which the intensity of radiant energy is assumed to be independent of direction. The integro-differential equation is then replaced by 2 N simultaneous differential equations in the 2 N unknown average intensities (or flux densities) of radiant energy. This would be called a 2 N- flux method of representation of the radiative transfer.

The differential equations in the unknown flux densities are similar in form to those which describe mass, momentum and energy transfer at a point, and can therefore be conveniently combined with them and solved numerically by utilising the techniques developed for solution of the flow field problem.

5.2 Two-flux solutions of simple problems.

Two-flux methods, which were originally devised for the solution of astrophysical problems on the assumption that radiative transfer is important in only one predominant direction, are all derived from original work by Schuster (19). The mathematical forms assumed by the two-flux equations depend upon the particular method of averaging the integrodifferential equation, and are known variously as the Schuster-Schwarzschild (20), Schuster-Hamaker (21) and Milne (22)-Eddington (23) approximations. The essential similarity of these approximations has been demonstrated recently (18). Two-flux methods have been successfully applied to the solution of several simple problems with prespecified flow and heat release patterns (21),(24),(25),(26),(27).

5.3 A two-flux solution for a combustion chamber.

Gibson and Monahan (28) have employed a two-flux Milne-Eddington method in cylindrical polar co-ordinates to predict the radiative transfer from a pulverised fuel flame in a cylindrical furnace (29) with simultaneous treatment of the flow and reaction. Simplifications were introduced by assuming axial symmetry (i.e. two dimensionality for mathematical purposes), by neglecting gas radiation, and neglecting radiative transfer in the axial direction, with the two flux components in the radial direction. Comparisons between predicted and measured values are shown in Fig.11. The authors suggest that the discrepancies are principally due to the oversimplified treatment of particle burning. Clearly further development and detailed experimental testing of this method is necessary before it can be recommended for application in industrial design.

5.4 Allowance for real gas behaviour.

Roesler (30) has extended the simple two-flux Schuster-Schwarzschild method to treat radiative transfer in co-current tube furnaces. Approximate allowance for real gas behaviour was made by treating the furnace gas as a one grey gas-one clear gas mixture, thus doubling the number of unknown flux components. Another novel feature of the solution method was the incorporation of the effect of radiative transfer normal to the predominant flux direction without the introduction of additional flux components.

5.5 Strongly scattering systems

For cases in which scattering is a major component of the total attenuation two-flux solutions do not provide an adequate treatment. Chu and Churchill (31) have employed a six-flux approximation to account for scattering in such situations, which leads to six simultaneous ordinary differential equations in the unknown fluxes, which are generally solved by finite difference methods. For the case in which no emission of radiation occurs, and scattering and absorption are the only radiative mechanisms, analytical solutions can be found. Brinkworth (32) has applied a six-flux method to the study of absorption of light in aerosols.

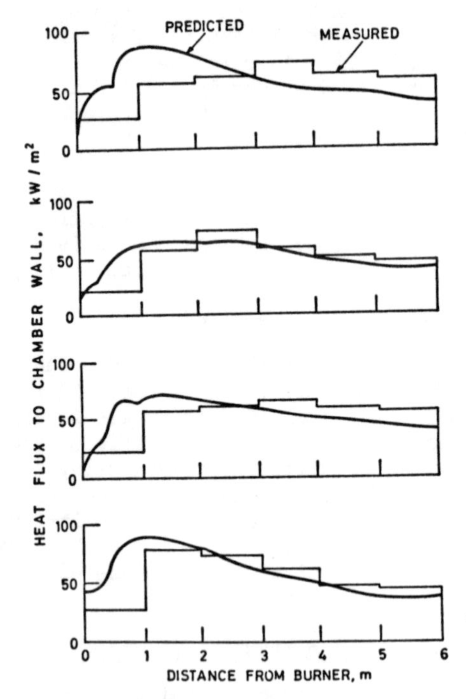

Fig.11. Comparison of predicted and measured heat flux to walls of p.f. combustion chamber (flux method) (after Gibson & Monahan (28)).

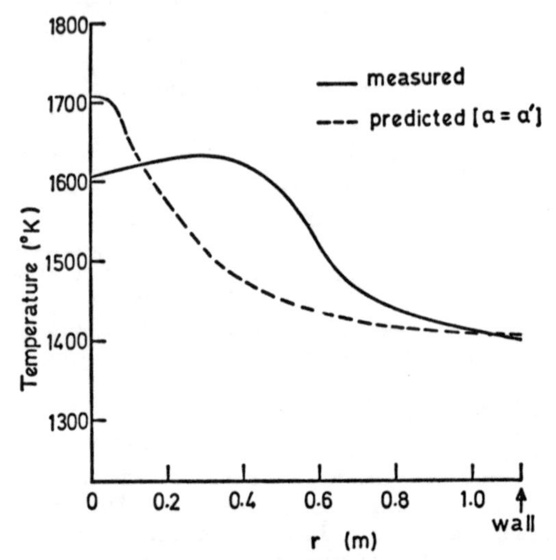

Fig.12. Comparison of predicted and measured radial temperature profiles (flux method). (after Gosman and Lockwood (33)).

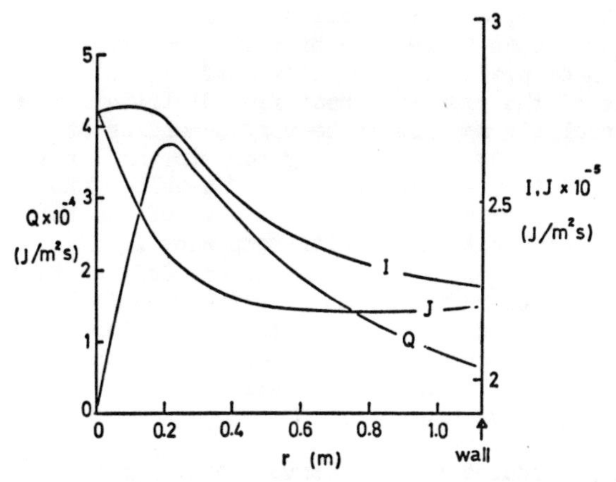

Fig.13. Predicted profiles of the radial fluxes of radiation and of heat (flux method).
(after Gosman and Lockwood (33)).

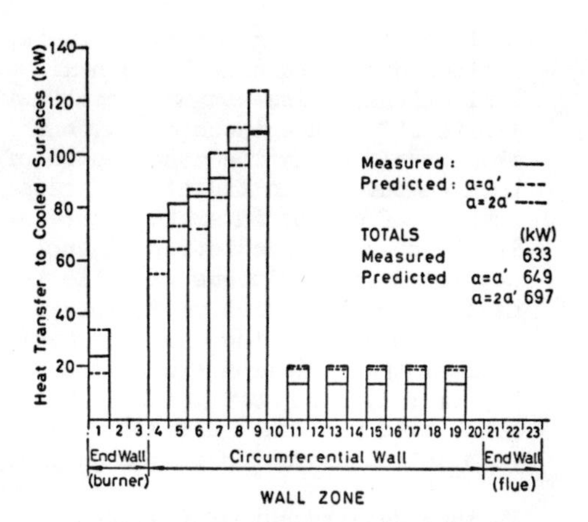

Fig.14. Comparison of predicted and measured heat transfer to water cooled surfaces (flux method).
(after Gosman and Lockwood (33)).

5.6 A four-flux solution for a combustion chamber.

Gosman and Lockwood (33) have recently employed a four-flux Schuster-Hamaker method in cylindrical polar co-ordinates to predict the radiative transfer from a gas flame in a furnace of square cross section (16) with simultaneous treatment of the flow and reaction, employing a more refined turbulence model than that utilised by Gibson and Monahan (28). Assuming axial symmetry and grey gas behaviour, the radiation field is represented by two fluxes in the axial and two fluxes in the radial direction. The predicted radial temperature distribution and the positive, negative and net radial fluxes for an axial station approximately midway between the end walls are shown in Figs. 12 and 13 respectively. The measured temperature distribution found by Johnson (2) is included in fig.12 for comparison. It can be seen that the four-flux method gives a reasonable prediction of temperature near to the chamber wall but is less satisfactory in the neighbourhood of the furnace axis. Fig.14 shows a comparison of predicted and measured heat transfer rates to the water-cooled surfaces at the furnace walls. The agreement is very satisfactory considering the simplicity of the flux model employed.

6. Conclusions

Techniques are now available for predicting radiative transfer in furnaces and combustors with allowance for variation in properties in any number of space dimensions, ranging from the well-stirred furnace model which assumes no variation with position within the enclosure, to the zone method of analysis and multi-flux methods which may be used for complete three dimensional predictions. The zone method is the most highly developed procedure now available, and has been rigorously tested against experimental data for a case in which complete information on the flow, composition and heat release patterns was available prior to the radiation calculation. It can also be conveniently combined with physical modelling of the flow and mixing pattern in the furnace or combustor, thus permitting application of the combined method of prediction to industrial furnaces of complex geometry and burner configuration. Prediction methods for determining flow and mixing patterns by computation are at present limited to systems of simple geometry. It is, however, expected that such methods will be further developed during the next decade so as to be free of their present limitations. The ultimate combination of predictions of flow patterns using the finite difference technique with those of the radiative heat flux distribution in the furnace using the zone method of analysis appears to be very demanding of both computer time and storage capacity, due to the necessity for storing large numbers of direct interchange areas. It is probable that future developments in the complete prediction of enclosure performance from a knowledge of the inlet and boundary conditions will lie in the application of multi-component flux methods to the radiative transfer. If 2 N component directions are defined and the emissivity of the enclosed medium is represented by an m term exponential series, the radiation field at any point can be completely described in terms of 2 N (m + 1) flux components. The improved angular and spectral resolution associated with the increased number of flux components should lead to substantial increases in the accuracy of solution.

Further development of the prediction procedures for radiative transfer must be accompanied by researches to provide further information on radiative properties and accurate data for testing the predictions:

I) Experimental information is required on the emissivity of gases at higher pressures to test the validity of the extrapolated values produced by use of the band models.

II) The complete prediction of luminous flame behaviour will require additional knowledge of the rates of formation and combustion of soot under various input conditions, and on how the optical properties of soot particles vary with their history. In the absence of such information the treatment of luminous radiation must be based on empirical data for absorption coefficients produced by measurement on a similar flame type.

III) Detailed experimental data on temperatures, flows, soot concentrations, gas compositions, heat release patterns and radiation flux densities for different flame types is desirable for an enclosure with a heat loading comparable with that encountered in current industrial practice, to provide a test basis for the calculation procedures.

IV) For the treatment of pulverised fuel flames it will be necessary to modify the flux methods to deal with strongly anisotropic scattering. Multi-component flux methods must also be compared with the zone method and experimental data to determine the minimum degree of angular and spectral resolution necessary to give acceptable accuracy (i.e. the lowest acceptable values of N and m).

REFERENCES

(1) Hottel, H.C., and Cohen, E.S. Radiant heat transfer in a gas filled enclosure: Allowance for non-uniformity of gas temperature. Amer. Inst. of Chem.Eng. J., 1958, 4, 3.
(2) Hottel, H.C., Section in heat transmission (McGraw-Hill Book Company, New York, 1954).
(3) Johnson, T.R. Application of the zone method of analysis to the calculation of heat transfer from luminous flames, Ph.D. Thesis, Sheffield University, 1971.
(4) Edwards, D.K. Radiative interchange in a non-grey enclosure containing an isothermal carbon dioxide/nitrogen gas mixture, J.Heat Trans., 1962, C84, 1-11.
(5) Boynton, F.P., and Ludwig, C.B. Total emissivity of hot water vapour - II. Semi-empirical charts deduced from long path spectral data. Int.J. Heat Mass Transfer, 1971, 14, 963-973.

(6) Hottel, H.C., and Sarofim, A.F. Radiative Transfer (McGraw-Hill Book Company, New York, 1967).
(7) Hadvig, S. Gas emissivity and absorptivity: A thermodynamic study, J. Inst.Fuel, 1970, 43, 129-135.
(8) Mie, G. Beiträge zur Optik trüber Medien, Speziall Kolloidaler Metallösungen. Annalen der Physik, 1908, 25, 377-445.
(9) Hawksley, P.G.W. The physics of particle size measurement - Part 2. Optical methods and light scattering. BCURA, Monthly Bulletin, April-May, 1952.
(10) Foster, P.J., and Howarth, C.R. Optical constants of carbons and coals in the infra-red. Carbon, 1968, 6, 719-729.
(11) Dalzell, H.W., and Sarofim, A.F. Optical constants of soot and their applications to heat flux calculations. Trans. ASME, J. Heat Trans. 1969, 91, 100-104.
(12) Hammond, E.G. Spatial distribution of spectral radiant energy in a small pressure jet oil flame, Ph.D. Thesis, Sheffield University, 1971.
(13) Foster, P.J. Calculation of the optical properties of dispersed phases. Combust. & Flame, 1963, 7, 277.
(14) Beér, J.M. and Claus, J. The 'traversing method' of radiation measurement in luminous oil flames. J.Inst.Fuel, 35, 1962, 437-443.
(15) Johnson, T.R., and Beér, J.M. The Zone Method of analysis of radiant heat transfer: a model for luminous radiation. Fourth Symposium on Flames and Industry, British Flame Res. Comm. and Inst.Fuel, Imperial College, Sept.1972.
(16) Johnson, T.R., and Beér, J.M. Further development of the Zone Method. Fourteenth Symposium (International) on Combustion, Penn. State Univ. Aug. 1972.
(17) Beér, J.M. and Siddall, R.G. Radiative heat transfer in furnaces and combustors. "Italian Flame Day", Pisa, 20/21 March, 1973. Italian Flame Res. Comm. Milano.
(18) Siddall, R.G. The flux method of furnace heat transfer analysis. Fourth Symposium on Flames and Industry. Brit. Flame Res. Comm. and Inst. Fuel at Imperial College, Sept. 1972.
(19) Schuster, A. Astrophys. J. 22, 1 (1905).
(20) Schwarzchild, K. Nachr. Akad. Wiss. Gottingen Math. - Physik Klass 1 (1906). Sitzber Preuss. Akad. Wiss. Physik-Math. Klasse 1183 (1941).
(21) Hamaker, H.C. Philips Research Repts. 2, 55, 103, 112 and 420 (1947).
(22) E.A. Milne. Handbuch der Astrophysik, 3, part 1, Springer. Berlin (1930).
(23) Eddington, A.S., The Internal Constitution of Stars, Dover Press, N.Y. (1959).
(24) Viskanta, R., Grosh R.J. T.A.S.M.E., J. Heat Transfer 84, 63 (1962).
(25) Larkin B.K., and Churchill, S.W. A.I.Ch.E. Journ. 5, 4, 467 (1959)
(26) Chen, J.C., and Churchill, S.W. A.I.Ch.E. Journ. 9, 1, 35 (1963).
(27) Chen, J.C. A.I.Ch.E. Journ. 10, 2, 253 (1964).
(28) Gibson, M.M., and Monahan, J.A. Int. J. Heat Mass Transfer 14, 141 (1971).
(29) Baker, P.C., Barker, M.H., D.J. Loveridge, G.G. Thurlow, and G.J. Wingfield, J.Inst.Fuel, 42, 371 (1969).
(30) Roesler, F.C., Chem.Eng. Science, 22, 1325 (1967).
(31) Chu, C.M., and Churchill, S.W. J.Phys.Chem., 59, 855 (1955).
(32) Brinckworth, B.J. Brit.J.Appl.Phys. 16, 1907 (1965).
(33) Gosman, A.D., and Lockwood, F.C. Paper presented at Fourteenth Symposium (International) on Combustion (1972).

Chapter 3

MATHEMATICAL SIMULATION OF AN INDUSTRIAL BOILER BY THE ZONE METHOD OF ANALYSIS

F. R. Steward and H. K. Gürüz

Fire Science Centre, University of New Brunswick
Frederiction, N.B., Canada

Abstract

A mathematical simulation of the heat transfer in a large modern boiler was carried out using the zone method of analysis. The input data required for this method were evaluated from the data normally available in an operating plant and the related literature by a combination of theoretical considerations and some simplifying assumptions. The predictions of the mathematical model were compared with the experimental data taken in a plant test and the level of agreement was reasonable.

NOMENCLATURE

a	:	weighting factor for various gray gases contributing to the absorptivity of the real gas, dimensionless
a'	:	weighting factor for various gray gases contributing to the emissivity of the real gas, dimensionless
A	:	area, ft^2
A_v, A_s	:	total radiant energy absorbed by a zone, Btu/hr.
B	:	center to center distance between two zones, ft
B_v	:	the total sensible heat content of the gas flowing into a volume zone, Btu/hr.
C_v, C_s	:	the net heat flow by convection to or from a zone, Btu/hr.
C_p	:	mean heat capacity of gases, Btu/lb°R
C_s'	:	mean soot particle concentration in the furnace lb/ft^3
d_o	:	outer diameter of a burner, ft
D_p	:	mean particle diameter, ft. (or microns)
D_v	:	heat generated by combustion within a volume zone, Btu/hr.
E_v	:	the total sensible heat content of gases flowing from a volume zone, Btu/hr.
f	:	fractional combustion in a volume zone
(f_{ij})	:	fraction of the radiant energy emitted by a zone i that is eventually absorbed by another zone j, dimensionless
F_v, F_s	:	the total radiant energy emitted by a zone, Btu/hr.
F_R	:	firing rate, Btu/hr. ft^2 sink
$(\overline{G_u S_s})$	:	the total radiative interchange area between two volume zones, ft^2
$(\overline{G_u S_u})$	:	the total radiative interchange area between a volume and a surface zone, ft^2
h	:	convective heat transfer coefficient, Btu/hr. ft^2 - °F
i	:	length, index, dimensionless
i	:	$\sqrt{-1}$
j	:	height index, dimensionless
k	:	width index, dimensionless

K_n	:	gray gas absorption coefficient, ft^{-1}
K_a	:	soot particle absorption coefficient, ft^{-1}
L_m	:	mean beam length of the enclosure, ft
m	:	mass flow rate, lb/hr
M	:	momentum flux parallel to jet axis, lb - ft/hr^2
n'	:	complex refractive index, dimensionless
n_1	:	real part, dimensionless
n_2	:	complex part, dimensionless
Nu	:	Nusselt number, dimensionless
Pr	:	Prandtl number, dimensionless
Q_s	:	total net heat flow through a surface zone, Btu/hr.
r'	:	equivalent nozzle radius, ft
Re	:	Reynolds number, dimensionless
$(\overline{S_rG_v})$	:	total radiative interchange area between a surface and a volume zone, ft^2
$(\overline{S_rS_s})$	:	total radiative interchange area between two surface zones, ft^2
T	:	temperature, °R
V	:	volume, ft^3
W	:	gas flow from one volume zone to another, lb/hr.

Subscripts

a	:	absorption
e	:	property of the mass being entrained by jet
g	:	gas
gp	:	gas plus solid particles
f	:	mean flame property
n	:	n^{th} gray gas
p	:	soot particle properties
r	:	a surface zone
r'	:	a neighboring surface zone
s	:	a surface zone
t	:	total
u	:	a volume zone
u'	:	a neighboring volume zone
u'→v	:	from a neighboring volume zone u' to the volume zone v
v	:	a volume zone
1	:	primary stream
2	:	secondary stream

1. Introduction

The analysis of radiative heat transfer within an industrial combustion chamber has in the past normally been restricted to two simplified cases; the well stirred enclosure which assumes perfectly mixed combustion products (1) and the long furnace model which assumes plug flow of combustion products with no radial temperature gradients (1, 2).

The desire to predict the detailed heat flux and temperature distributions in a furnace enclosure has led to the development of the zone method of analysis (3). For this method the enclosure and its surrounding surfaces are divided into a number of volume and surface zones each assumed uniform in properties. A steady state energy balance is written on each zone. The resulting set of simultaneous nonlinear algebraic equations can be solved given, the gas flow entering and leaving each volume zone, the fraction of the fuel combusted in each volume zone, convective heat transfer coefficients at the enclosing surfaces, descriptions of the radiative properties of the enclosed medium and its surrounding surfaces, distribution of the radiant species in the enclosure, and the total radiative interchange in the system. The development of high speed digital computers has made this type of analysis possible.

Because of the uncertainties encountered in the evaluation of the input data and the computational effort involved, few applications of the zone method of analysis are reported in literature. Hottel and Sarofim (1,4) used the zone method of analysis to investigate the effect of gas flow patterns on the radiative heat flux distributions in an idealized enclosure. In the above mentioned study, the total radiative interchange in the system was calculated by using the determinant method (1). Radiative heat flux distributions calculated using the Monte Carlo method to evaluate the total radiative interchange were found to compare favorably with those obtained by using the determinant method in this idealized enclosure (5,6). The Monte Carlo method was found to be flexible and easy to formlate. The accuracy of the zone method was later tested by experimental measurements taken in a laboratory furnace (7,8,9). The zone method of analysis was further tested by comparing its predictions with the experimentally measured heat flux and temperature distributions in a pilot scale furnace under more severe conditions (10). However, only a single application of this method to an industrial boiler is reported in literature (11). In this study the predicted heat flux distributions were found to be in qualitative agreement with experimental measurements taken in a gas burning boiler.

As a result of the above mentioned studies it can be concluded that the zone method of analysis is a reliable methematical description of the physical heat transfer occurring in a furnace enclosure. However, in so far as its application to an industrial boiler is concerned, the model is limited by the lack of data to accurately describe the flow and combustion patterns within the enclosure and the luminous radiation from soot particles within the flame.

On the other hand, a significant amount of experimental and theoretical effort has been spent in understanding the flow and the mixing behaviour of confined and free jet systems (12, 13, 14, 15,16). In addition to these studies, experimental data on the combustion patterns and soot particle concentration distributions, resulting from the combustion of fuel oils, have been taken under conditions similar to those that prevail in an industrial boiler (17, 18).

It was, therefore, felt that a mathematical simulation of the heat transfer in an industrial boiler could be carried out using the results of the above mentioned studies and the data normally available in an operating plant to determine the input data required for such a calculation.

Predictions of the mathematical model are compared to the experimental data taken in a plant test in order to assess the degree of confidence that can be placed in results obtained from such a calculation.

2. Description of the Boiler and Plant Operating Conditions

The furnace under consideration is shown in Figure 1 and is a Babcock and Wilcox radiant boiler with a design capacity of 700,000 lbs. steam (1925 psig)/hr., presently in operation at the Courtenay Bay Power Station in St. John, N.B.. The combustion chamber is approximately 90 feet in height, 30 feet in width and 22 feet in length with certain irregularities. The walls of the enclosure are covered by tube sheets made from 2.5" o.d. stainless steel tubes connected by 0.5" sheets with a center to center distance of 3.0" except for the superheater where the gases leave the combustion chamber. The furnace is fired with a residual oil through twelve burners in three rows of four located in the lower part of the chamber below the combustion gas exit to the super heater section.

A plant test of six hours duration was carried out on the boiler studied. During this time, in which operating conditions were held approximately constant, sufficient data was collected so that overall mass and energy balances could be made on the boiler. In the same period heat fluxes to the tube walls were measured through the various access ports located on the walls of the boiler. Under these operating conditions the firing rate, based on the lower heating value of the fuel and flat sink wall area, was 94,500 Btu/hr. ft^2 with part of the stack gases being recycled into the combustion chamber through an opening at the bottom of the chamber. Detailed information about this plant test can be found elsewhere (19).

3. Simplified Zoning of the Enclosure

For the purpose of the mathematical analysis the furnace was represented in the form of a rectangle with the same flat wall surface area as the actual furnace with an irregularity representing the superheater, as shown in Figure 2. Also shown in the same figure is the furnace subdivision into surface and volume zones obtained by dividing the height (j direction) into five equal increments, the length (i direction) into three equal increments, and the width (k direction) into four equal increments. Since the furnace is symmetric around its center line along its width only half of the system was considered in the analysis. Thus, a total of 45 surface and 26 volume zones were considered in the analysis. Each volume zone was assumed to have the same carbon dioxide and water vapor concentrations corresponding to those of complete combustion, and the same soot particle concentration obtained from the results of a previous study (18).

The complicated tube wall geometry was represented by an equivalent plane (1) with an emissivity of 0.85, by assuming it to be infinite in extent. To complete the system the superheater was also represented by an equivalent plane with an emissivity of 0.76.

4. Zone Method of Analysis

4.1 Steady State Energy Balances on Zones

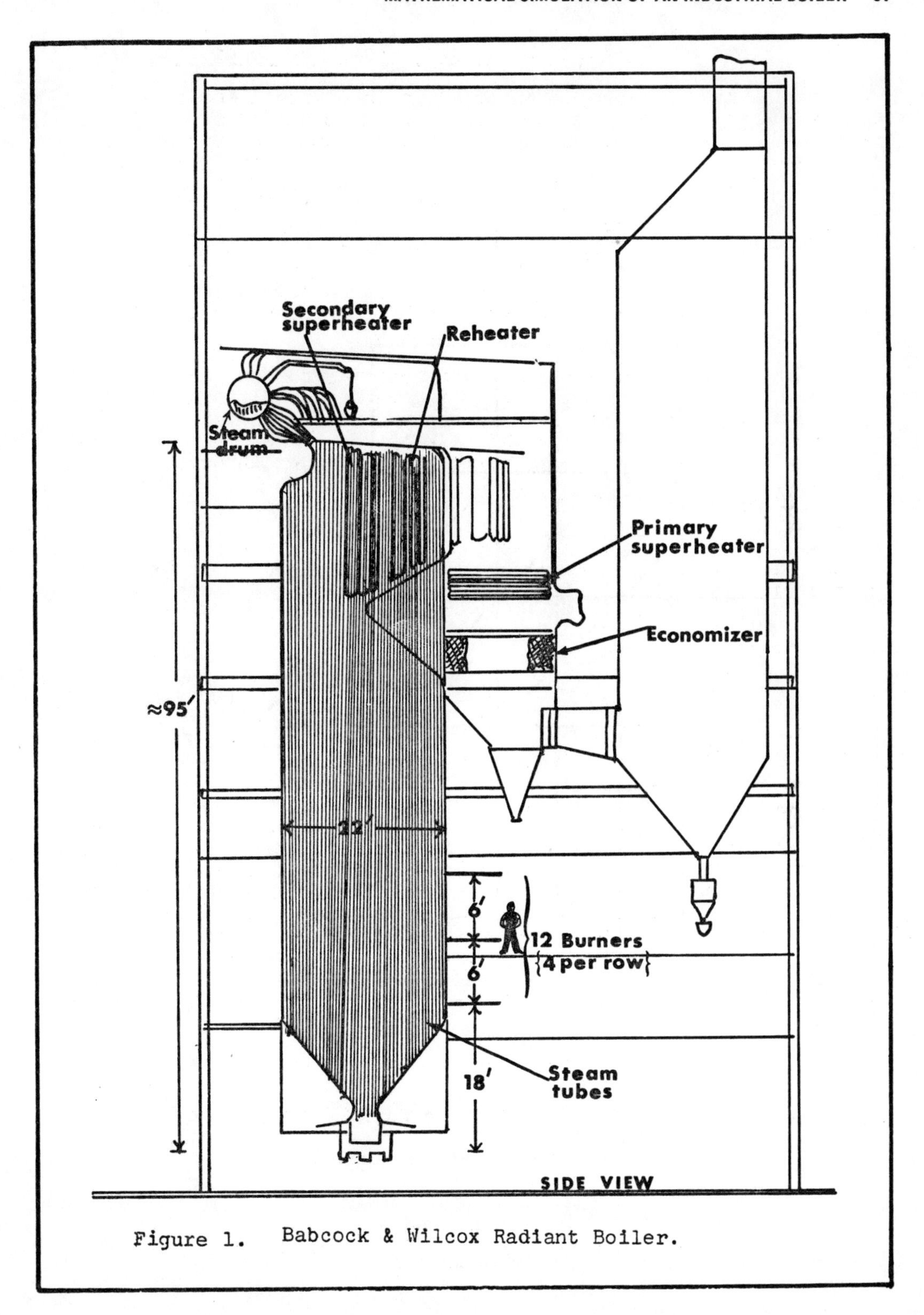

Figure 1. Babcock & Wilcox Radiant Boiler.

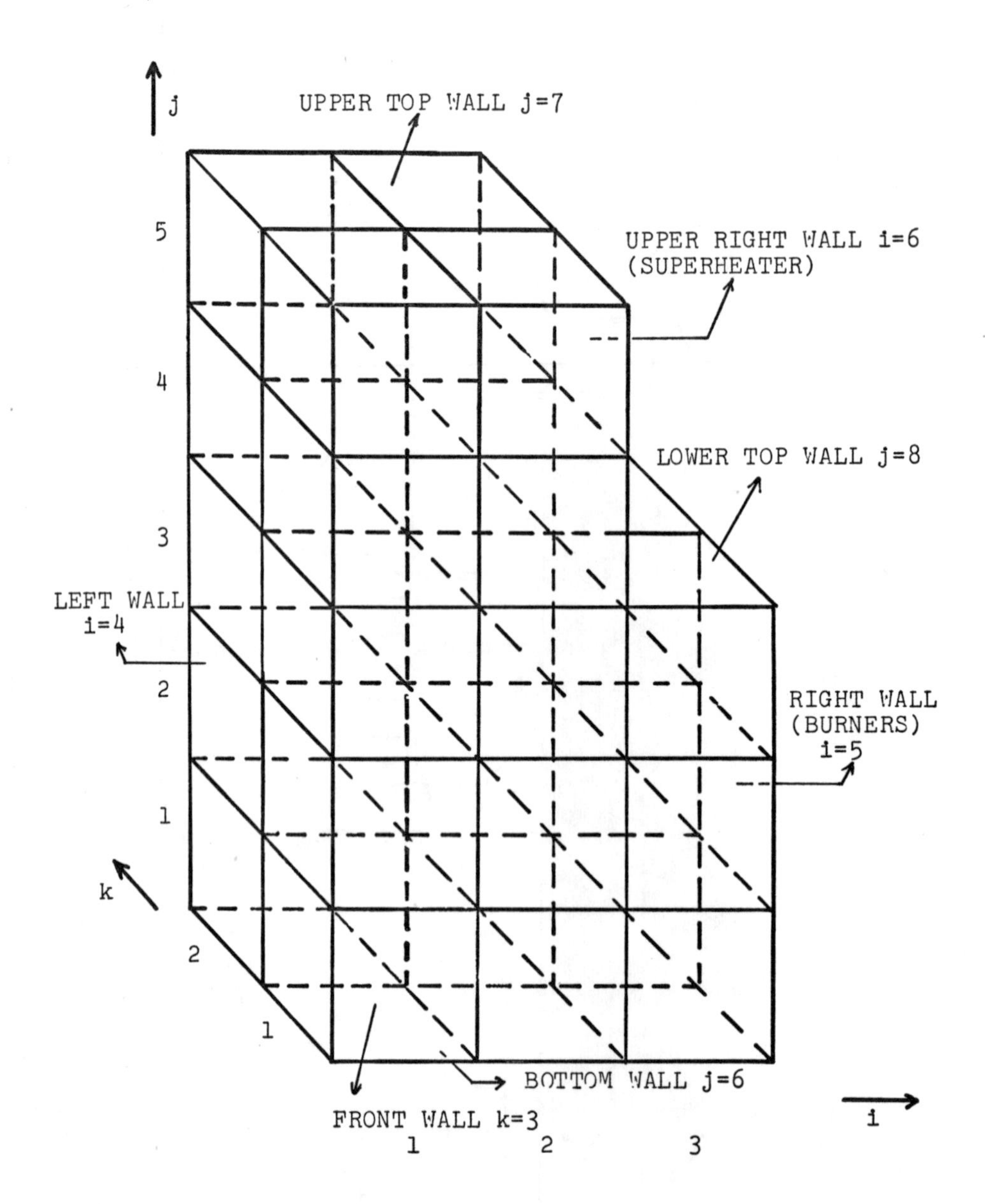

Figure 2 Boiler Subdivision into Surface and Volume Zones and the Indexing of Elements. Only one half of the Boiler in the k Direction is Shown.

A single volume zone is assumed to be perfectly mixed within itself and therefore has a single temperature which must be determined by a system of steady state energy balances written on each surface and volume zone in the system.

A steady state energy balance on each volume zone v can be broken down as follows (3, 4, 5, 6, 8):

A_v (the total radiant energy absorbed in v which was emitted by all the surface and volume zones in the system including v itself),
$+B_v$ (the total sensible heat content of the gas flowing into v),
$+C_v$ (the net heat flux by convection to or from surface zones adjacent to v),
$+D_v$ (the heat generated by combustion within v),
$-E_v$ (the total sensible heat content of the gas flowing from v),
$=F_v$ (the total radiant energy emitted by v).

In a similar fashion a steady state energy balance can be written on each surface zone s:

A_s (the total radiant energy absorbed by s which was emitted by all the surface and volume zones in the system including s itself),
$+C_s$ (the net heat flux by convection to or from volume zones adjacent to s),
$+Q_s$ (the total net heat flow through s),
$=F_s$ (the total radiant energy emitted by s).

The above energy balances result in a set of simultaneous equations equal in number to the total number of volume and surface zones in the system. The unknowns which cannot be determined by independent means are the temperatures of the volume zones and the total net heat flux through each zone, the temperature of the tube surfaces being fixed at the saturation temperature of the steam being generated and that of the superheater at a given superheat temperature. Therefore, a unique solution exists in principle, since the number of equations just equals the number of unknowns. Each of the other terms in the above equations can be determined by a combination of theoretical considerations and available experimental data.

4.2 Flow Pattern and Sensible Enthalpy Terms

The flow pattern resulting from a single burner is essentially a double concentric swirling jet, the swirl being generated by vanes in the annular space through which the secondary air enters the combustion chamber. At present there is little data available to calculate the flow and combustion patterns for a burner system of this kind. It is, however, believed that the effect of swirl lasts for a relatively short distance from the burner mouth and thereafter the flow behaves as a free jet. For this reason two of the flow patterns used in this study were based on the data available for such systems (15, 16).

The first flow pattern used was based on the data of Beer et al (15). The equivalent nozzle radius concept of Thring and Newby (14) was calculated from the combined masses and momenta of the primary and secondary streams as:

$$r' = \frac{(m_1 + m_2)}{[(M_1 + M_2)\ \rho_f \pi]^{1/2}} \tag{1}$$

where the momentum and mass flux through a single burner were evaluated at the measured inlet temperatures of fuel and air while the mean flame density within the jet, ρ_f, was evaluated at a mean flame temperature, T_f, corresponding to the arithmetic average of the adiabatic flame and measured exhaust gas temperatures. The ratio of the recirculating mass to the total mass flow was obtained from Figure 3 of the same reference for each subdivision along the lower part of the combustion chamber.

The free jet equation of Ricou and Spalding (16)

$$\frac{m}{m_t} = 0.32\ \frac{X}{d_o} \left(\frac{\rho_e}{\rho_f}\right)^{1/2} \tag{2}$$

gives the total mass flow rate in the jet, m, for a single burner at each cross section along the length of the furnace. The density of the gas being entrained, ρ_e, was evaluated at the measured exhaust gas temperature, while d_o was taken as the outside diameter of the double concentric burner.

In both cases the external recycle stream was assumed to supply part of the entrainment required by the jet for volume zones next to the bottom wall.

The third flow pattern used was a pseudo plug flow which does not involve any recirculation. This was obtained by superimposing a uniform burner flow and a uniform recycle flow.

In all cases the remainder of the flow pattern was fixed by individual total mass balances on each of the volume zones. Turbulent mixing between the volume zones was neglected as it can be shown that these flows are small compared to the bulk flow for volume elements of this size (20).

The flow pattern based on the data of Beer et al is shown in Figure 3. It must be pointed out that these flow patterns should only be considered as rough estimates of those in the actual furnace.

The specific heat function used is given by (4)

$$C_p = (7.171 + 0.1705 \times 10^{-3}\ T + 0.653 \times 10^{-7}\ T^2)/28.9 \tag{3}$$

This function gives the mean heat capacity in Btu/lb.°R between 0°R and T°R.

With these approximations the sensible enthalpy terms can be written as:

$$B_v = W_{u' \to v}\ (C_p\ T)_{u'} \tag{4}$$

$$E_v = W_{tv}\ (C_p\ T)_v \tag{5}$$

where $W_{u' \to v}$ is the mass flow rate of gas entering the volume zone v from a neighboring volume zone u', and W_{tv} is the total mass flow rate of gas leaving the volume zone v.

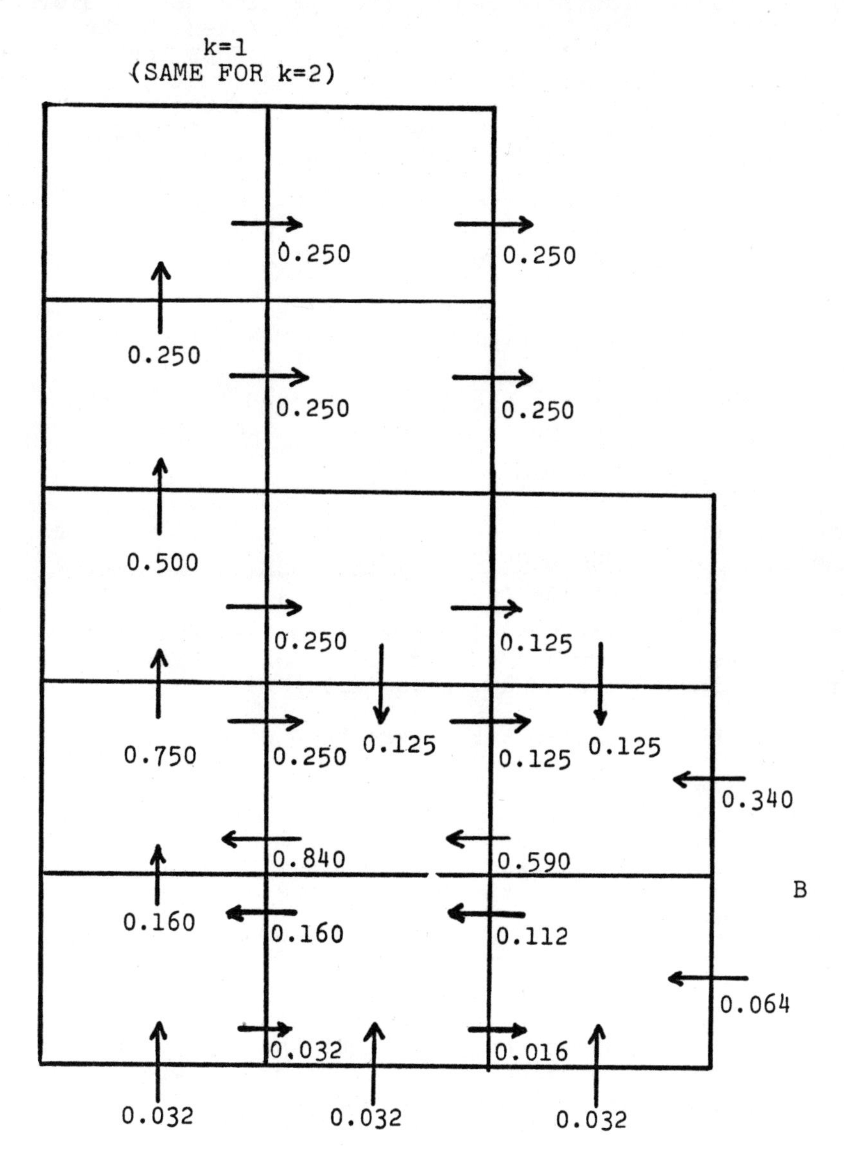

Figure 3 Flow Pattern Calculated from the Data of Beer et al (15). Flows as Fractions of Total Flow.
Firng Rate = 94,500 Btu/hr.ft.^{2}sink

4.3 Combustion Pattern

The calculation of the combustion pattern was based on the data of Maesawa et al (17). The fraction of the fuel combusted before a particular distance from the burner mouth was taken as the ratio of that distance to the total flame length defined as that length in which stoichiometric mixing is complete. This quantity was obtained from Figure 10 of the same reference in terms of the fuel flow rate and the momentum of the resulting jet parallel to its axis. Fractional combustion before any particular distance was then distributed among volume zones in front of the burners in proportion to the number of burners that a volume zone encompasses along its height. Calculations carried out in this fashion indicated that combustion was completed in volume zones adjacent to the wall on which burners are located.

The chemical enthalpy term, D_v, could then be written as:

$$D_v = f_v F_R A_t \tag{6}$$

where f_v is the fraction of the fuel combusted in the volume zone v, F_R is the firing rate and A_t is the total sink area.

4.4 Convective Heat Transfer Coefficients

An estimate of the convective heat transfer coefficient at the tube walls was obtained by assuming that (2)

$$N_u = 0.069\ Re^{0.8} Pr^{0.4} \tag{7}$$

while the convective heat transfer coefficient at the wall representing the superheater was obtained from manufacturer's data for flow perpendicular to a bank of tubes (21).

The convective heat transfer terms were then written as

$$C_v = \sum_{r'} h_{r'} A_{r'} (T_{r'} - T_v) \tag{8}$$

$$C_s = h_s A_s (T_{u'} - T_s) \tag{9}$$

where r' is a surface zone adjacent to the volume zone v, and u' is a volume zone adjacent to the surface zone s.

4.5 Radiative Properties of the Enclosed Medium

A technique for representing a real gas, in the absence of soot particles, by a mixture of gray gases each with its own absorption coefficient has been shown to approximate the emissivity and absorptivity of carbon dioxide and water vapor mixtures to a high degree of accuracy (1,4). The emissivity and absorptivity of the gas mixture formulated on this basis are given by

$$\varepsilon_g (T_g) = \sum_n a'_n (T_g) (1 - e^{-K_n L_m}) \tag{10}$$

$$\alpha_g (T_s) = \sum_n a_n (T_s) (1 - e^{-K_n L_m}) \tag{11}$$

where

$$\sum_{n+1} a'_n (T_g) = 1 \text{ and } \sum_{n+1} a_n (T_s) = 1$$

It has been shown that (4) a value n = 3 (three absorbing gases plus one clear gas) well represents the above radiative properties of the gas mixture under consideration if the variation of the weighting factors with temperature is given by third order polynomials such as:

$$a'_n (T_g) = a'_{no} + a'_{n1} T_g + a'_{n2} T_g^2 + a'_{n3} T_g^3$$
$$a_n (T_s) = a_{no} + a_{n1} T_s + a_{n2} T_s^2 + a_{n3} T_s^3 \qquad (12)$$

and the gray gas absorption coefficients for emissivity and absorptivity are taken as equal and independent of temperature. The numerical values of the various coefficients in the above relations have been determined for a carbon dioxide and water vapor mixture each with a partial pressure of 0.115 atm. (4). The values used in the present study are those recommended by Hottel and Sarofin (4) with some corrections (20).

The weighted gray gas mixture technique is a method of curve fitting based on Hottel's emissivity charts (1). Physically the absorbing gray gases represent combinations of absorption bands of different strengths while the clear component represents the windows in the combined spectrum of carbon dioxide and water vapor, and the weighting factors correspond to the fractional amount of the blackbody radiation associated with each gray gas.

The interaction of a solid particle with radiation incident on it depends on the dimensionless quantities: the complex refractive index, $n' = n_1 - in_2$, size parameter, the ratio of characteristic particle dimension to the wavelength of the incident radiation and the particle shape. The solution to this problem was obtained by Gustav Mie by solving Maxwell's wave equation with the appropriate boundary conditions for single spherical and cylinderical particles (1).

In so far as calculating the abosrption coefficient of a cloud of solid particles is concerned the quantity of interest is the absorption efficiency, X_a, defined as the ratio of the absorption cross section to the geometrical cross section of the solid particle. The total extinction efficiency of the cloud is the sum of the absorption and the scattering efficiencies

$$X_t = X_a + X_s \qquad (13)$$

For a cloud of spherical particles with a mean particle diameter, D_p, the absorption coefficient is given by

$$K_a = \frac{3}{2} \frac{X_a C'_s}{D_p \rho_p} \qquad (14)$$

where C'_s is the solid particle concentration in lb/cu.ft., and ρ_p is the density of the solid material.

From the point of view of calculating the total radiative interchange in the system there is considerable advantage in treating a cloud of solid particles as a gray gas although, the solid particle absorption coefficient is a strong function of wavelength.

For this purpose a mean wavelength, λ_m, was calculated from the equal energy division relation for blackbody radiation

$$\lambda_m T = 0.4011 \text{ (cm.°K)} \qquad (15)$$

The temperature for this purpose was taken as the previously mentioned mean flame temperature.

The quantities required for calculating the soot particle absorption coefficient, were taken from a recent experimental study carried out under conditions similar to those prevailing in the boiler studied (18). The quantities used in the present study were

Complex refractive index: $n' = 2.29 - 1.49\ i$
mean soot concentration : $C_s' = 3.09 \times 10^{-6}$ lb./cu.ft.
mean particle diameter : $D_p = 0.05$ microns
density of soot (ultimate): $\rho_p = 124.8$ lb./cu.ft.

The soot particle absorption coefficient, K_a, calculated by using the above and the Mie equations for spherical particles was 0.0184 ft.$^{-1}$. The scattering coefficient was negligible.

By treating a cloud of solid particles as a gray gas the emissivity, ε_p, and the absorptivity, α_p, of the cloud are both given by

$$\varepsilon_p = \alpha_p = (1 - e^{-K_a L_m}) \qquad (16)$$

Solid particles radiate over the entire spectrum as opposed to discontinuous radiation from polyatomic gas molecules. Since the absorption coefficients of radiant species whose spectra overlap are additive, the emissivity and absorptivity of a cloud of soot particles suspended in carbon dioxide and water vapor mixture is obtained by adding the soot particle absorption coefficient to the absorption coefficient of each of the gray gases including the clear component. Therefore, the luminous emissivity, ε_{gp} (T_g), and the luminous absorptivity, α_{gp} (T_s), formulated on the basis of the weighted gray gas mixture model are given by

$$\varepsilon_{gp}(T_g) = \sum_{n+1} a_n'(T_g)\ (1 - e^{-(K_n + K_a)L_m}) \qquad (17)$$

$$\alpha_{gp}(T_s) = \sum_{n+1} a_n(T_s)\ (1 - e^{-(K_n + K_a)L_m}) \qquad (18)$$

where the weighting factors were taken the same as those for a non-luminous mixture.

4.6 Radiant Emission and Absorption Terms

With the above approximations the rate of radiant energy emitted by a volume zone, F_v, is given by

$$F_v = 4\ V \sum_{n+1} a_n'(T_v)\ (K_n + K_a)\ \sigma T_v^4 \qquad (19)$$

and that for a surface zone is given by

$$F_s = A_s\ \varepsilon_s\ \sigma T_s^4 \qquad (20)$$

The total amount of radiant energy absorbed by a volume zone was expressed in terms of the total radiative interchange areas between the volume zone v and all other volume and surface zones in the system including v itself. Therefore,

$$A_v = \sum_u \sum_{n+1} a_n'(T_u)\ (\overline{G_u G_v})\ \sigma T_u^4 + \sum_r \sum_{n+1} a_n(T_r)\ (\overline{S_r G_v})_n\ \sigma T_r^4 \qquad (21)$$

The corresponding expression for a surface zone is

$$A_s = \sum_u \sum_{n+1} a'_n(T_u)\ (\overline{G_u G_s})_n\ \sigma T_u^4 + \sum_r \sum_{n+1} a_n(T_r)\ (S_r S_s)_n\ \sigma T_r^4 \qquad (22)$$

where the first summations are over the total number of volume or surface zones in the system and the second summations are over the number of gray gases.

In the present study the total radiative interchange areas were calculated by the Monte Carlo method (5, 6) for the first gray gas and the clear component where only soot particles radiate and by the diffusion approximation for the second and the third gray gases. In this application of the Monte Carlo method a prescribed number of bundles are emitted from each zone for each gray gas. Each bundle emitted in a gray gas from a given zone carried the same amount of radiant energy. All bundles are followed from initial points of emission through randonly determined paths until final points of absorption. The coordinates of the final absorption point can readily be related to the indices of the zone in which absorption has occurred.

After all bundles are released and followed the fraction of the radiant energy emitted in the nth gray gas from a volume zone u which is eventually absorbed in another volume zone, $(f_{uv})_n$, is readily calculated as the ratio of the number of bundles that are absorbed in v to the total number of bundles emitted from u in the nth gray gas. The total interchange area between the volume zone u and the volume zone v in the nth gray gas, $(\overline{G_u G_v})_n$, is then calculated from

$$(\overline{G_u G_v})_n = 4\ (K_n + K_a)\ V\ (f_{uv})_n \tag{23}$$

For radiation originating from a surface zone r and travelling in the nth gray gas, the total interchange area between two surface zones r and s is given by

$$(\overline{S_r S_s})_n = A_r \varepsilon_r\ \ (f_{rs})_n \tag{24}$$

In a similar fashion, the total interchange area between a volume and a surface zone in the nth gray gas, $(\overline{G_u S_s})_n$, and that between a surface and a volume zone, $(\overline{S_r G_v})_n$, are given by

$$(\overline{G_u S_s})_n = 4(K_n + K_a)\ V\ (f_{us})_n \tag{25}$$

$$(\overline{S_r G_v})_n = A_r \varepsilon_r\ (f_{rv})_n \tag{26}$$

In those cases where the product of center to center distance between two elements and the gas absorption coefficient is greater than 3.0 the total radiative interchange areas are readily calculated from the diffusion approximation as (1):

$$(\overline{G_u G_v})_n = \frac{4\ A}{3\ (K_n + K_a)\ B} \tag{27}$$

$$(G_u S_s)_n = \frac{8A}{3(K_n + K_a)\ B} \tag{28}$$

where A is the interfacial area and B is the center to center distance between two zones. In this approximation only neighboring zones are assumed to exchange radiation.

Since, according to the weighted gray gas mixture model, gas and soot absorption coefficients are taken as independent of temperature the total radiative interchange areas are independent of the temperature distribution in the enclosure.

4.7 Calculation Procedure

When the above relations are substituted into the original energy balances the following equations are obtained: for a volume zone v,

$$\sum_{u}\sum_{n+1} a_n'(T_u)(\overline{G_uG_v})_n\,\sigma T_u^4 + \sum_{r}\sum_{n+1} a_n(T_r)(\overline{S_rG_v})_n\,\sigma T_r^4$$

$$+ \sum_{u'} W_{u'\rightarrow v}(C_pT)_{u'} + \sum_{r'} h_{r'}A_{r'}(T_{r'} - T_v) + F_RA_tf_v$$

$$- W_{tv}(C_pT)_v = 4\,\sigma V \sum_{n+1} a_n'(T_v)(K_n + K_a)\;T_v^4 \qquad (29)$$

and for a surface zone s,

$$\sum_{u}\sum_{n+1} a_n'(T_u)(\overline{G_uS_s})_n\;\sigma T_u^4 + \sum_{r}\sum_{n+1} a_n(T_r)(\overline{S_rS_s})_n\;\sigma T_r^4$$

$$+ h_s A_s (T_{u'} - T_s) + Q_s = A_s\,\varepsilon_s\,\sigma T_s^4 \qquad (30)$$

Calculations start by feeding the variables describing the geometry and subdivision of the furnace together with surface and gas radiative properties. The total radiative interchange areas are calculated and stored in memory.

In the next step flow and combustion patterns, convective heat transfer coefficients, fixed surface temperatures and an initially assumed gas temperature distribution are supplied as input. The set of equations resulting from steady state energy balances on volume zones is simultaneously solved for zone temperatures and thereafter the total net heat flow to each surface zone, Q_s, is calculated. Intergration of the total net heat flux distribution over the tube walls gives the total rate of heat transfer to the feedwater in the tubes from which the rate of steam generation is calculated by dividing this quantity with the enthalpy required to produce one lb of saturated steam from the feedwater entering the boiler feed drum.

Computations were carried out on an IBM 370 Model 155 digital computer. A typical computation time was 574 seconds. A listing of the computer programme is shown elsewhere (20).

5. Results

5.1 Temperature Distributions

Calculations were carried out both with and without soot particle radiation being included in the description of gas radiative properties. Figures 4 and 5 show the predicted temperature distributions for different conditions with respect to flow pattern.

Gas temperatures rapidly rise in volume zones next to the burners where combustion is occurring. As a result of heat transfer by convection and radiation from the gas to the enclosing walls gas

Firing Rate = 94,500 Btu/hr.ft.^{2}sink

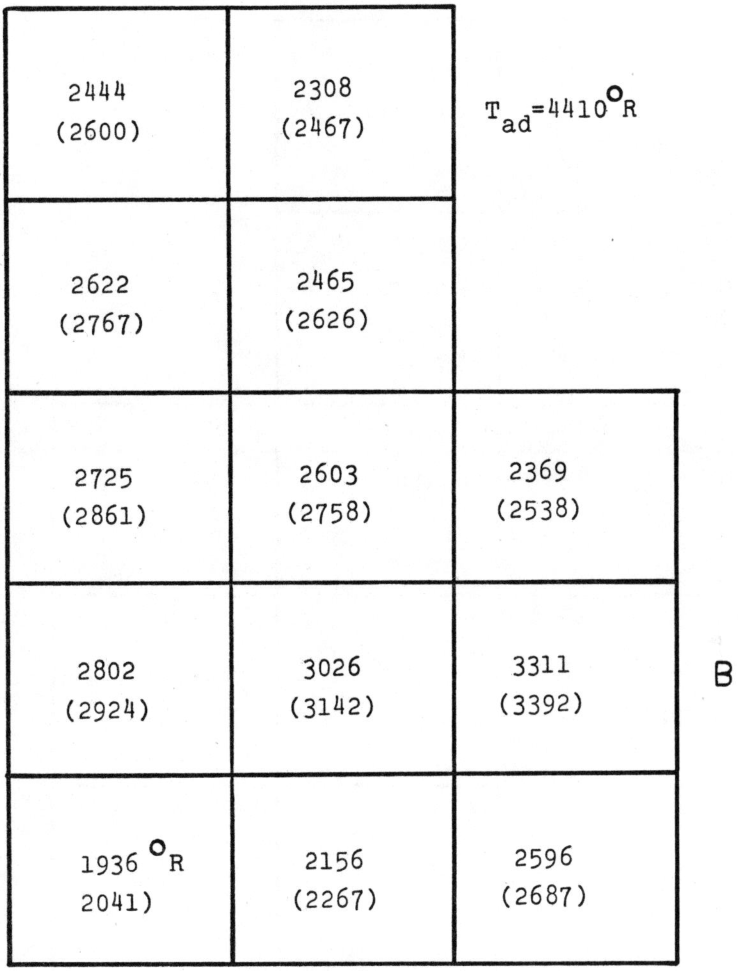

$T_{ad}=4410^oR$

2444 (2600)	2308 (2467)		
2622 (2767)	2465 (2626)		
2725 (2861)	2603 (2758)	2369 (2538)	
2802 (2924)	3026 (3142)	3311 (3392)	B
1936 ^oR 2041)	2156 (2267)	2596 (2687)	

Figure 4 Temperature Distribution Calculated by Using the Flow Pattern Calculated from the Data of Beer et al (15) and Including Particle Radiation. Temperatures in Parentheses are for k = 2 Volume Zones.

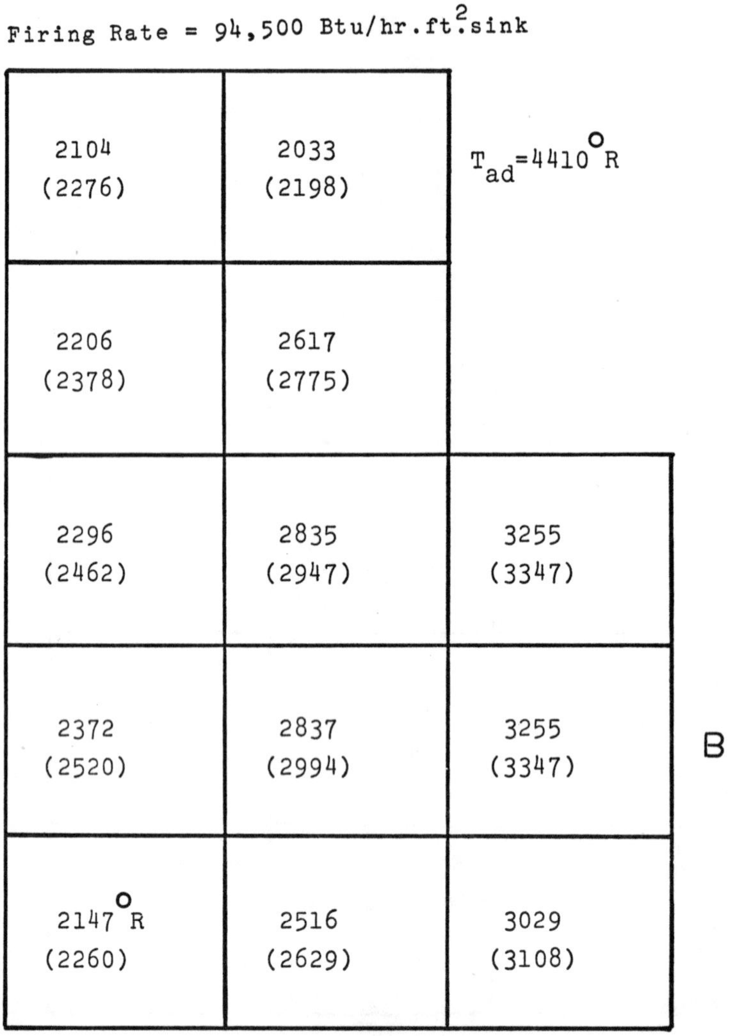

Firing Rate = 94,500 Btu/hr.ft.^{2}sink

2104 (2276)	2033 (2198)	T_{ad}=4410 $^\circ$R	
2206 (2378)	2617 (2775)		
2296 (2462)	2835 (2947)	3255 (3347)	
2372 (2520)	2837 (2994)	3255 (3347)	B
2147 $^\circ$R (2260)	2516 (2629)	3029 (3108)	

Figure 5 Temperature Distribution Calculated by Using the Pseudo Plug Flow Pattern and Including Particle Radiation. Temperatures in Parentheses are for k = 2 Volume Zones.

temperature decreases along the length and further along the height of the furnace. In all cases the temperatures of the inner volume zones (k=2) are higher than those of the outer volume zones (k=1) at a given height and length. This is due to the convective heat transfer from the latter to the front wall. The relatively lower temperatures of volume zones adjacent to the bottom wall result from the mixing of burner flow with the external recycle stream.

A comparison of Figure 4 with Figure 5 shows that the postulated flow pattern has an important effect on the predicted temperature distribution. In the lower part of the furnace temperatures calculated using the pseudo plug flow pattern are in general higher than those calculated using the recirculating flow pattern based on the data of Beer et al (15). This is because recirculation of relatively cooler combustion products mix with hotter gases coming from the burners and lower temperatures results.

A temperature measurement was taken at only one point in the furnace. This was at the top of the furnace about three feet in front of the superheater tubes. The temperature measured by a suction pyrometer equipped with a Pt/Pt-10%Rh thermocouple was 2534°R. This is in reasonable agreement with the exhaust gas temperature of 2463°R calculated from the predicted temperature distribution shown in Figure 4.

5.2 Heat Flux Distributions

Figure 6 shows the predicted total net heat flux distribution along the front wall. It is seen that the heat flux to the sink surface zones depends mainly on the temperature of adjacent volume zones. Figure 7 shows the effect of the postulated gas flow pattern on the total net heat flux distribution along the left wall. The general trend in the heat flux distribution, a maximum at a height of 36 feet from the furnace bottom, appears to be independent of the postulated gas flow pattern. However, the actual amount of heat flux to each surface zone does depend on the postulated gas flow pattern. The two recirculatory flow patterns give nearly the same heat transfer while the pseudo plug flow pattern yields lower heat fluxes to this particular wall. It is interesting to note that the maximum heat flux occurs at exactly the same height as the maximum in the temperature distribution of volume zones adjacent to this wall. The heat flux distributions shown in Figure 7 are very similar to that experimentally measured in a coal fired steam generator of similar geometry and burner arrangement (22).

A typical comparison of the predicted heat flux distributions obtained for the same postulated gas flow pattern, with and without the soot particle radiation being included, is shown in Figure 8. It is seen that the soot particle radiation does result in a significant increase in the radiative heat flux to surrounding walls.

5.3 Comparison of the Predicted Heat Fluxes with the Measured Values

During the plant test heat flux measurements were taken through the various access ports located on the walls of the furnace. The probe used for this purpose consisted of a 0.005" thick constantan foil soldered over the opening of a copper block which acted as a

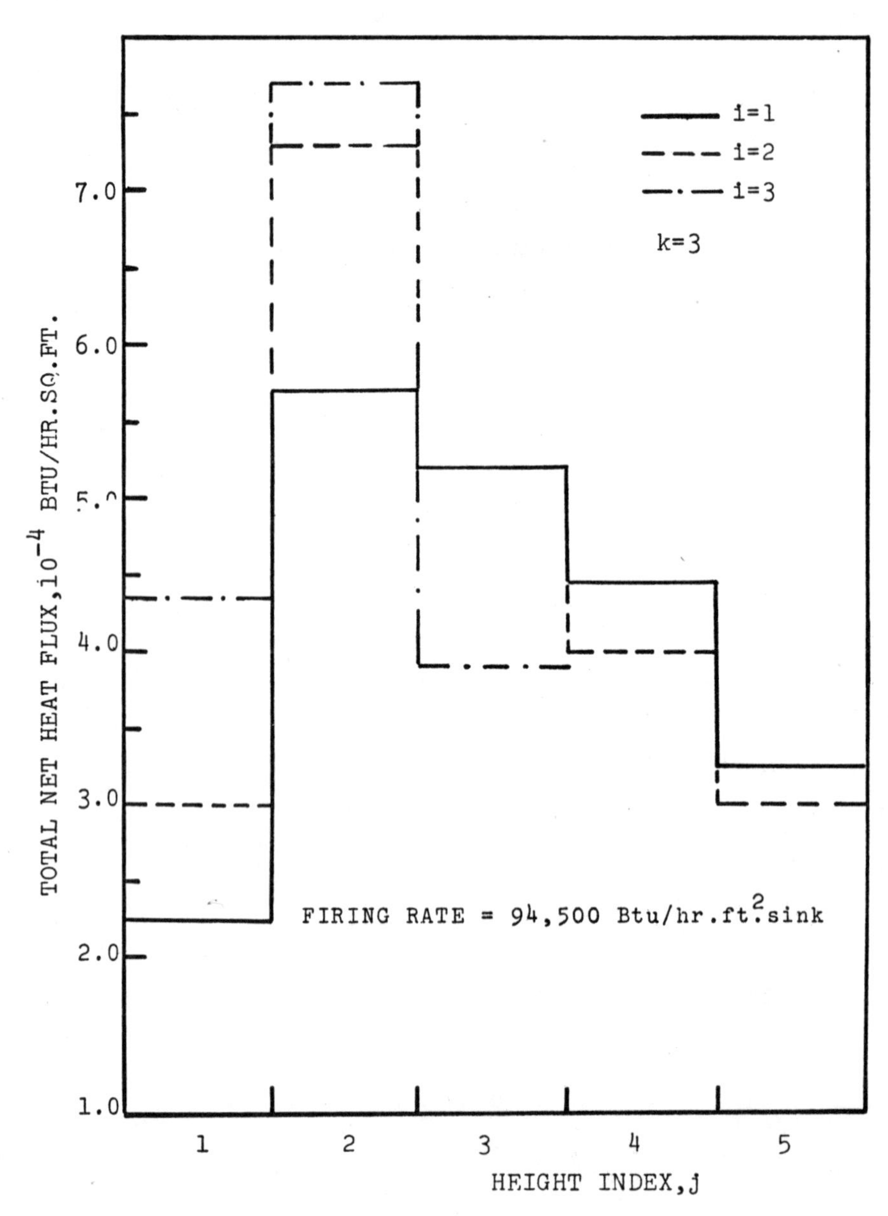

Figure 6 Total Net Heat Flux Distribution on the Front Wall. Calculated by Using the Flow Pattern Obtained from the Data of Beer et al (15) and Including Particle Radiation.

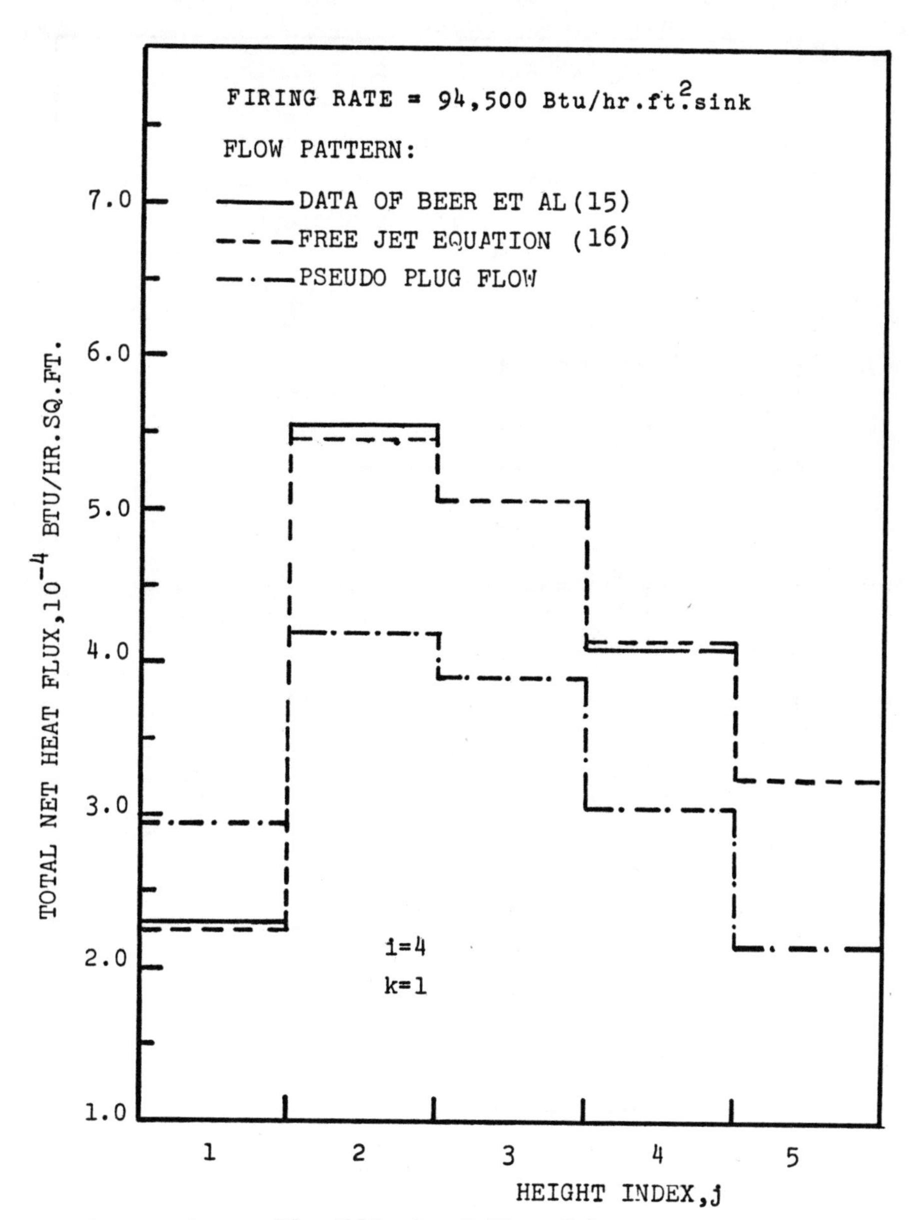

Figure 7 The Effect of Flow Pattern on the Total Net Heat Flux Distribution over the Left Wall. Particle Radiation Included.

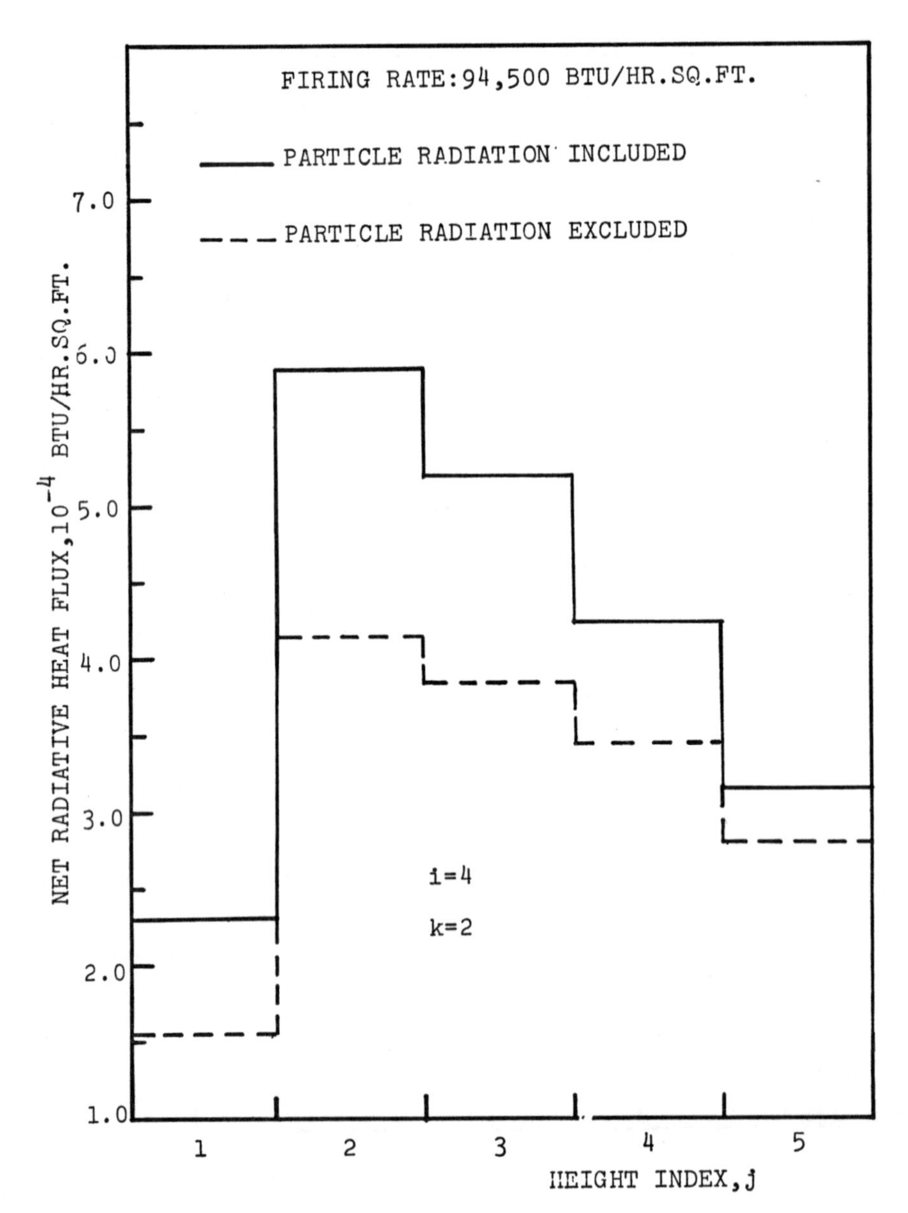

Figure 8 The Effect of Particle Radiation on the Net Radiative Heat Flux Distribution over the Left Wall.Flow Pattern Calculated from the Data of Beer et al (15).

heat sink. A thin copper wire was welded to the center of the foil while another thin copper wire was soldered to the base of the copper block. Thus a differential thermocouple was formed to measure the temperature rise at the center of the foil. This temperature rise can readily be related to the heat flux incident on the foil. The design of the probe was such that either a refractory collar or a hemispherical As_2S_3 lens could be fitted around the copper block. In the former case the response obtained from the probe corresponds to the sum of the incident radiative and net convective heat fluxes while in the latter case any convective heat flux is eliminated and the probe only measured the incident radiative heat flux.

Since the furnace is operated under slight pressure the gates of the access ports are equipped with an automatic mechanism which blows cold compressed air into the furnace as soon as the port gate is opened.

During the measurements the As_2S_3 lens could not withstand the thermal shock and all measurements were taken without the lens on the probe. However, because of the cold air blown in front of the probe the response obtained in this fashion is believed to be due to the incident radiative heat flux only. Since the sensing surface of the instrument was located approximately at the tube walls the incident radiative heat fluxes measured by the probe are nearly the same as those to the tube walls.

The heat flux probe without the As_2S_3 lens was calibrated against a Hy-cal Prime Standard Radiometer in front of a radiant panel. It must be pointed out that the conditions prevailing during the calibration were drastically different from those at the access ports of the furnaces.

Table 1 presents six measured incident radiative heat fluxes obtained in this manner located in zone 3, 2, 3 which is on the front wall near the burners. The measurements were taken on both sides of the furnace through symmetrically located ports so that the corresponding pairs separated by single lines should be equal if symmetry exits in the furnace. The average value of the six readings is also presented along with the calculated incident radiative flux for the same surface zone. The two values are in qualitative agreement and are typical of similar comparisons made in other zones which are reported in the original thesis (20).

Table 1

Comparison of the Predicted Incident Radiative Heat Fluxes with the Experimentally Measured Values

Measured Incident Radiative Heat Flux, Btu/hr. ft.^{2}	Average Btu/hr.ft.^{2}	Indices of the Sink Zone (i,j,k)	Calculated Incident Radiative Heat Flux Btu/hr.ft.^{2} *
55,425			
53,780			
86,427	69,615	(3,2,3,)	77,315
74,026			
79,088			
68,458			

* Calculated by using the flow pattern obtained from the data of Beer et al (15).

5.4 Comparison with Simpler Models in Terms of the Total Heat Transferred to the Feed Water

Table 2 gives a summary of the results in terms of the rate of steam generation obtained by different methods. It is seen that the total amount of steam generation as predicted by the zone method of analysis using recirculating flow patterns and including soot particle radiation are in good agreement with the value measured in the plant. Simpler models are out by a considerable amount, well-mixed models underestimating the heat transfer while plug flow models overestimate the heat transfer.

Table 2

Comparison of the Results Obtained by Different Methods

METHOD	FLOW PATTERN	STEAM GENTD., LBS/HR.
measured in the plant	-	627,000
well-mixed furnace, without soot radiation (1)	-	354,000
well-mixed furnace with soot radiation (1)	-	416,000
modified well-mixed furnace without soot radiation (1)	-	748,000
modified well-mixed furnace with soot radiation (1)	-	884,000
long furnace without soot radiation and convection (2)	-	871,000
long furnace with soot radiation without convection (2)	-	1,084,000
long furnace without soot radiation with convection (2)	-	942,000
long furnace with soot radiation and convection (2)	-	1,120,000
zone method without soot radiation	Data of Beer et al (15)	531,000
zone method with soot radiation	Pseudo Plug	692,000
zone method with soot radiation	Free Jet (16)	650,000
zone method with soot radiation	Data of Beer et al (15)	651,000

6. Discussion of Results

Calculations indicate that the postulated flow pattern is important in the analysis. The plug flow pattern gives the highest amount of steam generation while the two recirculating flow patterns result in lower but more realistic rates of heat transfer. This is due to the fact that the recirculation of relatively cool combustion products lowers the temperatures in the flame. This point is consistent with results obtained in a previous study by Hottel and Sarofim (1,4).

An important point is that soot particle radiation contributes substantially to the radiative heat flux even though its contribution to the absorption coefficient is small in the regions where the gas radiates. Most of the radiation originating from the gas is absorbed within the flame. As a consequence of the heat transfer from the combustion products to soot particles of small size the temperatures of the latter become nearly equal to the temperatures within the flame. It thus becomes possible for the soot particles to radiate appreciably over the entire spectrum, and most of the radiation emitted by the soot particles in the windows, where no gas absorption (or emission) occurs, penentrates to the enclosing surfaces.

It is believed that in a large boiler equipped with swirling burners, the representation of carbon dioxide and water vapor concentrations with single values corresponding to those of complete combustion is a realistic assumption. On the other hand, experimental data (18) show that the soot particle concentration distribution in a furnace is non-uniform, a higher concentration close to the burners and a lower concentration towards the exit. However, during the plant test, it was observed that a substantial amount of soot particles were present in the external recycle stream. This, therefore, justifies the representation of the non-uniform soot particles concentration distribution in the enclosure by a mean value.

The accuracy of the Monte Carlo method in calculating the total radiative interchange in the system was checked by noting the extent to which the reciprocity relation between the calculated total radiative interchange areas was satisfied. This was found to be quite satisfactory in spite of the statistical nature of the results obtained from a Monte Carlo calculation. The Monte Carlo method becomes superior to the determinant method as the level of complexity of the problem being studied increases. It is flexible, easy to formulate, and can readily be extended to account for concentration gradients (5,6), and anisotropic scattering of radiation by solid particles (20).

7. Conclusions

(1) It has been shown that a mathematical simulation of the heat transfer in an industrial boiler can be carried out using the data normally available in an operating plant and the related literature.
(2) Results obtained in this and other investigations have conclusively shown that the zone method of analysis is a useful mathematical description technique for evaluating the radiative heat transfer in a furnace enclosure. The accuracy of the method, however, is limited to the reliability of the input data which describe the other physical processes occurring within the enclosure.
(3) It is seen that a realistic description of the radiative properties of an oil flame must account for the contribution of soot particles in an appropriate fashion.
(4) A significant amount of the computational effort involved is in the calculation of the total radiative interchange in an enclosure with an irregular geometry like the boiler studied.

It is, however, believed that the generalized and simpler procedure of applying the zone method of analysis by considering the enclosure to be either rectangular parallelepiped or right cylindrical geometry can be developed.

References

(1) Hottel, H.C. and Sarofim, A.F., "Radiative Transfer", McGraw-Hill Book Co., New York (1967).
(2) Hadvig, S., "Heat Transmission by Gas Flow Including both Radiation and Convection", J. Inst. Fuel, 35, 129 (1970).
(3) Hottel, H.C. and Cohen, E.S., "Radiant Heat Exchange in a Gas Filled Enclosure, Allowance for Non-uniformity of Gas Temperature", A.I.Ch.E. J., 4, 3 (1958).
(4) Hottel, H.C. and Sarofim, A.F., "The Effect of Gas Flow Patterns on Radiative Heat Transfer in Cylindrical Enclosures", Int. J. Heat Mass Transfer, 8, 1153, (1965).
(5) Cannon, P., "The Calculation of Radiative Heat Flux in Furnace Enclosures Using the Monte Carlo Method", M.Sc. Thesis in Chem. Engineering, U.N.B., (1967).
(6) Steward, F.R. and Cannon, P., "The Calculation of Radiative Heat Flux in a Cylindrical Enclosure Using the Monte Carlo Method", Int. J. Heat Mass Transfer, 14, 245 (1971).
(7) Osuwan, S., "Radiative Heat Transfer in a Cylindrical Test Furnace", Ph.D. Thesis, Dept. of Chem. Eng. U.N.B., (1971).
(8) Steward, F.R. and Osuwan, S., "A Mathematical Simulation of Radiative Heat Transfer in a Cylindrical Test Furnace", Can. J. Chem. Eng., 50, 450 (1972).
(9) Steward, F.R., Osuwan, S. and Picot, J.J.C., "Heat Transfer Measurements in a Cylindrical Test Furnace", Fourteenth Symposium (International) on Combustion, (1972) (to be pub.).
(10) Johnson, T.R. and Beer, J.M., "Radiative Heat Transfer in Furnace Enclosures: Further Development of the Zone Method by Analysis", Fourteenth Symposium (International) on Combustion, (1972) (to be published).
(11) Hirose, T. and Mitunaga, A., "Investigation of Radiant Heat Exchange in a Boiler", Bull, J.S.M.E., 14, 829 (1971).
(12) Curtet, R., "Confined Jets and Recirculation Phenomena with Cold Air", Combustion and Flame, 2, 383 (1958).
(13) Becker, H.A., Hottel, H.C. and Williams, G.C., "Mixing and Flow in Ducted Turbulent Jets", Ninth Symposium (International) on Combustion, 7 (1963).
(14) Thring, M.W. and Newby, M.P., "Combustion Length of Enclosed Turbulent Jet Flames", Fourth Symposium (International) on Combustion, 789 (1953).
(15) Beer, J.M., Chigier, N.A. and Lee, K.B., "Modelling of Double Concentric Burning Jets", Ninth Symposium (International on Combustion, 892 (1963).
(16) Ricou, J.P. and Spalding, D.B., "Measurement of Entrainment by Axisymmetrical Turbulent Jets", J. Fluid Mech., 9 , 21 (1961)
(17) Maesawa, M., Tanaka, Y., Ogisu, Y., and Tsukamoto, Y., "Radiation from Luminous Flames of Liquid Fuel Jets in a Combustion Chamber", Twelfth Symposium (International) on Combustion, 1229 (1968).
(18) Godridge, A.M. and Hammond, G.E., "Emissivity of a Very Large Residual Fuel Oil Flame", Twelfth Symposium (International) on Combustion, 1219 (1968).

(19) Steward, F.R., "Industrial Boiler Test", Report Submitted to the Electric Power Commission of New Brunswick, Dept. of Chem. Eng., U.N.B., (1973).
(20) Guruz, H.K., "The Effect of Solid Particles on Radiant Transmission in Furnace Enclosures", Ph.D. Thesis, Dept. of Chem. Eng., U.N.B., (1973).
(21) The Babcock and Wilcox Co., "Steam", 37th Ed., Geo McKibbin and Sons, (1961).
(22) Winship, R.D. and Penner, G.R., "Heat Transfer Dynamics in Coal Fired Steam Generators", Paper presented at the Fourth Western Canadian Heat Transfer Conference, Winnipeg (1972).

Acknowledgement

The authors wish to express their appreciation to the New Brunswick Electric Power Commission for the use of the facilities at the Courteney Bay Power Station. They would also like to thank the operating personnel of the station who were cooperative in setting up test equipment, operating special services and explaining various aspects of plant operations.

The authors are particularly indebted to Mr. J. Jeffreys, Mr. E. Burton and Mr. L. Beaulieu of the Commission whose cooperation made the test possible.

Chapter 4

SIMULTANEOUS PREDICTIONS OF FLOW PATTERNS AND RADIATION FOR THREE-DIMENSIONAL FLAMES

Suhas Patankar and Brian Spalding

Imperial College of Science and Technology,
Mechanical Engineering Department, Exhibition Road, London, SW7

Abstract

The lecture describes a calculation method for the three-dimensional turbulent flow in a gas-turbine combustion chamber. The geometry considered involves the mixing of the streams of fuel and air in a confined space; additional air streams are used for film-cooling and dilution purposes. In addition to the equations of momentum and continuity, differential equations are solved for two turbulence quantities, for the concentrations of chemical species, and for the radiation fluxes.

1. Introduction

Contents:
- The problem.
- The solution procedure.
- A particular example.
- Results.
- Conclusions.

Fig. 1: Simultaneous prediction of flow pattern and radiation for 3D flames.

Methods can be found in the literature which purport to predict the heat-transfer rate from a flame when the temperature and composition distribution within it are specified beforehand [1]; but in practice these distributions rarely are specified beforehand. The engineer therefore needs a method which will predict the distributions simultaneously with the heat transfer.

The present paper supplies such a method, and exemplifies its use. The method is the result of a long process of development [2,3]; so it cannot truly be called a new one. However, it has only just reached the stage of practical implementation; so there has so far been little experience of its use. For this reason, I shall show no comparisons of predictions with experiment.

Fig. 1 explains how the lecture is constructed. First I shall clarify the nature of the problem, both from the physical and mathematical points of view. Then I shall describe the main features of the method which we are employing to solve the problem. The major contribution of the lecture then follows: the application of the method to a particular combustion chamber, that of a gas turbine; and the presentation and discussion of the results.

2. The problem

Given: • geometry of the combustion chamber
- fuel and air input conditions,
- thermal boundary conditions,
- thermodynamic, transport, radiative and chemical-kinetic properties;

Calculate: o velocity, temperature, composition, etc. throughout chamber.
- heat flux and temperature at wall.

Fig. 2: The problem: statement

The engineer knows the configuration of his furnace or combustion chamber, either because it exists or because he has a drawing of it; and he also knows for what conditions of operation he needs predictions of heat transfer. Text-books, scientific papers and his research advisers can inform him about the properties of the materials that he is concerned with. This information is his starting point; and it must be enough, in conjunction with a general prediction procedure,to lead to the information he seeks.

What information is that? Primarily it is the external characteristics of the flame that he wishes to know. What will be the heat flux at each point of the wall? What will be the resulting temperature distribution on the wall? What will be the temperature and composition distribution of the gases leaving the combustion chamber?

However, these external characteristics are the consequences of the phenomena within the combustion chamber. The prediction procedure must therefore perforce compute the whole three-dimensional fields of velocity, temperature and composition; so the engineer might as well take an interest in these also.

3. Interconnexions between phenomena

Variable	Depends on	Which depends on
q, T_{wall}	λ_{eff}	μ_{eff}, σ_{eff}
	emissivity	T, composition
	T	λ_{eff}, emiss., u,v,w
u,v,w	μ_{eff}	k, ϵ
k,ϵ	u,v,w,μ_{eff}	see above
composition		" " reaction rate
react,rate	T, composition	see above.

Fig. 3: The problem: why simultaneous prediction is needed.

If the last Fig. did not make it clear that it is pointless to try to calculate radiative heat transfer without regard to the flow pattern, this one may do so; for it draws attention to just a few of the many inter-connexions between the various aspects of the whole process.

The heat flux at a point of the wall, and so also the temperature taken up by the wall in general, depend on the whole distribution of temperature within the field; and also depend on the distributions of emissivity and of effective conductivity. These quantities depend on the distributions of effective viscosity and effective Prandtl number; and these cannot be dissociated from the distributions of velocity, turbulence, energy, etc.

The lesson is a simple one; yet it is often over-looked. To have a method of calculating radiation, which demands prior knowledge of the flow pattern in the furnace, is like having a plan for winning a football game, which demands prior knowledge of the tactics of the opposing side. It is of little use in practice.

4. The problem of calculating turbulent flow

Turbulence • Scale disparity entails use of turbulence models (TM).

- TM must be selected to give optimum balance of universality and simplicity.
- Fluctuations of concentration, temperature, must be computed, not just time-mean values.
- Influences of turbulence on reaction rate must be accounted for.

Fig. 4: The problem: physical aspects, 1.

Nearly all combustion-chamber flows are turbulent ones; so we must immediately face the difficulties which turbulent flows present the theoretician, even when chemical reaction is absent.

The first difficulty arises from "scale disparity": important phenomena like eddy decay take place on a scale which is small compared with the combustion-chamber dimension; therefore, finite-difference methods with current computers cannot follow them in detail. The way out is to use a "turbulence model", i.e. a set of equations believed to govern some important statistical properties of turbulence [4,5].

"Turbulence modelling" is fast becoming a branch of fluid mechanics in its own right. Here I mention just a few aspects, for example that there is a selection to be made, and that the criteria of choice are universality and simplicity.

The statistical properties of the turbulence, usually include velocity correlations of one kind or another; but concentration fluctuations are also to be considered when the chemical reaction is important. Methods now exist of calculating, for example, the root-mean-square fluctuations of concentration; and we have some knowledge of how these influence the time-mean reaction rates.

5. The problem of calculating the rate of chemical reaction

Chemical kinetics. • Fuel-air combustion proceeds by way of many intermediate steps.

- For separate fuel, air injection, main flame features can be predicted without knowledge of these steps.
- Chemical-kinetic knowledge is needed for prediction of carbon-content distribution.

Fig. 5: The problem:physical aspects, 2.

The general problem of calculating chemical reaction rates is a very difficult one; for what one may write as a simple transformation of fuel and air into combustion products proceeds in fact by way of innumerable intermediate steps, each one of which ought to be treated quantitatively if the whole process is to be correctly represented.

Fortunately, most furnace and engine flames are "physically controlled", by which is meant that their outward characteristics, such as temperature distribution, can be computed without knowledge of the detailed reaction-kinetic constants. This comes about because the fuel and air are injected separately; and their rate of mixing is small in comparison with the rate of reaction which chemical-kinetic processes could bring about.

However, even in furnace flames, two aspects of chemical-kinetics do require study. The first concerns carbon formation: small differences in operating conditions can turn a luminous flame into a non-luminous one; and this difference has large effects on radiative heat transfer. The second concerns the production of pollutants such as nitric oxide, and their possible escape from the combustion chamber with the exhaust gases. Although not the subject of the present paper, this topic is too important to be passed over in complete silence.

6. The problem of calculating radiative heat transfer

Radiation. • Properties (emissivity, scattering coeff.) depend on composition, temperature, wavelength, angle.

• Properties are not yet adequately documented.

• The most important composition feature is solid-carbon content, which is hard to predict.

Fig. 6: The problem; physical aspects, 3.

With this Fig. we come nearer to the centre of our topic. Radiation presents two difficulties, one physical and the other mathematical. Here we consider the first of these.

The radiative properties of materials are highly diverse. If we confine attention to a particle-free gas, we must reckon on its local absorptivity being a complex function of composition, temperature and wavelength. If particles are present, we must know the distribution of the scattering angle. Surfaces additionally have different properties according to their roughness, the presence of oxide films, etc.

It is no wonder that this information about properties is very far from having been fully collected and documented. Consequently we often have to make guesses, and are indeed glad of the relief from knowledge which would exceed our capacity to handle it.

We might as well admit that, until knowledge of the kinetics of solid-carbon formation is greatly improved, all calculations of radiative heat transfer from flames are liable to involve a good deal of guesswork. It follows that we should be careful not to allow other parts of the prediction procedure to become over-elaborate. There is no point in calculating everything to five significant figures when important input data are accurate only to one or two

7. The problem of solving the convection equations

Flow pattern. • Momentum and continuity equations are non-linear, simultaneous.

- TM's usually introduce additional non-linear simultaneous differential equations.
- In flames, density varies depending on temperature and composition.
- The non-uniform density links momentum, continuity with energy, concentration equations.
- Most practical flames are three-dimensional.

Fig. 7: The problem: mathematical aspects, 1.

Turning now to the mathematical part of the problem, let us note first that the hydrodynamic and convective-transfer aspects involve us in solving simultaneous, non-linear, partial differential equations. These may be parabolic, but are more often elliptic.

This would be true even if the flow were laminar; when a turbulence model has to be used, there are additional equations to be solved (usually two or three in number); and these are also linked with the other equations in several ways. For example, the equation for turbulence energy has a term containing velocity gradients; and the equations for velocity contain "effective viscosity", which depends upon turbulence energy. However, the inter-connectedness has already been illustrated by Fig. 3, so need not be additionally stressed.

The last point on the present Fig. is perhaps the most important: nearly all practical furnace and engine flames (and those in the environment also) are three-dimensional. This does not merely mean that three velocity components are important; for that can be true of two-dimensional phenomena, i. e. plane or axi-symmetrical ones. The significance is that there exists no space co-ordinate along which fluid properties are uniform; and this implies that much information must be supplied for the proper description of the flame.

8. The problem of solving the radiation equations

Radiation. • Radiation flux is 6-dimensional (3 position, 2 angle, 1 wavelength co-ordinate).

- Classical equations are integro-differential.
- Mathematical problem simplifies for
  - a) Very transparent media.
  - b) Very opaque media.
- Neither a) nor b) often apply to practical flames.

Fig. 8: The problem: mathematical aspects, 2.

It is bad enough that temperature, say, varies with three space co-ordinates; but the behaviour of radiation is even worse. For radiation must be regarded as six-dimensional, angle providing two dimensions and wavelength one. So the description of a radiant flux field is much more voluminous than that of a temperature field, and even exceeds that of a velocity field.

There is another thing. It is of the nature of radiation to penetrate; so, by its agency, all points in the field that can "see" each other participate in interactions, no matter how far away they are. The mathematical expression of this fact is by way of the class of equations called "integro-differential". They are especially difficult to solve.

It is true that there exist special situations in which the radiation equations greatly simplify. One of these is the thin-gas situation in which, perhaps because of small dimensions or low gas pressure, almost all the radiation emitted by the gas escapes to the wall without reabsorption. Then calculation is easy.

When the gas layer is very thick, another simplification is allowed: radiation reduces to the same form as thermal conduction.

Unfortunately, nearly all practical flames lie well between these two extremes; the engineer has to be able to handle flames of moderate thickness.

9. The turbulence-model

- The $k \sim \varepsilon$ (energy ~ dissipation-rate) turbulence model is used.
- The empirical constants and functions fit experimental data on 2D flows.
- Concentration fluctuations are taken as very small.
- It would however be easy to calculate the fluctuations.

Fig. 9: The solution procedure; 1, Turbulence

In the prediction procedure that is to be described, the $k \sim \varepsilon$ model of turbulence has been chosen, i.e. one in which the statistical properties for which special differential equations are solved are first the local energy of the fluctuating motion, and secondly the local rate of dissipation of that energy.

This choice has been made because this is currently the model which has been most completely investigated; it has much better claims to universality than any simpler one; and it is not itself too complicated for use.

Of course, every turbulence model requires an empirical input. The input to the $k \sim \varepsilon$ model is known to give good agreement with two-dimensional experiments; at present it is just a matter of faith that it will do the same for three-dimensional flows. However, we have no alternative except despair.

As the Fig. reveals, the following computations are based on the assumption that the fluctuations of concentration are small enough to be neglected. Although they are not small in reality, their neglect brings a minor simplification without damage to the main purpose of the paper.

Inclusion of the fluctuations could require us to solve one additional differential equation [6]. We are already solving eleven; so this is no very serious burden.

10. Thermodynamic and chemical-kinetic assumptions

- For simplicity, specific heat has been supposed not to depend on composition or temperature.
- Reaction is presumed to be: fuel + $O_2 \rightarrow$ product.
- Equilibrium prevails at all points (neither chemical kinetics nor turbulence delays reaction).

Fig. 10: The solution procedure; 2, Chemistry.

When one's purpose is demonstration rather than detailed realism, it is foolish to clutter one's calculations and thoughts with every refinement, regardless of relevance. For this reason, since heat transfer is in the forefront of attention, great liberties have been taken with the description of the thermodynamic and chemical-kinetic properties.

We make the simple-chemical-reaction assumption, according to which all specific heats are equal, there are no intermediate products, and fuel and oxygen combine in a unique proportion. Moreover, the reaction is supposed to proceed so rapidly that thermodynamic equilibrium prevails at every point.

Were we not to do this, we should simply have to solve more differential equations, one for each degree of freedom away from equilibrium. We could do this; and in some practical circumstances it may become necessary; but now is not the time.

11. The method adopted for radiative transfer

- The six-flux model is employed (leading to ordinary differential equations).
- Gas emissivity is taken as uniform.
- No division into wave-length bands is made.
- Neither of last two simplicifactions is essential.

Fig. 11: The solution procedure; 3, Radiation

There are two main methods of solving the radiative-transfer problem in practice. The most widely-discussed in the literature, that of Hottel and co-workers[1], can be regarded as a finite-difference representation of the integro-differential equations. It leads to a full matrix of coefficients in the equations, and so entails much computational work.

The second method, more approximate in principle but perhaps not so in practice, is the six-flux model [3], according to which attention is concentrated on six distinct bundles of radiation which are supposed to be representative of the whole. The six fluxes are the positive and negative elements of the components of radiation resolved in the three co-ordinate directions.

It is the six-flux method which we use. The reason is that its mathematical expression by way of finite-difference equations leads to a sparse matrix of co-efficients; and moreover the equations are of very similar form to those describing the convective part of theproblem.

The other simplifications mentioned on the Fig. are of less fundamental significance. We have used non-uniform gas emissivity in other computations; and we have also divided the wave-length range into a finite number of intervals. Neither of these things would however have brought any special advantage in our present study.

12. The method of solving the equations

- For hydrodynamics the algorithm is SIMPLE (semi implicit method for pressure-linked equations).
- For radiation, variables employed are (I+J), (K+L), (M+N).
- Hydrodynamic, energy, concentration, radiation equations are solved by program TRIC (three-dimensional recirculating-flow integrator in cartesian co-ordinates).

Fig.: 12 The solution procedure; 4, Mathematics.

This is not the place for describing the algorithm which has been employed; for earlier descriptions have been published [7, 8]. However, it may be useful to list some of its major features:- it is called SIMPLE (see Fig.); it employs a rectangular grid on which pressure, temperature and most other variables are stored, while the relevant velocity components are computed for the centre points of the links between the grid nodes; the difference equations are formulated implicitly; and they are solved line-by-line.

The variables appearing in the radiation calculation are not the six fluxes I,J,K,L,M,N (see previous Fig.), but rather the three flux sums (I + J), (K + L) and (M + N); for it so happens that the six first-order differential equations for the first set can be reduced to three second-order ones for the second set.

The prediction procedure therefore involves solution for the following complete list of variables:- $u,v,w,p,k,\epsilon,\tilde{h},f$, I + J, K + L, M + N, i.e. eleven in all.

The procedure has been incorporated into a Fortran 4 computer program, capable of being executed on CDC, IBM, Univac and other machines. The program, called TRIC (see Fig.), of course handles other problems beside those of furnace heat transfer.

13. The geometry of the particular combustor investigated

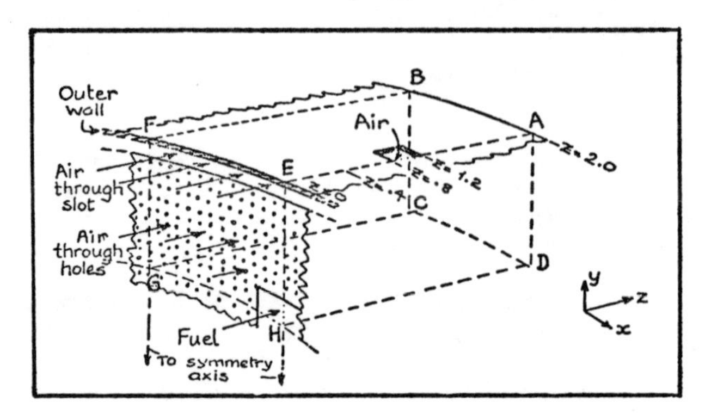

Fig. 13: A particular example; 1, geometry

We have chosen for our demonstration of the capabilities of the method a section of an idealised annular gas-turbine combustor. This is illustrated on the Fig., which shows segments of the inner and outer walls and a part of the air-inlet and fuel-inlet plate at the upstream end.

Because of the supposed symmetry of the combustor geometry, the cyclic repetition of air holes and fuel inlets around the annulus, and the large ratio of radius to combustor width, it suffices to confine attention to the nearly-rectangular region ABCDEFGH; in fact we treat it as fully rectangular.

The fuel, which is supposed to be gaseous, enters through the quadrant adjacent to H. Air enters through many small holes in the plate in the plane EFGH, through a gap between that plate and the wall in the vicinity of the line EF, and through a "dilution-air"hole in the wall ABFE.

The combustion gases pass out across the plane ABCD. The designer of the combustor is interested in knowing the distribution of gas temperature over that plane; and, since the mechanical strength of the flame-tube walldepends upon its temperature, he needs to know the temperature distribution over the wall ABEF.

14. Further details of the problem to be investigated

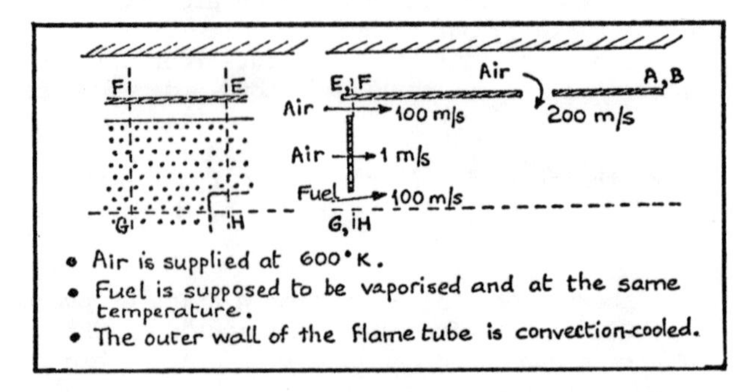

Fig. 14: A particular example; 2, input data

The Fig. shows some of the input data which have been held fixed for the majority of the computations. The fuel-air ratio is fixed at 1.5 times stoichiometric. Both fuel and air are supplied at 600K. The velocity of the fuel at inlet is 100 m/s, though its angle may vary. Always 47 % of the air enters through the dilution hole, 38 % through the slot EF, and 15 % through the holes in the inlet plate.

Heat is supposed to be lost from the wall ABEF to air at 600°K which flows outside it. The heat-transfer coefficient between the wall and that external air is taken as uniform; but various values are considered between 0 and 3000 $J/m^2 s^{\circ}K$. Obviously, the higher this coefficient is, the closer is the wall temperature to 600°K.

The other quantity which is varied over a wide range is the gas emissivity. Values between 0 and $2.0 m^{-1}$ have been used. The former corresponds to operation at extremely low pressures, the latter to operation at high pressure.

15. Some details of the computation

Grid employed:	7 x 7 x 7
Computer storage used:	37K words
Computer used:	CDC 6600
No. of iterations to give convergence:	50
Typical computer time:	60 seconds

Fig. 15: Aparticular example;
computer details

In order to make the prediction we need a computer program, a computer, and money to pay for the computation. I have said already that the TRIC computer program is available for solving the equations. This Fig. gives information about aspects of the computation which affect its user.

We need a large number of grid points if we are to represent the fine details of temperature distribution in our computation; however only 343 grid points were employed in the examples which I shall show. Were we to double the number of points in all three directions, we should of course need nearly 3000 grid points. The storage requirement would rise from 37K, about half of which is for the program, to about 180K.

The computer which we have used is a large and fast one. The present computations were accomplished in about one minute per run; this would rise to about 10 minutes per run for the fine grid computations. The cost is about £20 for the coarse-grid run and £200 for the fine-grid run at commercial rates in London, which are however untypically high.

It should be noted that the algorithm is a strongly convergent one. Only 50 iterations were needed to achieve acceptable convergence.

16. Distribution of longitudinal-directional velocity, w

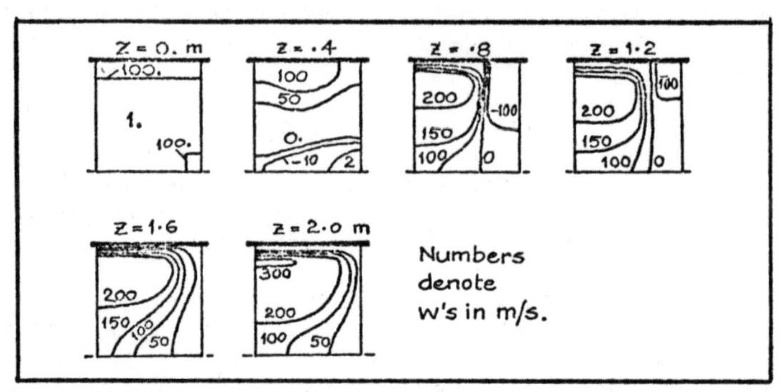

Fig. 16: Results of computation;
1, the w field.

Let us now examine some typical results. I exhibit first some Figures displaying the velocity distributions; they are all for the standard case with gas absorptivity of $2m^{-1}$ and an external heat-transfer coefficient of $300J/m^2{}^oK$. The rectangular contour diagrams are to be read from the left, which corresponds to the upstream end of the combustor, to the right, which corresponds to the downstream end.

Inspection of the first of these diagrams reveals the high-velocity regions along the top edge, representing the air-film cooling the wall, and in the bottom right-hand corner, representing the injection of fuel vapour; the low-velocity region is that of the porous end wall.

Recirculation, i.e. negative velocities, appears already at the second section; and it persists in the third and fourth. Undoubtedly a significant contribution is made by the fact that the air injected through the dilution hole has a strong upstream velocity component.

At the outlet section, the w component of velocity is positive at all points; but there are great non-uniformities. The highest velocities are still those near the wall, which shows that the locally injected reverse-flow air has only a local effect.

17. Velocity vectors in the x y planes

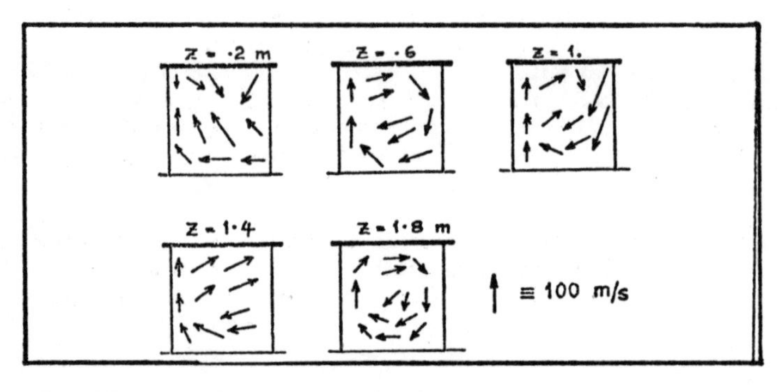

Fig. 17: Results of computation;
2, velocities in cross-stream planes.

The computer program prints out all three velocity components. We have not yet developed automatic graphical-display techniques; so the diagrams which you now see were made by hand. They represent the velocity field in the cross-stream plane by way of arrows, the length and direction of which correspond to the velocity vectors.

What is chiefly evident is that a significant swirling motion develops, induced mainly, no doubt, by the upstream injection of the dilution air. The absolute values of the velocity components are of course considerably lower than those in the z-direction; they are of the order of tens of metres per second, while the w-velocities are of the order of hundreds of metres per second.

It should perhaps be mentioned here that no particular combustor is in question; and the actual values of velocity are not necessarily all within the practically-occurring range. What we are demonstrating is that the computational ability now exists and is available for use, not that it has yet been used for design purposes, or even compared with particular experimental data.

18. Distributions of unburned fuel

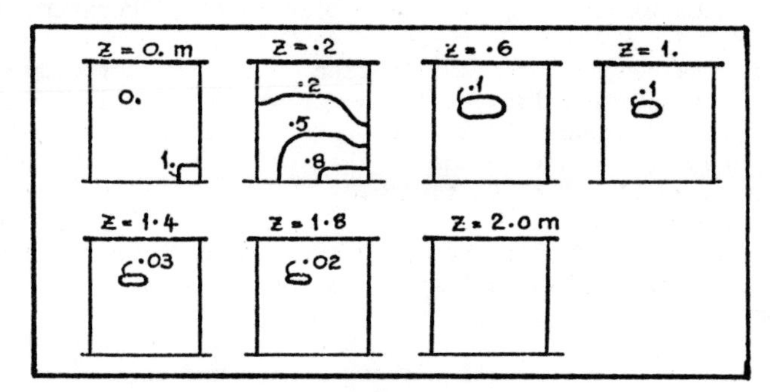

Fig. 18: Results of computation;
3, mass fraction of fuel.

As explained earlier, the chemical reaction is taken as being physically controlled; this means that it is impossible for both fuel and oxygen to have finite concentrations at the same point. The chemical reaction takes place at the outer envelope of the gas volume containing free fuel.

The Fig. shows the intersections of this volume with various planes along the length of the combustion chamber. At the first section, fuel is concentrated just at the corner where it is injected. It then spreads to cover all of the combustor cross-section except that area still swept by the air film.

Under the influence of the injection of dilution air, the fuel-rich region is greatly narrowed, and swept to the left. Finally, as mixing proceeds, it disappears entirely: the flame end has been reached; and all the fuel has been burned.

The oxygen concentrations have also been computed; but they provide little further information, and so are not represented here. Of course, in all the regions where unburned fuel exists, the oxygen concentrations are zero.

19. The distribution of temperature

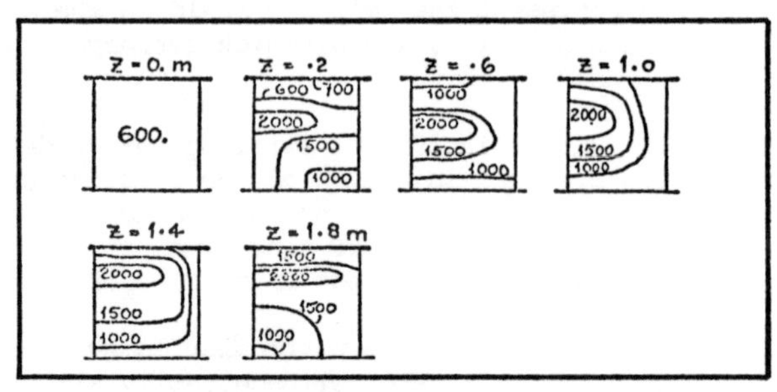

Fig. 19: Results of computation;
4, temperatures.

If this were an adiabatic combustor, there would be no need to represent the temperature distributions either; for the temperature would be uniquely dependent upon the fuel concentration. However, it is not; there is a convective heat loss to the wall, and a radiative heat loss and redistribution.

The Fig. represents the temperature distribution in the combustor by way of contour diagrams at the same sections as those for which we have already seen the fuel-concentration contours and the velocity vectors. It is evident that the high-temperature region corresponds to that of rapid reaction, where the fuel concentration is small but still finite.

The air film is fairly successful in reducing the temperatures close to the wall. However, I shall devote especial attention to the wall-temperature distribution in a moment.

Especially interesting to the combustion-chamber designer is the temperature distribution at the outlet. It can be seen from the farthest-down-stream temperature contours that this is very far from uniform. The combustor cannot yet be regarded as being at a stage of development that the engine designer could accept.

20. Distribution of radiation flux sums

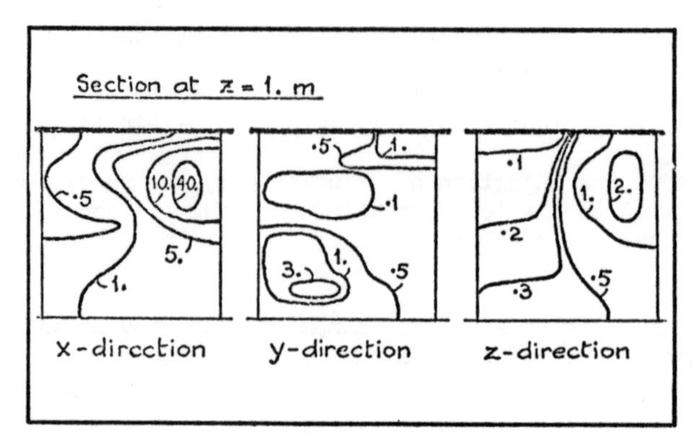

Fig. 20: Results of computation;
5, Radiation quantities.

It will be remembered that our radiation variables are the flux sums for the three directions: I + J, K + L, M + N. There is too much information for everything to be displayed; the Fig. therefore simply presents contour diagrams for these three flux sums for one particular central section across the combustor.

Strictly speaking it is not the flux sums themselves that are plotted but their values divided by the local emissive power of the gas. If the gas were very dense, the conduction approximation would be valid; then the values to be plotted would all equal unity.

It is evident that the conduction approximation is not valid; and values both above and below unity are to be seen.

The largest values are those for the x-direction which lies along the circumference of the annulus; the reason is of course that there are no sinks of this kind of radiation at the boundaries of the integration domain.

The smallest values are those for the y-direction; for radiation of this type can easily escape to the cooled wall. z-direction radiation can also escape; but it has a greater distance to travel.

This is an important Fig., worthy of detailed study. Of course, it does not exhaust the information that the program produces about the radiation fluxes.

21. The influence of the external convective heat-transfer coefficient

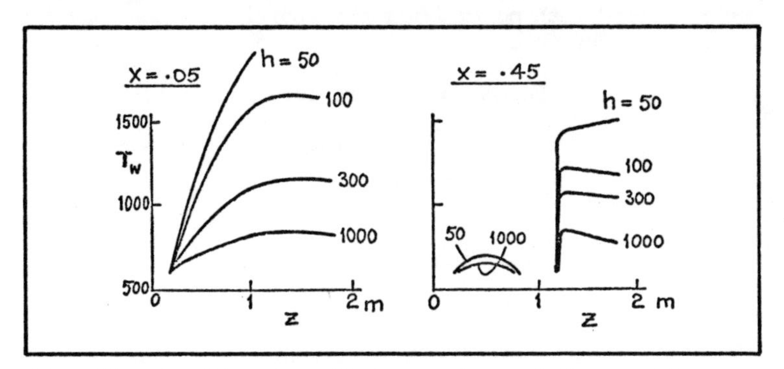

Fig. 21: Results; 6, wall temperatures influenced by external coefficient h.

We now turn to the predictions of the wall temperature. The Fig. displays the variation of this temperature with longitudinal distance z along two lines: for x = .05 m, and for x = .45 m. The second of these is near to the symmetry plane which passes through both the fuel inlet and the dilution-air inlet.

Several curves are shown, the parameter being the heat-transfer coefficient on the outside of the flame-tube wall, in J/m^2s^oK. As is to be expected, the larger this coefficient is, the lower is the temperature of the wall. In this connexion, you might like to be reminded that a coefficient of 1000 J/m^2s^oK is fairly easily attained when a high-velocity air stream is available.

The curves on the right have a shape which requires explanation: it is that the x = .45 line is interrupted by the dilution-air hole, through which enters air at 600^oK. Despite this, the flow pattern is such that there is little difference between the temperatures along the x = .05 and x = .45 m lines at the outlet of the chamber.

22. The influence of radiation on wall temperature

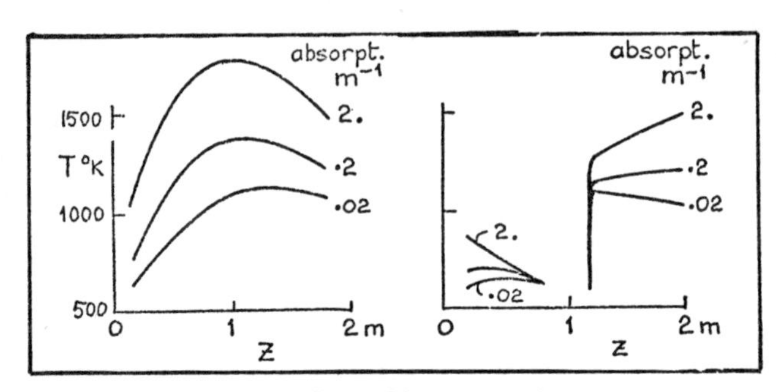

Fig. 22: Results; 7, wall temperatures influences by absorptivity.

It is worth considering the radiation problem, for a moment, from the point of view of the Hottel-Sarofim method [1]. There are one hundred and twenty-five gas volumes in question, and one hundred and fifty surfaces on the boundary of the integration domain. Therefore a 275 x 275 matrix would need to be inverted for the determination of the radiation flux even if the gas-temperature distribution were fixed. Since this would have to be done at least fifty times during the whole solution procedure, the task would be very formidable.

Had we been concerned with non-uniform absorptivities, the comparison would have been still more in the favour of the present method; for this is scarcely more troublesome when the absorptivities vary, whereas the interaction matrix of the Hottel-Sarofim method becomes much more troublesome to compute.

However, it must be admitted that the Hottel-Sarofim method is <u>in principle</u> the more satisfactory; for it does make better allowance for the angular distribution of radiation.

23. The influence of the velocity of the film-cooling air

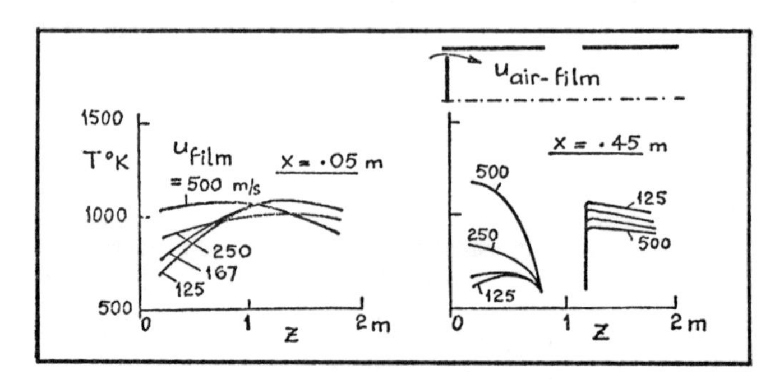

Fig. 23: Results; 8, wall temperatures influenced by air-slot width.

One of the dimensions at the easy disposal of the designer is the width of the slot separating the porous end-plate of the combustor from the flame-tube wall. If this is diminished, while the air-flow rate is maintained constant, the air-injection velocity is increased, with consequent effect on the flow pattern and on the wall-temperature distribution.

It will perhaps be expected that an increase in the air-injection velocity will lead to a reduction in the wall temperature; for high-velocity air may be expected to act as a more efficient coolant than low-velocity air.

The Fig. however shows that this expectation is not fulfilled. Along the x = .05m line, for example, the wall temperature becomes greater as the injection velocity rises. Detailed examinations of the computer output shows that this is not just because of the increased convective transfer coefficient on the flame side of the wall; it is mainly because the higher velocity of the injected air entrains more hot gas into the wall jet; and so, since the quantity of injected air has not increased, the temperature of the gases close to the wall must rise.

There is an important practical lesson to be learned: film-cooling air is best injected at such a velocity as to reduce entrainment; so the best velocity is that of the bulk of the hot gases, at the section in question. Of course, the designers of combustion chambers learned this by experience long ago, before there was any question of prediction by computer; however, now that we have computer models, we can expect them soon to be providing guidance where there is no experience to rely upon.

24. The influence of the velocity of injection of the fuel

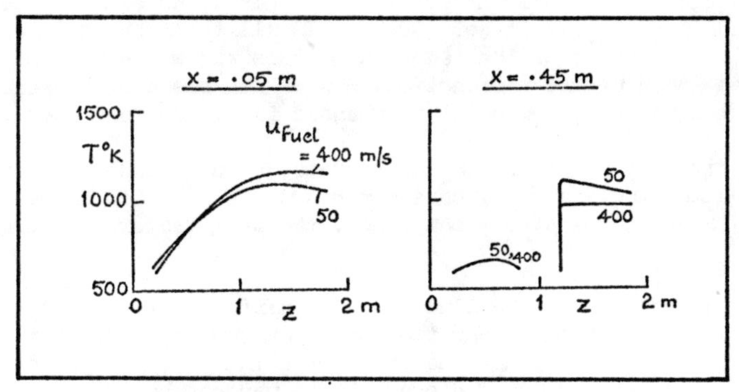

Fig. 24: Results; 9, Wall temperatures influenced by fuel velocity.

We are supposing that the fuel is injected as a vapour. Its mass flow rate is fixed; but we can inject it through a small hole or a large one, i.e. with high velocity or low. The present Fig. shows the results of calculations with fuel velocities of 50 m/s and 400 m/s.

The differences produced in the wall temperature are small. The reason is that the mass flow rate of fuel is much smaller than that of air; so even an eightfold change in the momentum of the fuel has little effect on the flow pattern as a whole.

How is the change of injection-hole size to be introduced into a computation for which the lateral dimension of a cell is 5 cm? The answer is: by the introduction of an artificial fuel density for the cell face through which the fuel enters; this is chosen so that, when the density-velocity product is multiplied by the cell-face area, the prescribed mass flow rate is obtained.

Of course, we cannot in this way account for all the effects of sub-grid-scale geometry. For example, we can make no distinction between a round hole, a square hole or a triangular hole; and the hole could be situated anywhere within the cell-face area without our being able to express the fact. If one wants to do better than this, one must employ a finer grid.

25. The influence of the axial velocity of the dilution air

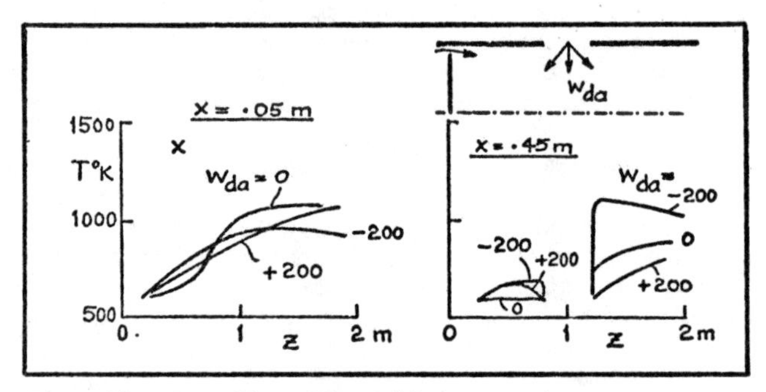

Fig. 25: Results; 10, Wall temperatures influenced by dilution-air angle.

This Fig. illustrates the influence on the wall temperature of the direction of the dilution air jet. In all calculations the air is caused to enter through the orifice with a normal velocity of 200 m/s. In all those shown in earlier Figures the axial component of the velocity of this air was -200 m/s, i.e. the air was injected in an upstream direction. Now we can see what happens when the value of the axial velocity component is changed first to 0 and then to +200 m/s.

The most noticeable effect is on the temperature distribution along the x = .45 line, which, incidentally, passes through the dilution hole. When the dilution air is given a downstream component, the wall temperature is appreciably lowered.

The effects along the x = .05 line are more difficult to understand; indeed there is perhaps little to say about the curves on the Fig. except that they are the end-result of the computations of all the interacting hydronamic, thermal and chemical-reaction processes. The predictions may not be correct in this instance; but at least they can be made; and, until experimental data are available, they are the best means we possess of indicating to the designer how the dilution-air angle will affect the wall temperature.

26. Influences on the temperature distribution at the outlet

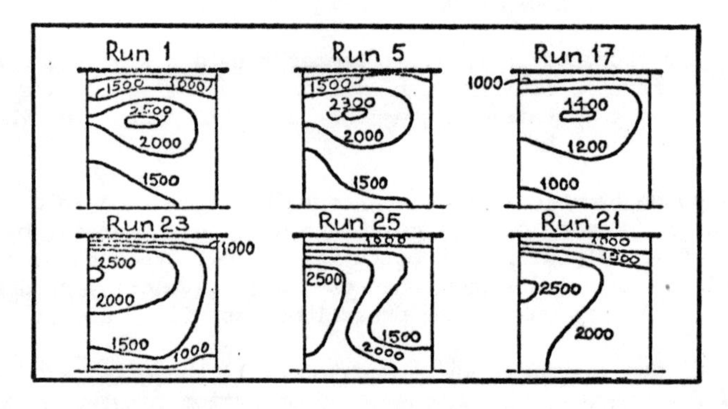

Fig. 26: Results; 11, outlet temperature distributions.

Let us now return our attention to the temperatures at the outlet plane. The contour diagram marked "Run 1" represents the standard situation, with absorptivity of .02m^{-1} and an external coefficient of 300 J/$m^2 s^oK$.

The "Run 5" diagram shows that an absorptivity two orders of magnitude greater has only a slight effect on the temperature distribution; but the upper and lower limits are somewhat nearer together.

The "Run 17" contours show what happens when the air film is injected at five times the normal velocity. The uniformity of the temperature is significantly greater.

Comparison of the "Run 23" and "Run 25" contours with those for "Run 1" reveals the the direction of entry of the dilution air radically alters the temperature pattern, and in an understandable direction. For "Run 23" the injection is normal to the wall; for "Run 25" it is at 200 m/s in the downstream direction.

Finally, inspection of the diagram for "Run 21" shows that the increase in fuel velocity also has a significant effect on the outlet temperature distribution.

It will be apparent that the ease with which predictions can be made of the influences of design variables on the maximum and minimum temperatures in the outlet stream can be of great help to the designer of the combustor; for it is his task to make the difference between the two temperatures as small as possible.

27. What has been achieved so far

- The combined flow + radiation problem has been formulated in terms of finite-difference equations.
- A computer program has been developed to solve the equations economically.
- The resulting predictions agree qualitatively with experiment.

Fig. 27: Conclusions; 1, Achievements

The TRIC computer program has been expensive to produce; and the funds available for it have been obtained from the earnings from less ambitious programs. It is therefore understandable that the program is still in a stage of promise rather than achievement. Nevertheless, some achievements can justly be claimed.

The main one is that a mathematical model of the combined fluid-flow and radiation processes has been constructed, and demonstrated to work for a fully three-dimensional turbulent combustor, the geometry of which is not without complication. This is new. A year ago, no such model existed anywhere in the world.

Moreover, the computer program is not prohibitively expensive to run. £20 is a large sum to spend on a meal or a theatre; but it is very small compared with what would be spent on experimentation on a physical model of a combustor of this kind; and it is negligible in comparison with the costs of full-scale experiments.

Finally, even though it is not possible to show that the predictions agree with experiments (because these have not been carried out), the predictions are qualitatively in accordance with all the expectations to which experience of practical combustors gives rise.

28. Improvements which can easily be made

- Use of more points in the grid.
- Better choice of emissivity distribution.
- Division of radiation fluxes into distinct wave bands.
- Computation of concentration fluctuations.
- Allowance for reaction rate limited by chemical kinetics and eddy breakdown.

Fig.: Conclusions:
2, improvements which can be easily made.

There are several ways in which the prediction procedure can be improved immediately; and some of them are mentioned on this Fig.

First of all, a finer grid is obviously needed if we are to predict the details of the flow accurately. This is just a matter of arranging the program so that it can employ auxiliary storage. Of course, the costs will increase; but the benefits should be more than commensurate to the design engineer.

Secondly, it is quite easy to introduce a more refined treatment of the radiation. More realistic emissivity values can be selected; and at least some account can be taken of the departure of the materials from greyness by splitting the radiation into distinct wave bands. Neither of these improvements is very expensive.

There are other relatively simple modifications that should bring more accurate predictions with little extra cost. Thus, the concentration-fluctuation equation can be introduced so as to give a more realistic representation of the thickness of the reaction zone.
Further, the reaction rate itself can be required to depend upon the chemical-kinetic properties of the materials, and on the local turbulence characteristics, rather than being infinitely fast.

29. The need for further research

- Computation of local radiation properties from soot concentration, etc.
- Computation of soot concentration from chemical-kinetic equations.
- Inclusion of other air-pollutants in the kinetic scheme.
- Validation of method by comparison with laboratory and full-scale experiments.

Fig. 29: Conclusions: 3, Improvements requiring further research

There are other improvements which, desirable as they are, we cannot immediately incorporate into the computer program; for the scientific knowledge to support them is not yet in our possession.

Partly this shortcoming is to be laid at the door of the chemical-kineticist who has been unable, so far, to develop a quantitative model for the production and decay of that most important of all radiating components - solid carbon. A completely accurate and general model of this kind will be a long time in coming; but I do believe that experience-based theoretical invention could provide us, in the next year or two, with a workable and useful model.

There are other chmical-kinetic features which will be needed in a comprehensive prediction procedure. Of particular importance is the incorporation of the kinetics of nitric oxide production. This is not straight-forward however, because the fluctuations of temperature and concentration play a major role; it is not the time-mean values alone that are important.

However, by far the greatest need is for the research which must be conducted, in collaboration, both by the developers of the method and by experimenters with equipment. Validation is what the method mostly needs, i.e. careful comparison of predictions with experiments, and insightful interpretation of the differences. Fortunately, an increasing number of organisations are recognising this need and are providing support for the work. Now that the method is available, the main point is to use it.

30. References

1. Hottel H.C. and Sarofim A.F.- "Radiative Transfer", McGraw Hill, New York, 1967

2. Patankar S.V. and Spalding D.B. - "A computer model for three-dimensional flow in furnaces"
Fourteenth International Symposium on Combustion.
To be published

3. Patankar S.V. and Spalding D.B. - "Mathematical models of fluid flow and heat transfer in furnaces; a review
Paper 2 of "Predictive Methods for Industrial Flames".
Published by Institute of Fuel, 1973

4. Launder B.E. and Spalding D.B. - "Mathematical models of turbulence"
Academic Press, London, 1972

5. Harlow F.H. (Ed.) - "Turbulence transport modelling"
AIAA Selected Reprint Series, Volume XIV, 1973

6. Spalding D.B. - "Concentration fluctuations in a round turbulent free jet".
Chem. Eng. Sci., Vol. 26, p. 95, 1971

7. Patankar S.V. and Spalding D.B.- "A calculation procedure for heat, mass and momentum transfer in three-dimensional parabolic flows".
Int. J. Heat Mass Transfer, p. 1787, Vol. 15, 1972

8. Caretto L.S., Gosman A.D., Patankar S.V. and Spalding D.B.- "Two calculation procedures for steady, three-dimensional flows with recirculation".
Proceedings of the Third International Conference on Numerical Methods in Fluid Mechanics. Vol. II, 1973.

Chapter 5

A MATHEMATICAL MODEL OF A LOW-VOLATILE PULVERISED FUEL FLAME

W. Richter and R. Quack

Institut für Verfahrenstechnik und Dampfkesselvesen der Universität Stuttgart, Federal Republic of Germany

Abstract

A two-dimensional mathematical model has been developed to describe the behaviour of axi-symmetrical non-swirling anthracite flames. The model is based on the SPALDING prediction technics for turbulent flows. A four-flux method is incorporated into the model to take into account the radiative heat exchange. The predictions of the model are compared with experimental results of the International Flame Research Foundation (IFRF) obtained for horizontal p.f. flames.

NOMENCLATURE

A	:	frequency factor	$kg/m^2 sec\ atm$
A,B	:	coefficients in the equations of radiant heat flux density	
a	:	local ash contents of the solids	kg_{ash}/kg_{solids}
$a_\varphi, b_\varphi, c_\varphi, d_\varphi$	:	coefficients of the transport equation for a property φ of the fluid	
D	:	diameter of the enclosure	m
d	:	inlet diameter	m
E	:	activation energy	kcal/kmole
I,J	:	radiant intensities in forward and backward directions	$kcal/m^2 sec\ sr$
i	:	radiant intensity	$kcal/m^2 sec\ sr$
K_D, K_S	:	diffusional and surface reaction rate coefficient	$kg/m^2 sec\ atm$
k	:	kinetic energy of turbulence	m^2/sec^2
k_a	:	absorption coefficient	1/m
m	:	mass concentration of a chemical species of the fluid	kg/kg_{fluid}
$\dot{m}, \dot{m}_r$	:	mass flow and recirculating mass flow	kg/sec
n	:	total number of size classes	
p_{O2}	:	partial pressure of O_2	atm
$\dot{Q}$	:	heat flux	kcal/sec
q	:	rate of removal of fixed carbon per unit external geometric surface area	$kg/m^2 sec$
$\dot{q}$	:	heat flux density	$kcal/m^2 sec$
R	:	source term of the equation for m	$kg/m^3 sec$
r,z	:	cylindrical polar co-ordinates	m
s	:	external geometric surface area per unit volume	m^2/m^3
T,t	:	gas temperature	K,C
U,1-U	:	unburnt and burnout	
$\vec{v}$	:	velocity vector	m/sec

$\sqrt{W}$	:	an average frequency of the turbulent fluctuations	1/sec
X_a	:	efficiency factor for absorption	
$x_{av,l}$	:	average particle diameter in the l th size class	m
ϵ_W	:	emissivity of the wall	
η_t	:	effective (turbulent) dynamic viscosity	kg/m sec
Γ_{eff}	:	effective (turbulent) exchange coefficient for mass	kg/m sec
λ	:	ratio of secondary to primary velocity at the inlet	
ω	:	component of the vorticity vector perpendicular to the z,r plane	1/sec
ϕ,ψ	:	polar angle,azimuthal angle	
φ	:	property of the fluid	
ψ	:	stream function	kg/sec
$\mathcal{R}$	:	universal gas constant	kcal/kmole K
ρ_c	:	apparent density of the carbon particles	kg/m^3
ρ_m	:	density of the suspension	kg/m^3
σ	:	STEFAN-BOLTZMANN constant	$kcal/m^2$ sec K^4
$\sigma_{t,z},\sigma_{t,r}$	:	turbulent normal stresses	kg/m sec^2
$\tau_{t,zr}$	:	turbulent shear stress	kg/m sec^2
θ	:	THRING-NEWBY parameter	

Subscripts

c	:	fixed carbon particles
l	:	l th size class of particles
m	:	condition on the axis
o	:	inlet condition
p,s	:	condition at primary inlet and secondary inlet
S	:	condition at the surface of the particles
W	:	condition at the wall
+,-	:	in forward and backward direction
$\bar{\ },{}'$	:	mean value and fluctuating value

1. Introduction

Mathematical flame models should be applied to the prediction of real industrial combustion processes.In order to produce realistic predictions of those processes one is often forced to start at least with a two-dimensional description of the flow and mixing pattern. A method for calculating two-dimensional turbulent flows was developed by SPALDING [1] .The purpose of the present paper is to describe a mathematical flame model in which the SPALDING-method combined with a four-flux model for the calculation of the radiative heat transfer forms the fundamental part.The presented flame model is valid for the prediction of axi-symmetrical non-swirling anthracite flames.Such flames were experimentally studied in the past by the International Flame Research Foundation (IFRF) [2],[3] and some of the experimental results will be used here to check the reliability of the developed flame model.Comparing the predictions of the present model with measurements discrepancies may arise mainly from three reasons:

(i) Deficiencies of the basic flow model,i.e. use of a non-appropriate turbulence model.
(ii) Insufficient kinetic datas for the reaction model.

(iii) Approximate evaluation of the radiative heat exchange by use of simple flux models.

To distinguish between the errors caused by the flow model and the other sub-models the flow model will be discussed separately.

2. Flow model

2.1 Geometry

Fig. 1 shows the geometry for which the mathematical flame model is built up. The fluid emerges steadily without swirl from a double-concentric burner into a cylindric enclosure which may be constricted at the end. As the equivalent burner diameter d_{eq} is smaller than the duct diameter D, recirculation occurs. Therefore it is necessary to solve the complete elliptic differential equations governing the fluid flow.

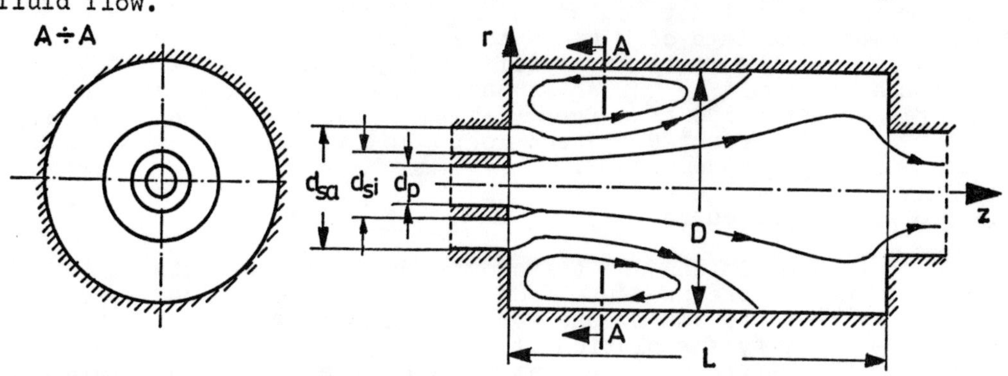

Fig. 1 The geometry considered

2.2 Calculation method

2.2.1 Effective viscosity concept

The method of calculating two-dimensional flows with recirculation is outlined in [1]. The basic equations to be considered are the equation of continuity and the equations for transport of momentum in axial and radial direction. Turbulence is accounted for by averaging the equations with regard to time. The double correlations of velocity fluctuations appearing in the momentum equations - the so-called turbulent stresses - are assumed to be proportional to gradients of the mean velocity like the viscous stresses according to STOKES's law of laminar friction.

$$\sigma_{t,z} = -\rho\overline{v_z'^2} = \eta_t\left(2\frac{\partial\bar{v}_z}{\partial z} - \frac{2}{3}\operatorname{div}\vec{\bar{v}}\right) \quad (1a)$$

$$\sigma_{t,r} = -\rho\overline{v_r'^2} = \eta_t\left(2\frac{\partial\bar{v}_r}{\partial r} - \frac{2}{3}\operatorname{div}\vec{\bar{v}}\right) \quad (1b)$$

$$\tau_{t,zr} = -\rho\overline{v_z'v_r'} = \eta_t\left(\frac{\partial\bar{v}_z}{\partial r} + \frac{\partial\bar{v}_r}{\partial z}\right) \quad (1c)$$

η_t in equation (1) is the turbulent or effective viscosity. η_t is no property of the fluid, but locally dependent and it is larger by an order of magnitude than the laminar viscosity except in near-wall regions.

2.2.2 Model of turbulence

The main difficulty of the effective viscosity concept is to find an expression for η_t that equation (1) is approximately satisfied. Several models of turbulence exist 4 .The model of turbulence incorporated in the present flame model is the k-W model invented by SPALDING 5 .This model states,that η_t is proportional to the local kinetic energy of turbulence k and inversely proportional to an average local frequency of the turbulent fluctuations denoted by $\sqrt{W}$:

$$\eta_t = \rho \frac{k}{\sqrt{W}} \tag{2}$$

k and W are determined assuming that they obey similar transport equations that are valid for a general property φ of the fluid (see 2.2.3).Six constants appear in the equations for k and W,two constants - the so-called turbulent SCHMIDT-numbers of k and W - refer the turbulent exchange of k and W to η_t and four constants are associated with the local dissipation and generation of k and W.The six constants should be universal ones,but calculations have shown that this is not quite true for axi-symmetrical flows.The values of the constants adopted for the present work don't differ very much from those ROBERTS [6] recommended.

2.2.3 Solution procedure

The basic flow equations are transformed by defining a stream function ψ and eliminating the pressure from the momentum equations.This leads to an equation for ψ and to one for the vorticity ω.Although the pressure - velocity formulation of the flow equations has numerical advantages,the ψ - ω formulation was chosen for ease of programming. Included the equations of the model of turbulence four equations, namely for ψ , ω ,k and W,have to be solved to predict the two-dimensional isothermal flow pattern in question.These four equations as well as the equations of mass concentrations and enthalpy which are necessary for the complete flame model (see 3.1) can be expressed in the mathematical form of equation (3):

$$a_\varphi\left[\frac{\partial}{\partial z}\left(\varphi\frac{\partial\psi}{\partial r}\right)-\frac{\partial}{\partial r}\left(\varphi\frac{\partial\psi}{\partial z}\right)\right]-\frac{\partial}{\partial z}\left[b_\varphi r\frac{\partial(c_\varphi\varphi)}{\partial z}\right]-\frac{\partial}{\partial r}\left[b_\varphi r\frac{\partial(c_\varphi\varphi)}{\partial r}\right]+rd_\varphi=0 \tag{3}$$

φ stands for a general property of the fluid which is transported by convection (the first two terms) and by turbulent diffusion (the following two terms) or which is created or destroyed at a certain rate (the last term).The solution procedure of the considered set of differential equations follows the finite-difference substitution technic described in [1] .

2.3 Prediction of isothermal flows with recirculation

Only few results of the calculations carried out for isothermal flows are mentioned here,a comprehensive report will be given in [7] . Three kinds of confined flows are considered:

(i) The single confined jet.
(ii) The double-concentric jet from a cement-kiln type burner.
(iii) The double-concentric jet from a boiler type burner.

The two latter flows occur in the flames for which the present mathematical model is constructed.

2.3.1 Single confined jet

Fig. 2 shows the calculated and measured axial velocity decay of some single confined jets. The parameter varied is the ratio of inlet to duct diameter d/D. In regions near the inlet the enclosed jet behaves like a free jet, but further downstream the confinement causes a more rapid decay of axial velocity. In Fig. 3 the calculated degrees

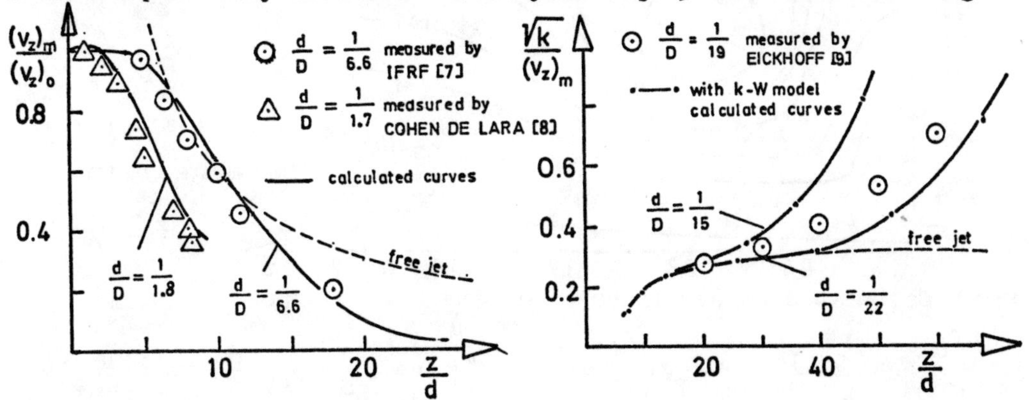

Fig. 2 Axial decay of velocity for single confined jets

Fig. 3 Degree of turbulence on the axis of single confined jets

of turbulence on the axis are plotted versus the axial distance from the inlet. The influence of the confinement leads to an increase of turbulence depending on the ratio d/D. Fig. 3 shows also the measured degree of turbulence of a confined jet [9], which follows the trend of the predicted curves. The position of the recirculating vortex and the amount of recirculating matter influence the stability of flames and should be known. Fig. 4 and Fig. 5 show that the mentioned characteristics of the recirculating flow are too quite well predicted.

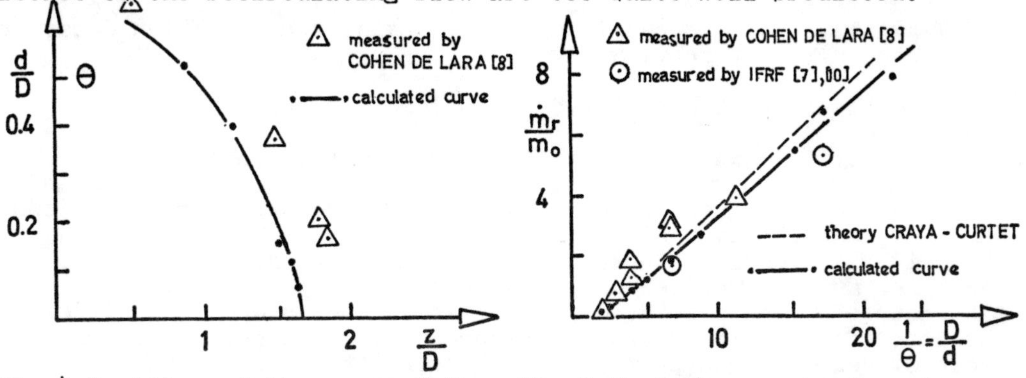

Fig. 4 Position of the core of the recirculation vortex

Fig. 5 Maximal amount of recirculating matter

2.3.2 Double-concentric jet from a cement-kiln type burner

In this case secondary air with low momentum is introduced around the primary jet. If the outer diameter of the secondary inlet equals the duct diameter, the flow pattern of Fig. 6 will be obtained. The primary jet first entrains the existing secondary air, if this is consumed the recirculation vortex is formed. The calculated position and size of the recirculation eddy and the calculated magnitude of the mass flow of recirculating fluid are again in agreement with measurements [10] as well as the predicted decay of velocity and primary mass concentration on the axis (Fig. 7 and 8).

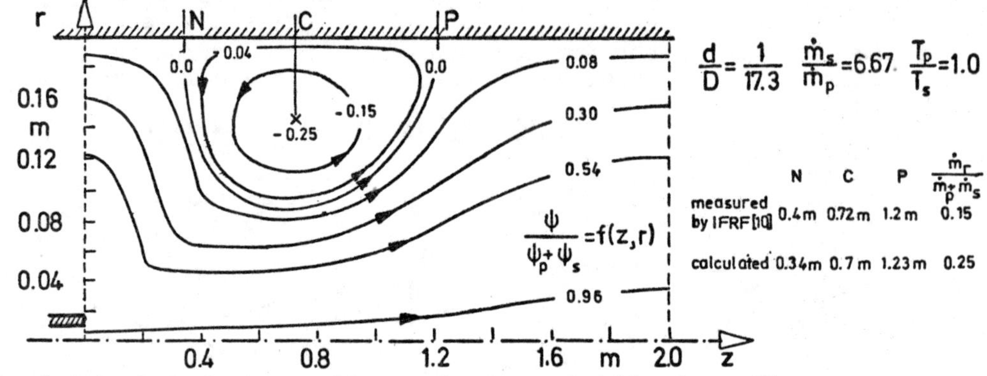

Fig.6 Calculated stream-lines of a cement-kiln type flow

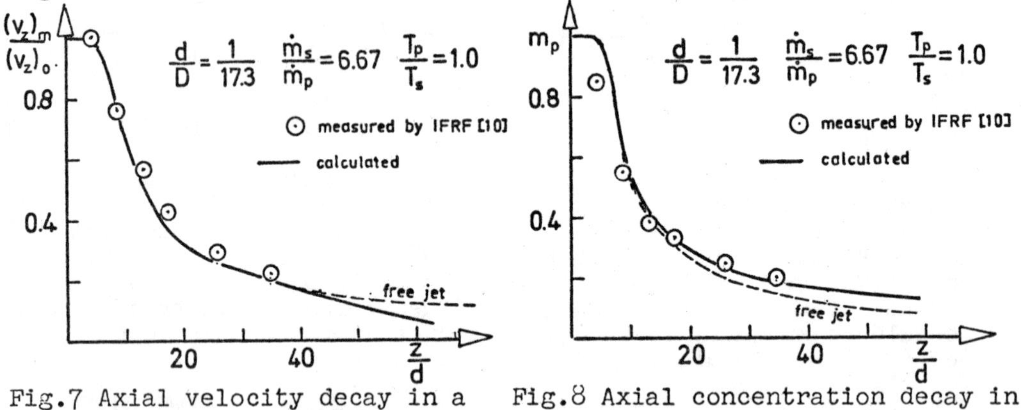

Fig.7 Axial velocity decay in a cement-kiln type flow

Fig.8 Axial concentration decay in cement-kiln type flow

2.3.3 Double-concentric jet from a boiler type burner

In boiler type flows the secondary annular jet is introduced into the furnace with high momentum. Depending on the velocity ratio and the interface thickness between the primary and secondary inlet two interior recirculation eddies are formed at the seperation lip as shown in Fig.9, where calculated and measured stream-lines are compared for a ratio λ of secondary to primary velocity of 1. Although the feature of the predicted recirculation eddies does not differ very much from the measured one, discrepancies are found for the decay of velocity and primary mass concentration on the axis (Fig.10 and Fig.11). Caused by the interior recirculation vortices the primary jet mixes very fast with the secondary jet, a steep velocity decay is found on the axis just behind the primary inlet. Further downstream the fluid on the axis is accelerated again by the predominant annular jet which is combining to a fully round jet. It can be seen from Fig. 10 that the predicted acceleration on the axis is too small. Two reasons may be responsible for this fact:

(i) RODI [11] found that in round wakes - the flow around a disc is similar to that out of an annular ring - the constants of the two-equation turbulence models have to be modified.

(ii) The effective viscosity concept implies that for example

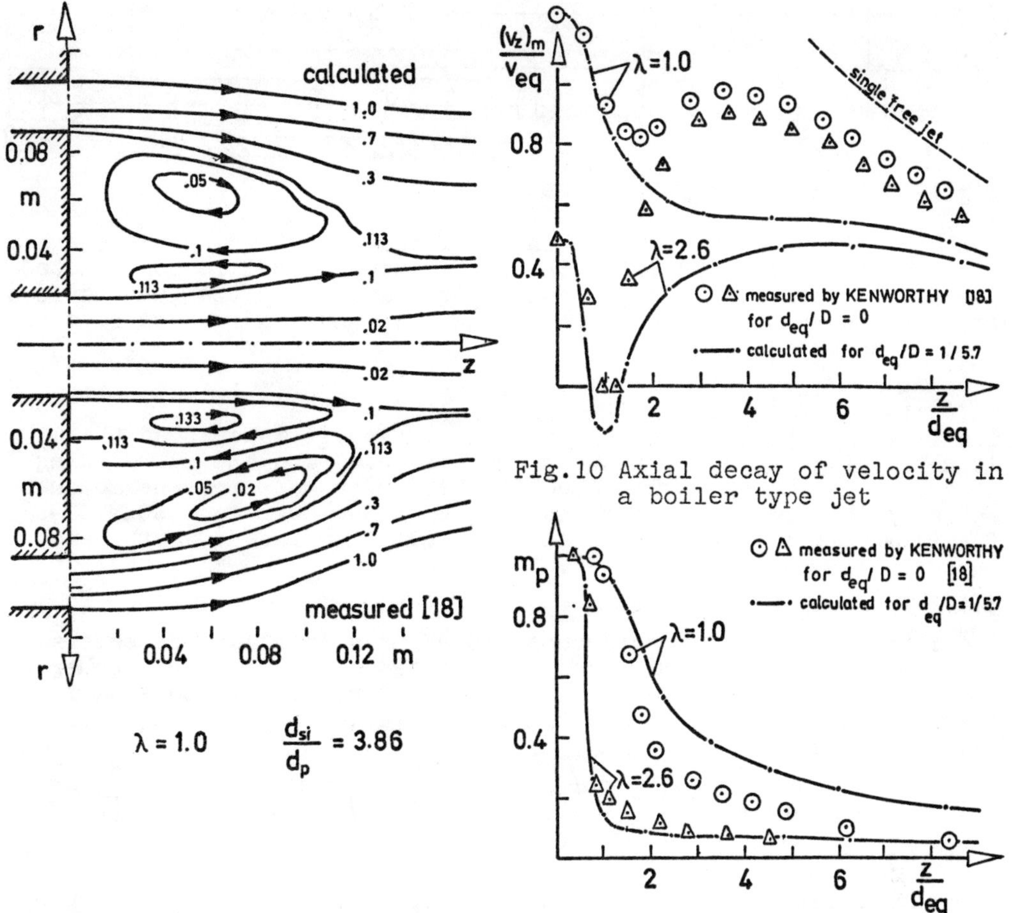

Fig.10 Axial decay of velocity in a boiler type jet

Fig.9 Flow pattern in a double-concentric boiler type jet

Fig.11 Axial decay of concentration in a boiler type jet

$\rho\, v_z'^2$ is replaced by $-2\eta_t \partial \overline{v}_z/\partial z$. As it is not possible for $\overline{v_z'^2}$ to become negative values, errors may arise in regions where $\partial\overline{v}_z/\partial z$ has considerable positive values and that is the region where the predictions were wrong.

3 . The complete mathematical flame model

The preceding flow model is extended to calculate the fields of concentrations and temperatures and the heat flux distributions in p.f. flames. For this purpose additional equations which describe the transport of the reacting species of the fluid and the transport of enthalpy are introduced into the system of the basic flow equations. An economic solution of the extended system of equations is only possible, if rigorous simplifications of the physical processes are made.

3.1 Some details of the flame model

3.1.1 Geometry of the furnace

The results of the flame model shall be compared with flame measure-

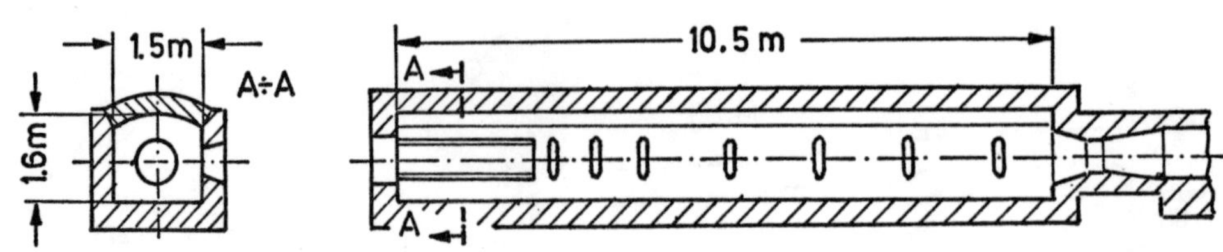

Fig.12 Furnace nr.2 of the IFRF

ments carried out by the IFRF in the furnace nr.2 which has a sqare cross-section (Fig.12).Since the mathematical model is constructed for cylindrical furnaces,an equivalent diameter of the furnace is determined so,that the real furnace and the modelled furnace have the same cross-section.

3.1.2 Flow pattern

The flow in a p.f. flame is a two-phase flow.It is assumed that there is no relative velocity between the particles and the surrounding gases.The turbulent exchange coefficients for mass concentrations and enthalpy are calculated assuming that the turbulent SCHMIDT- and PRANDTL-numbers are constant in the whole flow field.Gravity and buoyancy forces are neclected.

3.1.3 Property of the fuel

The initial pulverised coal consists of mass fractions of several size classes.Each size class is composed of fixed carbon spheres and seperated ash spheres.Transport equations similar to equation (3) are derived for the fixed carbon concentration $m_{c,1}$ of the various diameter classes:

$$\frac{\partial \psi}{\partial z}\left(m_{c,1}\frac{\partial \psi}{\partial r}\right) - \frac{\partial}{\partial r}\left(m_{c,1}\frac{\partial \psi}{\partial z}\right) - \frac{\partial}{\partial z}\left(r\Gamma_{c,1,eff}\frac{\partial m_{c,1}}{\partial z}\right) - \frac{\partial}{\partial r}\left(r\Gamma_{c,1,eff}\frac{\partial m_{c,1}}{\partial r}\right)$$
$$- rR_{c,1} = 0 \tag{4}$$

The ash particles are treated as an inert matter that is transported through the furnace without any physical changes.The contents of water and volatiles,which were about 8% for the anthracite studied, are neclected.However,the real composition of the coal is used to calculate the oxygen/fuel ratio of the reaction and the calorific value of combustion.The source term $R_{c,1}$ in equation (4) is splitted into three parts [12]:

$$R_{c,1} = - R_{1\rightarrow 1-1} + R_{1+1\rightarrow 1} - R_{1,f} \tag{5}$$

$R_{1\rightarrow 1-1}$ is the loss of mass of the 1 th size class,because particles of this class are burnt to the lower diameter limit,similar $R_{1+1\rightarrow 1}$ is the gain of mass from the upper diameter class.Finally,$R_{1,f}$ is the loss of mass of the 1 th size class due to reaction of fixed carbon.

3.1.4 Reaction mechanism and reaction kinetics

The only reaction considered is that between the carbon spheres and O_2 to CO that burns to CO_2 as soon as it has left the boundary layer around the particles.The reaction rate per unit external geometric surface is assumed to be of first order with respect to oxygen.The reaction rate is expressed as:

$$q = 1/(1/K_D + 1/K_S)\ p_{O2} \tag{6}$$

For combustion of pulverised anthracite the diffusional rate coefficient K_D is much larger than the surface reaction rate coefficient K_S, the overall reaction is controlled by the chemical surface reaction. K_S is evaluated using an ARRHENIUS-type formulation:

$$K_S = A \exp(-E/\mathcal{R}T_S) \tag{7}$$

T_S in equation (7) is the particle temperature which is calculated by means of a heat balance assuming the particles in thermodynamic equilibrium with their surroundings. In order to obtain the source term $R_{l,f}$ of the l th fraction due to combustion of fixed carbon the surface area s_l of that fraction locally available per unit volume must be known. If it is assumed that the particles burn with constant density of the coal, s_l can be evaluated by:

$$s_l = \rho_m m_{c,l} 6/(\rho_c x_{l,av}) \tag{8}$$

s_l is also needed to determine the absorption coefficient of the dust cloud.

$$k_a = X_a \left(\frac{1}{4}\sum_{l=1}^{n} s_l + \frac{1}{4} s_{ash}\right) \tag{9}$$

3.1.5 Evaluation of radiant heat exchange with a flux model

The mathematical formulation of the convective and diffusional transport of energy as a property of the fluid results in an equation similar to equation (3). The source term of this equation is the amount of radiant heat emitted or absorbed at a point of the flow field per unit time and unit volume. In order to obtain exact values for this term the complete integro-differential equations for radiant heat exchange in an enclosure should be solved. A numerical method for the solution of the equations governing the radiant heat exchange is that of HOTTEL et al. [13], but at the moment, the computational effort combining the complete HOTTEL-method with that of SPALDING is still too big. Therefore the calculation of radiant heat exchange is simplified by use of so-called flux models which allow for radiant heat fluxes only along few selected directions.

In the present flame model a four-flux method is used, net radiant heat fluxes are permitted only along the radial and axial directions. The following distribution of radiant intensity $i=f(\phi,\psi)$ is assumed around a small volume element of the furnace (Fig. 13):

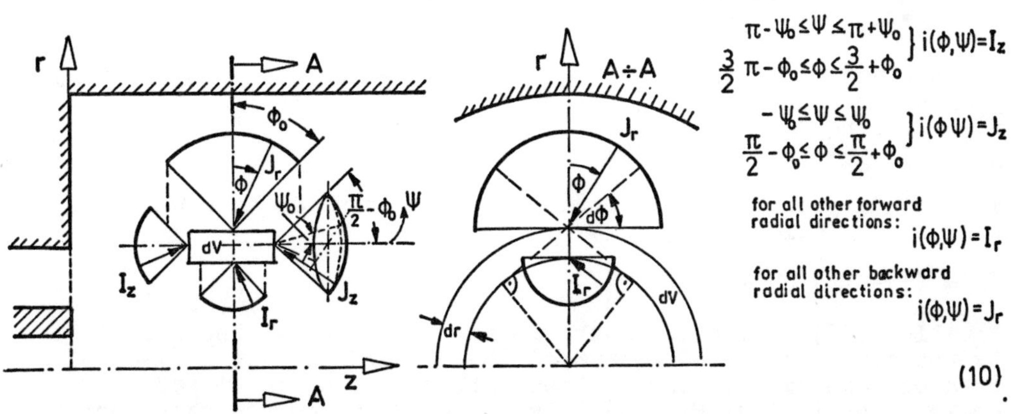

Fig. 13 Assumed distribution of intensity $i=f(\phi,\psi)$ in the furnace

The axial transport of radiant intensity takes place within a cone of the half-angle $\psi_o = \pi/2 - \phi_o$. Radiation from all other directions is combined to the net radial flux. The angle ψ_o can be adapted to the problem considered. If it is zero the four-flux model is reduced to a two-flux model. The present calculations were carried out with a value of 45° for ψ_o.

The following simplifications are made deriving the equations of radiant heat flux density for the assumed distribution of i:

(i) The solid body radiation of the particles predominates in relation to the gas radiation. Thus, it is justified to describe the directional transfer of radiant intensity in the suspension by a single equation as suggested by HEMSATH [14].

(ii) Scattering is either neclected or assumed to take place only in forward directions.

With these assumptions the equation of directional radiant energy transfer is seperately integrated over those ranges of the solid angle where i is constant. The integration procedure which is similar to that of SCHUSTER-SCHWARZSCHILD leads to 2 ordinary differential equations (11a and b) for the heat fluxes in forward and backward axial direction and to 2 corresponding equations (12a and b) for the radial direction.

$$\frac{d\dot{q}_z^+}{dz} = -A_z k_a \dot{q}_z^+ + B_z k_a \sigma T^4 \quad (11a) \qquad -\frac{d\dot{q}_z^-}{dz} = -A_z k_a \dot{q}_z^+ + B_z k_a \sigma T^4 \quad (11b)$$

with $A_z = 2/(1+\cos\psi_o)$, $B_z = 2(1-\cos\psi_o)$, $\dot{q}_z^+ = \pi(B_z/A_z)I_z$ and $\dot{q}_z^- = \pi(B_z/A_z)J_z$.

$$\frac{1}{r}\frac{d(\dot{q}_r^+ r)}{dr} = -A_r k_a \dot{q}_r^+ + B_r k_a \sigma T^4 + \frac{\dot{q}_r^-}{r} \quad (12a)$$

$$-\frac{1}{r}\frac{d(\dot{q}_r^- r)}{dr} = -A_r k_a \dot{q}_r^- + B_r k_a \sigma T^4 - \frac{\dot{q}_r^-}{r} \quad (12b)$$

with $A_r = 2\pi\cos\psi_o/(\pi - 2\psi_o + 2\cos\psi_o \sin\psi_o)$, $B_r = 2\cos\psi_o$,

$\dot{q}_r^+ = \pi(B_r/A_r)I_r$ and $\dot{q}_r^- = \pi(B_r/A_r)J_r$.

The equations (12a) and (12b) contain an additional term $\pm\,\dot{q}^-/r$ which takes the fact into account that a part of the radial inward flux does not cross the annular volume element for which the radiant energy balance is set up. The four heat fluxes in axial and radial direction are combined to net heat fluxes

$$\dot{q}_z = \dot{q}_z^+ - \dot{q}_z^- \qquad \dot{q}_r = \dot{q}_r^+ - \dot{q}_r^- \quad (13)$$

and the variation of $\dot{q}_z$ and $\dot{q}_r$ in z- and r-direction respectively yields the source term of the enthalpy equation:

$$\frac{d\dot{q}_z}{dz} + \frac{1}{r}\frac{d(r\dot{q}_r)}{dr} \quad (14)$$

The boundary conditions for radiant heat transfer at a wall are written as follows:

$$\dot{q}_W^- = (1-\epsilon_W)\,\dot{q}_W^+ + \frac{B}{A}\,\epsilon_W \sigma T_W^4 \quad (15)$$

Equation (15) implies, that the inward heat flux at the wall is formed by the total reflected amount of the outward radiation $(1-\epsilon_W)\dot{q}_W^+$ and by a certain part of the emitted wall radiation specified by the angle ψ_o or ϕ_o (see Fig.13). This formulation was necessary to obtain a correct radiant energy balance in case of thermal equilibrium.

3.1.6 Cooling of the furnace

During the p.f. trials for the flames studied the furnace was cooled by tubes introduced at the roof and the bottom.The side walls were uncooled.In order to consider the overall cooling effect in the flame model,the tube/refractory wall system was replaced by plane surfaces acting as equivalent heat sinks.This method was described by JOHNSON [15].

3.2 Results of the flame model

From all results obtained with the flame model so far,the results of two flames are discussed here.One flame (flame 1 C15 [2]) is produced with a cement-kiln type burner,the other one (flame 8 C11 [3]) with a boiler type burner.

3.2.1 Cement-kiln type flame

Fig. 14 shows some characteristic quantities like temperature,gas concentration,aerodynamic degree of mixing and burnout on the axis of the cement-kiln type flame.The overall agreement between the predicted curves and measured values is satisfying.However,discrepancies

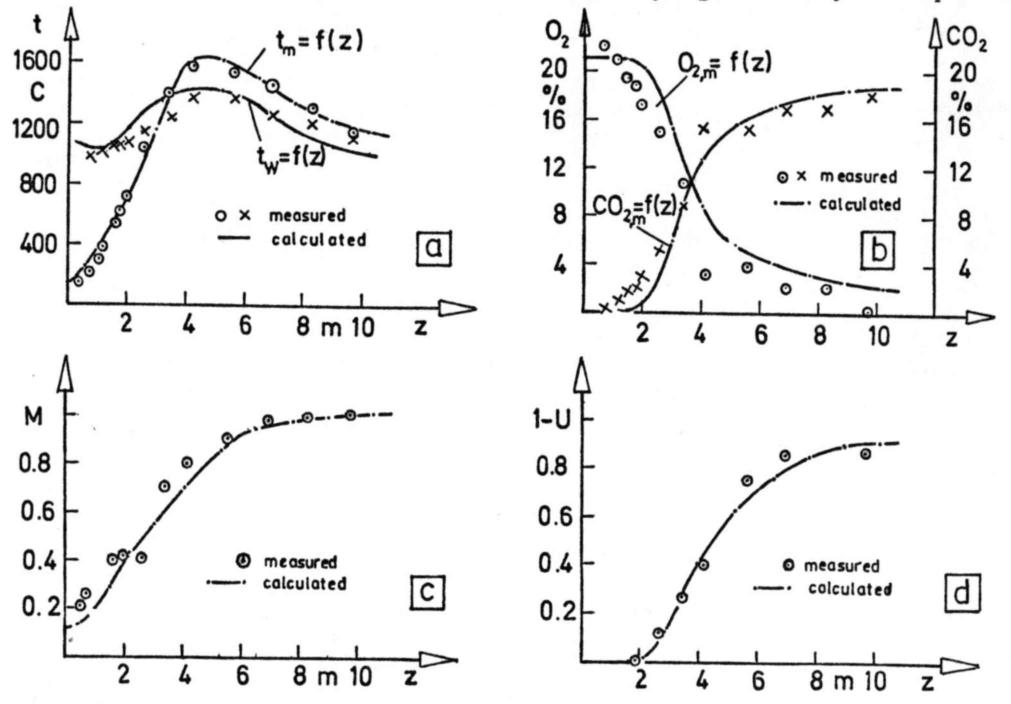

Fig. 14 Distribution of temperature (a), O_2 and CO_2 concentration (b), aerodynamic degree of mixing (c) and burnout along the furnace for flame 1 C15 [2]

typical for this kind of flame are found,if one looks at the complete temperature field (Fig.15).The measured isotherms show a distortion downstream the secondary inlet indicating a rapid heating of the secondary air which is not predicted to such an extent.On the other side,the evaluated temperature level of the recirculation zone is too high.A possible explanation for this fact is,that the slow moving secondary air stream,which is relatively cold compared with the recir-

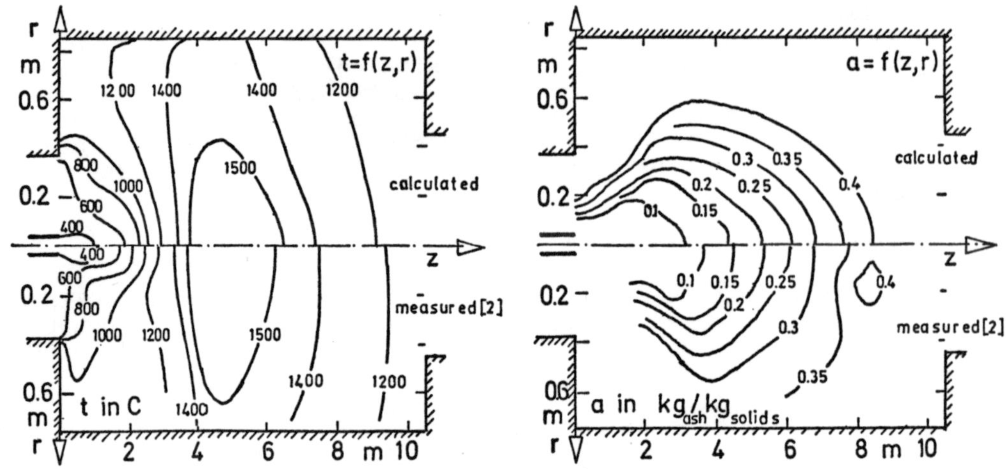

Fig.15 Temperature field of the ciment-kiln type flame

Fig.16 Local ash contents of the solids

culating gases,drops to the bottom of the furnace.Downstream,the secondary air mixes with the recirculating gases cooling them down somewhat.As the flow is horizontal,the two-dimensional flame model is not capable to take into account any gravity or buoyancy forces. Therefore,the dropping of the secondary air cannot be predicted.In the last section of the furnace,the measured temperature level of the near-wall regions surpasses the predicted one.This is probably caused by the fact,that the temperatures were measured in a plane perpendicular to the non-cooled side walls of the furnace,whereas the model gives only an average temperature level for all planes through the furnace axis.Fig.16 shows the distribution of the local ash contents of the solid particles in the furnace.Contrary to other measurements the local ash contents a can be measured without any big errors.In a low-volatile p.f. flame a is a function of the burnout 1-U of the particles provided the ash particles and the fuel particles travel along the same stream-lines:

$$a = a_o/[1 - (1 - U)(1 - a_o)] \qquad (16)$$

a_o is the initial ash contents of the solids at the inlet.Considering equation(16) it can be concluded from Fig.16 that the burnout in the whole flame is quite well predicted.

It was found,that from all parameters included in the mathematical model,the reaction rate coefficients had the strongest influence on the feature of the flame [16].Fig.17 shows the influence of the frequency factor Λ (equation (7)) on the calculated axial distribution of temperature and burnout.Λ affects especially the position of the ignition zone and the value of the maximal temperature,whereas the influence on the temperature of the flue gases is small.However, burnout is considerably affected along the whole furnace.The best fit of the measurements could be obtained by calculating with a frequency factor of 60 kg/m^2 sec atm and an activation energy of 19000 kcal/kmole.

Fig.18 shows the distribution of radial intensities along the furnace wall.The measured intensities were obtained along directions perpendicular to the non-cooled side walls through the axis of the furnace. The calculated intensities are the intensities averaged over that range of the solid angle that was assumed to give a contribution to the radial fluxes (see Fig.13).The difference between the predicted

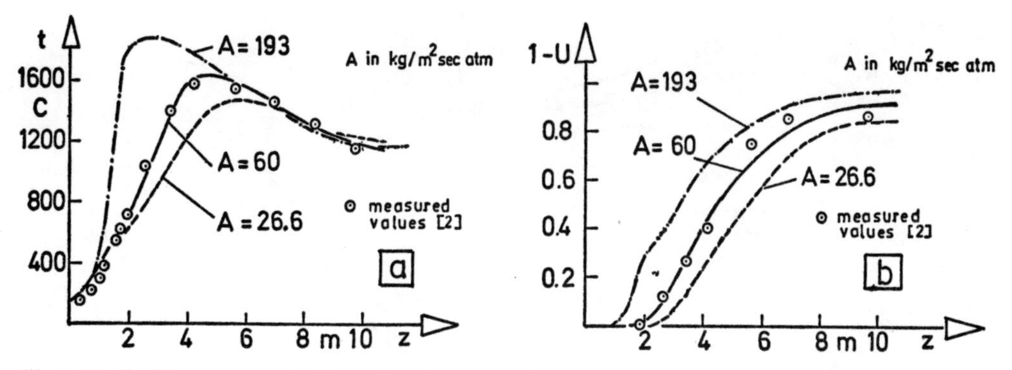

Fig.17 Influence of the frequency factor A on the temperature distribution (a) and on the burnout of combustible matter (b)

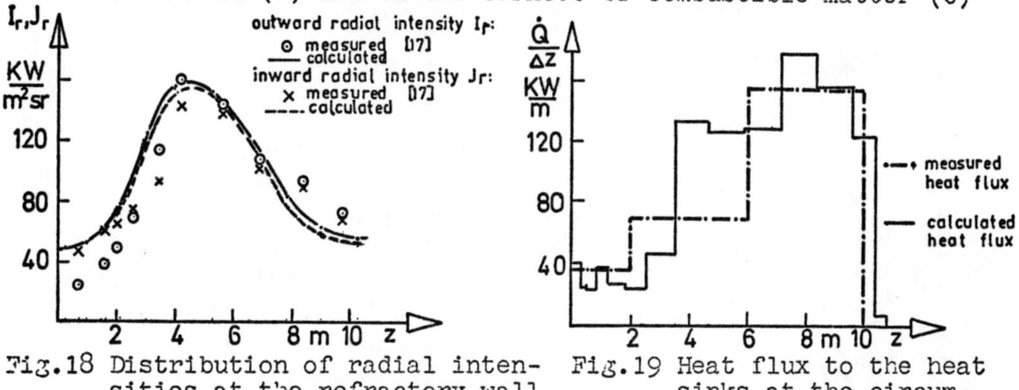

Fig.18 Distribution of radial intensities at the refractory wall

Fig.19 Heat flux to the heat sinks at the circumferential furnace walls

outward and inward radial intensities I_r - J_r has small positive values along the whole wall corresponding to small outward heat fluxes through the refractory.The measured difference I_r - J_r,however, is considerable in the first part of the furnace indicating that the distribution of intensities arriving the wall is non-isotropic in this region.Fig.19 shows the heat fluxes to the heat sinks along the circumferential walls of the furnace.The predicted total amount of heat absorbed by the heat sinks is about 6 % bigger than the measured one.As the calorific measurements were carried out dividing the furnace only in three sections,the comparison of the measured local heat fluxes with the predicted ones is somewhat doubtful.Nevertheless,one can conclude that the net heat flux opposite to the ignition zone (3m ÷ 5m) is overpredicted,a fact which corresponds to the too high calculated gas and wall temperatures in that region.

3.2.2 Boiler-type flame

In the figures 20 - 22 some results for the boiler type flame are shown.There is again a satisfying agreement between the predicted and measured burnout,discrepancies are found for the temperature distribution in the flame front.The temperature peak on the axis is predicted too high.This can probably attributed to a deficiency of the turbulence model,which gives a too poor mixing between the secondary and primary jet,if the velocity ratio is about 1 like in the present case.When the turbulent level on the axis was increased by assuming unrealistic high values of the length scale of turbulence at the primary inlet,the temperature peak on the axis could be lowered.

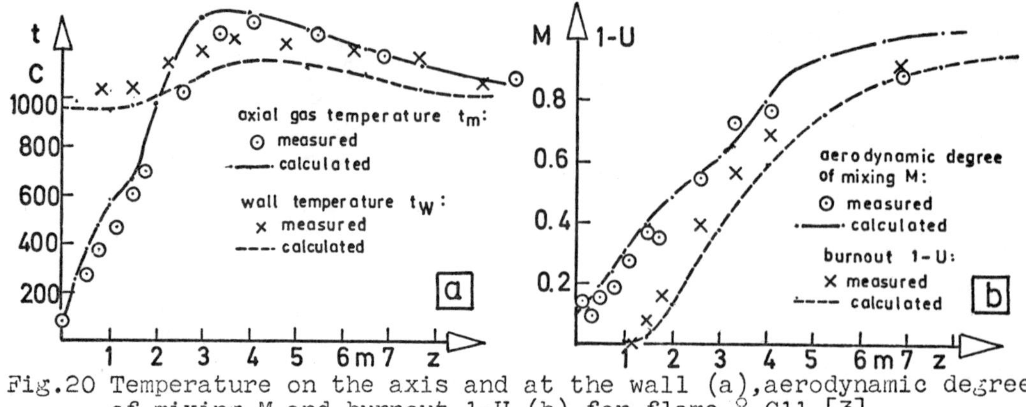

Fig.20 Temperature on the axis and at the wall (a), aerodynamic degree of mixing M and burnout 1-U (b) for flame 8 C11 [3]

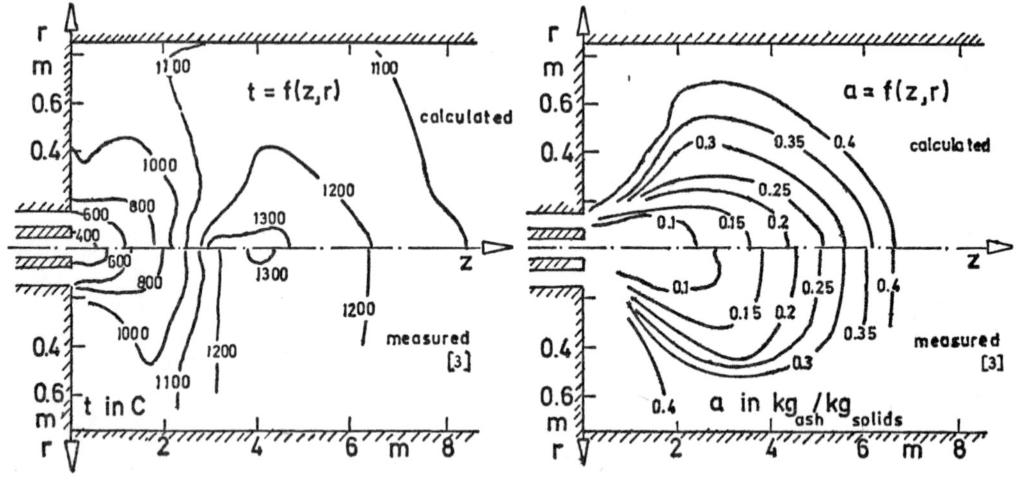

Fig.21 Temperature field of the boiler type flame

Fig.22 Local ash contents of the solids

4. Conclusion

In spite of the many simplifications that were necessary to construct the present mathematical flame model, satisfying predictions were achieved for non-swirling anthracite flames. It is now the task to extend the model to more complicated combustion processes as high-volatile coal combustion or oil combustion and probably to combustion with swirl. But for the present model some improvements of the implied sub-models are still desirable, for instance:

(i) A modified model of turbulence for the case of a double concentric jet when the secondary jet has a considerable momentum.

(ii) A flux model of radiant heat transfer, which takes into account fluxes not only in two perpendicular directions, but also in at least two oblique directions.

On the other side, some discrepancies between predictions and measurements can surely be attributed to a misuse of the mathematical model, which is only valid for two-dimensional flows, whereas the flow pattern of the measured flames was three-dimensional in certain parts of the furnace.

REFERENCES

[1] GOSMAN,A.D.,W.M.PUN,A.K.RUNCHAL,D.B.SPALDING and M.WOLFSHTEIN: "Heat and Mass Transfer in Recirculating Flows".Academic Press, London,(1969)
[2] HEIN,K.:"Investigation into the Combustion and Devolatilisation Behaviour of the Non-swirling Pulverised Coal Flames".IFRF, doc.nr. F32/a/41,(1971)
[3] CHEDAILLE,J. and K.H.HEMSATH:"Etude de mecanisme d'inflammation et de la recirculation des produits de combustion".IFRF, doc.nr.F33/a/35,(1967)
[4] LAUNDER,B.E. and D.B.SPALDING:"Mathematical models of turbulence". Academic Press,London,(1972)
[5] SPALDING,D.B.:"A two-equation model of turbulence".VDI-Forschungsheft 549,(1972)
[6] ROBERTS,L.W.:Unpublished work at Imperial College,London,(1971)
[7] PAI,B.R.,W.RICHTER and T.M.LOWES:"Flow and mixing in confined axial flows - second report".To be published in a report of the IFRF,(1973)
[8] COHEN DE LARA,G.,J.POUX and R.PERRIN:"Etude de la recirculation dans les modèles schématiques de fours à flammes de diffusion". SOGREAH,Grenoble,R.7.643,(1960)
[9] EICKHOFF,H.:"Statischer Druck und Turbulenz in drehsymmetrischen Freistrahlen und Freistrahlflammen".Thesis,Universität Karlsruhe (TH),(1968)
[10] PAI,B.R. and T.M.LOWES:"Flow and mixing in confined axial flows-first report".IFRF,doc.nr. G02/a/20,(1972)
[11] RODI,W.:"The Prediction of Free Turbulent Boundary Layers by Use of a Two-equation Model of Turbulence".Thesis,Imperial College, London,(1972)
[12] GIBSON,M.M. and B.B.MORGAN:"Mathematical Model of Combustion of Solid Particles in a Turbulent Stream with Recirculation". J.Inst.Fuel,Vol. 43,(1970)
[13] HOTTEL,H.C. and A.F.SAROFIM:"Radiative Transfer".Mc Graw-Hill Book Company,(1967)
[14] HEMSATH,K.H.:"Zur Berechnung der Flammenstrahlung".Thesis, Technische Hochschule Stuttgart,(1969)
[15] JOHNSON,T.R.,T.M.LOWES and J.M.BEER:"Comparison of Calculated Temperatures and Heat Flux Distributions with Measurements in the Ijmuiden Furnace".4th Symposium on Flames and Industry, London,(1972)
[16] RICHTER,W.:"Some Chemical Aspects in the Mathematical Modelling of Flames".6th Meeting of the Flame Chemistry Panel of the IFRF, (1973)
[17] HEIN,K.:"Beitrag zum Wärmeübergang in feststoffbeladenen Gasströmem technischer Staubfeuerungen".Thesis,Technische Hochschule Stuttgart,(1972)
[18] KENWORTHY,J.S.:"Flow and mixing in double concentric jets". IFRF,Internal Report nr. 156,(1971)

Chapter 6

THE PROBLEM OF FLAME AS A DISPERSE SYSTEM

A. Blokh

Central Boiler and Turbine Institute, Leningrad, USSR

Abstract

The lecture presents the principle results of the theoretical analysis, calculation and experimental research on radiative properties of soot particles in luminous flame and coal particles in coal powdered flame.

NOMENCLATURE

a_λ	: spectral absorptivity of particle cloud
c	: carbon content in fuel, %
$I_{\lambda,0} = c_1\lambda^{-5}(e^{c_2/\lambda T} - 1)^{-1}$	: Plank's function for black-body radiation
$\kappa_\lambda, \kappa_{s,\lambda}, \kappa_{\alpha,\lambda}$	: spectral extinction, scattering and absorption coefficients
$m = n - iH$	: complex refraction index
$M(\lambda) = \mathrm{Im}\left(\frac{1-m^2}{2+m^2}\right)$	: function of the complex refraction index
n	: refraction index
N(x)	: function of size particle distribution
T_0, T_a, T_F, T_w	: temperatures of exit gas, adiabatic burning, flame, screen wall
v	: volatile content in fuel, %
x	: particle diameter
x_m	: modal value of particle diameter in polydisperse system of particles
α	: coefficient of the air excess
β	: direction of scattering
$\gamma_\lambda(\beta), \tilde{\gamma}_\lambda(\beta)$	: scattering indicatrixes for mono- and polydisperse particle systems
H	: absorption index
λ	: radiation wave length
$\rho = \frac{\pi x}{\lambda}$	: diffraction parameter
$\rho_m = \frac{\pi x_m}{\lambda}$	: modal value of the diffraction parameter
$\tilde{\sigma}_{a,\lambda}, \tilde{\sigma}_{s,\lambda}$	: effective spectral absorption and scattering cross-sections for polydisperse particle system
$\sigma_{a,\lambda}, \sigma_{s,\lambda}$	: effective spectral absorption and scattering cross-sections for monodisperse particle system
$\tilde{\sigma}_{a,1}, \tilde{\sigma}_{a,R}$	: Plank's and Rosseland's mean absorption cross-sections for polydisperse system of soot particles

I. Introduction

Flame is one of the most complex nature phenomena. Over a long period of time its properties are studied in order to use them to the benefit of society.

Approximately 20 years ago A.Gaydon and H.Wolfhard wrote in their remarkable monograph/I/:"Almost without any exaggeration one can say that at present more is known about processes in the atmosphere of some stars than about processes in Bunsen burner".

At the same time famous German scientists V.Pepperhoff and A.Bahr wrote in the journal "Eisenhüttenwesen"/2/:"The processes of combustion in flame are so complex that any attempt of preliminary calculation for flame thermal radiation in certain conditions must be considered as hopeless".

From what point of view do we examine the flame? Have we learned to predict its properties and its behaviour?

In fact, the flame is extremely complex and varied, but to-day we already know enough to calculate its thermal radiation, even though without sufficient accuracy, but we can predict its properties. We have learned many things about physical processes in flames, their structure and radiation characteristics. During these years we have gone far forward from Bunsen burner and are engaged in investigation of large industrial flames.

In our lecture the main attention is focussed on the analysis of radiation characteristics of the solid disperse flame phase and on its influence upon radiant energy transfer.

I.I Radiant flame properties

Any industrial flame is an extremely complex multicomponent system which consists of gaseous combustion products, mainly of CO_2 and H_2O and suspended solid particles of different properties and sizes. We do not consider here the radiation characteristics of flame gaseous phase.

The greatest influence on flame heat radiation is exerted by soot, pulverized coal, ash, and coke particles. In reference to this solid phase the flame is a complex disperse system /3/.

The problem of flame as a disperse system includes the prediction of methods for heat radiation calculation of such system and of its effect on the condition of radiative energy transfer from flame to solid heating surfaces.

I.2 Particle radiation characteristics

A great part in the flame heat radiation is played by solid particles, suspended in the flow of gaseous combustion products. When burning gas and liquid fuel, finest carbon soot particles are formed in the flame. Due to a great quantity of such particles flame emissivity increases sharply. When burning solid fuels, particularly pulverized coal, flame radiation is defined by radiation properties of fuel, coke and ash particles. All these particles are large compared to carbon soot particles.

If the particle flow is considered as a disperse system, we can employ the known formula given by Mie /4/ which allows to determine the effective cross-sections of attenuation, scattering and absorption, and the scattering indicatrix. These data are basic for any calculation of heat exchange between a disperse system and solid wall.

For determination of the said radiation characteristics it is necessary to know first of all the complex index of refraction and its dependence on the radiation wave length. The second independent parameter which must be known for calculations is the diffraction parameter ρ .

I.3 Complex index of refraction

In our calculations for soot particles we use Stull's and Plass's data /5/about refraction and absorption indices for amorphous carbon. On the basis of these data we can adopt the following approximated relations describing satisfactorily the dispersion of n and H in the wave length band from 0,5 to 6 μ :

$$n^2 - H^2 = 2.63\lambda^{0.1} \tag{1}$$

$$nH = 0.4 + 1.55\lambda . \tag{2}$$

For particles of different solid fuels the indices of refraction and absorption were defined based on reflection coefficient measurements for optically smooth coal surface on the boundaries "coal-air" and "coal-immersion liquid" in a direction close to normal. As an immersion liquid, transparent in the infrared region, we used tetrachlorated carbon (CCl_4). The measurements covered the spectral range from 0.8 to 5.5 μ .Tests were performed on different coals from anthracite to brown coal.

The main results of these tests are presented in Figs.I.3-I ÷ I.3-4. The figures show the variation of n and H as a function of λ for each kind of coal. We shall examine these data in detail.

Anthracite and lean coal (Figs. I.3-I and I.3-2). Anthracite index of refraction n increases with increase in radiation wave length λ . However, as λ increases, this dependence gradually weakens. At $\lambda \geqslant 5\mu$ the index of refraction practically is independent of λ . The absorption index H augments with increasing of λ only in the range of 0.8 ÷ 2.5 μ . At $\lambda \geqslant 2.5\mu$, H no longer depends on λ .

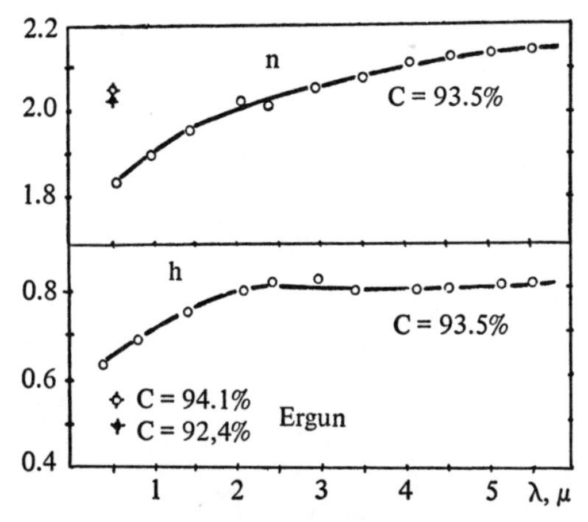

Fig. I.3-I Optical constants of anthracite

Lean coal index of refraction varies similarly to that of anthracite, but the dependence between n and λ is weaker. As to the absorption index H for lean coal in the range of 0.8-4 μ it even decreases slightly with increase in λ, and then starts growing.

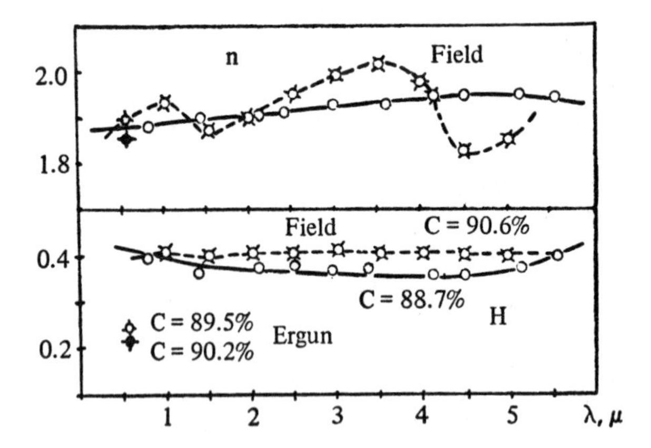

Fig. I.3-2 Optical constants of lean coal

On the presented plot the same data of Ergun and McCartney /6/ are shown for n and H at $\lambda = 0.546\mu$. For anthracite their data differ considerably from ours. For lean coal a satisfactory agreement is obtained only for the values of n.

For comparison we present Howarth's data published in /7/ for infrared region on the same plot. They are sufficiently close to our data for lean coal, but have some distinctive features. The first of them is connected with the normal dispersion of n found by Howarth in two spectral regions, (I.0 ÷ I.5 μ) and (3.5 ÷ 4.5μ) and the second one with the constancy of H within the entire studied range of radiation wave length.

Black coal (Fig. I.3-3). The tests performed do not reveal noticeable dispersion of n and H. The most important factor affecting n and H is the fuel elemental composition, particularly volatile (v) and carbon (C) contents.

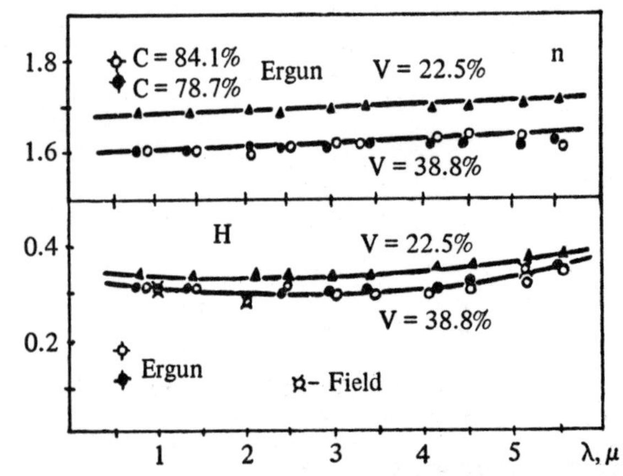

Fig. I.3-3 Optical constants of black coal

Brown coal (Fig. I.3-4). (v = 47-52 %, C = 66-7I %). The dependence of n and H on λ is characterized by the existence of a number of maxima and minima connected with the influence of absorption bands of water contained in fuel. The humidity of these coals reaches 30%.

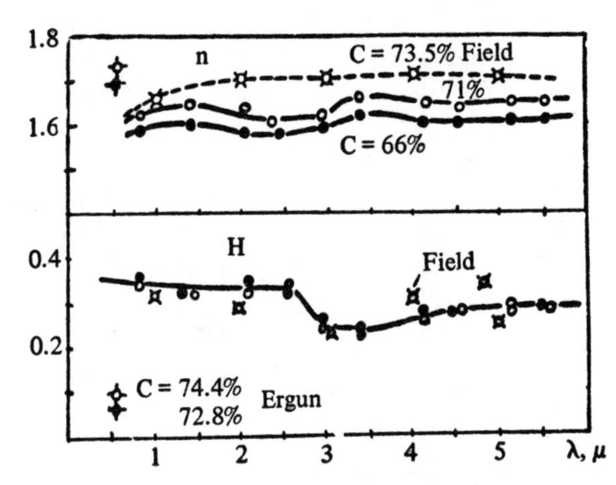

Fig. I.3-4 Optical constants of lean coal

From all cited data it follows that solid fuel optical constants depend substantially on the fuel carbon content. The most considerable variation of n and H is observed in the range of C > 85%. In the same range the dispersion of n and H shows up most vividly.

Concluding this section it should be noted that in our measurements the relative refraction index error does not exceed 6%, while the error in determination of H sometimes reaches I5%.

I.4 The influence of particle size and radiation wave length on particle radiative characteristics

On the basis of the known Mie's formulae /4/ and the above-mentioned soot carbon and solid fuel optical constants, such important radiation characteristics as the coefficients of absorption and scattering and the indicatrixes of scattering were determined. Using these data it is possible to accomplish different calculations connected with radiation energy transfer in disperse systems, specifically in flame.

a) Soot carbon particles. The graphs displayed in Fig. I.4-I show the variation of spectral absorption and scattering coefficients $\kappa_{a,\lambda}$ and $\kappa_{s,\lambda}$ as a function of ρ and λ for carbon soot particles.

One can see here three characteristic regions of absorption and scattering variation against parameter ρ : the small particle region; the region of extreme values for $\kappa_{a,\lambda}$ and $\kappa_{s,\lambda}$; the large particle region.

As applied to soot particles, we are interested in the small particle region. For this region, as may be seen from Fig. I.4-2, a linear relationship exists between $\kappa_{a,\lambda}$ and parameter ρ . Scattering can be neglected being extremely small. Thus, for this region

$$\kappa_{a,\lambda} = M(\lambda)\rho . \tag{3}$$

The function $M(\lambda)$ depends on the optical constants of particles.

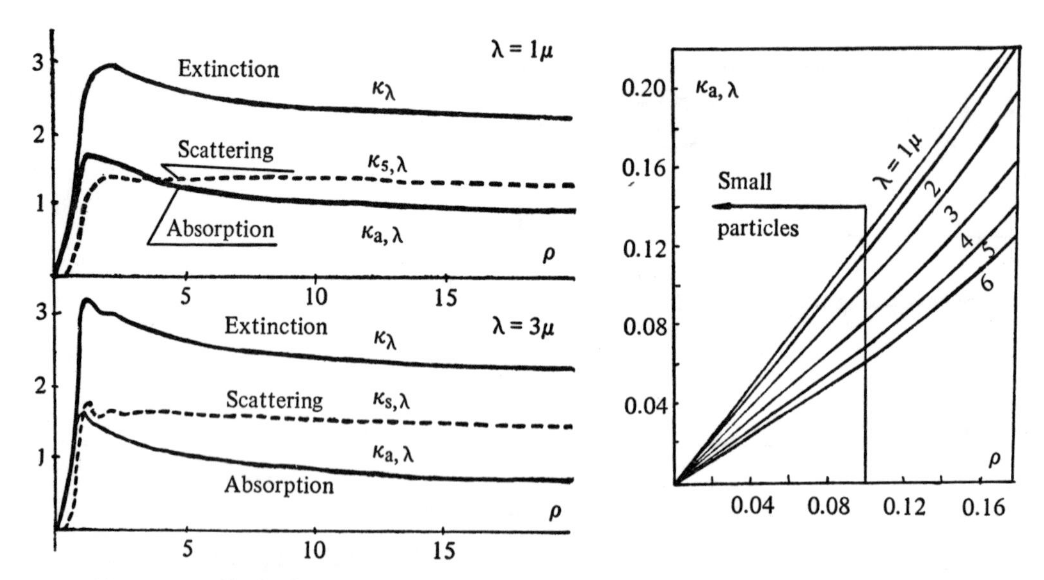

Fig. I.4-I Spectral absorption, scattering, and extinction coefficients for soot particles

Fig. I.4-2 The dependence of $\kappa_{a,\lambda}$ on ρ and λ for small soot particles

Fig. I.4-2 shows that due to the optical constant dispersion the boundary of the region in which $\kappa_{a,\lambda}$ is the linear function of ρ is determined by value of λ.

b) Pulverized coal particles. Figs. I.4-3 and I.4-4 show the radiation characteristics of different solid fuel particles. Fig. I.4-3 shows an immediate influence of the radiation wave length λ and the particle size x on the absorption $(\kappa_{a,\lambda})$ and scattering $(\kappa_{s,\lambda})$ coefficients.

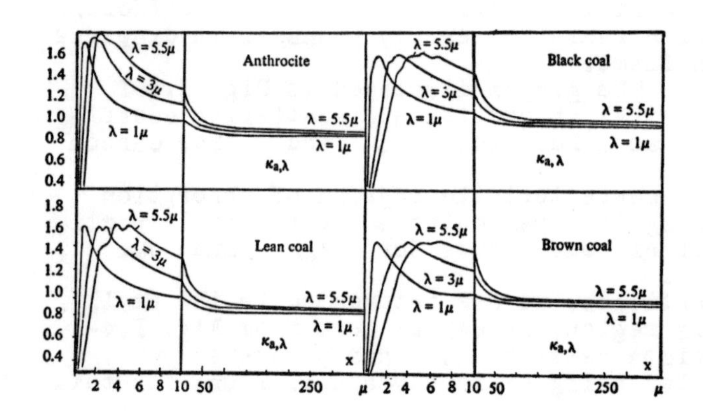

Absorption

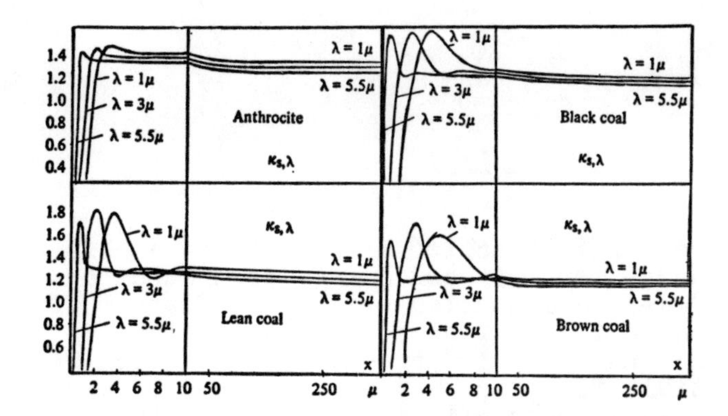

Scattering

Fig. I.4-3 Spectral absorption and scattering coefficients for different kinds of coal, as function of x and λ

In Fig. I.4-4 the spectral coefficients of absorption and scattering are given as a function of parameter ρ for one radiation wave length $\lambda = 1\mu$.

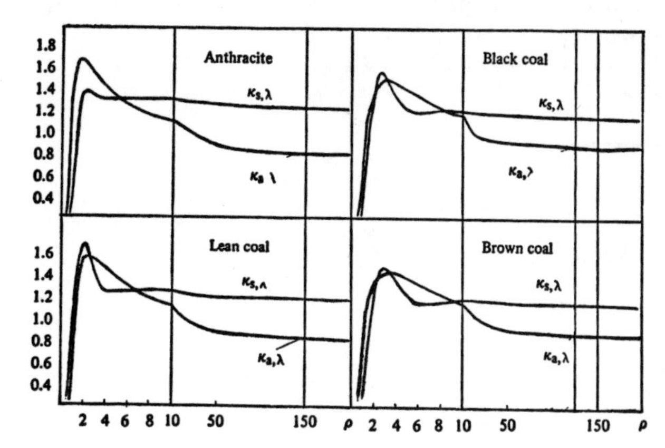

Fig. I.4-4 Spectral absorption and scattering coefficients for different kinds of coal as function of ρ at $\lambda = 1\mu$

From these data as well as from data on soot carbon one can see the extreme dependence of spectral absorption and scattering coefficients on parameter ρ . The correlation between scattering and absorption changes appreciably with ρ . Already at $\rho \geqslant$ IO the influence of scattering becomes prevailing. At $\rho \to \infty$ the coefficients of absorption and scattering cease to depend on ρ and tend to constant value, well determined for each fuel and each wave length. In this case the total attenuation coefficient ceases to depend on the fuel as well as on the wave length and tends to its asymptotic value equal to two.

For particles of all solid fuels at $\rho \geqslant 200$, it is possible to accept $\kappa_{a,\lambda} = 0.80 \div 0.95$ and $\kappa_{s,\lambda} =$ I.30 + I.I5.

Within the limits of $200 \geqslant \rho \geqslant$ IO there exists a linear relationship between the absorption and scattering coefficients and the inverse of diffraction parameter. For instance, for the anthracite particles over the radiation wave length range from I to

5.5 μ

$$\kappa_{a,\lambda} = 0.8 + \frac{2.7}{\rho} + \frac{2.5 \cdot 10^{-2}}{\lambda^2} \tag{4}$$

$$\kappa_{s,\lambda} = 1.28 + \frac{0.8}{\rho} - \frac{5 \cdot 10^{-2}}{\lambda} \tag{5}$$

For the other fuels these relationships are of the same character. The character of change in $\kappa_{a,\lambda}$ and $\kappa_{s,\lambda}$ as a function of the fuel kind (coal grade) is connected with the influence of elemental fuel composition on its optical constants (indices of refraction and absorption). As indicated above, the variation of carbon content in combustible fuel mass has a pronounced effect on the optical constants. It is naturally to expect that the particle radiation characteristics should substantially depend on C value. It should be noted that the effect of C on $\kappa_{a,\lambda}$ and $\kappa_{s,\lambda}$ is observed only at rather high carbon contents, i.e. C > 85% (anthracite and lean coal). At lower carbon contents (black coal and brown coal) the influence of C is very weak and may be practically neglected.

Finally, we have to examine the finest particles for which $\rho \leqslant 0.2$. The calculations on the approximate formula (3) agree well with the calculations on Mie's series. It is obvious from Fig. I.4-5. Here the calculation results from Mie's series are shown in the form of points.

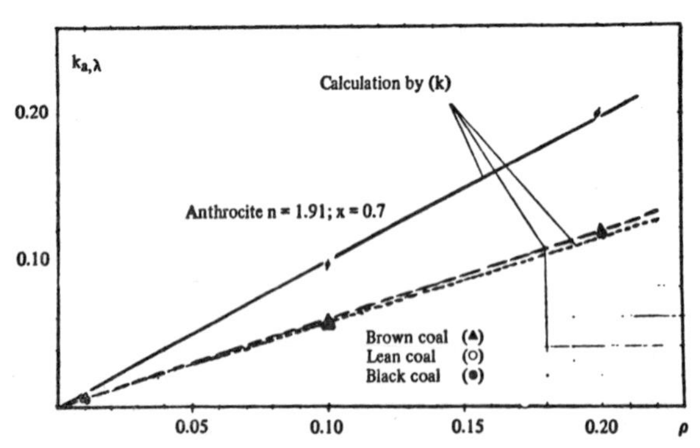

Fig. I.4-5 Spectral absorption coefficient for small coal particles at $\lambda = 1\mu$

Thus, in the range of $\rho \leqslant 0.2$ for all fuels a linear relationship between $\kappa_{a,\lambda}$ and ρ is observed. A tangent of the straight line inclination angle is defined by the values of optical constants n and H . This explains the fact that the data for anthracite (n = I.9I; H = 0.7) differ markedly from the data for all other fuels (n = I.89 ÷ I.63; H = 0.39 ÷ 0.33), mainly according to the differences in H .

At $\lambda = 1\mu$ for anthracite particles, $\kappa_{a,\lambda} \approx 0.97\rho$, and for particles of lean, black, and brown coals, $\kappa_{a,\lambda} \approx 0.58\rho$. Thus, for solid fuel particles at $\rho \leqslant 0.2$ it is possible to take into account only the first partial electrical oscillation, i.e. the first term in the expansion. Solid fuel particles can be considered as small already at $\rho \leqslant 0.2$ and for these particles the scattering can be neglected.

I.5 The angular distribution of scattered radiation

The question of scattered radiation angular distribution is considered for pulverized coal particles, as here we can observe all fundamental laws connected with the influence of the relative particle size variation ρ. For the pulverized coal it changes over considerably wider limits than for the soot particles. Fig. I.5-I shows the data indicating how the scattering indicatrix transforms for different fuel with the variation of ρ.

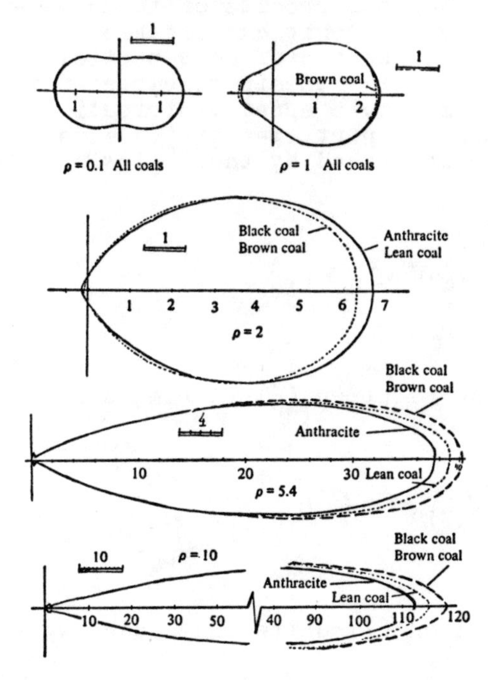

Fig. I.5-I Scattering indicatrixes for coal particles at $2 = 1\mu$

The graphs show that in the range of $\rho \leqslant 1$ the fuel grade change does not influence appreciably the character of scattering indicatrixes. But already at $\rho = 2$ we can notice a stratification of indicatrixes depending on the fuel grade. Later on, with increase in ρ this stratification becomes more significant.

From these data it is also obvious how the scattering indicatrix transforms with the change in parameter ρ. From Rayleigh type at $\rho = 0.1$ it changes to a forward-drawn type already at $\rho = 1$. At $\rho \geqslant 10$ the whole of scattered energy is practically directed forward with the incident radiation.

I.6 The law of particle size distribution

The experience shows that in flames we usually come across an irregular particle size distribution in solid phase. Based on the analysis of a great quantity of experimental data we propose some general relations describing the size distribution of soot particles in luminous flames.

From the data /II + I8/ as well as from our experiments we can conclude that for diffuse flames, when soot carbon is formed directly from hydrocarbon gas phase, the soot particle size is generally changed in the range from 50 to 800 A.

When burning liquid fuels, carbon particles of larger size are formed also as a result of pulverized fuel droplet carbonization. For such condition the soot particle size can attain I000 - 2000 A.

We have studied in detail the soot particle size in luminous flames of liquid fuel. Based on the results of these measurements we have established a relationship characteristic for such flames describing the distribution of soot particle sizes. According to its form, this relationship resembles Maxwell relation describing the gas molecula speed distribution at thermal equilibrium. The number of particles of the size range from x to $x + dx$ is described by the following dependence on the particle size

$$N(x)dx = \frac{4}{\sqrt{\pi}} \left(\frac{x}{x_m}\right)^2 e^{-(x/x_m)^2} d\left(\frac{x}{x_m}\right). \tag{6}$$

It is convenient to represent the relation (6) in the following form:

$$\tilde{N}(x) = \frac{N(x)}{N(x_m)} = \left(\frac{x}{x_m}\right)^2 e^{-[(x/x_m)^2 - 1]} \tag{7}$$

Fig. I.6-I shows how the relation (7) unites numerous experimental data of different authors.

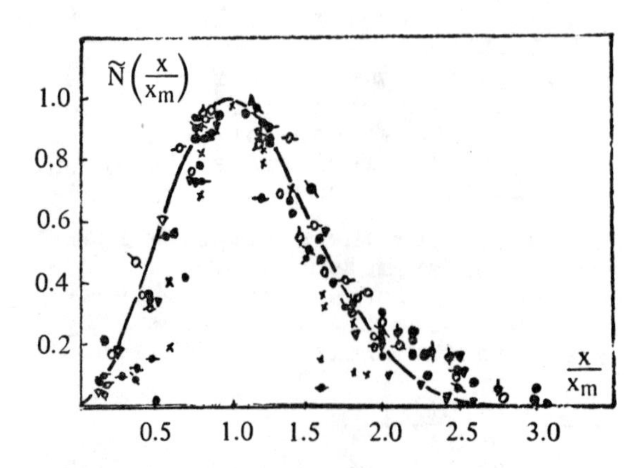

Fig. I.6-I The size distribution of soot particles

ɑ ϕ ○ ⊘ ○ ▽ △ - data of the Central Boiler and Turbine Institute, USSR (different burners)

● - data of Sato and Kunitomo /I3/

- data of Keltsev and Tesner /I6/
- data of Pepperhoff /2/
- data of Rossler /I8/
- data of Rogier and Leblanc /I2/
- data of Pribytkov and Ryzhkov /I7/

Specific distribution shape depends on the modal particle size x_m , which, in its turn, depends on fuel burning conditions, for instance, on the air excess α . The dependence of x_m on the air excess value α is shown in Fig. I.6-2.

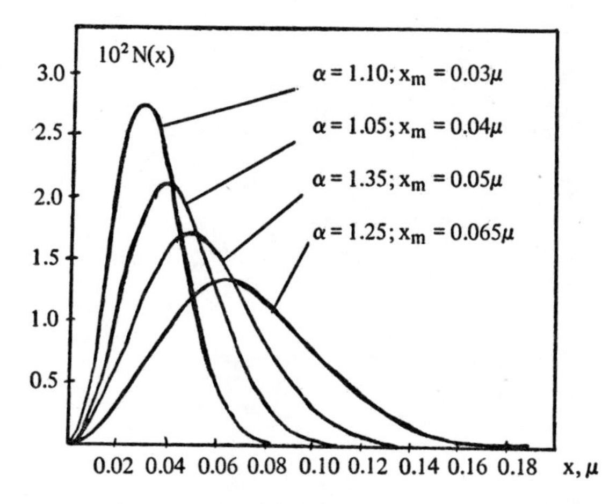

Fig. I.6-2 The influence of air excess α on the soot particle size distribution in luminous flame

As to the change of x_m in the direction of flame length, we have not noticed considerable variation of this value. At the same time Rogier's and Leblanc's data /I2/ indicate some tendency for a decrease of x_m .

I.7 Spectral radiation characteristics of polydisperse particle systems

Using the above data on radiation characteristics of single particles and their size distribution, it is not difficult to define all basic radiation characteristics of polydisperse soot carbon particle systems in luminous flames. One of the most important radiation characteristics of such systems is the spectral absorption cross-section

$$\widetilde{\sigma}_{a,\lambda} = \frac{\pi}{4} \int_0^\infty k_{a,\lambda}(x) x^2 N(x) dx . \tag{8}$$

Accounting for (8) and denoting

$$\left(\frac{x}{x_m}\right)^2 e^{(x/x_m)^2} = \Xi\left(\frac{x}{x_m}\right) = \Xi(z) , \tag{9}$$

where $z = x/x_m$, we can write

$$\tilde{\sigma}_{a,\lambda} = \sqrt{\pi}\, x_m^2 \int_0^\infty \kappa_{a,\lambda}(z x_m) z^2 \Xi(z)\, dz . \tag{10}$$

Fig. I.7-I shows the variation of $\tilde{\sigma}_{a,\lambda}$ for polydisperse systems as a function of modal particle size and radiation wave length.

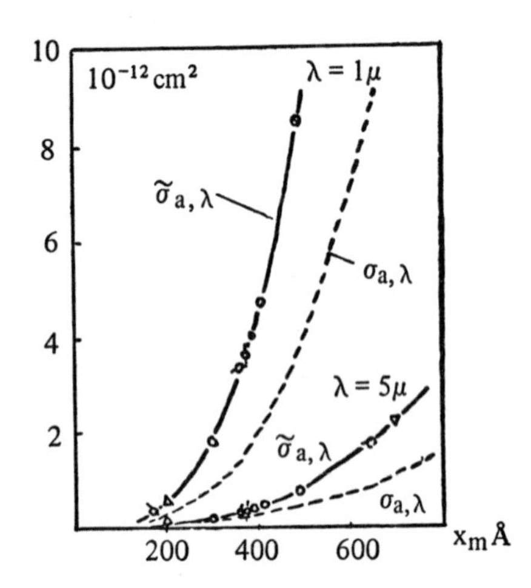

Fig. I.7-I Absorption cross-sections for poly- and monodisperse soot particle systems

Here for comparison the dotted curves show the absorption cross-sections for monodisperse systems where all particles are of the same size, equal to the modal particle size of corresponding polydisperse system. The value of absorption cross-section of such monodisperse systems is

$$\tilde{\sigma}_{a,\lambda} = \frac{\pi}{4} \kappa_{a,\lambda}(x_m) x_m^2 . \tag{11}$$

As may be seen from the graph, the polydisperse system absorption cross-sections $\tilde{\sigma}_{a,\lambda}$ are far in excess of monodisperse absorption cross-sections $\sigma_{a,\lambda}$ with corresponding value of x_m. The value of $\tilde{\sigma}_{a,\lambda}/\sigma_{a,\lambda}$ ratio changes with x_m and λ from I.8 to 2.4. This ratio reaches the greatest values at small and high x_m. When passing from $\lambda = 1\mu$ to $\lambda = 5\mu$ the value of this ratio decreases approximately by I5%, which is mainly connected with prevailing influence of large particles having $x > x_m$.

For small soot particles, when $\rho \ll 1$, due to relation (3) we can write for absorption cross-section

$$\tilde{\sigma}_{a,\lambda} = \pi\sqrt{\pi}\, \frac{M(\lambda)}{\lambda} x_m^3 \int_0^\infty z\Xi(z)\, dz . \tag{12}$$

Taking into account that

$$\int_0^\infty z^3 \Xi(z)\, dz = 1 ,$$

we can write

$$\widetilde{\sigma}_{a,\lambda} = \sqrt{\pi} M(\lambda) x_m^2 \rho_m . \tag{13}$$

If all particles under these conditions (I) had the same size x_m, their absorption cross-section should have been

$$\sigma_{a,\lambda} = \frac{\pi}{4} M(\lambda) x_m^2 \rho_m . \tag{14}$$

In this case the ratio

$$\frac{\widetilde{\sigma}_{a,\lambda}}{\sigma_{a,\lambda}} = \frac{4}{\sqrt{\pi}} . \tag{15}$$

Thus, for a polydisperse system of small soot particles the real absorption cross-section exceeds the absorption cross-section of a monodisperse system with equivalent x_m value by a factor of more than 2.

If the absorption cross-section of polydisperse systems exceeds the absorption cross-section of monodisperse systems approximately by a factor of 2, the scattering cross-section of poly- and monodisperse systems differ approximately by an order of magnitude.

I.8 Angular distribution of radiation scattered by polydisperse particle systems

Using the above relations, which define particle size distribution, it is not difficult to calculate how the radiation, scattered by a polydisperse particle system, is distributed in different directions β. The effective scattering indicatrix for polydisperse particle system is defined by the following formula

$$= \quad \widetilde{\gamma}_\lambda(\beta) = \frac{4}{\sqrt{\pi}} \int_0^\infty \gamma_\lambda(\beta) \Xi(z)\, dz . \tag{16}$$

The calculations show that for $\rho_m \leqslant 0.1$, the scattering indicatrixes $\widetilde{\gamma}_\lambda(\beta)$ of soot particle polydisperse systems do not practically differ from the indicatrixes for monodisperse systems with $\rho \leqslant 0.1$. Both are of Rayleigh type. But already at $\rho_m \geqslant 0.6$, as is seen from Fig. I.8-I, the scattering indicatrixes for polydisperse particle compositions are formed with a marked degree of asymmetry.

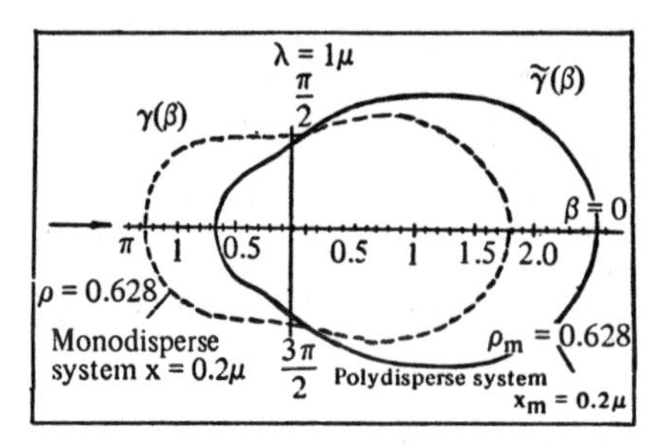

Fig. I.8-I Scattering indicatrixes for poly- and monodisperse systems at $\rho = \rho_m =$ 0.628

Polydispersed systems are characterized by a more forward-drawn scattering indicatrix compared to monodisperse systems. In these conditions the transformation of scattering indicatrixes can be characterized by the value of relation between the forward and the backward scattering. For ρ_m = 2 the forward scattering exceeds the backward scattering approximately by an order of magnitude, and at ρ_m = 50 by four orders of magnitude.

Note that the scattering indicatrix transformation during transition from poly- to monodisperse systems is mainly connected with the contribution to forward scattering made by particles with the size exceeding the value of .

I.9 Absorptivity of polydisperse particle system

Above we have shown the variation of absorption and scattering cross-sections and scattering indicatrixes depending on disperse composition of soot carbon particles. It is convenient to consider the question of system absorptivity as applied to a flat layer in the limits of the known Schuster's two-flow approximation, developed by V.N.Adrianov /I9/. It is not necessary to cite here the primary formulae. They are known very well. We consider only the main results of analysis.

Fig. I.9-I shows the influence of scattering indicatrix character on spectral absorptivity of an optically infinite layer. From the presented data it is evident that if the real scattering indicatrix is not taken into account, this can cause errors in a_λ value up to 20%.

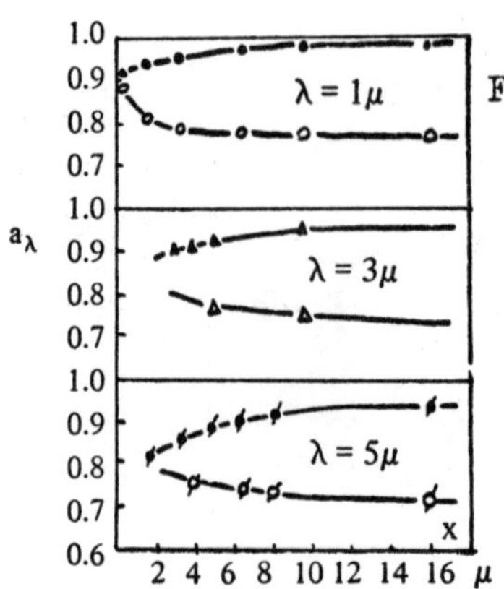

Fig. I.9-I The spectral absorptivity of infinitely thin layer of particles

● Symmetrical scattering indicatrix
○ Real scattering indicatrix

to page 126

of radiation heat transfer between flame and a wall tube covered with deposits. In this case a method is optimal if it reflects correctly the physical nature of heat transfer process, gives the best correlation with a large number of experimental data, acquired in real working conditions, and allows to predict reliability and effectiveness of new, original furnace constructions. This is possible only in such a case, when the effect of basic process parameters on heat transfer conditions is completely taken into account.

2.I The method of Central Boiler and Turbine Institute

In the USSR an engineering method of furnace heat calculation developed at Central Boiler and Turbine Institute is the most widely used. This empirical method, based on a large number of experimental data, considers the relation between nondimensional gas temperature at furnace exit $\theta_0 = T_0/T_a$ and nondimensional complex Bo/a_F. The value Bo in this complex, named Boltzmann criterion, is the relation of the change of gas heat content in furnaces to the black body heat radiation at the adiabatic burning temperature /20, 2I/. The furnace emissivity a_F is the function of radiation properties of flame and furnace screen and it depends on the furnace volume and configuration.

For the conditions of instantaneous fuel burning, the form of this relation can be approximately established analytically for one-dimensional scheme /22, 23/

$$\theta_0 = \sqrt[3]{1 + 3\frac{a_F}{Bo}} \tag{21}$$

The formula (2I) is very approximate and, naturally, cannot be used for the furnace design. However, it indicates that it is necessary to find a relation between the value of θ_0 and the complex a_F/Bo. Such relation was found in works performed by Central Boiler and Turbine Institute. In this relation an empirical coefficient A is taken into account which characterizes the influence of the relative level of burner disposition on the heat transfer in the furnaces. Changing this coefficient according to the burner level and to the kind of fuel we can obtain an essential correlations between calculation results and experimental data on furnace exit temperature.

Fig. 2.I-I shows the relation between $\theta_0/(1-\theta_0)$ and the complex $\Pi = Bo/Aa_F$ for three kinds of fuel: anthracite, mazut, and gas. This figure illustrates a great number of experimental data for furnaces of various constructions. The experiments involve power boilers of different output, from 200 to 800 Mwt. The data for 800 Mwt unit are obtained in the USSR at Slavianskaya power station where anthracite is burned. The data for gas include the experiments of American investigators at Sterlington power station.

As one can see from these data, for all fuels and for various constructions of furnaces the values of $\theta_0/(1-\theta_0)$ and Π are related by a single dependence. In logarithmic coordinates it is a straight line passing through the origin. The tangent of straight line inclination angle is equal to 0.6 for all fuels and furnaces.

to page 128

Some words can be said about the influence of particle size on absorptivity value a_λ . It is well known that this dependence is extremal. In the range of small particles a_λ does not depend on x ; then as x increases, the a_λ value increases too, reaches the maximum and drops rapidly.

In the range of small particles the radiation is selective, and beyond the maximum it can be assumed grey, as α_λ value is independent of λ .

I.IO Integral absorption cross-sections

When calculating radiation energy transfer, integral values of absorption cross-section are often used. Plank's and Rosseland's mean absorption cross-sections are used as such values.

We have performed the calculations as applied to polydisperse systems of soot carbon particles.

In this case the mean Plank's and Rosseland's absorption cross-sections were determined from the formulas

$$\tilde{\sigma}_{a,P} = \frac{\int_0^\infty \tilde{\sigma}_{a,\lambda} I_{\lambda,0} \, d\lambda}{\int_0^\infty I_{\lambda,0} \, d\lambda} \tag{17}$$

and

$$\frac{1}{\tilde{\sigma}_{a,R}} = \frac{\int_0^\infty \frac{1}{\tilde{\sigma}_{a,\lambda}} \frac{\partial I_{\lambda,0}}{\partial T} \, d\lambda}{\int_0^\infty \frac{\partial I_{\lambda,0}}{\partial T} \, d\lambda} \tag{18}$$

Fig. I.IO-I shows the computation results of Plank's mean absorption cross-section for a polydisperse soot particle system.

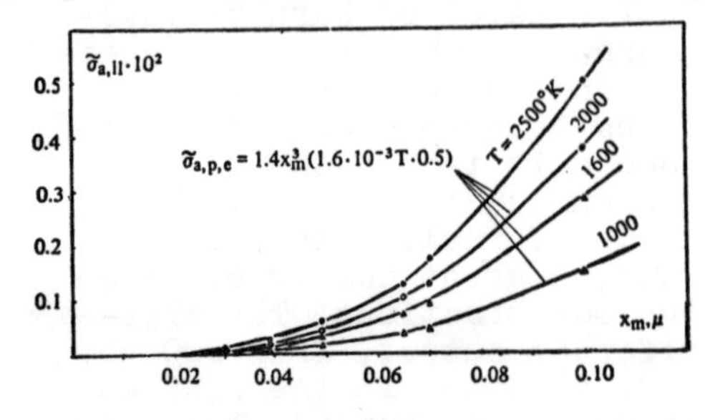

Fig. I.IO-I Average Plank's absorption cross-section for soot particles

The influence of particle disperse composition is taken into account by the value of modal particle size x_m . All curves are satisfactorily described by a generalized relation

$$\tilde{\sigma}_{a,P} = 1.4x_m^3(1.6 \cdot 10^{-3}T - 0.5) \tag{19}$$

The dependence of Rosseland's mean absorption cross-section

on temperature T and modal particle size x_m is similar to the above relation

$$\tilde{\sigma}_{a,R} = x_m^3 (1.6 \cdot 10^{-3} T - 0.5) \tag{20}$$

For comparison, Fig. I.10-2 shows the curves describing the variation of the mean absorption cross-sections $\tilde{\sigma}_{a,P}$ and $\tilde{\sigma}_{a,R}$ as a function of x_m .

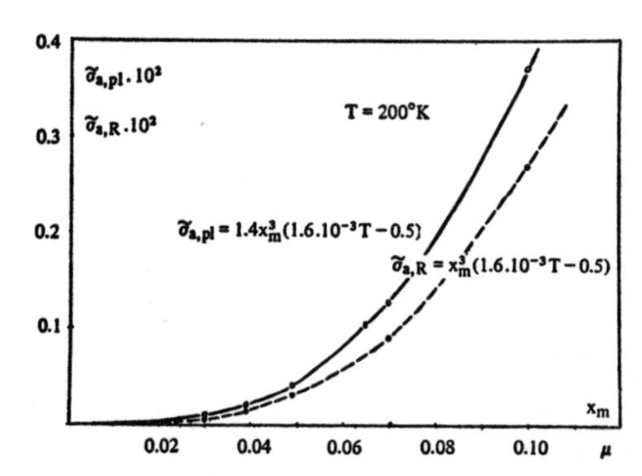

Fig. I.10-2 Average absorption cross-sections $\tilde{\sigma}_{a,P}$ and $\tilde{\sigma}_{a,R}$

It is interesting to note that the values of $\tilde{\sigma}_{a,P}$ and $\tilde{\sigma}_{a,R}$ are in a constant relation which is independent of x_m and T .

2. Engineering methods for the computation of heat transfer in furnaces

The problem of heat transfer computation in furnaces is one of the most complex problems in the heat transfer theory. It is the problem of heat transfer between a moving, burning, radiating, absorbing, and scattering medium and furnace wall surfaces, insulated from the flame by a layer of external fouling on tubes. The medium is, in turn, nonisothermal, multicomponent and nonhomogeneous in different points of flame.

It is precisely by complexity of the problem and by inadequate knowledge of combustion chamber processes that up to now there is no analytical computation method accounting for all features of process and well correlated with the experimental data obtained for different combustion chambers and burners and for different operating conditions.

In a general case, at high velocities of gas flow, for instance, in high intensity combustion chambers, the energy transfer between flame and a wall is achieved both by radiation and convection.

As to the furnaces of steam power boilers, here the convective heat transfer does not play an important role, and the most part of energy is transferred by radiation. It is explained by the fact that the gas velocity in such chambers is not high, and the temperature of external deposit layer on tubes is close to the temperature of gas layer near the wall.

Thus, in reference to steam power boilers, it is a question

to page 125

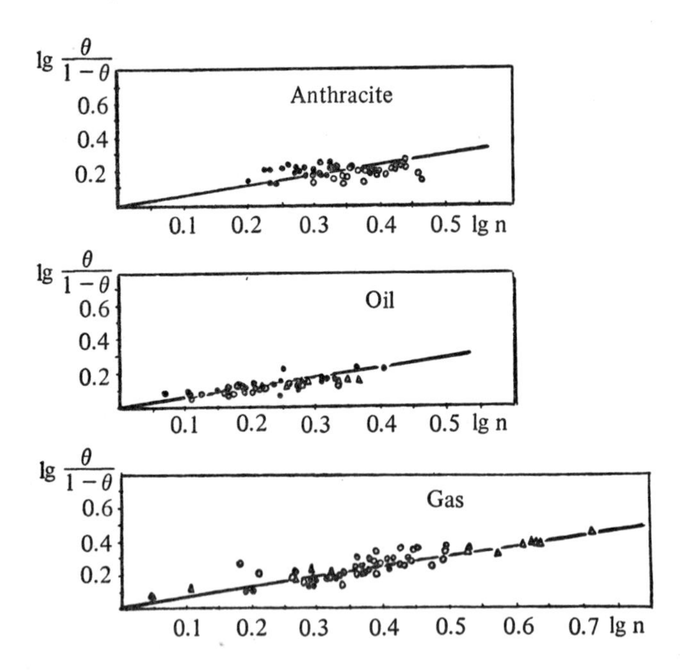

Fig. 2.I-I Experimental data on the heat transfer in furnaces

All experimental data are well described by the following empirical relation

$$\frac{\theta_0}{1 - \theta_0} = \Pi^{0.6} \tag{22}$$

The formula (22) is used in the USSR by all boiler manufacturing works and construction bureaus for the furnace heat transfer calculations as the basis for a new improved method /2I/. The formula is supported by extensive experimental studies including more than 2000 tests. The mean-square error of computations from this formula does not exceed 50°C. It is very important that for modern large boilers the computation from this formula gives very good results.

The influence of the parameter A in /2I/ is connected with the character of furnace temperature field, more exactly, with the relative height of maximum flame temperature zone. In modern large furnaces with horizontal burner arrangement this zone is commonly located at the level of burners.

It was thought possible to eliminate the flame and wall temperature from direct consideration. In essence, these values are in themselves functions of furnace process and depend on the same defining parameter as the gas temperature at furnace exit.

The distinctive feature of the said method is that the existence of a deposit layer on tubes is directly taken into account /24/. These deposits have a very low thermal conductivity as a result of which the temperature of their external surface reaches very high values, commensurate with flame temperature. The wall covered with deposits sends about 50% of incident radiation flow back to the furnace volume. The calculation method uses the numerical values of these quantities characterized by the wall thermal effectiveness coefficient /2I, 24/.

2.2 The method of All-Union Heat Engineering Institute

Lately in the USSR another method of furnace heat transfer calculation is developed. This method is published in /2I/. It is based on three fundamental equations: the equation of radiation heat transfer between flame and furnace wall; the equation of furnace heat balance; and the equation of conductive heat transfer through the wall covered with deposits.

From these equations the basic computation formula is deduced:

$$\theta_0 = 1 - \frac{a_F}{Bo} \frac{T_F^4 - T_w^4}{T_a^4}. \tag{23}$$

The mean effective flame temperature T_F is determined from Polak's and Shorin's formula /25/:

$$\frac{T_F}{T_a} = \tau \left(\frac{T_0}{T_a}\right)^{\eta}. \tag{24}$$

At η = I, it follows from this formula that the mean effective flame temperature is proportional to the least flame temperature

$$T_F = \tau T_0. \tag{25}$$

At η = 0, the mean effective flame temperature is proportional to the highest possible emitter temperature

$$T_F = \tau T_a \tag{26}$$

Real flame temperature is restricted by the above-mentioned limits.

Some numerical values of the coefficients τ and η in the formula (24) are proposed for calculations in /2I/.

We do not examine this method in detail. Note only, that it is convenient in the cases where it is necessary to account for both radiative and convective heat transfer.

REFERENCES

/I/ Gaydon A.G., Wolfhard H.C. "Flames, Their Structure, Radiation and Temperature".(I953)

/2/ Pepperhoff W., Bahe A. "Archiv Eisenhüttenwesen", No.9/IO, (I952).

/3/ Beer J.M. Journal of the Inst. of Fuel, No 252,(I962)

/4/ Mie A. Ann. d. Phys., 25, 377,(I908)

/5/ Stull V.R. and Plass G.N. Journal of the Opt. Soc. of Am., No.2,(I960)

/6/ Ergun, McCartnes T.T., Walline R.E. "Fuel", No.2, (I96I)

/7/ Field M.A., Gill D.B., Morgan B.B., Hawsklev P.G."Combustion of Pulverized Coal".(I967)

/8/ Parker W.G., Wolfhard H.C. Journal Chem. Soc., 2038 (I959)
/9/ Mc.Grath I.A. Ph. D. Thesis, Sheffield University (I960)
/I0/ Ibiricu M. Ph. D. Thesis, Sheffield University (I962)
/II/ Thring M.W. Journal of the Inst. of Fuel, No.250 (I96I)
/I2/ Rogier I., Leblanc B. Revue Generale de Thermique, No.96 (I969)
/I3/ Sato T., Kunitomo T. Bulletin of JSME, No.36 (I966)
/I4/ Jagi S. Journal of the Soc. of Chem. Ind., Japan, No.2 (I937)
/I5/ Foster P.G. Ph. D. Thesis (I96I)
/I6/ Keltsev V.V., Tesner P.A. "Soot" (I952)
/I7/ Pribytkov I.A., Ryzhkov L.N. Izvestiya VUZOV "Ferrous Metallurgy", No.3 (I968)
/I8/ Rössler F. "Optik". Heft I0 (I953)
/I9/ Adrianov V.N. "Heat Transfer and Hydrodynamics". Nauka (I968)
/20/ "Standard Method of Power Boiler Heat Calculation". Gos-energoizdat (I957)
/2I/ "Standard Method of Power Boiler Heat Calculation". Energia (I973)
/22/ Gurvich A.M., Blokh A.G. Energomashinostroyeniye, No.6 (I956)
/23/ Blokh A.G. "Heat Radiation in Boiler Plants". Energia (I967)
/24/ Mitor V.V. "Heat Transfer in Steam Boiler Furnaces". Mash-giz (I963)
/25/ Polak G.L., Shorin S.N. Izvestiya AN SSSR, OTN, No.I2 (I949)

Chapter 7

SOLID/GAS PHASE HEAT EXCHANGE IN COMBUSTION OF POWDERED FUEL

V. I. Babiy

All-Union Heat Engineering Institute, USSR

Processing of experimental evidence on combustion of carbon pellets/1/ and particles of coal powder /2/ has revealed that an equation of the type

$$Nu_d = 2 + 0.17Re^{0.66}$$

is equally good for mass exchange in combustion of large and small carbon particles, droplet evaporation /3/ and mass exchange between a gas flux and a sphere is isothermic conditions/4/ within Re = 0.01 to 1000.

Because there is an analogy between differential equations of diffusion and heat conductivity, empirical mass exchange equations are frequently used for heat exchange calculations. This approach was theoretically justified by Nusselt /6/, extended by Rubinshtein/7/ and applied in actual research,/3,4,5/

The findings of these studies are often used in calculation of convective heat exchange between burning particles of coal powder and a gas medium. There is, however, no clear-cut analogy between heat and mass exchange in combustion of carbon particles. Furthermore, this seems very dubious when heat sources generate in the gas boundary layer at the particle surface as a result of carbon monoxide oxidation in diffusion from the hot coal surface into the gas medium.

In view of the complex nature of heat transfer from burning coal particles into the gas medium, the equations should rather be built around experimental evidence on heat exchange.

This paper will be an attempt to establish an experimentally justified relation for calculation of convective heat exchange between the gas medium and coal powder particles in quasisteady-state combustion.

Heat exchange in ignition of coal powder particles which heat to give liquid and gas products of coal thermal decomposition is an independent problem per se and will not be discussed here. Depending on the CO/CO_2 ration in the gases leaving a burning carbon particle the heat evolution from its surface may vary by a factor of 3.6.

There is no reliable method for calculation of the CO/CO_2 ratio in carbon combustion products as a function of the temperature and composition of the gas medium or of the coke particle surface structure. This fact is a hindrance to calculation of heat evolution on the surface of burning particles.

Heat evolution in the boundary layer of the gas around the particle in oxidation of carbon monoxide is also rather complicated. In combustion of carbon powder particles the heat exchange may be greatly influenced also by the final rate at which thermodynamic equilibrium is established in the boundary layer over the particle surface. Accurate theoretical assessment of the above factors in calculation of heat exchange between burning carbon particles and the gas medium has been impossible thus far.

For engineering calculations, however, an empirical equation may suffice. Certain assumptions should be made; in particular, carbon combustion and all heat evolution are assumed to be completed on the particle surface. In this case an experimental convective heat exchange coefficient denoted as α_c^* will also include heat transfer by heat and chemical energy and other processes in the boundary layer of gas at a burning coal particle. In the final analysis the problem reduced to finding experimentally the dependence of this coefficient on temperature and composition of the gas medium and on the size of coal particles.

To calculate the coefficient of heat exchange between a burning coal particle and gas, it is necessary to have experimental knowledge of carbon burning rate, temperature of the particle, gas and the radiating region around the particle, and the relative rate of the gas flux.

Using the Stefan-Boltzmann equation for calculation of radiation heat transfer, the convective component of heat exchange between a particle and the gas may be determined through the difference between heat evolution and radiation losses.

This was the method used to process the findings of Khitrin and his co-workers/8/ who studied the burning of dia. 14-16 mm carbon pellets at flux rates of 0.1 to 20 m/sec and gas temperature in the chamber of 900 to 1400°K. Oxygen contents in the gas medium amounted to 21 and 4%. Calculations revealed that the value of the Nusselt heat criterion for burning carbon pellets,

$$Nu^* = \frac{\alpha_c^* \delta}{\lambda},$$

as a rule exceeded the diffusional counterpart of this criterion,

$$Nu_d = \frac{d\delta}{D}$$

At higher temperatures their relation was

$$\frac{Nu^*}{Nu_d} = 3 \text{ to } 5$$

which is the evidence that in combustion of carbon particles the heat exchange with the gas medium is much more active than could be expected from the mass exchange equations. The notation is:

α_c^* = convective heat transfer coefficient for burning carbon particle surface;
δ = particle size;
λ = gas heat conductivity at the incident flow.

The reactive gas exchange coefficient was calculated by the equation

$$\alpha = \frac{K_s^c}{BCO_2}$$

The mass concentration of oxygen CO_2 in kg/m^3 and the carbon burning rate per unit surface, K_s^c in kg/m^2 sec, had been obtained experimentally.

The stoichiometric coefficient was assumed to be $\beta = 0.375$. The diffusion coefficient of oxygen and the burning carbon surface was calculated following Ref. 2 by the equation

$$D = 0.16 \times 10^{-4} \left(\frac{T_g}{273}\right)^{1.9} \quad m^2/sec$$

$$d_r = 4.9 \cdot 10^{-8} a_p \frac{T_p^4 - \psi T_{ch}^4}{T_p - T_g} \left[\frac{\text{k cal}}{\text{m}^2 \text{ hr deg}}\right] \qquad (5)$$

The coal particle blackness coefficient a_p was assumed to be 0.8, the screening of particles by chamber walls, ψ = 0.9 to 0.95.

In the experiments the reaction chamber wall temperature T_{ch} was 10 to 30°K in excess of the inside gas temperature T_g. Adding the possible inaccuracy in selection of a_p, the overall error in calculation of α_r could not exceed 10 to 20%.

Figure 1 shows the values of the convective heat exchange coefficient obtained by the above method for anthracite powder particles. Curves corresponding to the heat conductivity heat transfer coefficient without combustion are also given,

$$\alpha_c = \frac{2\lambda}{\delta}$$

In this case it was assumed that Nu = 2 because the Reynolds number in these experiments was $Re < 0.1$.

For comparison the graphs show the dependence of α_r on the particle size calculated by eq. (5) at O_2 = 21%. With decreasing oxygen content and T_g = const, α_r decreases. The experimental points in Fig. 1 are grouped by temperatures of the gas medium. The various characters denote variances of oxygen concentration within O_2 = 5 to 21%. The X-line is the size of coal particles. The experimental values of the convective heat exchange coefficients in combustion, α_r^*, are about five to six times the calculated value, α_c, at T_g = 1600°K. The difference of α_c^* and α_c decreases with temperature of the gas medium.

It is interesting to note that decrease of oxygen content in the gas medium from 21 to 5% and the associated increase of CO_2 content up to 16% at T_g = const under these conditions does not greatly affect α_c^*.

This conclusion is somewhat unexpected and seems to suggest that in the boundary layer of gas at the surface of burning coal particles, and at high temperatures the heating processes are influenced by the CO_2 content as well as O_2 content.

The specifics of heat exchange in combustion of coal particles will be better understood if we consider in more detail the phenomena in the boundary gas layer at the burning surface of carbon.

Khitrin and Tsukhanova /8/ have noted the existence of a luminescent layer over the surface of burning carbon particles. That layer forms owing to the oxidation of carbon monoxide and its thickness at small velocities of the flux was about 0.1 to 0.2 of the particle radius.

Wicke and Wurzbacher /11/ used capillaries to sample the gas from over the burning carbon particle surfaces and also found that carbon monoxide oxidizes near the particle surface. Bukhman and Nurekenov /12/ have measured the field of temperatures in the gas boundary layer near the surface of burning spherical coal particles. These measurements have proved the existence of a temperature maximum at a certain distance from the burning carbon surface. The excess of the gas temperature over the particle temperature and the position of the maximal temperature were found to be a function of many variables.

where T_g °K is the temperature of gas far from the particle surface.

The processing of experimental data on heat exchange in combustion of coal powder has its own specific features.

Direct measurement of carbon burning on the particle surface in terms of weight as done in studying the combustion of large carbon pellets was impossible. In experiments with coal powder particles/9/ the initial particle sizes, their change during burning and complete time of particle burning were measured. With this data available the specific rate of carbon burning from particle surface was calculated by the equation

$$K_S^C = -\frac{100 - A_k}{100}\frac{\rho_k}{2}\frac{d\delta}{d\tau} \tag{1}$$

where A_k in per cent and $_k$ in kg/m^3 are the ash content and density of the coal particle coke residue.

The rate of particle size change, $d\delta/d\tau$, can be found either directly from cine pictures or calculated through the known time of particle burning.

In further experiments the second version was preferred. In the experiments under consideration the coal particles burned diffusionally. In this case their burning out extent is proportioanl to their squared initial size and, as shown in Ref. 9, is well described by the equation

$$\tau_{rk} = K_{rk} \cdot 2.21 \cdot 10^8 \frac{100 - A_k}{100} \frac{\rho_k \delta^2}{T_g^{0.9} O_2} \tag{2}$$

where T_g = the gas flux temperature in °K;
O_2 = volume concentration of oxygen in the gas, in per cent;
δ = medium size of particles in m;
K_{2k} = a coefficient characteristic of a particular kind of coal.

Under these conditions the rate of particle size change is given by the equation

$$\frac{d\delta}{d\tau} = -\frac{\delta}{2\tau_{rk}} \tag{3}$$

The average initial particle size δ is given as the geometrical mean of three microscopic measurements of a particle in different planes. The overall coefficient of heat transfer from the burning particle surface into the gas medium is then

$$d_s = d_c^* + d_r = \frac{K_S^C Q}{\Delta T} = \frac{100 - A_k}{100}\frac{\rho_k \delta}{4\tau_{rk}}\frac{Q}{\Delta T} \tag{4}$$

where Q = the carbon heating value in kcal/kg;
α_c^* = the heat transfer coefficient characterizing convective heat exchange in combustion of coal particles;
d_r = the radiation heat transfer coefficient.

The temperature drop between the burning particle and the gas medium $\Delta T = T_p - T_g$ was found through direct measurement of experimental values of T_p and T_g. In a number of cases the difference was calculated by an equation based on these measurements./10/

Once the overall heat transfer coefficient is found, the convective heat exchange coefficient may be found as the difference $\alpha_c^* = \alpha_s - \alpha_r$. In this case the coefficient of radiation heat exchange between a burning coal particle and the medium, α_r, was calculated by the equation

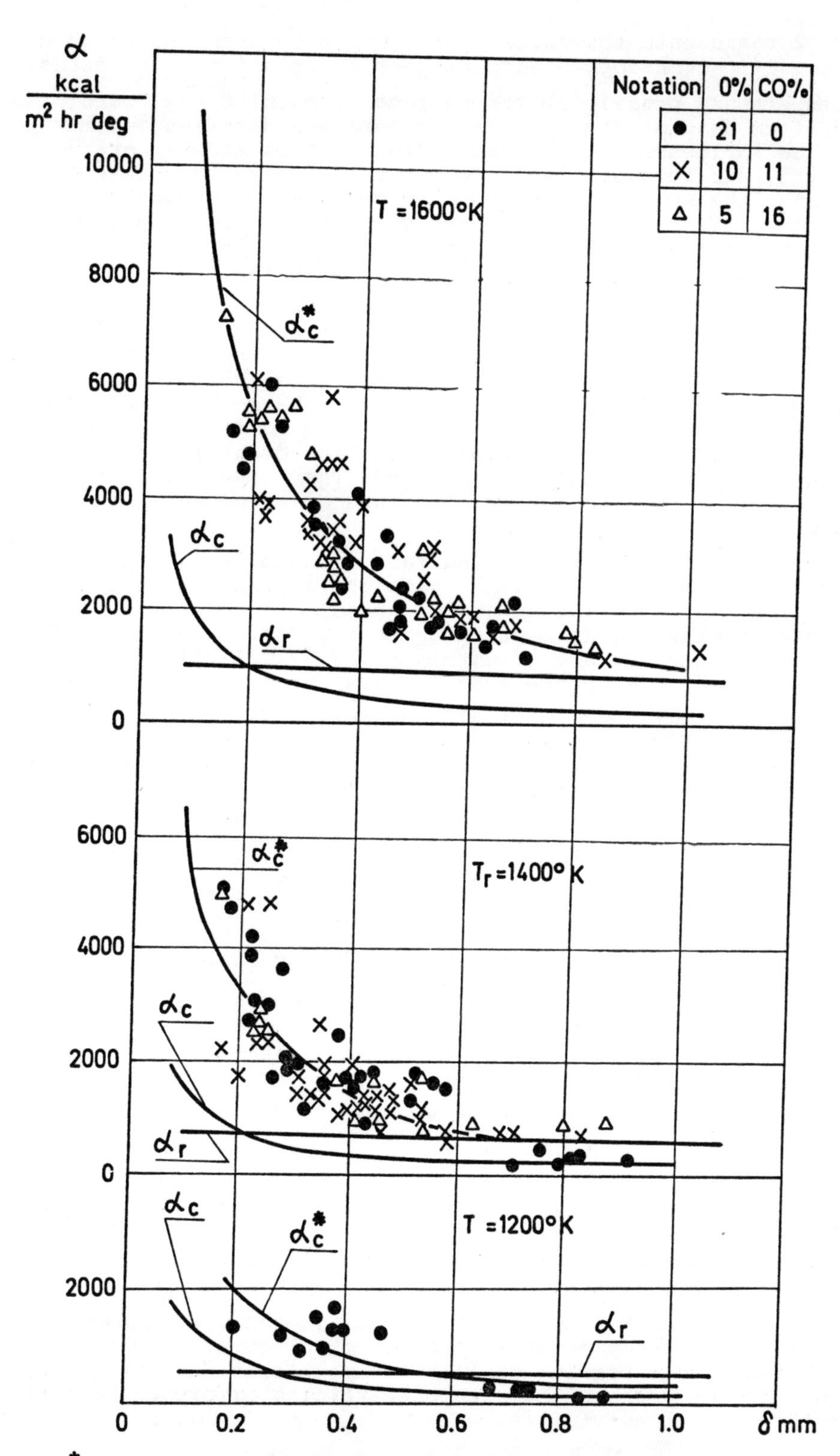

Fig. 1 α^*_c, α_c and α_r (O_2 = 21%) versus particle size at various gas medium temperatures and oxygen contents.

Figure 2 represents the experimental evidence obtained by Bukman and Nurekenov/12/ transformed with respect to relative coordinates.

The temperature profile at the surface of burning coal particles may differ from that of Fig. 2. The temperature distribution may generally be influenced by the gas medium content and temperature, the flux rate, particle size and type of coal and by the position of the point over the surface size in relation to the incident flux. Despite its specific nature the above graph is of profound interest in that it shows that processes in the boundary layer over the surface of burning coal particles may greatly affect the heat exchange between these particles and the gas medium.

As evidenced by Fig. 2, the gas may take the temperature of the particle at a considerable distance from its surface. In this case the heat evolution proceeds, as it were, not from the particle surface but from the surface of some conditional sphere whose radius may be much larger than that of the particle. For comparison, the dotted line shows temperature variation in cooling of an inert sphere (δ = 6 mm) in an immobile gas medium. The temperature of the gas and sphere far from the surface are assumed to be equal to those for a burning carbon 6 mm particle.

There is also another interesting fact. Assuming that the temperature curve maximum is associated with the surface where the carbon monoxide completes oxidation (which is in good agreement with experiments of Refs. 8, 11 mentioned above), Fig. 2 is the evidence of a much stronger effect of carbon monoxide oxidation in the boundary layer on heat exchange than on mass exchange.

With heat transfer from chemical reactions and the thermodynamic equilibration of gas assumed completed inside a sphere of a radius $R > r_o$ on whose surface the gas temperature is equal to the particle temperature, for steady state we can write

$$\frac{Nu^{*}\lambda}{2ro} \; 4\pi ro^2 \; (T_p - T_g) = \frac{Nu\lambda}{2R} \; 4\pi R^2 \; (T_p - T_g)$$

where Nu^* = Nusselt criterion for heat transfer from a burning carbon surface;
Nu = Nusselt criterion in the absence of burning;
R = the radius of a sphere on whose surface $T_g = T_p$;
r_o = particle radius.

Simplifications give:

$$Nu^* = Nu \frac{R}{ro} = Nu \; K \tag{6}$$

Equation (6) can be reduced to the form

$$\alpha^*_c = \alpha_c \cdot K \tag{7}$$

where α^*_c = convective heat transfer coefficient for the coal particle surface;
α_c = convective heat transfer coefficient from particle surface in the absence of burning.

Analysis of the above experimental evidence shows that the value of the coefficient K representing the effect of burning on the convective heat exchange between coal powder particles and the gas medium does not depend much on the size of particles or oxygen content under the conditions studies.

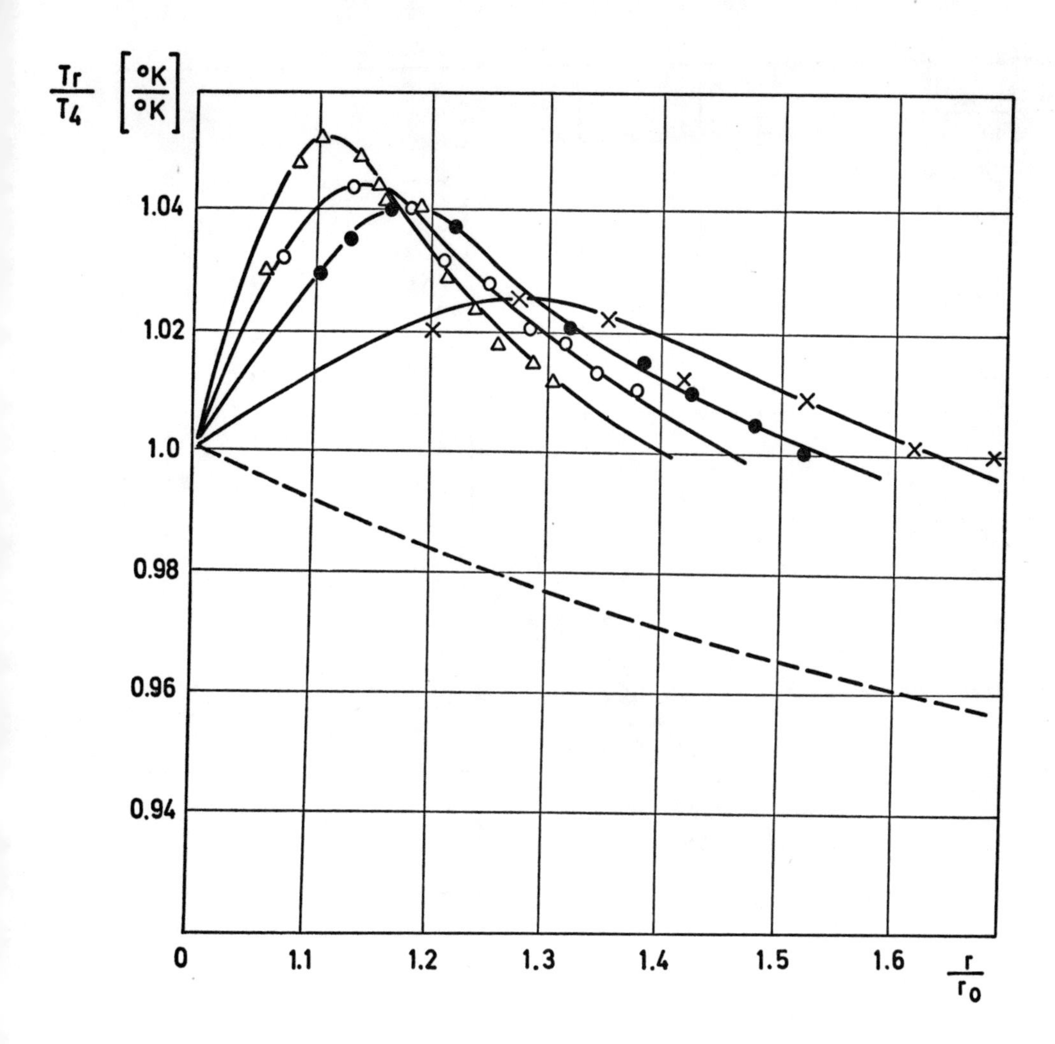

Fig. 2 Gas temperature in the boundary layer of a burning particle on the side of the incident flux (following Buchman, Ref. 11)

Tr = 1443°K, 0_2 = 21%, W = 0.15 m/sec.

x - δ = 6 mm. ● - δ = 10 mm. o- δ = 15 mm. △ - δ = 20 mm.

— — — calculated values, insert particle, δ = 6 mm.

W = 0 m/sec. T_4 = 1610°K, T_r = 1443°K

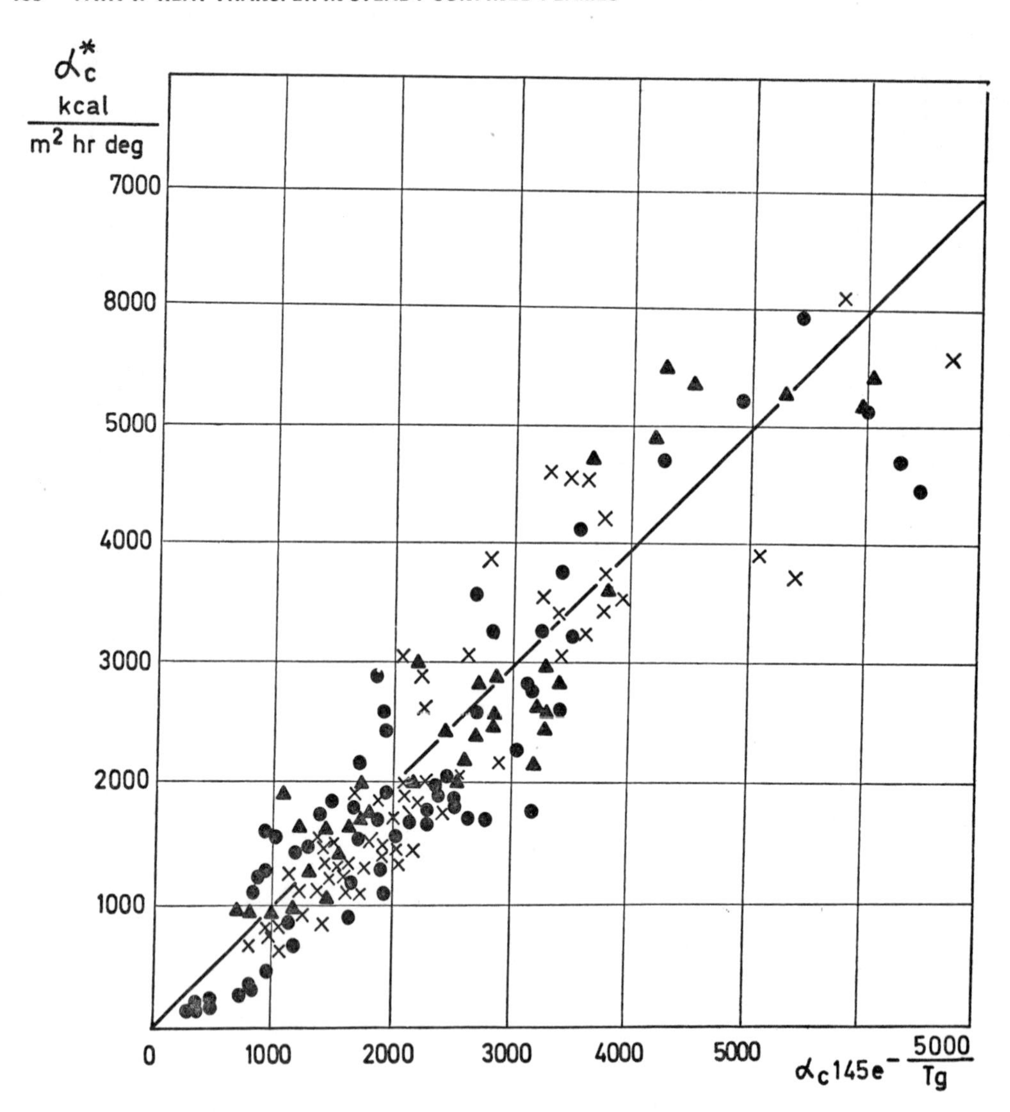

Fig. 3 Comparison of experimental values of convective heat exchange in combustion of coal powder particles, with the relation $\alpha_c^* = \alpha_c$ 145 exp $(-5{,}000/T_g)$ at different gas compositions.

● – O_2 = 21%, CO_2 = 0%

x – O_2 = 10%, CO_2 = 11%

▲ – O_2 = 5% , CO_2 = 16%

The effect of the medium temperature in the range T_g = 1200 to 1600°K on K was found to be rather strong and can be given in the form:

$$K = 145 \exp(-5{,}000/T_g) \tag{8}$$

Figure 3 is a comparison of the relation

$$\alpha_c^* = \alpha_c\, 145 \exp(-5{,}000/T_g) \tag{9}$$

with experimental data. Most points fit the relation (9) with a scatter of ± 35%, which can be regarded as satisfactory for experiments with coal powder particles of irregular form. Equations (6) and (9) lead to an empirical equation for heat exchange between burning particles of coal dust and the gas medium in the form

$$Nu^* = Nu\, 145 \exp(-5{,}000/T_g)$$

This equation applies at:

T_g = 1200 - 1600°K, O_2 = 5 - 21%;
δ = 150 - 1000 m, $Re < 1$.

A conclusion can be made that heat exchange calculation bur burning coal particles on the experimental evidence of mass exchange with the specifics of carbon combustion disregarded may lead to large errors. The calculated combustion temperature may prove to be much higher than the actual value.

REFERENCES

/1/ Sugawara, S., S. Kikkawa, K. Soda, Bulletin of JSME, Vol. 4, No. 15, 1961.
/2/ Babiy, V.I., A.G. Serebriakova, "Teploenergetika," No. 2, 1971.
/3/ Sokol´sky, A.L., F.A. Timofeeva in "Studies of Natural Fuel Combustion Processes," G.F. Knorre, ed., Gosenergoizdat, 1948 (in Russian).
/4/ Vyrubov, D.I., Zh.TF, Vol. IX, issue 21, 1939.
/5/ Katsnel´son, B.D., F.A.Timofeeva, "Kotlotrubostroenie," No. 5, 1948.
/6/ Nusselt, W., "ZaMM," Bd. 10, Hf. 2, 1930.
/7/ Rubinshtein, Ya.M. in "Studies of Heat Regulation, Transfer and Recooling," GONTI, 1938.
/8/ Predvoditelev, A.C., L.N. Khitrin, O.A. Tsukhanova, Kh.I. Kolodtsev, M.K. Grodzovsky, "Carbon Combustion," Academy of Sciences Publishing House, Moscow, 1949 (in Russian).
/9/ Ivanova, I.P., V.I. Babiy, "Teploenergetika," Nos. 4 and 5, 1966.
/10/ Babiy, V.I., I.P. Ivanova, "Teploenergetika," No. 12, 1968.
/11/ Wicke, E., G. Wurzbacher, Int. J. Heat Mass Transfer, Vol. 5, Nos. 3 - 4, 1962.
/12/ Rezniakov, A.B., I.P. Basina, S.V. Bukhman, M.I. Vdovenko, B.P. Ustimenko, "Natural Solid Fuel Combustion," Nauka Publishing House, Alma-Ata, 1968 (in Russian).

Chapter 8

GEOMETRICAL-OPTICAL CHARACTERISTICS AND CALCULATION OF RADIANT HEAT TRANSFER BETWEEN A FLAME AND A WALL

I. Mikk

The Tallinn Polytechnical Institute, USSR

Abstract

Some geometrical-optical characteristics of radiating systems are dealt with. Especially the problem of nonisothermal volume is considered using the Taylor series expansion of radiation flux function. Data for practical calculation of coefficients of nonisothermality are presented. Some cases of radiation exchange of nonisothermal furnace chambers are discussed by numerical examples.

SYMBOLS

a	:	absorbtivity; dimensionless distance
B	:	intensity on a beam
d	:	transmissivity
F	:	surface, area
h	:	dimensionless distance
I	:	hemispherical radiation flux
k	:	extinction (absorption) coefficient
L, l	:	beam length and dimensionless beam length
M	:	point
N	:	point on a surface
P	:	point in the volume
p	:	relative coordinate
Q	:	radiant energy per unit time
q	:	radiant heat flux per unit time and unit area
R	:	radius (Fig. 3)
r	:	reflectivity
s	:	dimensionless coordinate in the direction of a beam
T	:	temperature
X, Y, Z	:	coordinates
x, y, z	:	dimensionless coordinates

Greek Letters

α, β	:	angles (Fig. 2)
ϵ	:	emissivity; coefficients of nonisothermality in formulae (17) - (22)
η	:	dimensionless length (15)
θ	:	angle between a surface normal and the direction of a beam
H	:	coefficient in formulae (28), (29), (33)
μ	:	coefficient in formulae (22) - (26)
ν	:	reduced coefficient of nonisothermality (23)
ξ	:	coordinate in the direction of a beam
ρ	:	dimensionless length of a beam
ϕ	:	function in formulae (2), (3)
ω	:	solid angle

Subscripts

ef	:	refers to the effective
f	:	refers to the incoming radiation
g	:	refers to the gaseous
o	:	refers to the centre
res	:	refers to the resulting
sc	:	refers to the scattering
w	:	refers to a wall
α	:	refers to the absorption
β	:	refers to the scattering
λ	:	refers to the wavelength of monochromatic radiation

1. Introduction

A flame together with a gaseous medium and surrounding walls form a complicated radiating system. The spreading of radiant energy in such a system is accompanied by the weakening of the beam as a result of absorbing and scattering processes and by intensifying of the beam as a result of spontaneous emission of the medium. It is common knowledge that the intensity of the beam B in an arbitrary point P (Fig. 1) can be expressed by the equation

$$B^{f}_{(\lambda)}(P) = B_{(\lambda)}(N)e^{-\int_0^L k_{(\lambda)} d\xi} + \frac{1}{\pi} \int_0^L \phi^{ef}_{(\lambda)}(P')e^{-\int_0^{\xi} k_{(\lambda)} d\xi^*} k_{(\lambda)} d\xi \qquad (1)$$

where the subscript (λ) is used when the wavelength of monochromatic radiation is referred to and is left out when total radiation is meant. The extinction coefficient $k_{(\lambda)}$ is generally presented as a sum of absorption and scattering coefficients $k_{(\lambda)} = k_{\alpha(\lambda)} + k_{\beta(\lambda)}$

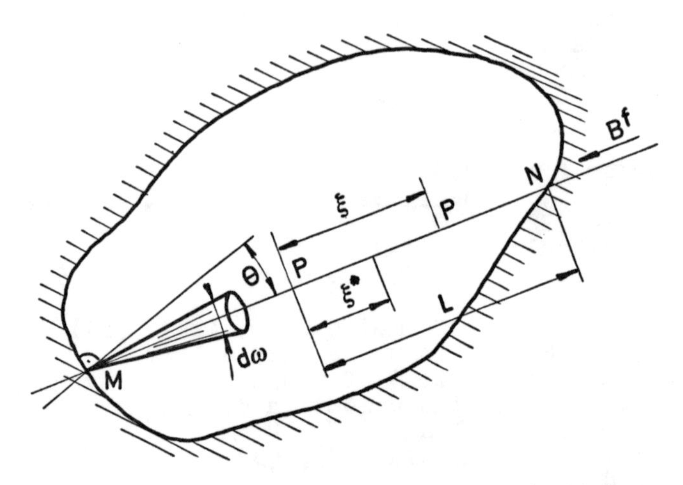

Fig. 1. Scheme to equations of the spreading of radiant energy in a system

The function $\phi^{ef}_{(\lambda)}$ is proportional to the volume density of the radiation and is also expressed as a sum

$$\phi^{ef} = \phi + \phi^{sc}, \tag{2}$$

where ϕ^{sc} is determined by the scattering of the radiation reaching into the point $-\phi^{sc} = k_\beta \phi^f$. By the function ϕ the self-emission of the medium is determined. To a monochromatic radiation ϕ_λ is expressed by the Planck law $\phi_\lambda = (k_{\alpha\lambda}/k_\lambda)I_\lambda$ and if total radiation is meant, by Stefan-Boltzman's Law

$$\phi = \frac{k_\alpha}{k} I = \frac{k_\alpha}{k} \sigma T^4. \tag{3}$$

The flux of incoming radiation is given as a sum of all beams projected on the straight line under consideration. In a volume it corresponds to the integration of $B^f(P) \cos\theta$ over the complete solid angle 4π. If the point M is on a solid surface (Fig. 1) the integral is to be taken over the hemispherical solid angle of 2π.

$$I^f(M) = \int_{2\pi} B(M) \cos\theta \, d\omega . \tag{4}$$

The hemispherical effective radiation flux in M is given as a sum of self-emission $\epsilon(M)\,I(M)$ and of reflected radiation flux $r(M)I^f(M)$:

$$I^{ef}(M) = \epsilon(M)I(M) + r(M)I^f(M). \tag{5}$$

From (1), (4) and (5) follows an integral equation

$$I^{ef}(M) = \epsilon(M)I(M) + r(M) \int_{2\pi} I^{ef}(N) e^{-\int_0^L k\,d\xi} \frac{\cos\theta\, d\omega}{\pi}$$

$$+ r(M) \int_{2\pi} \int_0^L \phi^{ef}(P') e^{-\int_0^\xi k\,d\xi^*} k\,d\xi \frac{\cos\theta\, d\omega}{\pi} . \tag{6}$$

The equation (6) and another analogous equation, which may be derived for the function of volume radiation $\phi^{ef}(P)$, form a system of integral equations, determining principally all effective radiation fluxes in the system. However, trying to find a practical solution to these equations one meets with various difficulties. One of the ways to simplify the solving of the problem is to give the function to be integrated in the form of two factors - one of them depending on the temperature and the other one on the optical and geometrical properties of the system - the "geometrical-optical characteristics".

In the present paper some geometrical-optical characteristics of radiating systems are dealt with. Some corresponding calculation methods and application of the characteristics by analysing various problems have been taken into consideration, too - especially the case of nonisothermal volumes because the geometrical-optical characteristics of nonisothermal volumes, are rarely available.

Later on some simplifying assumptions are applied. Firstly, the exponential law and the diffusivity of the radiation are assumed. Secondly, the refractive index of the medium is taken to be equal to unity. It is assumed, too, that the scattering of the radiation is negligible and therefore we can say

$$k_{\alpha(\lambda)} = k_{(\lambda)}, \quad \text{or} \quad \phi^{ef}_{(\lambda)} = \phi_{(\lambda)} .$$

2. Radiant energy transfer in a nonisothermal volume

The nonisothermity of the medium in a volume can be taken into account in the following way. Let us rewrite the equation (1) in a simplified form using dimensionless coordinates and lengths ($x = kX$; $y = kY$; $l = kL$; $s = k\xi$, etc.):

$$B(P) = B(N)e^{-\int_0^l ds} + \frac{1}{\pi}\int_0^l \phi(P') \, e^{-\int_0^s ds^*} ds. \qquad (7)$$

The function ϕ , describing the temperature field, can be presented in Taylor series expansion in the direction of a beam /s/ around the point M on the wall of a radiating system:

$$\phi(P') = \phi(M) + s\,\phi'(M) + \frac{1}{2}\,s^2\,\phi''(M) + \frac{1}{3!}\,s^3\,\phi'''(M) + .. \qquad (8)$$

By the comparison of (7) and (8), and integration between the limits (0;1) we obtain the following formula for the intensity of a beam of incoming radiation in point M

$$B(M) = B(N)e^{-1} + \frac{1}{\pi}\sum \phi^{(n)}(M)(1 - e^{-1}\sum_{m=0}^{n} \frac{1^m}{m!}). \qquad (9)$$

By means of (9) the integral equation (6) is transformed as follows:

$$I_w^{ef}(M) = \epsilon_w(M)I_w(M) + r_w(M) \int_{2\pi} I_w^{ef}(N)e^{-l}\frac{\cos\theta\, d\omega}{\pi}$$

$$+ r_w(M) \int_{2\pi}\sum I^{(n)}(M)\left(1 - e^{-l} \sum_{m=0}^{n} \frac{l^m}{m!}\right) \frac{\cos\theta\, d\omega}{\pi} \quad . \qquad (10)$$

The subscript "w" is used to refer to the walls because the temperature and other properties in the point M could have different values for the medium and for the walls. Note, that the distribution of the temperature function I(P), more exactly its derivatives $I^{(n)}(M)$, on the boundaries of the volume, are taken as predetermined. In this way the equation (10) can be applied immediately to solve the problem in the fundamental statement. The selfemission of the volume is expressed by the integral in the last term of the equation (10). This term can be replaced by a reduced product of emissivity of the volume ϵ by the blackbody radiation flux I. In this case one of the multipliers can be chosen arbitrarily $(\epsilon I)^* = \epsilon^* I = \epsilon I^*$. For example, we can take the emissivity ϵ as for an isothermal volume and the result is

$$\epsilon(M)I^*(M) = \int_{2\pi}\sum I^{(n)}(M)\left(1 - e^{-l} \sum_{m=0}^{n} \frac{l^m}{m!}\right) \frac{\cos\theta\, d\omega}{\pi} \quad . \qquad (11)$$

In this manner the selfemission of a volume is to be calculated as the selfemission of an isothermal volume having an effective temperature $T^* = \sqrt[4]{I^*/\sigma}$ (if the total radiation is considered). In this case the integral equation (10) can be written as

$$I_w^{ef}(M) = \epsilon_w(M)I_w(M) + r_w(M)\ \epsilon(M)I^*(M)$$

$$+ r_w(M) \int_{2\pi} I_w^{ef}(N)e^{-l} \frac{\cos\theta\, d\omega}{\pi} \quad . \qquad (12)$$

The equation (12) can be approximately replaced by a system of algebraic equations if the zoning method is to be applied. If we restrict ourselves to a case of only two zones (volume - boundary walls), then the equation (12) reduces to the wellknown formula

$$q^{res} = \frac{\frac{\epsilon}{a} I^* - \frac{\epsilon_w}{a_w} I_w}{\frac{1}{a} + \frac{1}{a_w} - 1} . \qquad (13)$$

Now let us consider cylindrical systems. According to Fig. 2 we can write

$$\frac{\cos\theta\, d\omega}{\pi} = \frac{1}{\pi} \qquad \cos^2\beta\ \cos\alpha\ d\beta d\alpha, \tag{14}$$

$$\eta = 1\cos\beta\ . \tag{15}$$

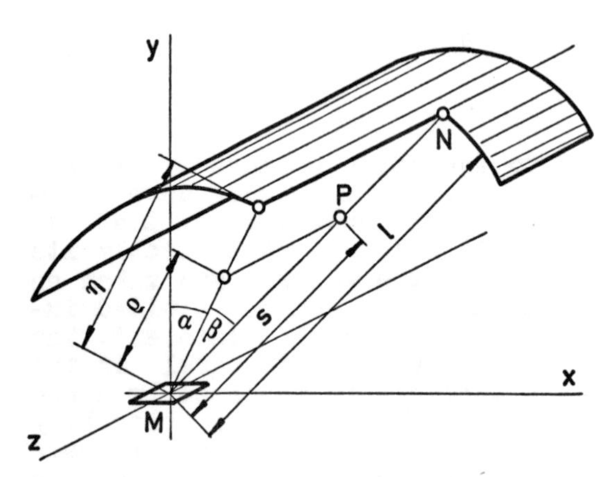

Fig. 2. A scheme of a cylindrical system

Taking into consideration that $d\rho/ds = \cos\beta$ and $dz/ds = \sin\beta$, the derivatives $I^{(n)}(M)$ in direction of a beam could be expressed as sums from the derivatives of I with respect to and z. For example

$$I^{(2)} = \frac{d^2I}{ds^2} = \cos^2\beta\frac{\partial^2 I}{\partial\rho^2} + 2\cos\beta\ \sin\beta\frac{\partial^2 I}{\partial\rho\partial z} + \sin^2\beta\frac{\partial^2 I}{\partial z^2}\ . \tag{16}$$

and using the equation (11) results in

$$\epsilon(M)I^{*}(M) = \frac{1}{2}\int_{-\frac{\pi}{2}}^{+\frac{\pi}{2}}\left[\epsilon^{(oo)}(\eta)\ I_M + \sum C^{j}_{i+j}\ \epsilon^{(ij)}(\eta)\ \left(\frac{\partial^{i+j} I}{\partial_\rho i\partial_z j}\right)_M\right]\cos\alpha\, d\alpha\ , \tag{17}$$

where C^{j}_{i+j} are binomial coefficients and

$$\epsilon^{(ij)}(\eta) = \frac{4}{\pi}\int_{-\frac{\pi}{2}}^{+\frac{\pi}{2}} \left(1 - e^{-1}\sum_{m=0}^{i+j}\frac{1^m}{m!}\right)\cos^{2+i}\beta\ \sin^j\beta\ d\beta\ . \tag{18}$$

The integrals (18) can be expressed by means of special functions $Ki_n(\eta)$. In case $j = 0$ the equation becomes

$$\epsilon^{(i0)}(\eta) = \frac{4}{\pi}Ki_{3+i}(0) - \frac{4}{\pi}\sum_{m=0}^{i}\frac{\eta^m}{m!}Ki_{3+i-m}(\eta), \tag{19}$$

$$\epsilon^{(00)}(\eta) = 1 - \frac{4}{\pi}Ki_3(\eta). \tag{20}$$

In case /j/ is an odd number, it gives $\epsilon^{(ij)} = 0$. To the even numbers of /j/ we can use the formula $\sin^2\beta = 1 - \cos^2\beta$ and obtain again for $\epsilon^{(ij)}(\eta)$ formulae by means of functions $Ki_n(\eta)$. For example

$$\epsilon^{(i4)}(\eta) = \frac{4}{\pi} Ki_{3+i}(0) - \frac{4}{\pi} \sum_{m=0}^{i+4} \frac{\eta^m}{m!} Ki_{3+i-m}(\eta)$$

$$- 2\,\epsilon^{(i+2,2)}(\eta) - \epsilon^{(i+4,0)}(\eta). \tag{21}$$

In this way the equation (17) contains the integrals of $\epsilon^{(ij)}(\eta)$, being considered as the geometrical-optical characteristics of a nonisothermal volume. Their concrete expressions depend on the shape of the volume and on the kind of the function I(x, y, z).

3. Geometrical-optical characteristics of a nonisothermal volume

The following determination of integrals (17) is connected with determination of the cross-section of the duct. In Fig. 3 4 schemes of cross-sections considered in this paper are pictured.

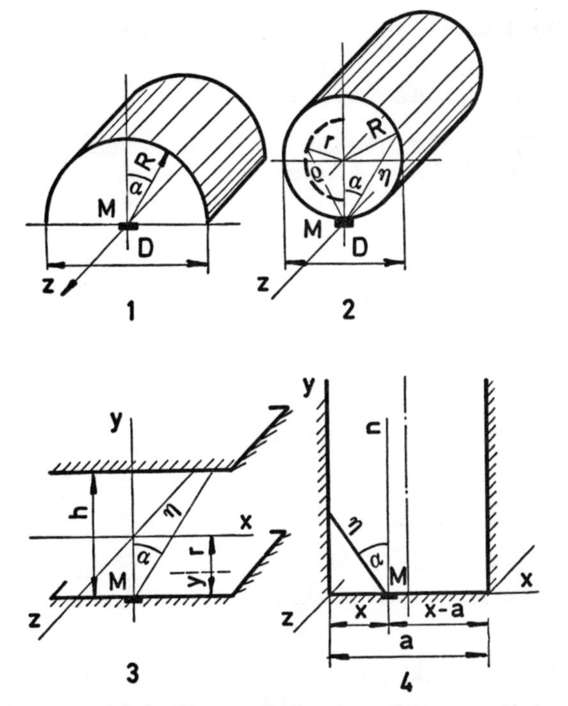

Fig. 3. Schemes of ducts with a nonisothermal volume

The most simplified case we have for the cross-section in Fig. 3, No. 1. In this case $\eta = R = \text{const.}$ and the values of $\epsilon^{(ij)}(\eta)$ do not depend on the angle α. Taking a radial temperature field, the derivatives of I do not depend on α, too, and in eq. (17) we obtain $\frac{1}{2}\int \cos\alpha \, d\alpha = 1$. Hence the formulae (19) - (21) can be applied without further integration. The final form of the formula depends on that, in what variables the function $I(\rho, z)$ is given. If we apply the relative coordinates $p = \rho/R$, $p_3 = Z/R$ for the temperature field, then a rational presentation of the formula (17) corresponding to the scheme 1 (Fig.3) can be written

$$\epsilon(M)I^{*}(M) = \epsilon_{oo} I_M + \sum \mu_{ij} \epsilon_{ij} R^{-(i+j)} I_M^{[ij]} , \tag{22}$$

where $\epsilon_{oo} = \epsilon(M) = \epsilon^{(oo)}(\eta)$ is emissivity of the isothermal volume;

$\epsilon_{ij} = \epsilon^{(ij)}(\eta)/\epsilon^{(ij)}(\infty)$ is coefficient of nonisothermity, dependent on geometrical and optical properties of the system;

$\mu_{ij} = C_{i+j}^{j} \epsilon^{(ij)}(\infty)$ is coefficient, dependent on geometrical variables and the shape of the volume as well as on the kind of function I;

$I_M^{[ij]}$ is the partial derivative of I with respect to coordinates p, p_3.

Dividing all terms of the equation (22) by the ϵ_{oo}, the equation reduces to

$$I^{*}(M) = I_M + \sum \mu_{ij} \nu_{ij} R^{-(i+j)} I_M^{[ij]} , \tag{23}$$

where $\nu_{ij} = \epsilon_{ij}/\epsilon_{oo}$ is reduced coefficients of nonisothermity of the volume.

For schemes 2, 3 and 4 (Fig. 3) formulae (22) and (23) are applicable too. However the terms μ_{ij}, ϵ_{ij} and ν_{ij} are more complicated to determine. For scheme 2 the numerical data are obtained by a digital computer.
For a plane layer (scheme 3) relatively simple formulae can be obtained using the special functions $E_n(x)$. If we consider one-dimensional temperature field depending on a relative coordinate p = 2r/h and take R = h/2 in formulae (22), (23), it gives

$$\mu_i \epsilon_i = (-1)^i \, 2\left[\frac{1}{2+i} - \sum_{m=0}^{i} \frac{h^m}{m!} E_{3+i-m}(h)\right] . \tag{24}$$

For scheme 4 the values of μ_{ij} ϵ_{ij} are relatively simply obtainable, if the temperature field is depending on relative coordinates $p_1 = \frac{2x}{a} - 1$; $p_2 = (\frac{2y}{a})^2$; $p_3 = (\frac{2z}{a})$. Let us consider the two-dimensional case (subscript i refers to p_1 and subscript j to p_2). Then for example for μ_{io} ϵ_{io} we can give the following equation:

$$\mu_{io}(x) \, \epsilon_{io}(x) = \frac{1}{1+i}\left[\epsilon^{(io)}(x) + \frac{x^i}{i!} \frac{4}{\pi}(2\,Ki_3(x) - Ki_1(x))\right] . \tag{25}$$

However, it should be borne in mind, that in the point M we have $dp_2/dy = 0$; $d^2p^2/dy^2 = 2\left(\frac{2}{a}\right)^2$. Hence, in the index of power of $R = a/2$ must be doubled in the part of subscript /j/ in formulae (22) and (23), and we obtain

$$I^*(x) = I(x) + \sum \mu_{ij}(x)\ \nu_{ij}(x)\left(\frac{a}{2}\right)^{-(i+2j)} I(x)^{[ij]} . \qquad (26)$$

Table 1

Some Values of Coefficient μ_{ij} in Formulae (22) and (23)

i	j	Scheme 1 $p_1 = \rho/R$; $p_3 = Z/R$	Scheme 2 $p_1 = (r/R)^2$; $p_3 = Z/R$	Scheme 3 $p = 2r/h$	Scheme 4 $p_1 = \frac{2x}{a} - 1$; $p_2 = \left(\frac{2y}{a}\right)^2$
1	0	$8/3\pi$	− 4/3	− 2/3	$-4/3\pi$
2	0	3/4	+ 2	+ 1/2	+ 1/4
3	0	$32/15\pi$	− 16/5	− 2/5	$-8/15\pi$
4	0	5/8	+ 16/3	+ 1/3	+ 1/8
5	0	$64/35\pi$	− 64/7	− 2/7	$-32/105\pi$
6	0	35/64	+ 16	+ 1/4	+ 5/64
0	1	-	-	-	-
1	1	-	-	-	$-16/5\pi$
2	1	-	-	-	+ 1
3	1	-	-	-	$-64/21\pi$
4	1	-	-	-	+ 15/16
0	2	1/4	+ 1/4	-	+ 4
1	2	$8/5\pi$	− 4/5	-	$-384/35\pi$
2	2	3/4	+ 2	-	+ 11/2
3	2	$64/35\pi$	-	-	-
4	2	75/64	-	-	-
0	4	1/8	+ 1/8	-	-
1	4	$8/7\pi$	-	-	-
2	4	45/64	-	-	-
0	6	5/64	5/64	-	-
0	8	7/128	7/128	-	-

It should be taken into consideration, too, that in scheme 4 the normal n divides the region of the integration into two parts and the formula (39) gives double value for one part. The total value for both parts can be written as follows

$$\epsilon(M)I^*(M) = \frac{1}{2}\left[\epsilon(x)I^*(x) + \epsilon(a-x)I^*(a-x)\right] . \qquad (27)$$

Some values of μ_{ij} for schemes in Fig. 3 are given in Table 1. In Fig. 4–7 there are plotted curves of coefficients ν_{ij} and of emis-

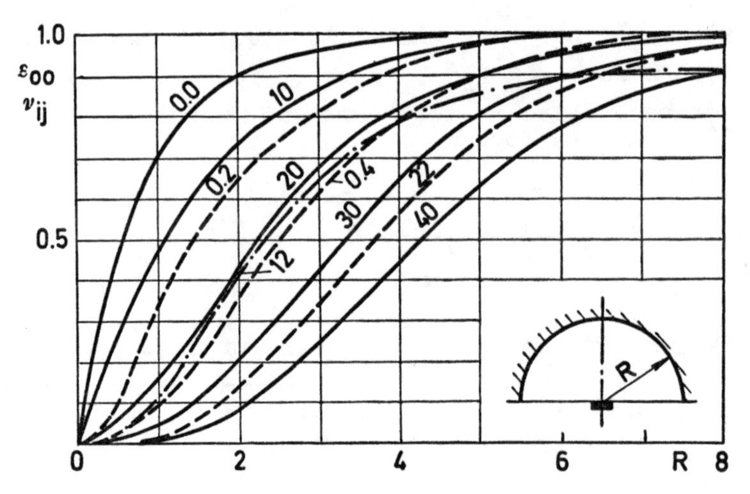

Fig. 4. Values of ν_{ij} and ϵ_{oo} to the scheme 1, Fig. 3

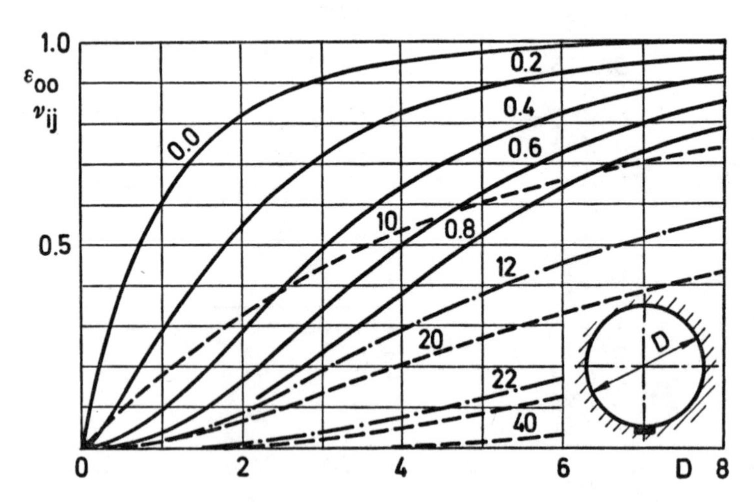

Fig. 5. Values of ν_{ij} and ϵ_{oo} to a cylinder scheme 2, Fig. 3

sivity of the isothermal volume ϵ_{oo} by some first numbers of subscripts i, j.

4. Application of the characteristics of nonisothermal volume when calculating heat exchange between a flame and a wall

Later on the considered problems are illustrated by some examples. Let us assume, that heat exchange is to be analysed dependent on the filling of a furnace chamber by the flame and on the nonisothermality of the furnace. The temperature field is considered in general as known and can be expressed in the first case (1) (the volume is accurately filled with the flame) by an approximate formula

$$T/T_o = \sqrt{1 - H p^2}\,, \tag{28}$$

and in the second case (2) (the flame fills the volume badly) by a formula

$$T/T_o = (1 + H p^2)^{-1}\,. \tag{29}$$

In both cases the T_o is the temperature in the centre of the volume and the term H determines the temperature on the boundary.

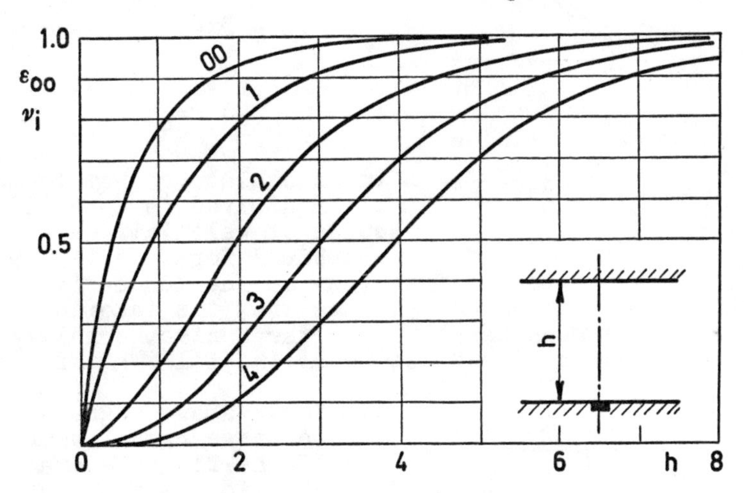

Fig. 6. Values of ν_i and ϵ_o to a plane layer, scheme 3, Fig. 3.

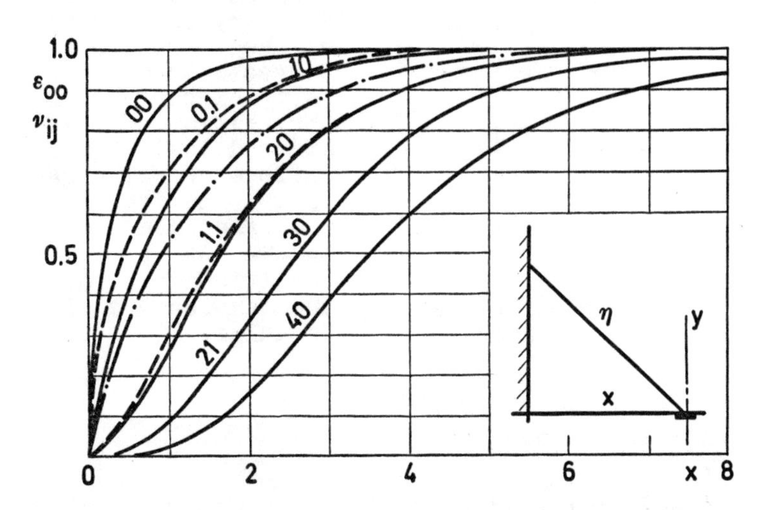

Fig. 7. Values of ν_{ij} and ϵ_{oo} to the scheme 4, Fig. 3.

Let us carry out the calculation for a plane layer (scheme 3, Fig.3). The temperature in the centre is chosen, e.g. T_o = 2000 °K. The corresponding temperature fields are shown in Fig. 9. The values

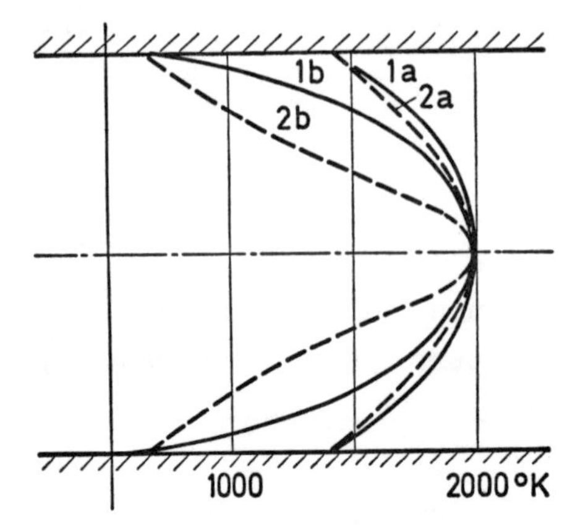

Fig. 8. Temperature fields corresponding to formulae (28) and (29)

of H are chosen in the first case 1a) H = 0.5; 1b) H = 0.9 and in the second one 2a) H = 0.4 and 2b) H = 2. We get the partial derivatives from formulae (28) and (29). Using these values and Table 1 the formula (23) can be written as

$$I^{*}/I_{o} = (1 - H)^{2} + \frac{8}{3}H(1 - H)\left(\frac{2}{h}\right)\nu_{1}$$

$$- 2H(1 - 3H)\left(\frac{2}{h}\right)^{2}\nu_{2} - \frac{48}{5}H^{2}\left(\frac{2}{h}\right)^{3}\nu_{3}$$

$$+ 8H^{2}\left(\frac{2}{h}\right)^{4}\nu_{4}\,. \qquad (30)$$

In an analogous way formulae are to be derived in case (2). The results of calculations are given in a graphical form. In Fig. 9 the relations of fluxes of the selfemission on the boundaries of the volume are plotted. These curves demonstrate that the incoming radiation on the wall increases considerably when the nonisothermality of the volume decreases (curve 1) and also when the filling of the volume with the flame (curve 2) increases.

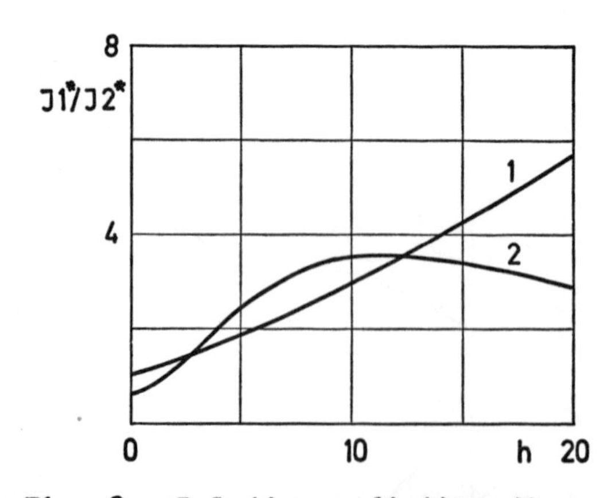

Fig. 9. Relative radiation fluxes on the boundaries of the volume

1 - I_a^*/I_b^*, the dependence on nonisothermality,

2 - I_{1b}^*/I_{2b}^*, the dependence on filling-in degree of volume by the flame

The above graphs correspond to these cases where the centre of the flame is characterized by a uniform temperature. Hence the enthalpy of the medium has various values for each case. However, the temperature level of the process is frequently characterized by means of enthalpy of combustion products. In the given example it is considered that the speed field is uniform (or the gaseous medium is not moving at all) and the specific heat capacity of the medium is constant. In these conditions the average value of the temperature is given by the formulae:

$$\frac{\bar{T}}{T_o} = \frac{1}{T_o}\int_0^1 T dp \begin{cases} = \frac{1}{2}\sqrt{1-H} + \frac{1}{2}\frac{\arcsin\sqrt{H}}{\sqrt{H}} \text{ (to case (1))}, \\ = \frac{1}{H}\operatorname{arctg}\sqrt{H} \text{ (to case)2)).} \end{cases} \tag{31}$$

The ratio of the effective temperature T^* to the average temperature $\bar{T}$ can be calculated by the formula

$$T^*/\bar{T} = (\bar{T}/T_o)^{-1}\sqrt[4]{I^*/I_o}, \tag{32}$$

and is plotted in Fig. 10 for the cases discussed above. It is interesting to note, that the effective temperature value, required for the calculation of the selfemission of the volume, may considerably differ from the average temperature obtained by means of enthalpy of the combustion products. We have small values of the ratio $T^*/\bar{T}$ just in case of large furnaces having considerable nonisothermality. Therefore the methods of heat transfer calculation of furnaces, which are based on the use of the average temperature of a gaseous medium, need refinement if a furnace with a very different nonisothermality and optical thickness is dealt with. For the above described method we must preliminarily set up an approximate equation of temperature field, e.g. (28); (29), using some additional data (e.g. analysing combustion process, aerodynamics of the flame, or using experimental data, etc.). It should be taken into consideration, that

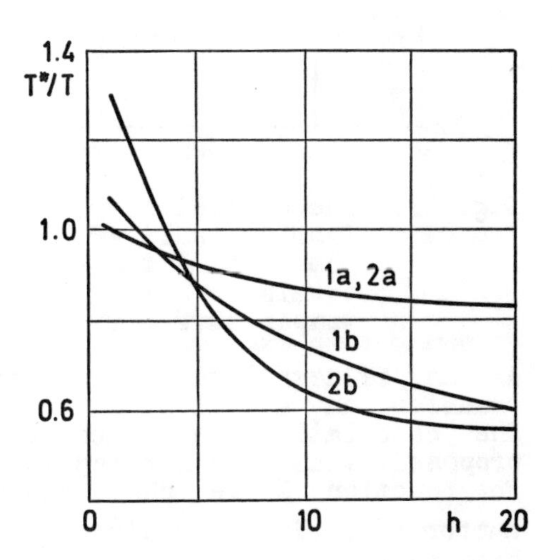

Fig. 10. Relation of the effective temperature to the average temperature of the medium $T^*/\bar{T}$

1) Function $T(p_1, p_2, p_3)$ must be continuous and single-valued;
2) The series expansions (22) or (23) must converge by not very large number of terms.

To restrict the number of the terms the simplest way is, if the derivatives $I^{[ij]}$ would become zero beginning with some subscript numbers i and j. That is what happened in the case (1) of the example (see (30)) and it may be obtained in other cases, if the formula $I(p_1, P_2, p_3)$ is chosen suitably. It follows, for example, from Fig. 8 that in some conditions different formulae may give practically identical distributions of the temperature (curves 1a and 2a). The series in formulae (22), (23) may be better converged, if the optical thickness of the volume and the isothermality are increasing.

Note, that heat transfer is mainly determined by the selfemission of the boundary layer with dimensionless thickness approximately one unit. In Fig. 11 curves of dimensionless distance y^* from the wall, on which the real temperature T equals to the effective temperature T^*, are plotted. These curves are obtained from the data of the example given above. It shows, that the values of y^* are located in an interval of $0 \div 1$ and for the large values of h the curves of y^* are tending to a limit value. These circumstances can be useful because for volumes of large optical thickness we can set up approximate formulae describing the temperature distribution correctly only in the boundary layer, and their accuracy is not required at distances very far from the walls.

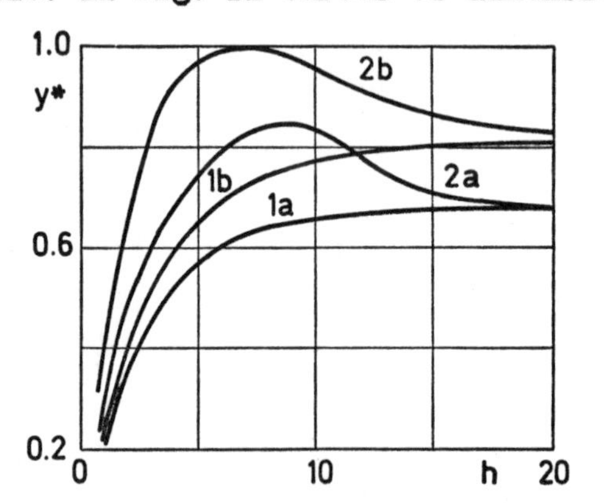

Fig. 11. The dimensionless distance from the wall y^*, by which the real temperature T equals to the effective temperature T^*

As an illustration let us consider the nonuniformity of the heat transfer on a wall of a furnace with rectangular cross-section. The calculation is based on the scheme 4 (Fig. 3) and it is proposed, that the furnace should be optically very thick (a = 10). For function I we shall choose such a function which has a small number of derivatives $I^{[ij]}$ differing from zero. A function of this kind is

$$I/I_o = H + (1 - H)(\frac{a}{2})^2 (1 - p_1^2)p_2. \tag{33}$$

Let $p_1 = 0$ and $(\frac{a}{p})^2 p_2 = y^2 = 1$ to give $T/T_o = 1$. It means that the temperature in the centre of the furnace at the distance of $y = 1$ will be equal to T_o. The temperature of the gaseous medium contacting with the walls will be $T/T_o = \sqrt[4]{H}$. In this example we put $T_o = 1500\ ^oK$ and carry out the calculation for three cases:

1) Isothermal furnace - H = 1;
2) Small nonisothermality - H = 0.4 ($T_{y=0} \approx 1200\ ^oK$);
3) Considerable nonisothermality - H = 0.025 ($T_{y=0} \approx 600\ ^oK$).

The temperature field for the case (3) is pictured on the left-hand side of Fig. 12. On the right hand side of Fig. 12 curves of incoming radiation flux ϵI^* ratio to its values in the centre of the wall $(\epsilon I^*)_{p_1=0}$ are plotted, being calculated by formulae (27) and (33). Note that in the case (3) the absolute value of ϵI^* in the centre of the wall equals 0.93 and correspondingly the value of effective temperature is $T^* = 1470\ ^oK$. So the temperature determining the incoming radiation stays in expected limits in spite of infinite increase of actual temperature at large distances from the wall in accordance with formula (33).

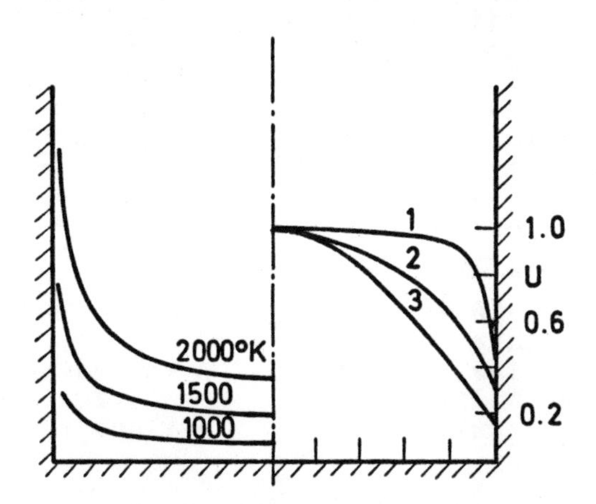

Fig. 12. The dependence of the heat flux of the furnace walls on the corner regions. On the left hand side - temperature field of a nonisothermal furnace (case 3). On the right hand side - ratio $u = \epsilon I^{*}/(\epsilon I^{*})_{p_1=0}$ for three cases: 1 - H = 1; 2 - H = 0.4; 3 - H = 0.025.

5. Conclusion

In this paper the geometrical-optical characteristics of radiating systems are discussed not only by means of the well-known zoning method, but the problem has been developed with regard to nonisothermal volumes and zones. The applied principle of expansion of the function I in Taylor-series is not new and has been used in various papers before /1, 2, 3/. However, it appears that in connection with the determination of geometrical-optical characteristics, it is as yet applied only in /4/.

In the limits of the above discussed matter radiant heat transfer could be calculated for the monochromatic radiation and for the grey approximation of total radiation. The problem of the selective total radiation is strictly connected with the integration of (ϵI^{*}) over the parts of the spectrum.

The above discussed principles of the calculation of radiation of nonisothermal volumes could be also useful in the measurement of the flame radiation fluxes and radiation characteristics of gaseous medium. For the study of a nonisothermal flame by means of various radiometric and optical instruments the formulae (22) and (23) could be recommended, too. However, taking into consideration the radiation of the background $\epsilon_w I_w$ these equations are transformed and written as

$$I^{ef} = \epsilon_w I_w (1 - \epsilon_o) + \epsilon_o I + \sum \mu_i \epsilon_i I^{[i]}$$

$$= \epsilon_w I_w (1 - \epsilon_o) + \epsilon_o I (1 + \sum \mu_i \nu_i I^{[i]}/I). \qquad (34)$$

If the radiation fluxes are to be measured at various distances $x = kX$ (Fig. 13) we can have a system of equations for all characteristics we are interested in from (34). In this case $I(x)$ is considered to be a function of the coordinate x and ϵ_i and ν_i dependent on $l - x$. If the temperature field is measured by any other method and we know the value of background radiation $(\epsilon_w I_w)$, then only the values of ϵ_i remain unknown in (34). In this case it is sufficient to measure I^{ef} only in one point, e.g. $x = 0$ and $l = kL$ is determined by formula (34). For the instruments, having a wide field of vision (20) the functions ϵ_i, ν_i and μ_i could be found as for a plane layer (Table 1 and Fig. 6). For the instruments of a narrow field of vision the one-dimensional formula (9) may be recommended. In this case $\mu_i = (-1)^i$ and the functions ϵ_o, ν_i are to be taken from Fig. 14.

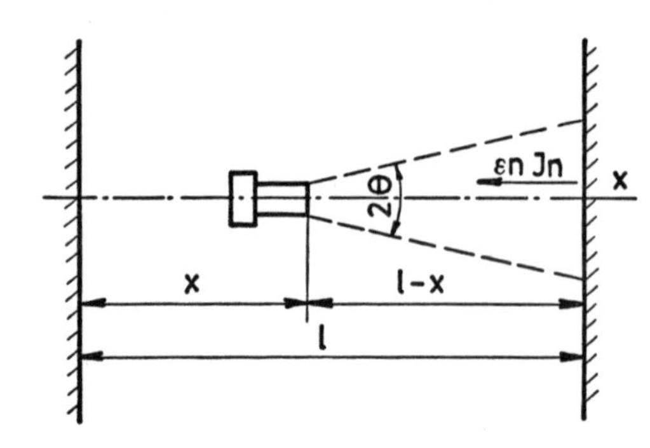

Fig. 13. A scheme of measurement of the radiation characteristics of a flame

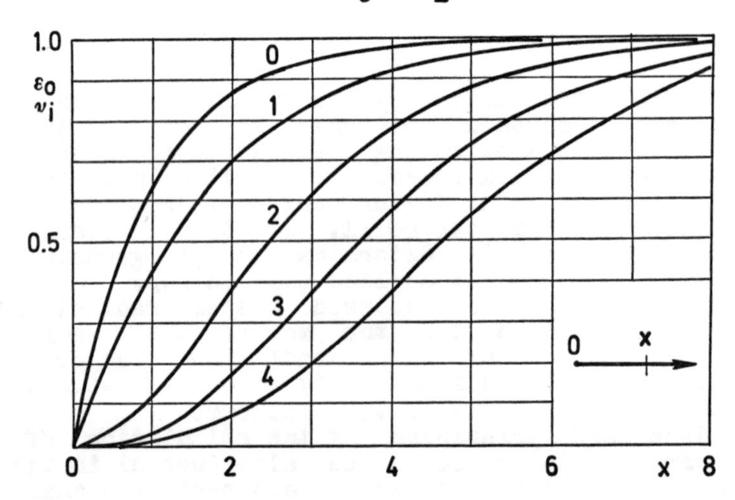

Fig. 14. Values of ν_i and ϵ_o to a one-dimensional model of radiation

REFERENCES

/1/ R. Viskanta, Interaction of Heat Transfer by Condition, Convection and Radiation in a Radiating Fluid. "Heat Transfer" Ser. C, 85, 4 (1963).

/2/ E. E. Anderson, R. Viskanta, Effective Conductivity for Conduction-Radiation by Taylor Series Expansion. "Int. J. Heat and Mass Transfer, 14, 8 (1971).
/3/ Y.Taitel, J.D. Hartnett, Application of Rosseland Approximation and Solution Based on Series Expansions of the Emission Power to radiation Problems. Rocket Techn. and Cosmonautics, 6, 1 (1968).
/4/ I. Mikk, V. Pomerantsev, On the Radiant Heat Transfer in a Duct with a Variable Temperature of the Medium. Izv. Akad. Nauk ESSR, Fiz. Mat. 19, 1 (1970).

Chapter 9

FLAME AS A PROBLEM OF THE GENERAL THEORY OF FURNACES

M. A. Glinkov

Moscow Steel and Alloys Institute, Moscow, USSR

An industrial flame is a controlled fire element, and a flame is a form of fire convenient for practical use. In contrast to combustion products, heat is continuously generated from some other kind of energy, i.e. from chemical energy of a fuel or from electrical energy of plasma. Thus, flame is a typical heat generator, and from this point of view it ranks with a bed of a burning fuel, electric arc or a heater of electrical resistance furnaces. The equipment may be of two kinds: power or technological. In the first case one type of energy is obtained from another source of energy or from a change of energy carrier or from a change of parameters in the system; in the second case a given kind of energy, which we call "working energy," is used to overcome forces hindering the technological process.

The efficiency of power equipment is characterized by minimum energy consumption, whereas the efficiency of technological equipment is characterized by maximum energy consumption for the needs of the technological process. The loss of energy in the power equipment is usually less than in the technological equipment, and this factor determines the efficiency difference. Flame may be an element of both power and technological equipment. However, there is a great difference in flame functions in both cases.

To illustrate this, it is quite enough to compare flame functions in the cylinder of an internal-combustion engine and the functions of flame in a flame furnace for metal heating. In the first case the problem is concerned with combustion effect, whereas in the second case it deals with a complex of interdependent processes of heat and mass transfer. In an internal-combustion engine the low efficiency (less than 0.5) is the result of imperfect process of the transformation of heat into power on the engine shaft. In flame furnaces the same or lower efficiency is the result of imperfect heat transfer from flame to the material being heated. Whereas the power equipment works in a steady regime, an unsteady regime in furnaces is a usual phenomenon.

Summing up, we may say that the flame problem in industrial furnaces is especially complicated. Therefore, in spite of numerous studies in this field the calculation technique is very imperfect.

The furnace is thermal equipment in which heat is a kind of energy. The definition refers to combustion furnaces and to all kinds of electric furnaces as well, since transformation of electrical energy into heat always takes place before the technological process starts.

So, any furnace may consist of two zones - a zone of technological process and a zone of heat generation. The zone of technological process is the main zone since any furnace is technological equipment. The technological zone can be the bath of an open-hearth furnace, or converter, ingots in a soaking pit or billets in a continuous furnace. The zone of heat generation is an auxilary zone in furnaces, as it is designed to create certain energy conditions on the boundaries of the technological zone. For example, the volume of a soaking pit between ingots and the space above billets in a continuous furnace are heat generation zones. Furnaces with a separated zone of technological process and a zone of heat generation are called heat exchanging furnaces. There are furnaces in which these zones coincide. We may call them heat generating furnaces. A convertor and kiln with a fluidized bed are examples of heat generating furnaces. Many furnaces must be put into a group of furnaces of a mixed type where these two zones are partly combined. There is a zone of heat generation in these furnaces but its functions are less important since a part of the heat is generated directly in the zone of technological process. Open-hearth furnaces working on liquid iron where oxygen is blown through the bath are examples of furnaces of a mixed type.

The schematic presentation of a furnace consisting of two zones permits us to single out two groups of processes, namely determining and determined. The former provides heat input into the zone of technological process, while the latter provides heat distribution through the zone. It is evident that various kinds of heat and mass transfer may be both determining and determined. In order to correctly organize the thermal work of furnaces and to obtain maximum efficiency it is very important to determine the limiting link in a complex of interdependent processes, which is very difficult to do in unsteady regimes which often occur in furnaces.

The efficiency of various kinds of energy may be characterized by the value of the virtual thermal equivalent Q, the actual value of which depends on the coefficient of heat comsumption (η cuh)

$$Q_{ve} = Q_{ae} \cdot \eta_{cuh}$$

where Q_{ve} - virtual thermal equivalent and Q_{ae} - amount of energy in thermal units.

The virtual thermal equivalent characterizes that portion of energy which may remain in the zone of heat generation as heat. On account of what has just been said, the virtual thermal equivalent depends on the heat content of energy carrier leaving the zone and carrying the energy away. Thus, for a natural gas (98% methane, its calorific power is 8400 kcal/m^3). When burning it without any excess of oxygen the values of the virtual thermal equivalent are as follows.

Temperature of outgoing gases	300°C	800°C	1600°C
Burning in cold air, kcal/m^3	7420	5650	1730
Burning in oxygen, kcal/m^3	8095	7540	6670

The temperature of characteristic outgoing gases are: 300°C for steam boilers; 800°C for heating furnaces, 1600°C for open-hearth furnaces.

From the above we can see the difference in flame efficiency of a natural gas in boilers and in different furnaces and the significance of air enrichment with oxygen in high temperature furnaces. Now let us imagine the conditions of natural gas burning in oxygen in the zone of technological process in a steel melting bath. Under these conditions the hydrogen of a natural gas will not oxidize, the carbon will oxidize to CO, and the virtual thermal equivalent will be negative (-560 kcal/m^3) if it is taken into account that the temperature of the products of incomplete combustion of a natural gas from the bath cannot be less than 1600°C. In this respect electric energy possesses remarkable advantages, and its virtual thermal equivalent retains its value (860 kcal/kwh) under any conditions.

The fuel efficiency is the first important characteristic of flame used as a heat generator in industrial furnaces. It is this feature that demonstrates the difference between the conditions for furnaces and the conditions for flame in various power installations where the temperature of outgoing gases is low.

Furnaces are efficient at high temperatures in the zone of technological process and the zone of heat generation at relatively low rates of combustion products. Therefore, the convective component of mass and heat transfer is of limited value and the radiant component is predominant. At the same time it is impossible to neglect the convective component in mass and heat transfer, for instance the influence of the furnace atmosphere on metal in the zone of technological process.

If flame radiation is regarded as such in steam boilers with screened walls, in internal-combustion engines and for flames, in flame furnaces it is the result of interaction with surrounding hot surfaces. Formerly, all flame furnaces were called reverberatory, thus emphasizing the significance of brickwork in heat exchange. Let us consider a simplified model of a flame furnace as space of a cylindric shape, full of lame and characterized by permanent temperatures in the field of heating surface area (M) and brickwork (B). If prolonged radiation is neglected, the following equation of heat balance for any furnace section may be written for the unsteady regime

$$Q_f = q_M F_M + q_B F_B \tag{1}$$

$$q_M = Q_f^M + Q_f^B (1-\bar{E}_f) - Q_M\left[1-(1-\bar{E}_f)^2 \frac{1}{K}\right] - q_B (1-\bar{E}_f) \tag{2}$$

where Q_f - heat generation in flame per 1m of furnace length, kcal/mh; q_M and q_B - resulting heat flow for surfaces of material and brickwork, kcal/m^{2}h; Q_f^M and Q_f^B - flame radiation to surfaces of material and brickwork, kcal/m^{2}h; Q_M - effective radiation of material surface, kcal/m^{2}h; F_M and F_B - heating surface area both of material and brickwork per 1m along the length; k - coefficient of relative brickwork surface - ratio of brickwork surface to heating surface area; $\bar{E}_f$- average emissivity value.

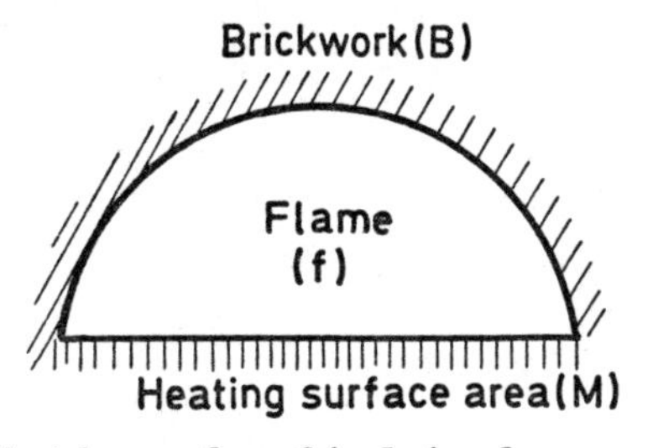

Fig. 1 Schematic depiction of cylindrical space cross-section

Analysis of the equation (2) shows that the intensity of heat emission is defined by flame radiation in the direction of material surface (Q_f^M) and brickwork surface (Q_f^B) as well. Thus Q_f^M characterizes direct flame radiation onto the heating surface area and Q_f^B is indirect flame radiation using brickwork as a medium. The values of Q_f^M and Q_f^B may vary within wide limits depending upon temperature distribution and may characterize various regimes of radiation in flame furnaces.

Let us call the radiation regime

uniformly distributed if $Q_f^M = Q_f^B$

obtained directly if $Q_f^M > Q_f^B$

obtained indirectly if $Q_f^M < Q_f^B$

The solution of equation (2) shows that the value of q_M at $Q_f^M > Q_f^B$ increases if the value $\bar{E}_f$ increases to the limit $\bar{E}_f = 1$. In the case of $Q_f^M < Q_f^B$ the maximum value of q_M is reached at $\bar{E}_f$ in the range of 0.5 - 1.0 depending upon the ratio of Q_f^M and Q_f^B.

In equation (2) one member shows the effective radiation of heating surface area (M). After deciphering we get

$$Q_M = Q_M^{OW} + Q_M = 4.9 \cdot 10^{-8} \cdot E_M \cdot T_M^4 + (1-E) \cdot Q_M^{in} \tag{3}$$

The second member of the right part of the equation (3) is always considerably greater than the first, especially at low values of T_M. The fact accounts for increasing q_M as the degree of emissivity of heating surface area (E_M) grows, other conditions being equal. It also explains why radiation is lower when heating non-ferrous metals and the convective component is correspondingly increasing. The availability of two components in radiation between flame and heating surface area enables substitution of one of the components for another retaining the same or equal radiation intensity and thus transforming one regime to another.

Real maximum of radiation intensity is reached when $Q_f^M \gg Q_f^B$ with $E_B = 1$. If we assume that brickwork with $E_B = 1$ is heated up to the flame temperature in some other way, but not by flame radiation, then the same intensity radiation is evidently obtained. Flame emissivity is considerably less than 1. Therefore, almost complete substitution of a direct leading regime for an indirect one is quite real and is used in the so-called infrared heaters.

Such a regime is called the maximum case of an indirect leading regime. It is quite natural that in this case equation (2) is not used. Furnaces of rapid heating are often designed taking into account this heat exchange principle and are distinguished by high efficiency.

From equation (2) we see that the brickwork emissivity and its temperature do not influence heat emission (only by means q_B) whereas the ratio of brickwork surface to heating surface area (K) has little effect and the greater is flame emissivity, the less the effect. To define the influence of brickwork emissivity on the brickwork temperature and that of a flame, let us write an equation for brickwork radiation

$$Q_B^{OW} = \left[Q_f^B + Q_M(1-\bar{E}_f) \right] E_B - q_B \tag{4}$$

$$\text{when } E_M = 1,\ T_B^4 \leqslant \bar{E}_f T_f^4 + T_M(1-\bar{E}_f) - \frac{q_B \cdot 10^8}{E_B \cdot 4.9} \tag{5}$$

By means of equation (5) it is easy to conclude that an essential change of brickwork emissivity (from 0.8 to 0.4) decreases the brickwork temperature and increases the flame temperature only by some degrees. From equation (5) we see that E_B does not influence the brickwork temperature and the temperature of flame at all when $q_B = 0$. Thus, the organization of heat radiation in furnaces really does not depend upon the brickwork material if its heat resistance remains unchanged, other conditions being equal. The organization depends on flame emissivity and emissivity of heating surface area (except the extreme case of an indirect leading regime).

All the above conclusions are based on the assumption that flame, brickwork and heating surface area radiate as "grey" bodies. It is known that the above-mentioned does not correspond to reality, but this method is widely used in calculations, though the magnitude of the error in unknown. At the Department of Theory and Automation of Furnaces of the Moscow Steel and Alloys Institute the studies of the problem have been carried on for a number of years by Professor Krivandin and Assistant Professor Mastrukov. The considerations given below are based on the results of the studies as well as on the investigations of other scientists. First of all, we are interested in the spectral structure of incident radiation. From the balance equation it follows

$$Q_M^{in} = Q_f^M + (Q_B^{OW} + Q_B^r)(1-\bar{E}_f)\frac{F_M}{F_B} \quad (6)$$

In screened boiler fire chambers the incident radiation on the heating surface area is defined on the whole by flame radiation. As follows from equation (6) the problem becomes more complicated in furnaces. Three spectra are present in the incident rediation spectrum, namely: flame spectrum, brickwork spectrum and spectrum of heating surface area; the specific value of individual parts of the spectrum depends upon the temperature distribution in the working space and absorption effects.

The spectrum of heating surface area is less represented in the incident radiation spectrum, as the temperature of the latter is extremely low. The weaker the flame spectrum in the incident radiation, the lower the flame emissivity and the higher brickwork emissivity on the surface where the flame spectrum is somewhat transformed into the spectrum of brickwork radiation.

The working temperatures of flame and brickwork in various types of furnaces do not exceed 1200-2200°K in practice. Therefore, we may conclude that the range of wavelengths from 1 to 5 μ and in a lesser degree up to 8 μ is a decisive factor for furnaces. It is known that the flame spectrum in contrast to the spectrum of pure gases is of an uninterrupted character and determined by the radiation of suspended solid particles, soot in general.

Physico-chemical processes in flames are connected with diminishing the number and size of soot particles or pulverized coal and with the formation of ash particles CO_2 and H_2O. They result in a considerable change of the spectral flame emissivity along its length, and at the end of the flame spectral emissivity approaches that for pure gases, especially if a natural gas or heavy oil is burnt, the flame of which contains few ash particles. In outward appearance it is expressed by the weakening of flame luminosity. However, this feature is not similarly connected with the change in flame emissivity, since it refers to the visible part of the spectrum. The spectral absorbing capacity of soot particles is studied in detail for the visible and nearest infrared part (up to 1.5) of the spectrum. The maximum absorption coefficient for the widest range in size of soot particles in flame (from 0.03 to 0.3μ) is in the range of wavelengths (from 0.15 to 0.7μ), and then the absorption coefficient is rapidly decreasing while the wavelength is increasing. However, the radiation of heterogenous flame is of a complex structure which represents the radiation of pure gases depending not only on the number of particles per unit volume but also on their distribution according to size, not to mention the temperature regime of flame.

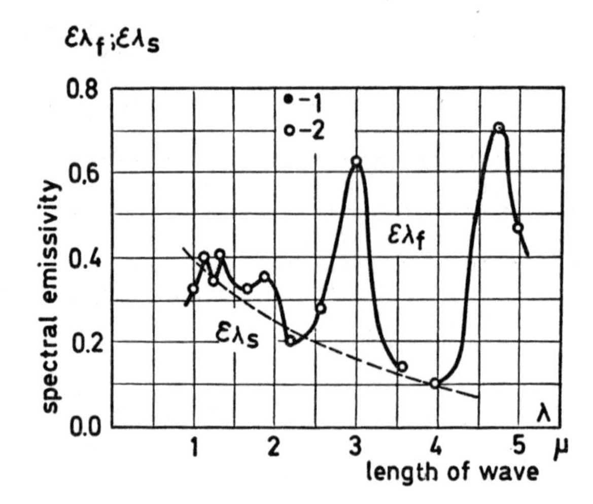

Fig. 2 Spectral emissivity of luminous flame (1) and soot carbon particles (2) at a distance of 450 mm from the burner, the emitting layer being 400 mm (2).

In Fig. 2 the curves (2) for the spectral emissivity of flame and soot are given. They illustrate the monotonous character of decreasing the soot emissivity whereas the wavelength increases.

The curve in Fig. 2 represents the thickness of the flame layer (400mm). The maximums in a solid curve which characterizes the summarized radiation of soot and pure gases demonstrate the spectral emissivity of combustion products for wavelengths close to about 3 and 4.75μ of the order 0.42 and 0.65 respectively. It may be said that the gas radiation of c ombustion products brings the luminous flame radiation nearer to the "grey" one (average for Λ = 1-5μ) depending on the thickness of the flame layer.

It accounts for the fact that all the points in Fig. 2 will go upwards but in a different degree. The given data enable us to come to the conclusion that the flame as a weakening medium retains its properties while the wavelength of penetrating radiation increases. Fig. 3 and 4 give the data obtained in the Moscow Steel and Alloys Institute for various refractories during radiation in the atmosphere and in the neutral medium respectively. On the whole these data correspond with the results of the studies of other authors. The same dependence is seen for alumine-silica and magnesite refractories: relatively slow decrease of the spectral emissivity in the range of 1-3μ and its more rapid increase in the range of 4-6μ for the maximum value. Neither the working temperature of refractories nor the character of the atmosphere changes this dependence though the left branches of the curves are somewhat inverted with respect to the right ones. The exception is chrome magnesite refractories where the spectral emissivity increases while the wavelength increases too.

The radiation intensity maximum in the range of temperatures characteristic of furnace flames (1000°C - 2000°C) correspond with the wavelength 2.26-1.32μ , the intensity of its radiation being determined by the spectral emissivity of the given material near the corresponding maximum.

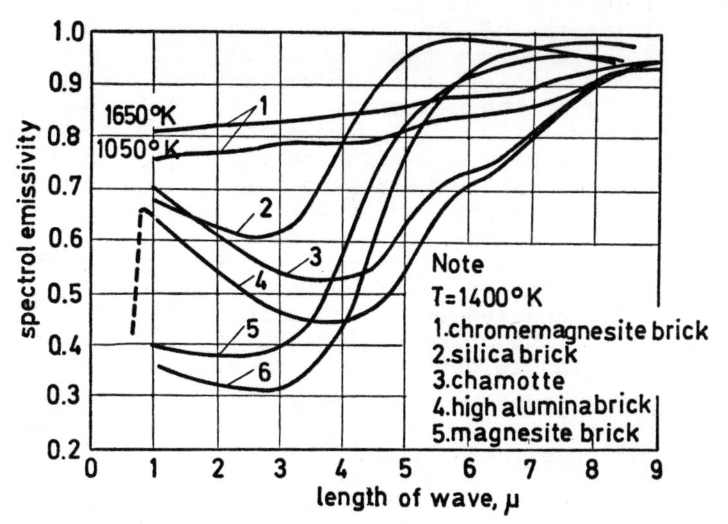

Fig. 3 Spectral emissivity of different refractories (in open air)/3/

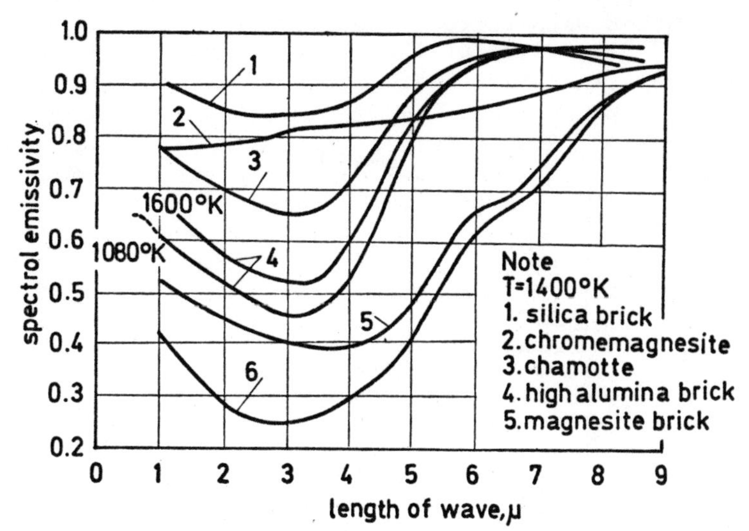

Fig. 4 Spectral emissivity of different refractories (in a neutral medium) /3/

As we see from Fig. 3, the spectral emissivity at the radiation maximum for magnesite refractories in the oxidizing medium lies in the range of 0.30-0.40 for silica brick, 0.5-0.65 for alumina-silicous refractories, and 0.76-0.83 for chrome magnesite refractories. From the given figures we may conclude that magnesite refractories as well as silica brick and alumina-silicous refractories in flame furnaces reflect a grey radiation of flame, whereas chrome magnesite refractories possess their own radiation approaching grey radiation. Consequently, the incident radiation onto the heating surface area in flame furnaces must be practically grey or approaching it, and therefore this approximation without any error will be allowed in the calculations of furnaces. N.P. Kuznetsova /3/ made detailed calculations of the radiation in conformity with a two-dimensional radiating system which consists of four zones: heating surface area, electric resistors, brickwork between them, and unoccupied brickwork. Comparison of the "grey" calculating approximation and the "quasi-grey" one is the basis of the calculations. At the "quasi-grey" approximation the whole spectrum in the range of 1.5-8 was divided into 8-10 bands 0.5-2 mm wide depending on the region of the spectrum. Within each band the emissivity value was averaged on the basis of the experimental data. The results of the calculations are given in Fig. 5 and 6 for different zone temperatures and for different spectrum characteristics. The curves (Fig. 5 and 6) demonstrate the essential difference in the value of absorbed radiation ($J_{\lambda,a}$), incident radiation (J_{λ}^{in}) and resulting radiation ($J_{\lambda,r}$) for heating surface areas as well as the value of incident radiation, which is virtually equal according to both methods of calculations.

Some difference may be found only in the region of wavelengths adjacent to the maximum radiation. It is important to note that the lack of dependence between the incident radiation and the calculating approximation is obtained at various regularities of the change of spectral emissivity of the heater from wavelength.

The reason for this is the averaging of the emissivity value of the incident radiation in the process of reradiations as well as the dominating role of the emissivity values in the narrow region of the maximum radiation with virtual equality of this value and the integral value of emissivity for the given refractory at the same temperature (Table I).

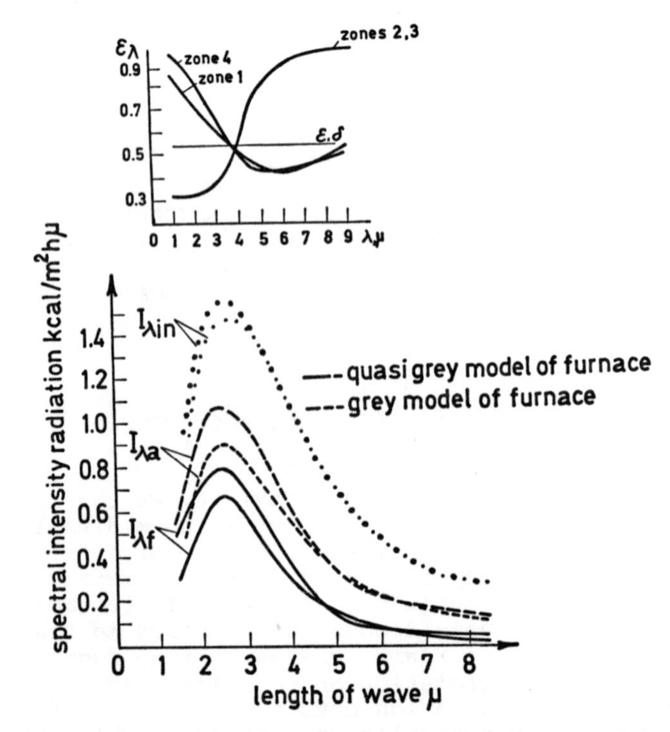

Fig. 5 Radiation intensity dependence on wavelength for zone temperatures T_1=850°K, T_2=1000°K, T_3=1050°K, T_4=1200°K /3/

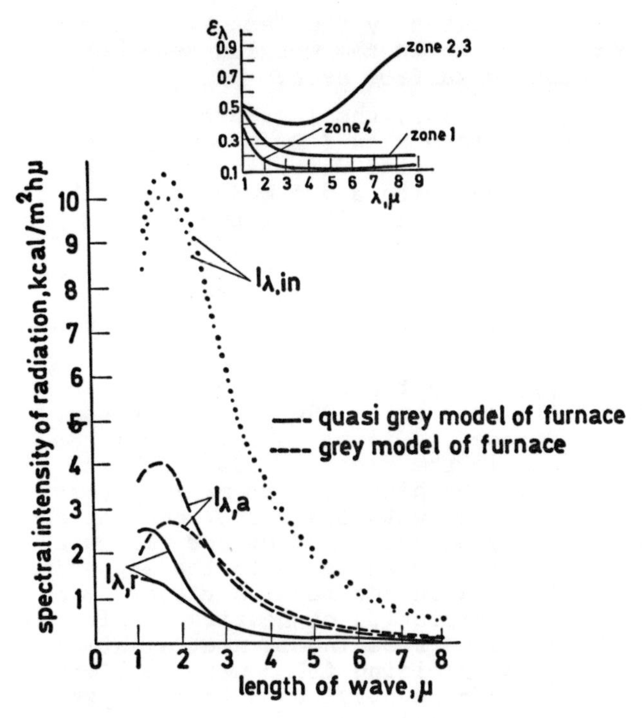

Fig. 6 Radiation intensity dependence on wavelength for zone temperatures T_1=1400°K, T_2=1550°K, T_3=1566°K, T_4= 1800°K /3/

Table 1 Comparison of integral emissivity of refractories and spectral emissivity corresponding to the maximum radiation wavelength when T=1650°K

(Neutral medium)

Refractories	Integral Emissivity	Spectral Emissivity (for Λ_{max}=1.8μ)
Silica brick	0.86	0.85
Chamotte	0.69	0.7
High-alumina brick	0.61	0.55
Magnesite brick	0.46	0.45
Chrome magnesite brick	0.80	0.81

The given results confirm the previous conclusion of the "grey" character of the incident radiation of flame furnaces. These results, however, may be applied to all the cases of incident radiation in furnaces, since this regularity if observed even under radiation conditions without an intermediate weakening medium. The calculated data (Fig. 5 and 6) show that the resulting radiation ($J_{\lambda,r}$), which unequivocally characterizes transfer by radiation, depends upon the character of the selected calculating pattern, and the results are obtained in all the calculations according to the "grey" pattern. In some cases the error may be 25-30% and more. From the standpoint of physics this is explained by the fact that some part of the incident radiation circulating in the working space of the furnace is not absorbed by the heating surface area.

As follows from equation (3), the increase of Q_M at a given temperature of the heating surface area (E_M = const) is connected with the increase of incident radiation; whereas for the permanent incident radiation it is connected with a decrease of E_M, that is with the increase of the reflecting capacity of the heating surface area. Thus, in all the cases one must increase the emissivity of the heating surface area in order to intensify heat transfer. In Fig. 7 and 8 there are three groups of experimental curves for steel with 0.5% carbon content, for some alloying steels and for copper and its alloys, respectively.

The same regularity, namely the low value of the integral emissivity (< 0.3) for temperatures below 300°C may be noted. This emissivity depends upon the surface condition (smooth, polished, rough) and upon the emissivity growth in the range of 300°-900°C up to the maximum value of the order 0.9 and higher, with maximum emissivity value being reached at 400°C for steel with low carbon content and at 500°C for carbon steel, for alloying steel and copper alloys at higher temperatures. The nature of this phenomenon is the appearance of exidizing film of various thickness on the metal surface. From the given data we can see a sharp deterioration of heating conditions of the same metal if it is heated in a reducing or neutral atmosphere. It should be noted that the thin oxidizing films were mentioned above, but not the scale with low heat conductivity, a thick layer of which deteriorates the conditions of metal heating.

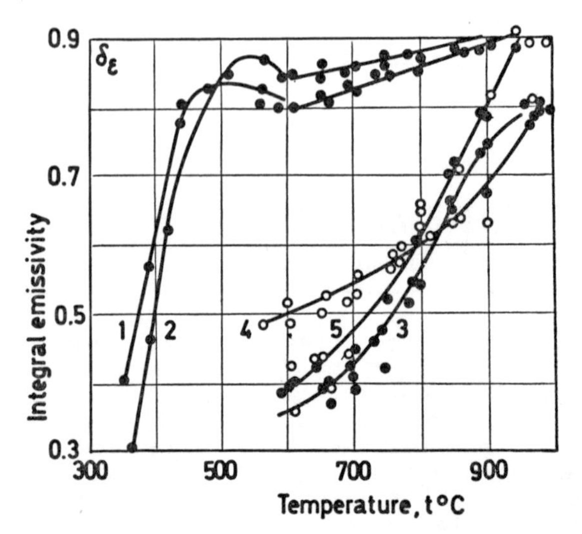

Fig. 7 Emissivity variation for preliminary ground steels at heating with the speed of 100°C/h: 1 - steel 50; 2 - 33XH3MA, 3 - 1X18H9T; 4 - 4X8B2; 5 - 481 /4/

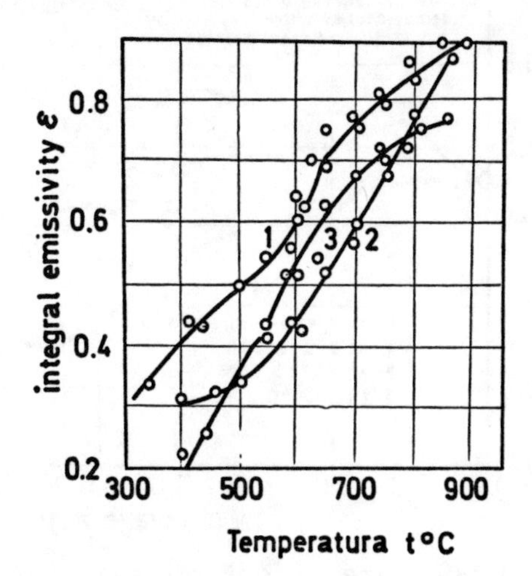

Fig. 8 Emissivity variation for preliminary ground copper and brass at heating with the speed of 100°C/h: 1 - copper; 2 - brass LOTO-1; 3 - brass 2 /4/

In Fig. 9 and 10 the experimental data for the spectral emissivity of some alloying steels and alloys are given.

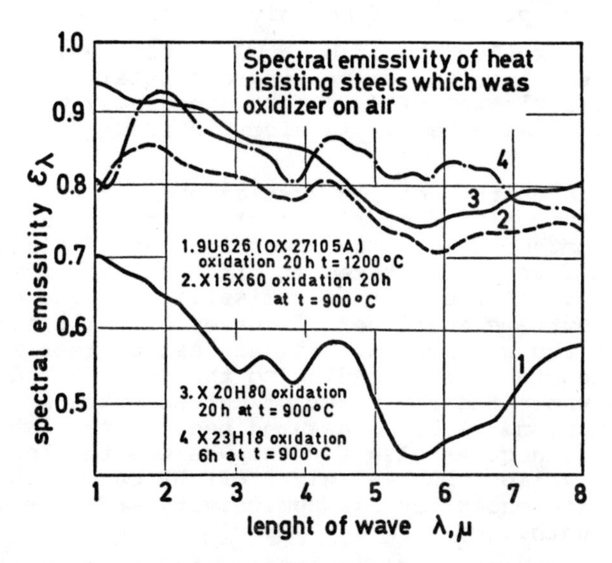

Fig. 9 Spectral emissivity of heat resistant alloys exidized in the open air. 1-3H626(OX2705A) oxidized during 20 hours at 1200°C, 2-X15H60 oxidized during 20 hours at 900°C, 3-X20H80 oxidized during 20 hours at 900°C, 4-X23H18 oxidized during 6hours at 900°C. /5/

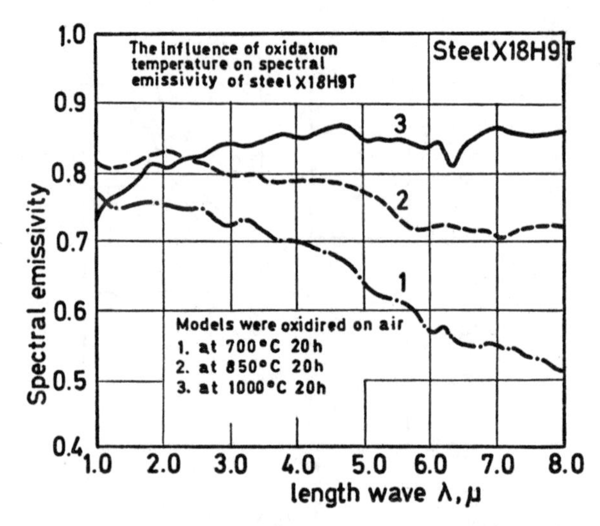

Fig. 10 Oxidation temperature effect on spectral emissivity of steel X18H9T. 1 - in the open air during 20 hours at 700°C, 2 - in the open air during 20 hours at 850°C, 3 - in the open air during 20 hours at 1800°C. /5/

From the analysis of the curves (Fig 9 and 10) we can observe some decrease of spectral emissivity, while the wavelength increases. However, in most cases the absolute value is retained in the range of 0.75-0.9.

As a rule, the increase of oxidation temperature equalizes the curves, bringing the radiation nearer to the "grey" radiation (Fig. 10). The influence of steel composition and the character of oxidizing film formation are probably of decisive importance.

On the whole the problem should be studied further. However, the right value of the integral emissivity essentially excludes serious errors in the calculation of furnaces, especially when the radiation of the given alloy as a heater in electric furnaces is taken into account.

In smelting furnaces the heating surface area is the surface of molten slag. The composition of liquid slags is very diverse and depends upon the method used. The investigation of slags (Fig. 11) in the open-hearth furnace /6/ taken at different moments of heat indicates the growth of the integral emissivity to the end of the heat as well as the extreme character of curves of the spectral emissivity dependence upon the wavelength. The highest extreme is seen in the region of wavelengths corresponding to the maximum radiation intensity, but as a whole the character of curves makes it possible to use "grey" approximation. This is assumed because the laboratory experiments referred to a quiet slag surface whereas the real slag surface is covered with various bubbles given off by carbon oxide. This factor apparently increases the slag emissivity and averaged it along the spectrum length.

Synthetic slags were studied simultaneously to clear up the influence of slag composition. It was determined that the growth increase of slag basicity ($CaO : SiO_2$) and the content of iron and manganese oxides increase the emissivity. No temperature influence was found in the range of 1500°C - 1650°C.

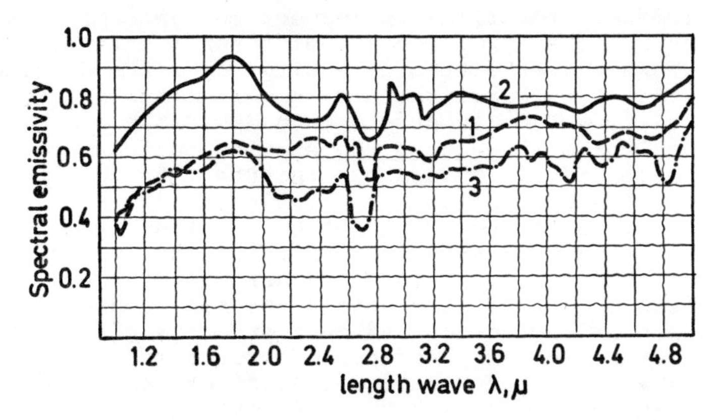

Fig. 11 Dependence of spectral emissivity of open-hearth slags taken at different periods of a heat. 1 - melting period FeO+ MnO=21, 66% CaO:SiO_2=1.39; 2 - "clean boiling" period 32.28% and 1.98%; 3 - "end of melt" period 15.16% and 2.94 /6/

The above-mentioned analysis of the influence of the spectral character of radiation of flame, brickwork and heating surface area was aimed at supporting the point of view of the lecturer on the possibility of restricting a "grey" pattern of radiation, according to the general theory of furnaces. As far as the calculations of furnaces are concerned, it is necessary to take into account that spectral radiation character of the heating surface area, but only in those cases when other elements of the calculations are carried on with comparable precision.

In the alternative coase complex calculations with regard to spectral radiation are virtually unjustified. First of all, it refers to heating surface areas with high values of the integral emissivity, since variations of spectrum values are not considerable in this case.

Let us consider the flame problem in light of the problem that work intensification of industrial furnaces is, as is know, the main trend in furnace heat engineering.

While in internal combustion engines the intensification problem leads to acceleration of fuel combustion, in boilers it leads to the improvement of heat emission from flame to heating surface area. This problem is more complicated in industrial furnaces.

In practice, in modern furnaces heat generation in flame is not a restricting moment, since the combustion process is smoothly organized; it may reach 250.10^6 μcal/m^3 per hour and even more, whereas heat emission does not exceed 800-900 μcal/m^2 per hour.

It is the organization of heat generation in the zone of technological process that is the main principle in furnace construction in order to eliminate or bring to a minimum everything hindering heat exchange. An example of such intensification is the possibility of completing the entire process of liquid iron refining in a converter within three minutes, with heat generation from the chemical energy of iron in a converter bath reaching $3 \cdot 10^6$ μcal/t per hour and the mass transfer intensity being provided by creation of a large interphase surface by slag emulsification in metal. The more heat for a technological process is generated in the zone of technological process, the less are the requirements for flame in respect to heat emission. In this case, the requirements for flame are different; that is, the flame should be uniform but of little intensity along the heating surface area. But it is often complicated by requirements for a flame as an oxygen carrier, which can create both desirable and undesirable effects of chemical influence on the heating surface area. Similar conditions are created in modern open-hearth furnaces in the "liquid" period (oxygen blowing) and in many heating furnaces during the "soaking" period. It is necessary to get a higher value of the incident radiation together with uniformity of the heating surface area in heat exchanging furnaces where heat generation in the zone of technological process is impossible for some reason or other. The organization of combustion and heat exchange plays a decisive role in this case. However, for high temperature furnaces, taking into account the fact that the organization of heat exchange from flame is closely connected with the refractory service, this problem should be considered in detail. Bearing in mind what has been said above, the division of all possible regimes of radiant heat exchange in furnaces into uniformly distributed, direct leading, and indirect leading is of great importance. The main difference between them lies in the brickwork in the course of heat exchange of flame and the heating surface area.

The uniformly distributed regime of radiant heat exchange is characterized by the equality of specific flame heat flows falling onto the heating surface area and brickwork. The similar case of heat exchange can be imagined only theoretically but in reality we can approach it. The heating surface area (especially in its volumetric arrangement) is arranged in such a manner that its individual elements are oriented unequally in respect to the incident radiation. Such are the conditions for ingots in soaking pits, kilns, etc.

Uniform use of the heating surface area is best approached in furnaces with the uniformly distributed regime.

Such a regime is created by obtaining a uniform temperature field and radiant constants in flame; the higher flame emissivity, the easier to get uniform heating of a surface. All kinds of fuels with a luminous flame are given for this regime. Fuel combustion should be performed by numerous small burners which are located in respect to outlets to provide rapid mixing with the surrounding atmosphere. From the point of view of aerodynamics there are furnaces with intensive flame circulation in the working chamber. The brickwork is subjected to service under severe conditions as it is in close contact with flame, and therefore the choice of refractories is very important.

The regimes of radiant leading heat exchange (direct or indirect) are the effects of the volumetrical character of flame radiation.

According to the theory of volumetrical radiation, the radiation of an elementary volume itself weakens in passing through other volumes of medium without any dependence of their temperatures. If there is a volume of medium with permanent temperature and radiant properties, its radiation in all directions is of the same intensity. On the other hand, if the radiant properties of the medium are permanent and we create uniform distribution of temperatures in the volume, its radiation will be more in the direction of high temperatures, since the radiation in the direction of low temperatures will be screened by cold gases.

The possible intensification of furnace work is in the eccentricity of flame temperatures, the intensification depending upon working conditions. To illustrate this, let us take a one-dimensional problem using the equation (2) when K=1. The first curve (Fig. 12) has to do with the two-layer flame, the second and the third with the three-layer flame. The temperature of flame, a middle layer, an upper layer and heating surface area are permanent and equal to 1800°C, 1700°C and 1600°C respectively. The integral emissivity for an upper flame layer and that of heating surface area are permanent and equal to 0.2 and 0.6 respectively. The curves (Fig. 12) show the temperatures of a bottom layer adjacent to heating surface area. Our attention is drawn to a sharp increase of heat transfer and heating surface area when the temperature gradient is created and to a special role of the high temperature of the bottom layer emissivity. The middle layer emissivity, which is the bottom layer in a two-layer regime, is of decisive significance in regime I. It is of almost no significance in regime II, it is very important in regime III, but it works in the opposite direction from regime I. This is explained by the diminishing role of brickwork in heat exchange.

To clear up the working conditions of brickwork let us analyze the equation for effective brickwork radiation in the one-dimensional problem.

$$Q_B = \frac{1}{\overline{E}_f(2-\overline{E}_f)} \left[Q_f^B + (Q_f^M - q_M)(1 - \overline{E}_f) - q_B \right] \qquad (7)$$

Let us simplify the problem taking into account that the flame is of the same emissivity everywhere ($\overline{E}_f = 0.5$), the brickwork does not irradiate heat into surrounding space ($q_B = 0$) and the brickwork emissivity is $E_B = 1$, so $Q_f^B \neq Q_f^M$. Thus, the given flame temperatures in the direction of heating and those of flame cannot be equal $T_f^B \neq T_f^M$.

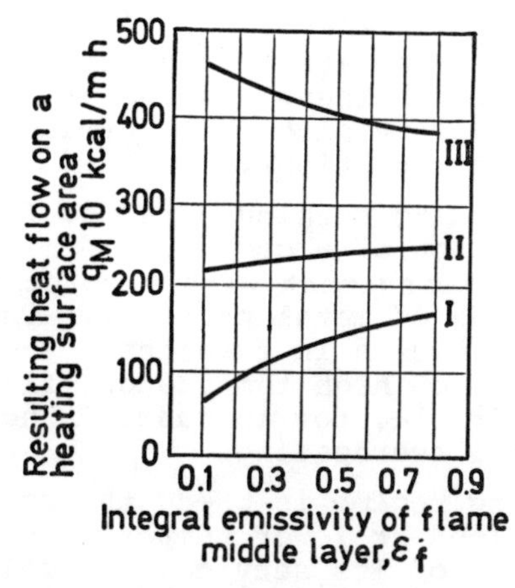

Fig. 12 Dependence of resulting thermal flow per heating surface area on emissivity of the next flame layer for three working regimes

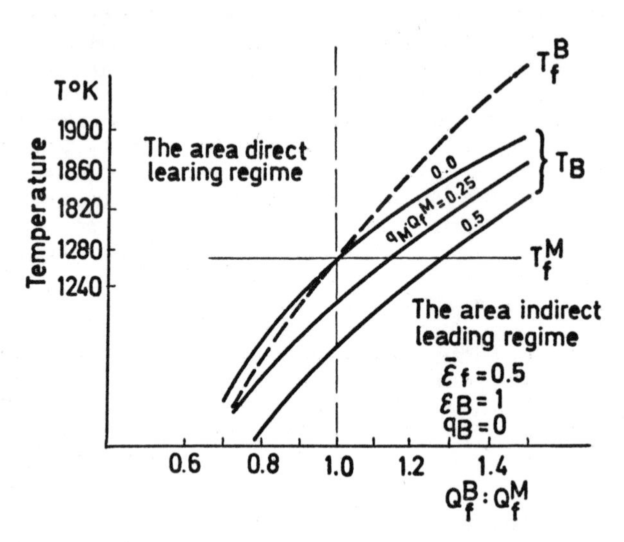

Fig. 13 Brickwork temperature for different variants of the leading regimes (direct and indirect)

The curves (Fig. 13) are given according to the equation 7, the above-mentioned facts being taken into account. The variables in this case are the ratio of radiation from flame onto the brickwork and heating surface area ($Q_f^B : Q_f^M$) and the ratio of the resulting heat flow for heating surface area and radiation from flame ($q_M : Q_f^M$), which depends upon the temperature distribution in flame and the radiant constants as follows from Fig. 12. By choice of the given values at various temperatures, working conditions of brickwork may be created. It is evident that the working conditions are better in the direct leading regime (see the left side of diagram in Fig. 13) than in the indirect leading regime. This is true in the case when there is no convective heat transfer from flame to brickwork or it is poorly expressed. The equation (7) refers to the case when the temperature regime of the indirect leading regime arises only when the brickwork is heated not by volumetric flame radiation in the furnace working chamber but in some other way.

In this case the flame temperatures near to the limit temperature of fuel combustion (infrared burners) can be attained. The intensity of heat transfer may be very high, but refractories are subjected to service under severe conditions.

This case may be applied to burning purified fuels of high efficiency with preliminary formation of a combustible mixture in furnaces with symmetrical brickwork arrangement with respect to the heating surface area (high-speed methods of metal heating). The conventional indirect leading regime is widely used in furnaces, especially in heating ones, when the direct contact of high temperature flame with the heating surface area is undesirable, for instance in heating a thin metal layer disposed to local overheating.

When the direct leading regime is used, the equivalent intensity of heat transfer is attained at lower temperatures of brickwork, i.e. under easier conditions of its service. In melting furnaces when there is no danger of a direct contact of the high temperature part of the flame with the heating surface area, high temperatures and slag aggression make the conditions of brickwork service especially severe, the direct leading regime being one of the best decisions. In some cases it is absolutely necessary when the brickwork, as an effective factor of heat transfer ($T_B \cong T_M$), is excluded and the thermal work entirely depends on the eccentricity of flame radiation. This is the case, for instance, in some moments of melting in open-hearth furnaces with silica roof. A good luminous flame created with the help of a small number of high capacity burners is characteristic of the direct leading regime as well as of the indirect leading one.

The flame should be "layless" in relation to heating surface area and retain high luminosity along its length. Therefore, gas recirculation in the working space of such furnaces should be decreased, for it promotes temperature equalizing along the volume.

With account taken of the above-mentioned, the aimlessness of the controversy becomes quite clear -- whether a luminous or a non-luminous flame is better and whether a short or a long flame is better. These questions cannot be answered synonymously; everything depends upon the character and relative arrangement of the heating surface area, properties of a fuel, refractories and some other factors.

All these factors allow us to choose the rational regime of the organization of heat transfer, the consequences of which are the other parameters of furnace thermal work.

One of the most important problems of flame as a heat generator in industrial furnaces is an increase of the flame working temperature. Let us dwell on this problem in detail.

The flame working temperatures in furnaces is something indefinite which varies in wide limits depending upon the intensity of heat generation and the conditions of heat exchange with surrounding surfaces. Therefore, the concept "furnace temperature" has no real sense. There is a complex field of temperatures which is calculated by means of balance methods. Therefore, when we speak about an increase of the flame working temperature we mean a temperature increase of flame and brickwork, i.e. elements which define the value of incident radiation. Other conditions being equal, the value of incident radiation depends upon the theoretical temperature of fuel combustion.

When generator gases with low calorific value are used as a fuel, the theoretical combustion temperature (in cold air) does not exceed $1800^{\circ}K$. For a natural gas (8500 $kcal/m^3$) it is of the value $2350^{\circ}K$. The simplest and a well-known method of increasing the theoretical combustion temperature is the use of air preheating due to the physical or chemical heat of outgoing gases. But it is always connected with complication of furnace design and furnace exploitation as well. Therefore, nowadays we do not use slightly preheated air, since the resulting economy of the fuel is not justified by the other expenses. The exception is blast furnaces using very deficit coke as a fuel. As is known, some economy in coke can be obtained in a blast furnace by increasing the balst temperature or by using enriched air if we substitute a natural gas for a part of coke. The use of preheated air is justified in furnaces for scaleless heating. HAwever, when the recuperative heating is used it is difficult to obtain sufficient air temperature, whereas the regenerative heating makes the furance design more complicated. The use of oxygen is some sort of a revolution in the steel industry, the result of which is slow disappearance of Siemens-Martin regenerative furnaces.

A new type of open-hearth furnace with oxygen blown through a bath has emerged. These furnaces work almost without fuel due to the chemical energy of iron which cannot be completely realized in converters, as is known. As far as the use of exygen in heating furnaces is concerned, there are some short-comings due to a sharp increase of oxidizing ability of combustion products and as a consequence an increase of metal losses with exides. Besides that, overburning takes place at lower temperatures. In the near future it will be possible to speak about the use of oxygen for small enrichment of air (O_2 up to 24-25%) in heating furnaces and in furnaces for scaleless heating of a continuous type, i.e. with the zone where it is possible to burn up products of incomplete combustion. In Fig. 14 the curve for combustion of a natural gas in air with O_2=20% crosses the absciss axes at 2350°K.

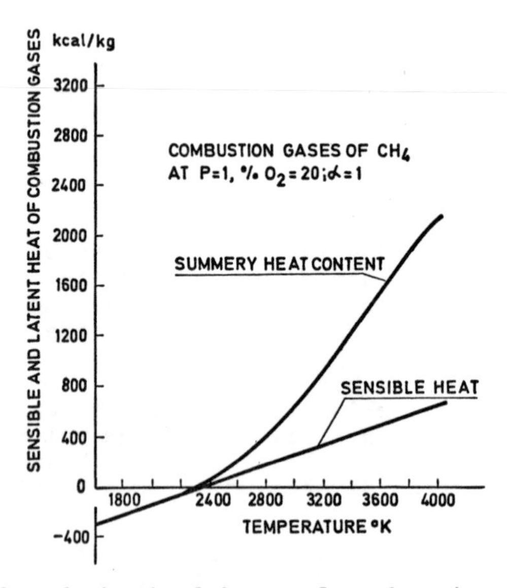

Fig. 14 Physical and chemical heat of combustion products of a natural gas in the range of temperatures up to 4000°K.

This means that it is necessary to introduce heat from another kind of energy to increase the temperature of combustion products owing to the development of endothermic dissociation processes.

At 3000°K the chemical energy share is 55.7%, whereas at 4000°K it is already 71% when a natural gas is burnt in enriched air with O_2=40%, 60% and 80%, the maximum temperature is 2701°K, 2887°K and 2987°k. In constast to a fuel, electrical energy has limitless possibilities of increasing flame temperature.

The methane combustion products about $6 \cdot 10^{-4}$ kcal/kg are used up for ionization at 4000°K, i.e. even at these temperatures the electroconductivity of combustion products is small. However, while introducing potassium salts in amount of up to 1 g/kg the energy consumption for ionization increases ~ 150 times and it becomes possible to realize a distributed electrical charge in flame. On account of the above, it is possible to increase flame temperature up to 3000°K without any essential change of oxidizing ability of combustion products.

I believe that in the nearest future this method of intensification of furnace work may become remarkable.

REFERENCES

/1/ Glinkov, M.A., "Fundamentals of the General Theory of Furnaces," Metallurgy, USSR, 1962.

/2/ Bloh, A.G., "Heat Radiation in Boilers", Energia, USSR, 1967.

/3/ Kuznetsova, N.P., Thesis 1973, USSR.

/4/ Bloh, A.G., "Principles of Heat Exchange by Radiation," Gosenergoisdat, USSR.

/5/ Zubov, V.V., Thesis 1973, USSR.

/6/ Kairov, E.A., Thesis 1969, USSR.

/7/ Krivandin, V.A., "Luminous Flame of Natural Gas," USSR, 1973.

/8/ Kluchnikov, A.D., G.P. Ivantsov, "Heat Transfer by Radiation in Fire Technical Installations," Energia, USSR, 1970.

/9/ Lisienko, V.G., Thesis 1972.

/10/ Glinkov, M.A., "General Theory of Furnaces," Journal of the Iron and Steel Institute, Vol. 206, June, 1968.

Chapter 10

PREDICTION OF RADIANT HEAT FLUX DISTRIBUTION

T. M. Lowes, H. Bartelds, M. P. Heap, S. Michelfelder and B. R. Pai

International Flame Research Foundation, Ijmuiden, Holland

Abstract

Methods for the prediction of radiative heat flux distribution are reviewed. Development work on the "Flux methods" is indicated. A four flux model for an axisymmetrical system which can accommodate non gray radiation, is developed and tested against practical and theoretical situations. The necessity of the correct formulation of the emission/attenuation characteristics of the furnace volume is demonstrated. Recommendations for further development of the flux model are made.

NOMENCLATURE

I - intensity of radiation
$I_{b\lambda}$ - black body radiation intensity at wavelength, λ
I_{λ} - intensity of radiation at wavelength, λ
$I_{r,n}$ - intensity of radiation in +ve r direction in nth gray gas
$I_{z,n}$ - intensity of radiation in +ve z direction in nth gray gas
$J_{r,n}$ - intensity of radiation in -ve r direction in nth gray gas
$J_{z,n}$ - intensity of radiation in -ve z direction in nth gray gas
K_a - gray gas attenuation coefficient
K_e - gray gas emission coefficient
$K_{g,s,n}$ - emission/attenuation coefficient for gas and solid particles in nth gray gas
$K_{\lambda a}$ - monochromatic attenuation coefficient
$K_{\lambda s}$ - monochromatic scattering coefficient
L - beam length
$T_{g,s}$ - effective gas/solid particle temperature
$a_{g,s,n}$ - weighting factor gas and solid radiation
c_s - soot concentration
$k_{g,n}$ - specific emission/attenuation coefficient in nth gray gas for CO, CO_2 and H_2O
$k_{s,n}$ - specific emission/attenuation coefficient in nth gray gas for soot
p - partial pressure
$q_{in,w}$ - incident radiative heat flux at wall
$\alpha_{g,s}$ - absorptivity of gas and solid particles
γ, ϑ, ϕ - angle defined in fig. 4
$\varepsilon_{g,s}$ - emissivity of gas and solid particles
μ - direction of radiation intensity
τ - optical depth
ω - solid angle

1. Introduction

One of the important parameters in assessing the performance of a furnace is the heat flux distribution to its thermal load. Methods based on fundamental principles are now available using numerical techniques and the digital computer, that permit predictions to be made. These prediction procedures require empirical inputs to describe turbulent transport, chemical kinetics and thermodynamic properties. The quality of the predictions essentially depends on the validity and relative importance of these factors.
An unsuccessful attempt /1/ has been made to predict the heat flux distribution in the IJmuiden furnace No. 1, from a natural gas flame produced by a simple double concentric burner. The flame aerodynamics were simple, but a high degree of unmixedness and steep temperature gradients existed in the combustion chamber. The prediction procedure used was similar to that suggested by Spalding et al /2/ with a simple turbulence model, one step chemically controlled combustion and a Milne-Eddington two flux radiation model. Although the flow and mixing were predicted with a reasonable degree of accuracy, the heat flux distribution was substantially in error. The coupled solution procedure made it difficult to assess the principle cause of the error. Work on decoupled models is therefore necessary. The paper describes the current work of the I.F.R.F. in the assessment of the complexity of the radiation model and its emission/attenuation characteristics that are necessary to predict a range of axisymmetrical situations.

2. Prediction of radiation exchange

The problems associated with the prediction of radiation heat transfer within a combustion chamber can be divided in three categories:

a) Evaluation of radiation exchange to all points within an enclosure if the temperature, absorption, emission, scattering, emissivity and reflectivity characteristics are known.
b) Evaluation of the absorption, emission and scattering characteristics from a knowledge of temperature and radiating species concentration.
c) Evaluation of temperature and radiating species concentration.

In the development of a radiation exchange model only points a) and b) need be considered, as c) can be obtained: either by iteration when coupled with a prediction procedure such as that suggested by Spalding /2/; or by iteration when coupled with information obtained from a physical model on flow and chemical heat release pattern /3/.

2.1 Evaluation of radiant heat exchange in an enclosure

Ideally the accurate solution of radiation exchange within an enclosure at equilibrium is obtained by the spectral summation of the multiple reflection, absorption and scattering of a semi infinite number of limited monochromatic radiation beams. However, several problems arise in the application of this technique, i.e.:

- spectral information does not exist for the emission, absorption or reflection from solid surfaces /4/;
- insufficient spectral information is available for the emission, absorption and scattering of radiation from soot, pulverised coal and flame gases /4/ /5/ /6/;

- even if all the spectral information was available, the computer storage capacity required would be excessive. Additionally, even if the storage capacity was available, the computational time for only one iteration of the total spectrum would be excessive.

With these physical and computational limitations in view, several techniques for the evaluation of radiation exchange have been developed. These techniques are commonly referred to as "Monte Carlo method" /7/ /8/, "Zone method" /9/ and "Flux method" /10/ /11/ /12/ /13/ /14/. All methods are fundamentally based and in their limits return to the ideal solution. The Monte Carlo method is potentially the most rigorous and flexible, however it has a statistical uncertainty unless excessive computing times are used. With a reasonable use of computing time, the Monte Carlo method has little merit over the Zone method in terms of accuracy. Although little development work has been done on either of these techniques, the Zone method has been applied successfully to the prediction of radiation heat transfer in enclosures, e.g. /15/ /16/ /17/. Acceptable predictions of radiation heat transfer to the combustion chamber walls and heat sink have been obtained, but the prediction of temperature profiles has not been good. Several contributory factors have given rise to the error, coarse zone sizing and incorrect allowance for emission/absorption of radiation in the flame, being factors associated with the radiation model. The incorrect allowance for emission/absorption and scattering within the flame is common to all modelling techniques and will be dealt with in section 2.2. Coarse zoning is a problem associated with the Zone method. In principle the Zone method is not limited to coarse zoning. However, the calculation of the direct and total exchange areas that would be necessary for its coupling with the technique suggested in section 1, renders the Zone method computationally undesirable. Flux methods, as discussed in references /10/ - /14/, replace the exact integro-differential equations on radiative heat transfer, by approximate differential ones, similar in form to the heat conduction equation. The advantages which this transformation brings are twofold: firstly the differential equations can be solved with relative ease by standard finite difference techniques and secondly the finite difference equations are of a sparse matrix kind, where a zone temperature is linked only to its immediate neighbour. Unfortunately the physical property approximation is only valid in a combustion system for monochromatic radiation in a non scattering medium. Furthermore the limited directionality of the radiation exchange (2 flux), used by Gibson and Morgan /18/ and Pai and Lowes /1/, although producing reasonable results, is open to question. Despite these physical disadvantages, the computational features of the Flux method appears to make it the most promising area for development.

2.2 Radiative properties of combustion products

The basis of the flux methods of radiation heat transfer prediction is the assumption that the radiation exchange within the combustion chamber is gray. Unfortunately the majority of combustion chambers contain both a mixture of gases, which will only absorb and emit in certain wavebands and suspended particulate material such as soot, pulverised coal or fly ash, which have a continuous radiative contribution (not necessarily gray) over the infrared spectrum. Any flux method developed must be able to account for the simultaneous occurrence of banded and continuous radiation. This is possible provided the radiation can be separated into pseudo gray spectral components. However, a knowledge of the banded and continuous radiation characteristics is required before the separation into pseudo gray spectral components can be attempted. The present state of this knowledge as it applies to engineering solutions of radiative heat flux prediction, has recently been reviewed by the I.F.R.F. /19/. They conclude that while a gray waveband approximation is the most rigorous, lack of physical information and computational efficiency makes a pseudo physical clear/gray approximation of the type suggested by Johnson /20/ more desirable. The representation equation being:

$$\varepsilon_{g,s} = \alpha_{g,s} = \sum_{n=1}^{n=m} a_{g,s,n} \quad f(T) \quad (1 - e^{-K_{g,sn,L}}) \tag{1}$$

where

$$K_{g,s,n} = k_{s,n}\, c_s + k_{CH_4,n}\, P_{CH_4} + k_{g,n}\, (P_{CO} + P_{CO_2} + P_{H_2O})$$

Using equation (1) to evaluate the emission and attenuation of unidirectional radiation, the operation equation becomes:

$$(\hat{n}\,\nabla)\, I = \sum_{n=1}^{n=m} (\hat{n}\,\nabla)\, I_n = \sum_{n=1}^{n=m} \left(-K_{g,s,n}\, I_n + a_{g,s,n}\, K_{g,s,n} \frac{\sigma T^4_{g,s}}{\pi} \right) \tag{2}$$

Consequently the emission/attenuation of unidirectional radiation through a luminous flame can be assessed, providing information on $K_{g,s,n}$ and $a_{g,s,n}$ is available. Johnson /20/ has suggested values for the constants, but both their numerical values and the validity of equation (1) and (2) still require substantial theoretical and practical evaluation.

3. Flux methods

The basis of all methods for the solution of radiation problems is the equation of radiant energy transfer, which is derived by formulating a balance on the monochromatic radiant energy passing in a specified direction through a small volume element in an emitting-absorbing-scattering medium. This equation can be written in the form /21/ :

$$(\hat{n}.\nabla)\, I_\lambda = -\, (K_{\lambda a} + K_{\lambda s})\, I_\lambda + K_{\lambda a}.I_{b\lambda} + \frac{K_{\lambda s}}{4\,\pi} \int_{\omega=4\pi} I.d\omega \tag{3}$$

For a gray medium, in which there is no scattering, the equation reduces to

$$(\hat{n}.\nabla.)\, I = -\, (K_a)\, I + K_a \frac{\sigma T^4}{\pi} \tag{4}$$

The assumptions involved in moving from equation (3) to (4) is one of the physical generalisations that are made in most flux methods of analysis, as applied to combustion systems. The assumption that $K_{\lambda a} = K_a = K_e$ and that K_a or K_e is not $f(T,\tau)$ removes the integral nature of radiation exchange from the flux formulation. However this physical generalisation is not correct for combustion systems. Therefore the computational advantages of removing the integral nature of radiation exchange, can be offset by poor predictions. Equation (2) shows an alternative formation of the intensity equation. The integral nature of radiation exchange has been removed by the pseudo gray approximation. Therefore flux models based on equation (2) can potentially account for unidirectional radiation exchange in combustion systems.

The other physical generalisation associated with the flux methods, is the division of the solid angle surrounding the point into a number (2N) of solid angles in which the intensity is assumed to be a specific function. The integro-differential equation is then replaced by 2N simultaneous differential equations in the 2N unknown average intensities of radiant energy. The problems associated with this generalisation are discussed in sections 3.1 and 3.2.

3.1 Two flux (N = 1) solutions

Two flux methods, which were originally devised for the solution of astrophysical problems on the assumption that radiative transfer is important in only one predominant direction, are all derived from the original work of Schuster /10/. The mathematical forms assumed by the two flux equations depend upon the particular method of averaging the integro-differential equation and are known variously as Schuster-Schwarzschild /22/, Schuster-Hamaker /12/ and Milne /23/ - Eddington /13/ approximations. The Schuster-Schwarzschild approximation assumes that the radiation intensity is constant within a hemisphere, while the Milne-Eddington approximation assumes it to be a function of Cos ϑ. The Schuster-Hamaker approximation is derived from the Schuster-Schwarzschild approximation to achieve mathematical simplicity and is obtained by omitting the mean obliquity factor of $\frac{1}{2}$. Siddall /24/ has demonstrated the similarity between the mathematical formulation of the two flux approximations, and compared their predictions for an emitting gray gas between parallel plates. He demonstrates that the Schuster-Schwarzschild and Milne-Eddington approximations give the best agreement with the exact solution, but they are inferior in their prediction to a coarse Zone method. Siddall in his recommendations for the further development of flux methods suggests that the prediction characteristics would be improved by increasing the number of flux directions, i.e. N = 2,3.

In the limit at $N\rightarrow\infty$ all three approximations tend towards the exact solution. However, an increase in the number of flux directions to an infinite number is obviously undesirable computationally. Consequently the approximation which has to be further developed is that which can give acceptable predictions with minimum mathematical complications and a minimum of flux directions.

3.2 Multi-flux models

Gosman and Lockwood /25/ have employed the radiation intensity formulation of the Schuster-Hamaker approximation to develop a four flux model for an axisymmetrical system. Although the predictions can be termed reasonable, the validity of the four flux model cannot be evaluated as it has been coupled with a procedure similar to that suggested in ref. /2/. The choice of the Schuster-Hamaker approximation for a four flux model can be justified on the basis of mathematical simplicity. However the lack of interaction between the axial and radial radiation, would eventually necessitate more flux directions to be employed to obtain the accuracy that could be achieved by a four flux model based on the radiation intensity formulations in Schuster-Schwarzschild or Milne-Eddington approximations. The work of Siddall /24/ partially confirms this hypothesis on the limitations of the Schuster-Hamaker approximation. Unfortunately an alternative choice of radiation intensity formulation is not apparent. Viskanta /21/ in his review of both the "Spherical Harmonics" method and the "Discrete Coordinate" method of radiation approximations, has considered that the latter method

is the more convenient mathematically. The Spherical Harmonics method achieves a solution of radiation exchange by expressing the variation of $I\ (\tau,\mu)$ in a series of Legendre Polynomials $P_n\ (\mu)$. The Milne-Eddington two flux approximation is a simplified form of this approach. Difficulties are encountered however in extending this approximation to a multiflux model, due to discontinuous functions such as $I\ (o,\mu)$ not easily being expressed by continuous functions of μ. The Discrete Coordinate method achieves a solution by dividing the radiation field $I\ (\tau,\mu)$ at depth τ into 2N streams in direction μ_i ($i = \pm 1, \pm 2 \ldots \pm N$). The Schuster-Schwarzschild two flux approximation is a simplified form of the approach. Chandrasekhar/26/ has successfully used this approach to produce a multi-flux formulation.
The available evidence /24/ indicates that the two flux Schuster-Schwarz-schild and Milne-Eddington approximations give approximately the same errors when compared to the exact solution. No experimental or theoretical evidence exists to suggest which form of Legendre Polynomial will be suitable for a combustion system. Therefore an approach similar to that of Chandrasekhar/26/ can be justified for a combustion system.

4. A four flux model for an axisymmetrical system

The discussions in section 3 have indicated that the development line of a flux model for an axisymmetrical system, should be via a four flux model based on equation (2). Furthermore the intensity should be considered constant in each quadrant, and characterised by the axial and radial intensities.
A four flux model has been produced with these characteristics by integrating equation (5) over all solid angles, i.e. $\omega = 4\pi$, or $\phi = \pi$ and γ s 2π. Equations (1) and (2) define ϕ, γ, ϑ, I_r, I_z, J_r and J_z. Equations (6) - (9) are the resulting four flux model.

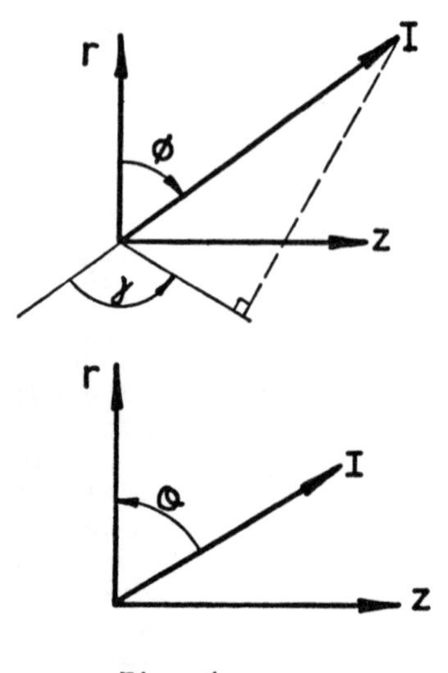

Fig. 1.
Definition of γ, ϕ, ϑ

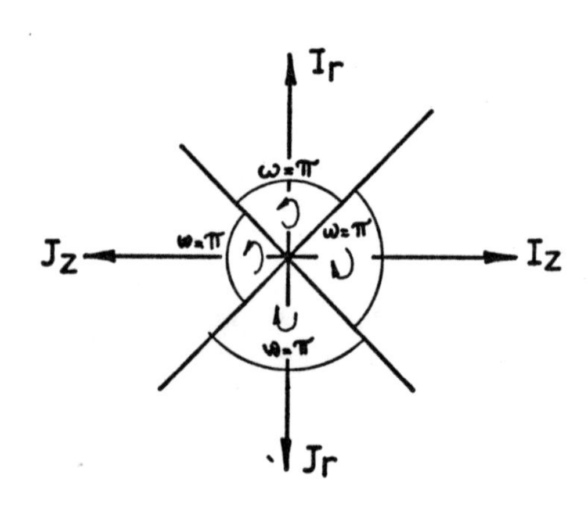

Fig. 2.
Definition of I_r, I_z, J_r, J_z

$$\int_0^{4\pi} (\bar{n}.\nabla)I.d\omega = \int_0^{4\pi} \sum_{n=1}^{n=m} (\hat{n}.\nabla)I_n.d\omega = \int_0^{4\pi} \sum_{n=1}^{n=m} \left(\cos\phi \left(\frac{\partial I_n}{\partial r}\right)_{\sigma,Z} \right.$$

$$- \frac{1}{r} \sin\phi \ \cos^2\gamma \left(\frac{\partial I_n}{\partial \phi}\right)_{\gamma,r,Z} + \frac{1}{2r} \cos\phi \ \sin\gamma \left(\frac{\partial I_n}{\partial \gamma}\right)_{\phi,r,Z}$$

$$\left. + \sin\phi \ \sin\gamma \left(\frac{\partial I_n}{\partial Z}\right)_{r,\sigma} \right).d\omega = \sum_{n=1}^{n=m} \left(-\int_0^{4\pi} K_{g,s,n}.I_n.d\omega + \int_0^{4\pi} a_{g,s,n}K_{g,s,n} \frac{\sigma T_{g,s}^4}{\pi}.d\omega \right) \quad (5)$$

$$\frac{d}{dr}\frac{\sqrt{2\pi}}{2}(I_{r,n}-J_{r,n}) + \frac{I_{r,n}-J_{r,n}}{r} . \frac{\sqrt{2\pi}}{4} = -K_{g,s,n}.\pi(I_{r,n}+J_{r,n})$$

$$+ 2a_{g,s,n}K_{g,s,n}\sigma T_{g,s}^4 \quad (6)$$

$$\frac{d}{dr}\frac{\sqrt{2\pi}}{2}(I_{r,n}+J_{r,n}) + \frac{(I_{r,n}+J_{r,n}-I_{z,n}-J_{z,n})}{r} . \frac{\sqrt{2\pi}}{4} = -K_{g,s,n}.\pi(I_{r,n}-J_{r,n}) \quad (7)$$

$$\frac{d}{dZ}\frac{\sqrt{2\pi}}{2}(I_{Z,n}-J_{Z,n}) + \frac{I_{r,n}-J_{r,n}}{r} . \frac{\sqrt{2\pi}}{4} = -K_{g,s,n}.\pi(I_{Z,n}+J_{Z,n})$$

$$+ 2a_{g,s,n}K_{g,s,n}\sigma T_{g,s}^4 \quad (8)$$

$$\frac{d}{dZ}\frac{\sqrt{2\pi}}{2}(I_{Z,n}+J_{Z,n}) = -K_{g,s,n}.\pi(I_{Z,n}-J_{Z,n}) \quad (9)$$

The radial flux is equal to $q_R = \sum_{n=1}^{n=m} \frac{\sqrt{2\pi}}{2}(I_{r,n}-J_{r,n})$

with $I_{r,n} = J_{r,n}$ when $r = 0$.

The axial flux is equal to $q_Z = \sum_{n=1}^{n=m} \frac{\sqrt{2\pi}}{2}(I_{Z,n}-J_{Z,n})$

The incoming flux at the wall (r=R) is equal to

$$q_{in,w} = \sum_{n=1}^{n=m} \frac{\sqrt{2}}{2} \cdot \pi \cdot I_{r,n} + \left(\frac{1 - \frac{\sqrt{2}}{2}}{2}\right)\pi \cdot (I_{z,n} + J_{z,n})$$

The potential superiority of this four flux model over that used by Gosman and Lockwood is demonstrated by equations (7) and (8), in which the axial and radial fluxes are coupled. Equations (6) - (9) can be solved conveniently by a finite difference technique to give information on the radiation heat flux distribution within a combustion chamber. A programme with a central differencing procedure has been developed to predict this radiation heat flux distribution. The programme requires as inputs the temperature field, wall boundary conditions $a_{g,s,n}$ and $k_{g,s,n}$, p_g, c_s or $K_{g,s,n}$. The programme can be applied to both gas and oil flames, providing information on the emission/attenuation coefficient characteristics is available.

5. Testing of the heat flux model

In the development of any model for radiation heat transfer, its eventual testing produces difficulties due to:
a) Exact solutions for realistic practical situations not being available.
b) Radiation measurements on practical situations not being available.
c) If b) is available, suitable emission/attenuation laws may not.

Detailed information on b) is available at the I.F.R.F. for natural gas and fuel oil flames /27/, therefore the necessity for a) is part removed. However all flames have complex emission/attenuation characteristics, therefore the sparse matrix of the four flux model cannot be tested if the emission/attenuation laws are not known. The validity of this statement can be seen by reference to fig. 3, where a Milne-Eddington two flux approximation has been used to predict the radiation incident on the wall for the flame situation referred to in ref. /1/. The measured gas concentrations and temperatures were used (with appropriate data reduction and interpretation and extrapolation procedures) as inputs to a Milne-Eddington two flux approximation. For a range of emission/attenuation formulations, these were:

a) A gray gas situation where K_e=K_a throughout the furnace, and assigned a value K=0.2 m^{-1} based on the mean beam length for the furnace.
b) A gray gas situation where K_e=K_a, but can vary throughout the furnace depending on local concentration, temperature and beam length.
c) K_e defined as in (b), but K_a being specified by wall temperature and not local temperature.
d) A one clear/one gray gas approximation, with CO_2 and H_2O emissivity fitted at path lengths of 0.1 to 6 m.
e) A one clear/two gray gas approximation, using the constants suggested by Johnson /20/.

All these formulations have been suggested in the literature. All the emission/attenuation formulations predict the correct trends with axial distance. However, the predictions vary by over 100 %, depending on the emission/attenuation relationship chosen. This not only emphasises the difficulties of testing the sparse matrix of the four flux model, but indicates the necessity of the correct formulation of emission/attenuation

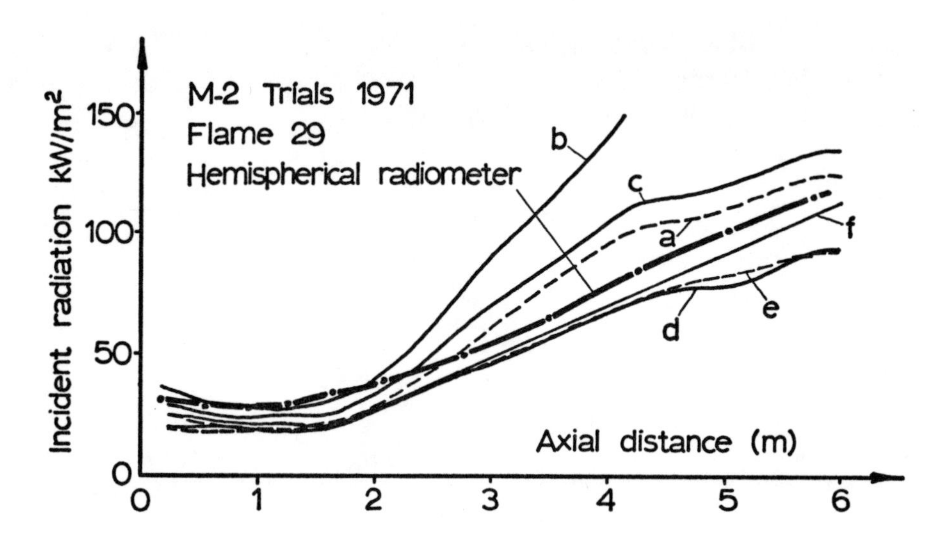

Fig. 3.

Prediction of incident radiant flux using a two flux model, with different attenuation coefficient formulations.

characteristics if realistic predictions are to be made. The better predictions came from relationships a) and e), the worst from relationship b) (relationship b) was used with the four flux model developed in ref. /25/). Therefore a) and e) can be used to provide some information on the validity of the sparse matrix of the four flux model. Fig. 4 shows four flux model predictions of the same situation as was shown in fig. 3, using three emission/attenuation formulations, two gray gas (K=0.1, and K=0.2) and the clear/gray formation of ref. /20/. The predictions are better than those shown in fig. 3, particularly within the first four metres. After this position the shape of the curve is incorrect. The reason for this error is difficult to substantiate due to:

- interaction of axial and radial radiation;
- representation of boundary conditions and emission/attenuation characteristics.

The latter problem has been overcome by a Zone method prediction of the same situation, which is shown in fig. 5. The prediction of the Zone method compares well with the measurements, throughout the length of the combustion chamber. If predictions with K = 0.15 could be justified, a prediction to ± 5 % could be achieved at any position. However, it should be noted that the clear/gray formulation always underpredicts the incident radiative flux. The Zone method can be considered to be an exact solution of the situation providing the temperature and concentration have been adequately averaged and the boundary conditions and emission/attenuation characteristics have been formulated correctly. The inputs of boundary conditions and emission/attenuation characteristics are the same for the four flux model and Zone method, therefore the only potential differences lie in the validity of the average temperatures produced from the interpolation and extrapolation procedures. Comparison of fig. 4 and fig. 5 can therefore indicate that the errors produced by the sparse matrix of the four flux model are not as high as would be expected from only a study of fig. 4. However, while the matrix

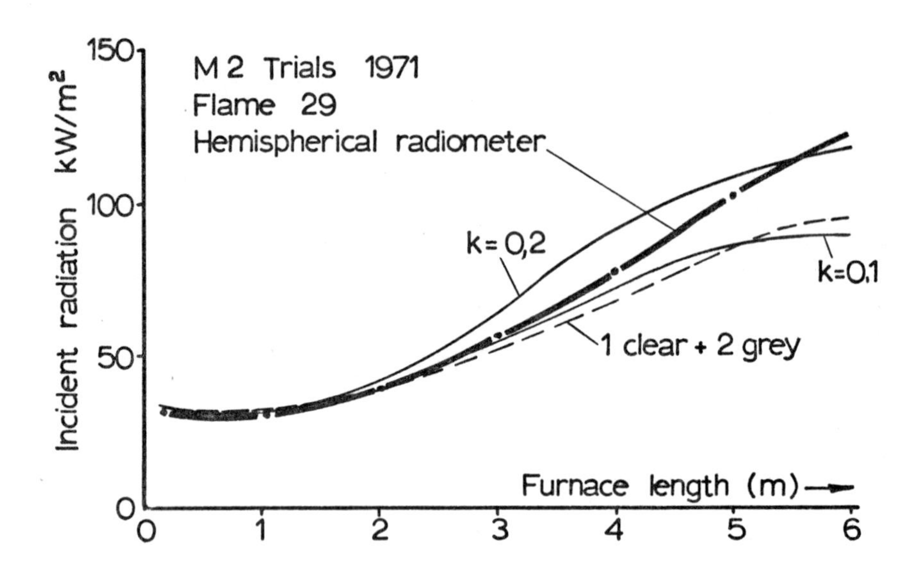

Fig. 4.

Prediction of incident radiant flux using the four flux model with different attenuation coefficient formulations.

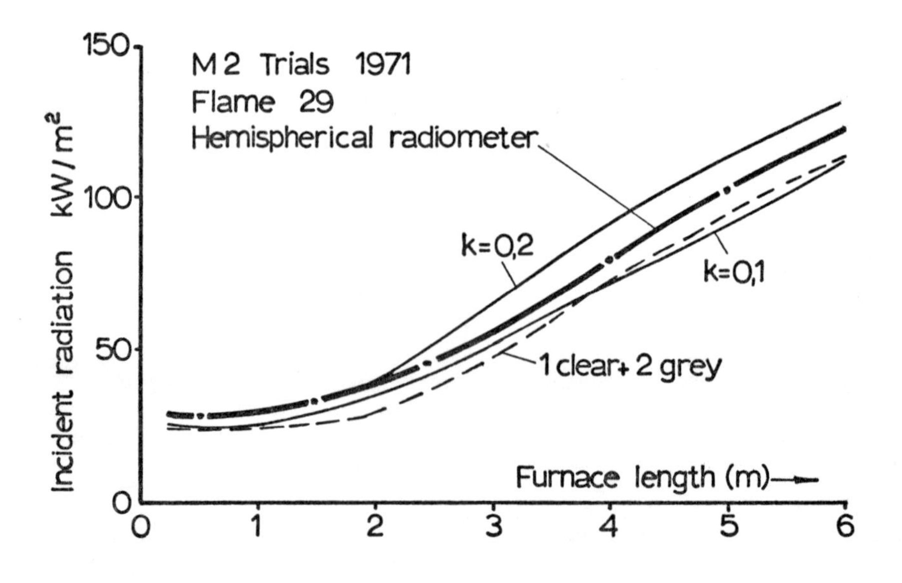

Fig. 5.

Prediction of incident radiant flux using the Zone method with different attenuation coefficient formulations.

of the four flux model allows for axial radiation exchange to a certain extent, its differences from the Zone method indicate that axial radiation is not fully accommodated. Therefore the effect of the cold and hot end walls cannot be adequate, the resulting errors are the overpredictions near the burner and underpredictions near the chimney. In order to correct this situation, time routes are possible:

- introduction of more flux directions;
- modification of the intensity distribution within the quadrant.

The former possibility is not desirable due to mathematical and computational problems associated with the representation of intensities not in the purely axial or radial directions. The latter possibility is the most promising route, however the choice of intensity distribution is open to question. Similar curves to a) and e) in fig. 3 have been produced using the Schuster-Schwarzschild two flux approximation. Therefore it is reasonable to conclude that if the four flux model was modified to allow the intensity to vary as a function of Cos σ within its quadrant, it would not greatly improve its predictive capacity. Another formulation is therefore required. This formulation has been assessed /19/, by carrying out a series of theoretical tests. The Zone method has been used as an absolute standard, by chosing the temperature profiles to coincide with the zone boundaries and postulating gray gas situations. The conclusions show that in many circumstances the four flux model represented by equations (6) - (9) agree with the Zone method to $\pm$ 5 % (K=0.1, 0.2 and 0.5), however in certain circumstances when K=0.1 errors of up to 35 % can occur. The situations where high errors occurred were somewhat unrealistic practical situations. However these errors could be substantially reduced if sudden transition of intensity from I_r to I_Z at a quadrant boundary could be removed. A relationship such as $I = I_r \cos^2\alpha + I_Z \sin^2 \alpha$, will produce this desired effect. Work is now in progress at the I.F.R.F. te re-integrate equation (5) with this distribution of intensity.

5. Conclusions

A four flux model for gas and fuel oil flames, in an axisymmetrical system has been justified, developed and tested.
The testing indicates that the model gives a prediction of radiation heat flux distribution to better than 10 % in many circumstances. However, errors can arise due to its sparse matrix and the distribution of intensity with the matrix.

Appropriate recommendations have been made concerning the further development of the model, while maintaining its sparse matrix.

Although the model developed gives acceptable predictions in many circumstances, these predictions are only as good as the representation of the emission/attenuation characteristics of the combustion chamber volume. Differences in predictions of over 100 % have been obtained by using the range of emission/attenuation formulations for non luminous radiation, that has been suggested in the literature.

REFERENCES

/1/ Pai, B.R. and T.M. Lowes: "The prediction of flow, mixing and heat transfer in the IJmuiden furnace". I.F.R.F., Doc.nr. G 02/a/22 (1972)
/2/ Spalding, D.B. et al: "Heat and mass transfer in recirculating flows". Academic Press (1969)
/3/ Hottel, H.C. and A.F. Sarofim: Int. J. Heat Mass Transfer, Vol. 8 (1965), 1153
/4/ McAdams, W.H.: "Heat transmission". New York, MacGraw Hill (1954)
/5/ Sparrow, E.M. and R.D. Cess: "Radiation heat transfer". Belmont, Brooks/Cole Publishing Co (1970)
/6/ Lowes, T.M. and M.P. Heap: "Emission/attenuation coefficients of luminous radiation". I.F.R.F., Proceedings of 2nd Members' Conference (1971)
/7/ Howell, J.R.: "Application of Monte-Carlo to heat transfer problems". Advances in Heat Transfer, New York, Academic Press (1968)
/8/ Steward, F.R. and P. Cannon: Int. J. Heat Mass Transfer, Vol. 14 (1971), 245
/9/ Hottel, H.C. and A.F. Sarofim: "Radiative transfer". New York, MacGraw Hill (1967)
/10/ Schuster, A.: Astrophys. J., Vol. 22 (1905), 1
/11/ Milne, E.A.: Handbuch der Astrophysik, Vol. 3 (1930), 1
/12/ Hamaker, H.C.: Philips Research Reports 2, 55, 103, 112, 420 (1947)
/13/ Eddington, A.S.: "The internal constitution of stars". New York, Dover Press (1954)
/14/ Patankar, S.V. and D.B. Spalding: Imperial College Rep. C/TN/A/8 (1972)
/15/ Fitzgerald, F. and A.T. Sheridan: "The heating of slab in a furnace". Conference on Mathematical Modelling in Metallurgical Process Development. Iron and Steel Institute (1970)
/16/ Latch, R.: Gas Wärme Int., Vol. 20 (1971) : 3, 106
/17/ Wall, T.F.: "The calculation of temperature and radiative transfer in an industrial pulverised fuel furnace". I.F.R.F., Italian Flame Day, Pisa (1973)
/18/ Gibson, M.M. and B.A. Morgan: J. Inst. Fuel, Vol. 23 (1970), 517
/19/ Lowes, T.M. et al: "The prediction of radiant heat transfer in axi-symmetrical systems". I.F.R.F., Doc.nr. G 02/a/25 (1973)
/20/ Johnson, T.R. and J.M. Beér: "Comparison of calculated temperatures and heat flux distributions with measurements in the IJmuiden furnace. Institute of Fuel and B.F.R.C., 4th Flames and Industry Symposium (1972)
/21/ Viskanta, R.: Advances in Heat Transfer, Vol. 3 (1966), 175
/22/ Schwarzschild, H.: Nachr. Akad. Wiss., Göttingen, Math. Physik-Klasse 1 (1906); Sitzber. Preuss. Akad. Wiss. Physik-Math. Klasse 1183 (1941)
/23/ Milne, E.A.: Handbuch der Astrophysik, Vol. 3 (1930), 1
/24/ Siddall, R.G.: "Flux methods for the analysis of radiant heat transfer". Institute of Fuel and B.F.R.C., 4th Flames and Industry Symposium (1972)
/25/ Gosman, A.D. and E.C. Lockwood: Imperial College Rep. CCK/TN/A9 (1972)
/26/ Chandrasekhar, S.: "Radiative transfer". New York, Dover Publications (1960)
/27/ Michelfelder, S. and T.M. Lowes: "Report on M-2 trials". I.F.R.F. Doc.nr. F 36/a/4 (1973)

Chapter 11

THE APPLICATION OF FLUX METHODS TO PREDICITION OF THE BEHAVIOR OF A PROCESS GAS HEATER

Richard G. Siddall and Nevin Selcuk

Department of Chemcial Engineering and Fuel Technology, University of Sheffield, Sheffield, England

Abstract

The prediction of temperatures and radiative heat transfer by means of simple flux methods is considered. To test the accuracy of two flux approximations, the exact solution of a parallel plate problem is compared with approximate analytical and iterative numerical solutions. An extended two flux method is finally applied to prediction of the behaviour of a process gas heater.

NOMENCLATURE

a	:	coefficient in the generalised two flux equation	
A_G	:	cross-sectional area of the furnace gas space normal to main flow direction	(m^2)
b	:	coefficient in the generalised two flux equation	
c	:	mass specific heat of gas	$(Jkg^{-1}K^{-1})$
D	:	overall thermal resistance between the outer surface of a tube and the process gas based on unit area of the outer surface	(Km^2W^{-1})
e_b	:	black emissive power of the outer surface of a tube $(=\sigma T_s^4)$	(Wm^{-2})
e_{bs}	:	black emissive power of plate surfaces	(Wm^{-2})
E	:	dimensionless black body emissive power	
E_b	:	black emissive power of the medium $(=\sigma T^4)$	(Wm^{-2})
H	:	rate of energy generation per unit volume	(Wm^{-3})
K_a	:	volumetric absorption coefficient of the medium	(m^{-1})
K_s	:	volumetric scattering coefficient of the medium	(m^{-1})
L	:	distance between parallel plates or height of process gas heater	(m)
m	:	number of grey gas components used to represent a real gas	
$\dot{m}$	:	mass flow rate	(kgs^{-1})
n	:	number of intervals in x direction	
q	:	flux density of a gas based on unit cross sectional area of furnace gas space	(Wm^{-2})
q^+	:	radiant flux density in the direction of increasing x	(Wm^{-2})
q^-	:	radiant flux density in the direction of decreasing x	(Wm^{-2})
Q^+	:	dimensionless radiant flux density in the direction of increasing X	
Q^-	:	dimensionless radiant flux density in the direction of decreasing X	
$2R$	:	area of sidewall refractory surface per unit volume of furnace gas space	(m^{-1})
$2S$	:	area of tube surface per unit volume of furnace gas space	(m^{-1})
T	:	absolute temperature of the medium	(K)
T_o	:	datum temperature for measuring quantities of heat	(K)
x	:	distance measured in the predominant direction	(m)
X	:	dimensionless distance $(= x/L)$	
ε	:	surface emissivity	
σ	:	Stefan-Boltzman constant	$(Wm^{-2}K^{-4})$
τ	:	optical thickness of gas layer of actual thickness x	
τ_o	:	optical distance between two plates $(= K_aL)$	

ω_o : albedo for single scattering = $K_s/(K_a+K_s)$

Subscripts
G : furnace gas
i : number of pass
p : process gas
R : refractory
S : tube surface

1. Introduction

In recent years, industrial combustion chambers and process fluid heaters have been designed for progressively increasing throughputs and heat loadings. To avoid the resulting possibility of overheating the tube surfaces and the enclosed process fluid, the maximum local heat flux and tube surface temperature must be kept below predetermined limits. Since different furnaces and burner arrangements will lead to considerable variations in the locations and magnitudes of the maximum tube surface heat flux and temperature, it is essential for design purposes to use an accurate method for predicting detailed radiative heat flux and temperature distributions.

During the past decade the detailed evaluation of radiant heat transfer in enclosures containing an emitting-absorbing medium has been based principally on the zone method of analysis (1), (2), (3) devised by Hottel and his co-workers. This method relies upon the subdivision of all bounding surfaces and the gas space into small regions (zones) in each of which the temperature and all other properties are assumed to be uniform. In carrying out the solution by the zone method it is assumed that compositions, velocities of flow, and all physical properties other than temperature and radiant energy flux are known in advance for each zone. Even with this simplification, the iterative solution of the algebraic equations to give the unknown zone temperatures involves considerable numerical work. For cases where it is required to predict flow patterns and the progress of reaction simultaneously with temperature and radiative heat flux distributions, flux methods for representing the radiative field possess several advantages. Such methods replace the exact integro-differential equations for radiative transfer at a point by a set of approximate differential equations which are similar in form to those representing the conservation of mass, momentum and energy, and can therefore be conveniently solved simultaneously with them. Although more approximate than the zone method of analysis, they usually lead to results of acceptable accuracy for engineering purposes with considerably reduced computational effort.

This paper considers various flux methods, and describes their application to the prediction of temperature and radiative heat flux distributions in a process gas heater.

2. Flux methods for the solution of radiant heat transfer problems

The mathematical formulation of the radiant energy balance in a specified direction on a small radiatively grey volume element in thermodynamic equilibrium within an enclosure results in an integro-differential equation for the unknown intensity of radiation (4). Solution of this equation is complicated by the fact that the intensity is a function of both direction and position. The basis of all flux methods is to simplify the problem by making an approximate allowance for the variation of intensity with direction by subdividing the total solid angle at any point into a number of smaller solid angles, in each of which the intensity is assumed to be independent of direction. The integro-differential equation is then averaged over each of the smaller solid angles in turn, leading to a set of differential equations for the unknown intensities. Simultaneous solution of these equations in conjunction with a total energy balance and appropriate boundary conditions leads to values of intensity for each solid angle at all points within the enclosure. The accuracy of solution of any problem can be improved at the expense of increased computational effort by using a finer subdivision of the angle range.

For systems in which the radiation field varies rapidly in one predominant direction a two flux approximation may be used. A considerable amount of published work on the use of flux methods is based on the application of two flux solutions to essentially one-dimensional transfer situations (5), (6), (7), (8). The basis of such approximate two flux solution methods is to divide the solid angle surrounding any point into two hemispheres whose common diametric plane is normal to the predominant direction, and to assume uniform but different intensities of radiation over each hemisphere. By suitable averaging of the integro-differential equation over the two hemispheres, two ordinary differential equations are produced giving the variation of the two intensities with position. Two methods of this type, which are produced by different forms of averaging, are the Milne-Eddington (9), (10) and the Schuster-Schwarzschild (11) approximations. A third method of a similar type was proposed by Hamaker (8) on the assumption that all radiation is parallel to the predominant direction (rather than isotropic over two hemispheres). All three methods, which are based on original work by Schuster (12) and are reviewed in a recent publication (13), lead to ordinary differential equations in terms of the forward and backward dimensionless radiant flux densities in the predominant direction of the form

$$\frac{1}{\tau_o} \cdot \frac{dQ^+}{dX} = 2aE - (a+b)\,Q^+ + (b-a)\,Q^- \qquad (1)$$

$$-\frac{1}{\tau_o} \cdot \frac{dQ^-}{dX} = 2aE + (b-a)\,Q^+ - (a+b)\,Q^- \qquad (2)$$

where $Q^+ \left(= \frac{q^+ - e_{bs}}{H/2K_a} \right)$ is the dimensionless forward flux density, $Q^- \left(= \frac{q^- - e_{bs}}{H/2K_a} \right)$ is the dimensionless backward flux density and $E \left(= \frac{E_b - e_{bs}}{H/2K_a} \right)$ is the dimensionless black body emissive power of the enclosed medium. Values of coefficients 'a' and 'b' are given in Table 1. Each of the equations is a statement of the fact that in travelling a short distance in either the positive or negative X direction, the flux density of radiation in that direction is increased by emission of radiation and by forward scattering and is reduced by absorption and back scattering.

Table 1. Values of coefficients in the generalised two flux equation.

	Milne-Eddington	Schuster-Schwarzschild	Schuster-Hamaker
a	$1 - \omega_o$	$1 - \omega_o$	$\frac{1 - \omega_o}{2}$
b	$\frac{3}{4}$	1	$\frac{1 + \omega_o}{2}$

3. Solution of a problem by use of two flux methods

Consideration is given to a system of two large parallel black plates at the same temperature separated by a grey emitting-absorbing medium in which heat generation occurs at a uniform rate. Heat transfer by conduction and convection is taken to be negligible. An exact solution of this problem has been presented by Usiskin and Sparrow (14).

As the dimensionless emissive power occurring in equations (1) and (2) is unknown, an additional relationship involving this quantity must be found before complete solution of the problem is possible. This is provided by writing a total energy balance for an elementary volume of the enclosed medium, which takes the form

$$\frac{d}{dX}\left(Q^+ - Q^-\right) = 2\tau_o \tag{3}$$

Mathematical formulation of the problem is completed by the radiation boundary conditions at the plate surfaces, which take the simple forms

$$Q^+ = 0 \quad \text{at} \quad X = 0 \tag{4}$$

and

$$Q^- = 0 \quad \text{at} \quad X = 1 \tag{5}$$

It should be noted that equations (1), (2) and (3) may be combined to give a relationship between the variables which may be used in place of equation (3) for the purposes of solution of the problem

$$E = \frac{1}{2a}\left[1 + a\left(Q^+ + Q^-\right)\right] \tag{6}$$

3.1 Approximate analytical solution

As equations (1), (2) and (6) are linear in Q^+, Q^- and E, analytical solution is possible to give the distribution of the dimensionless emissive power between the plates as

$$E = b\,\tau_o^{\,2}\left(X - X^2\right) + \frac{1}{2}\left(\frac{1}{a} + \tau_o\right) \tag{7}$$

3.2 Numerical solution of the differential equations

The solution of the approximate differential equations associated with the flux methods may be regarded as a boundary value problem, as the fluxes occurring in the differential equations are required to satisfy boundary conditions at both ends of the range of integration. However, for the purposes of numerical solution, such a problem may also be conveniently treated as a pseudo initial value problem by combining step-by-step integration of the differential equations with an iterative procedure. This may be illustrated by considering a scheme for the numerical solution of equations (1), (2) and (6). The first step is to subdivide the range of integration ($0 \leq X \leq 1$) into n equal intervals of length ΔX, the (n+1) end points of the intervals being treated as grid points. Initial guesses are made for Q^- at all grid points, and equation (1) is integrated step-by-step in the positive X direction starting from the known zero value for Q^+ at $X = 0$. The values of E occurring in the equation are evaluated simultaneously with Q^+ at each grid point in turn by using the guessed Q^- value and the calculated Q^+ value in equation (6). On completing the integration (i.e. calculating Q^+ at $X = 1$), equation (2) is integrated in the negative X direction starting from the known zero value of Q^- at $X = 1$, utilising the newly calculated values of Q^+. On reaching $X = 0$ the calculated value of Q^- is compared with the original guess. If these do not agree within specified limits the whole integration cycle is repeated, starting with the newly calculated Q^- values in place of the original guesses. Iteration continues until the desired agreement is obtained between successively calculated Q^- values at $X = 0$.

Before utilising two flux methods widely for the solution of radiation problems, two important questions must be answered: (i) which set of coefficients (Milne-Eddington, Schuster-Schwarzschild or Schuster-Hamaker) yields approximate differential equations which most closely represent the actual radiation field, and (ii) in solving the approximate differential equations, which method of numerical integration will produce results which give best agreement with their exact solution (i.e. which method gives the highest degree of convergence).

The first question can be answered by solving a problem for which analytical solution of the approximate differential equations is possible and for which an exact solution exists. The second question can be answered by comparing analytical and numerical solutions of the differential equations for the same problem.

3.3 Results of calculations and conclusions

Approximate analytical solutions of the differential equations evaluated by use of equation (7) with the Milne-Eddington, Schuster-Schwarzschild and Schuster-Hamaker coefficients are compared with the exact solutions derived by Usiskin and Sparrow (14) for $\tau_o = 2$ in Fig.1. It can be seen that the Schuster-Schwarzschild coefficients lead to the greatest accuracy for the problem under consideration.

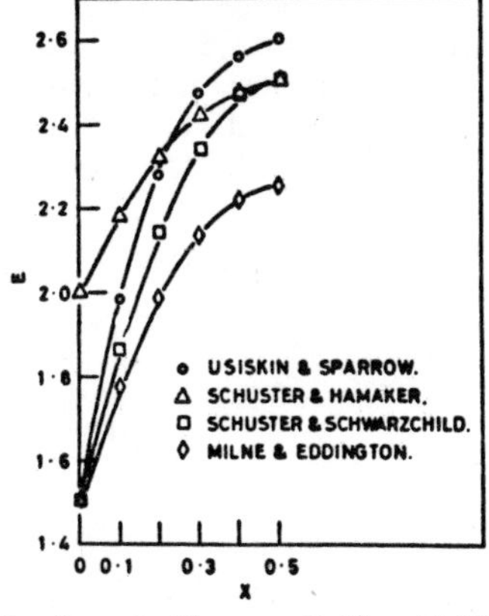

Fig.1. Distribution of emissive power in a heat generating and radiating gas between parallel black plates ($\tau_o = 2$)

Results of numerical solutions of the differential equations obtained by the use of the iterative scheme described in section (3.2) and a two-point and a combined two and three-point predictor-corrector method for the numerical integrations (15) have been compared with the corresponding approximate analytical solutions. It was found that the use of the combined two-point and three-point predictor-corrector methods produced significantly greater accuracy than the use of the simpler two-point method.

Additional calculations to test the effect of grid size have shown that halving the grid size reduces the percentage errors by an order of magnitude, at the expense of doubling the computing time.

It appears from the results of the calculations for the particular problem investigated that the use of the Schuster-Schwarzschild coefficients leads to greatest accuracy of solution, and that, when necessary, the numerical solution of the ordinary differential equations representing the radiation field can be conveniently and accurately carried out by an iterative scheme coupled with a combined two and three-point predictor-corrector method of numerical integration. Such a computing scheme has the advantages of ease of computer programming, low computing time, rapidity and accuracy of convergence, and shows no signs of instability. In any problem the degree of convergence can be improved by reduction of the grid size but only at the expense of increased computing time.

4. A two flux method for two dimensional radiation

Two flux methods are based on the assumption that the transfer of radiant energy takes place predominantly in one direction. However, in many practical situations, significant radiative transfer occurs at right angles to the predominant direction, and neglect of this effect can lead to serious errors.

Roesler (16) has suggested a simple modification of the Schuster-Schwarzschild two flux method which makes an approximate allowance for two dimensional transfer without the introduction of two additional and distinct fluxes, by assuming that the q^+ and q^- flux densities in the main flux direction are associated with corresponding flux densities of αq^+ and αq^- respectively at right angles to the main flux direction. In order that all balances should be correct when every point of the system is at the same temperature, it is found that the value of α must be taken to be $\frac{1}{2}$. The two dimensional radiative field is then finally represented by two ordinary differential equations in the two unknown flux densities q^+ and q^-.

4.1 Application of the method to a process gas heater

The Roesler two flux method has been applied in the present work to predict the behaviour of a multipass process gas heater in which a furnace gas exchanges radiative energy with bounding refractory surfaces and tubes containing the process fluid. The burners located at the bottom of the furnace, produce the heat supplying furnace gas which exits at the top. Tubes carrying the process gas are arranged vertically along the walls and each has eight passes, the fluid entering and leaving at the top of the heater. A simple representation of the radiative situation is shown in Fig.2.

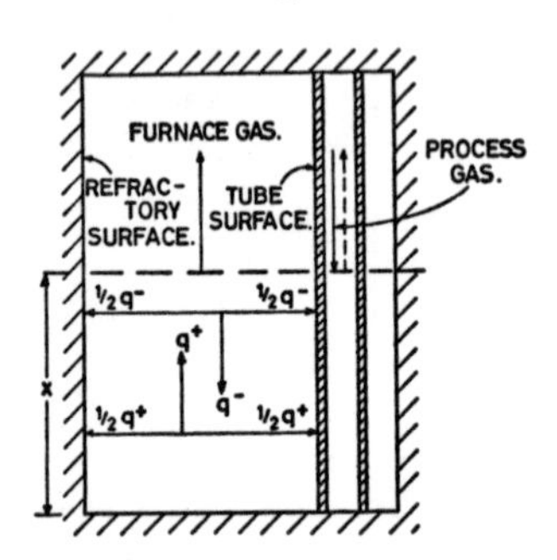

Fig.2. Physical set-up for the extended two flux method

To simplify the formulation and solution of the radiant and total energy balances, the following assumptions have been made: (i) refractory surfaces are grey and radiatively adiabatic, (ii) in the furnace gas space the effects of conductive and convective heat transfer are negligible, (iii) within the process gas tubes heat is transferred from the inner wall to the fluid by convection alone, (iv) in both the furnace gas and process gas there is no variation of properties in any cross section normal to the main flow direction, (v) the heater operates adiabatically under steady state conditions, (vi) the furnace gas is radiatively grey and non-scattering, (vii) the specific heats of the furnace and process gases are independent of temperature, and remain constant in their passage through the heater, (viii)instantaneous combustion of the furnace gas occurs at the bottom of the heater.

With the assumptions the radiant energy balance on the flux in the positive x direction becomes

$$\frac{dq^+}{dx} = 2K_a E_b + \frac{\varepsilon_S S}{8} \sum_{i=1}^{8} (e_b)_i - \left(2K_a + \frac{\varepsilon_R R}{2} + \varepsilon_S S\right) q^+ + \frac{\varepsilon_R R}{2} q^- \tag{8}$$

The corresponding balance on the flux in the negative x direction is

$$-\frac{dq^-}{dx} = 2K_a E_b + \frac{\varepsilon_S S}{8} \sum_{i=1}^{8} (e_b)_i + \frac{\varepsilon_R R}{2} q^+ - \left(2K_a + \frac{\varepsilon_R R}{2} + \varepsilon_S S\right) q^- \tag{9}$$

Total energy balances for the furnace gas and the outer tube surfaces can also be written in terms of the unknown radiative flux densities and emissive powers

$$\frac{dq_G}{dx} = 2K_a \, (q^+ + q^- - 2E_b) \tag{10}$$

$$(-1)^i \, \frac{d}{dx} \, [(q_p)_i] = \frac{\varepsilon_S S}{8} \, [q^+ + q^- - 2(e_b)_i] \tag{11}$$

The flux densities of furnace and process gases are related to their respective temperatures by the relationships

$$q_G = \frac{\dot{m}_G}{A_G} \cdot C_G \, (T_G - T_o) \tag{12}$$

$$(q_p)_i = \frac{\dot{m}_p}{A_G} \cdot C_p \, [(T_p)_i - T_o] \tag{13}$$

where $(T_p)_i$ is the absolute temperature of the process gas on its ith pass. The process gas and tube surface temperatures can finally be related by an energy balance on the tube wall of the form

$$\frac{(T_S)_i - (T_p)_i}{D} = \frac{\varepsilon_S}{2} \, [q^+ + q^- - 2\,(e_b)_i] \tag{14}$$

The twenty eight equations [(8), (9), (10), (12) and (11), (13), (14) with $i = 1,2,\ldots,8$] involving the twenty eight unknown variables q^+, q^-, q_G, T_G and $(q_p)_i$, $(T_p)_i$, $(T_S)_i$ with $i=1,\ldots,8$, can be solved simultaneously once initial and boundary conditions are specified at the top and bottom of the heater. If the end walls of the heater are treated as radiatively adiabatic, then the boundary conditions become

$$q^+ = q^- \quad \text{at} \quad x = 0 \tag{15}$$

$$q^- = q^+ \quad \text{at} \quad x = L \tag{16}$$

The necessary initial conditions are the inlet temperatures of the furnace gas and the process gas.

4.2 Computational procedure

The problem under consideration involves the solution of eleven ordinary differential equations most of which are nonlinear due to the presence of the T^4 terms. Under such conditions analytical solution is impossible and a numerical solution is obligatory. The numerical approach suggested by the solution of the parallel plate problem has been adopted i.e. an iterative solution using a combined two-point and three-point predictor-corrector method for the numerical integrations.

4.3 Results of calculations and discussion

With 24 space increments in the vertical direction convergence of the solution to five decimal places was obtained in about 25 iterative cycles, taking approximately 50 seconds computing time on an ICL 1907 computer. The use of the

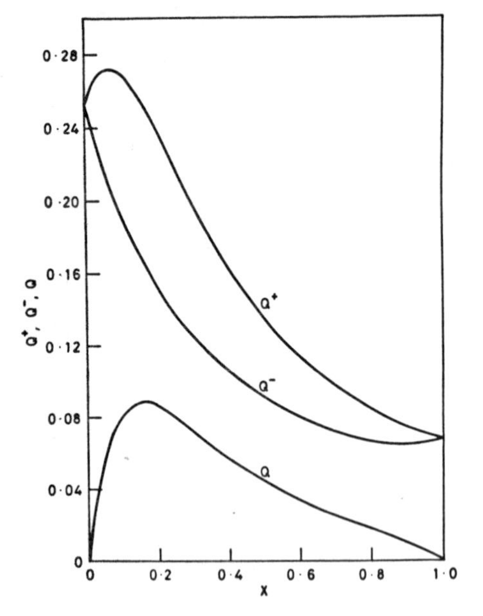

Fig.3. Distributions of forward, backward and net dimensionless radiant flux densities in a process heater.

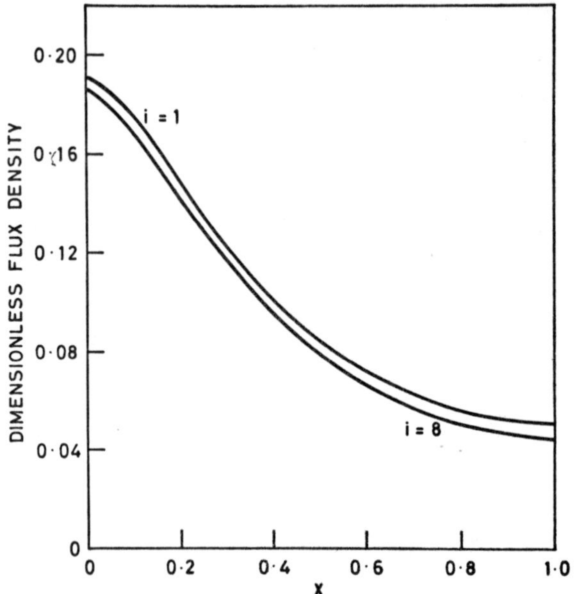

Fig.4. Distributions of dimensionless radiant flux densities to the tubes.

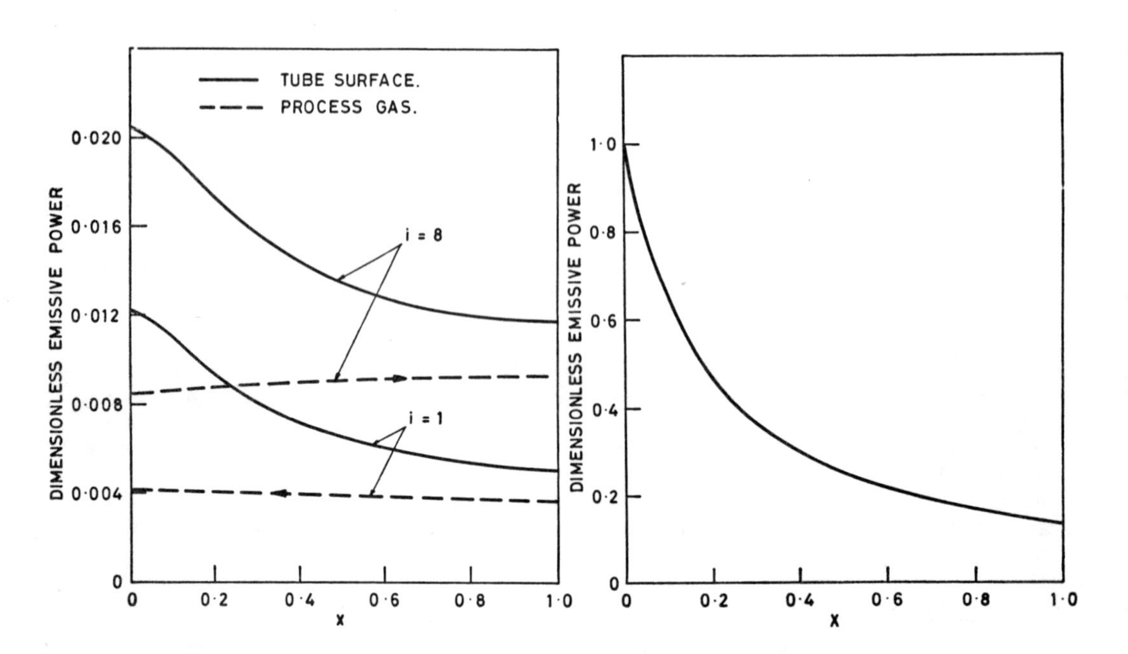

Fig.5. Distributions of tube surface and process gas dimensionless emissive powers.

Fig.6. Distribution of furnace gas dimensionless emissive power.

combined predictor-corrector methods for integration with a Newton-Raphson technique for the solution of equation (14) provided no computational difficulties, and the iterative solution proved to be extremely stable, appearing to be relatively insensitive to the value of the initial guess for q^-.

For greatest generality the results of the calculations for a particular heater are presented in dimensionless form; all fluxes and emissive powers being divided by the initial black body emissive power of the furnace gas, and distance from the base of the heater being divided by the heater height.

Dimensionless forward, backward and net radiant flux densities are shown in Fig.3. The net dimensionless radiant flux density Q, calculated as the difference $Q^+ - Q^-$, increases rapidly near to the bottom of the heater, passes through a peak, and then decreases continuously to the top of the heater. It is expected that if a more realistic heat release pattern were assumed within the furnace gas, the position of the peak would shift towards the top of the heater.

The distributions of dimensionless heat flux density at the tube surfaces for the first and last passes are shown in Fig.4. It can be seen that the difference in magnitude between the flux densities to the two passes at any distance from the base of the heater is approximately the same as the difference between the values at the two ends of the same pass.

Dimensionless emissive powers for the tube surfaces and process gas for the first and last passes are shown in Fig.5, and the dimensionless emissive power of the furnace gas in Fig.6. Although these exit temperatures of the furnace and process gases agree very approximately with those measured in practice on the heater in question, it is felt that significant improvement in the reality of the model can be achieved in future calculations by modifying assumptions (vi), (vii) and (viii) of section 4.1.

The assumption of grey gas behaviour for the furnace gas is a gross oversimplification. A better description of the radiating characteristics of a real gas can be achieved by treating it as a mixture of m grey gases and a clear gas. The flux of radiation in the forward or backward direction is then subdivided into (m+1) components; m grey gas components (band radiation) and the clear gas component (window radiation), and each of equations (8) and (9) is replaced by (m+1) equations. Incorporation of the additional differential equations into the existing computation scheme should provide no difficulties, leading only to an increase in computing time.

Further improvements in accuracy should also be achieved by allowing for variation in the specific heats of the gases with temperature and by assuming a more realistic heat release pattern. The modifications in the existing equations necessary to incorporate these effects are to define q_G and q_p in terms of the integral of specific heat over the temperature range involved in equations (12) and (13), and the addition of a term accounting for heat release per unit distance in equation (10). Both changes should involve a minimal increase in computational effort.

REFERENCES

(1) Hottel,H.C., and Sarofim, A.F. "Radiative Transfer", McGraw Hill, New York, 1967.

(2) Hottel,H.C., and Cohen,E.S., Radiant heat exchange in a gas-filled enclosure: allowance for nonuniformity of gas temperature. A.I.Ch.E. Journal, Vol.4. No.1. (1958).

(3) Hottel,H.C., and Sarofim,A.F. The effect of gas flow patterns on radiative transfer in cylindrical furnaces. Int.J.Heat and Mass Transfer, vol.8 (1965).

(4) Viskanta,R. Radiation transfer and interaction of convection with radiation heat transfer. Chapter in "Advances in heat transfer Vol.3." (Editors: Irvine,T.F., and Hartnett,J.P.) (1966).

(5) Viskanta,R., and Grosh,R.J. Heat transfer by simultaneous conduction and radiation in an absorbing medium. T.A.S.M.E., J. Heat Transfer (1962).
(6) Larkin,B.K., and Churchill,S.W. Heat transfer by radiation through porous insulations. A.I.Ch.E. Journal, Vol.5. No.4. (1959).
(7) Chen,J.C., and Churchill,S.W. Radiant heat transfer in packed beds. A.I.Ch.E. Journal, Vol.9. No.1. (1963).
(8) Hamaker,H.C. Philips Research Repts. Vol.2. (1947).
(9) Milne,E.A., Handbuch der Astrophysik. Vol.3.pt.1. Springer, Berlin (1930).
(10) Eddington,A.S. "The Internal Constitution of Stars", Dover Press, New York, (1959).
(11) Schwarzschild,K. Nachr. Akad. Wiss. Gottingen Math-Physik Klasse 1, 1906; Sitzber, Preuss. Akad. Wiss. Physik-Math Klasse (1941).
(12) Schuster,A. Astrophys.J., Vol.22 (1905).
(13) Siddall,R.G., Flux methods for the analysis of radiant heat transfer. Fourth Symposium on Flames and Industry, London. (1972).
(14) Usiskin,C.M., and Sparrow,E.M. Thermal radiation between parallel plates separated by an absorbing-emitting nonisothermal gas. Int.J.Heat Mass Transfer, Vol.1 (1960).
(15) Lapidus,L., and Seinfeld,J.H. "Numerical Solution of Ordinary Differential Equations", Academic Press, New York (1971).
(16) Roesler,F.C. Theory of radiative heat transfer in co-current tube furnaces. Chem.Eng.Science, Vol.22 (1967).

Chapter 12

A NEW FORMULA FOR DETERMINING THE EFFECTIVE BEAM LENGTH OF GAS LAYER OR FLAME

Milos Gulic

University of Novi Sad, Yugoslavia

Abstract

In this work a new mathematical expression is given for effective beam length, capable of producing satisfactory results at higher values of optical density. The new formula has advantages when determining the flame emissivity because it enables the physically correct Buger's formula to be used.

NOMENCLATURE

a : total absorption coefficient of gas radiation, m^{-1}
a_λ : mono-chromatic absorption of gas radiation, m^{-1}
b : segment of the ordinate,
d : diameter, m
E_s : semi-spherical radiant heat flow of absolute black body at the temperature T, W/m^2
E_{s_λ} : semi-spherical radiant heat flow of absolute black body at the temperature T, within the portion of the spectrum λ to $\lambda + \Delta\lambda$, W/m^2
F : area of enclosure surface, m^2
k : coefficient of direction (tg α)
ℓ : mean beam length or effective beam length, m
ℓ_o : mean beam length of gas, m
ℓ_G : mean beam length according to the author, m
q : elementary heat flow with definite direction with respect to the surface dF, in the area of the solid angle $d\omega$, W
Q : semi-spherical heat flow, W
Q_λ : semi-spherical heat flow within the portion of the spectrum from λ to $\lambda + \Delta\lambda$, W
V : volume of gas, m^3
ε : emissivity of gas
ε_G : emissivity according to the author
τ : optical density of the gas

The emissivity of the flame or of the combustion products in industrial furnaces considerably influences radiant heat transfer. For the present, the emissivity of the flame may be expressed in the gormula given by Buger:

$$\varepsilon = 1 - \exp(a\ell) = 1 - \exp(\tau) \qquad (1)$$

and is dependent on the optical density of the absorbing medium,i.e. absorption coefficient of gas radiation and effective beam length.

In the last few years it has been possible to precisely determine the value of the absorption coefficient, in spite of the great number of influential factors involved, due to the great quantity of theoretical and experimental research. This is not the case with the effective beam length (or mean beam length) of the gas layer which was determined for over thirty years by means of an insufficiently accurate formula:

$$\ell = \frac{3{,}6\ V}{F} = \frac{3{,}6}{f} \qquad (2)$$

The effective beam length, as suggested by H.C. Hottel, represents the radius of the hemispheric gas volume irradiating to the center of the base, the quantity of energy being equal to the energy of any other shape of gas volume irradiated to a small area at its base under the same remaining conditions. This formula could be applied to a furnace with relatively low absorbability of gas volume. At the time the formula was established, the furnaces were relatively small in size, and determination of the absorption coefficient of gas radiation, which was made empirically, was unreliable. In the last few decades the size of furances has been greatly enlarged, especially furances of modern steam generators. Consequently, the effective beam length of the absorptive medium increased considerably, therefore augmenting the absorbability of the flame, i.e. of combustion products. For this reason the formula for the effective beam length in the aborptive medium (2) could not be more usefully applied, and the purpose of this work is to establish a new mathematical expression capable of producing satisfactory results at higher values of optical density.

In order to effectuate our aim, let us consider the diathermic volume bound by an absolute black sphere of a definite temperature, as represented in Fig. 1.

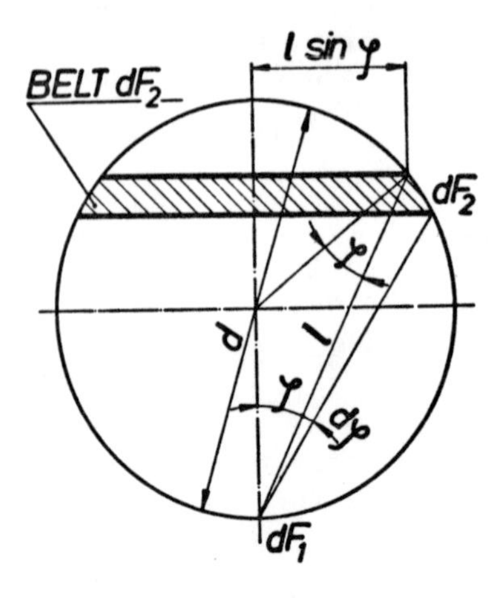

Fig. 1

At thermal equilibrium a certain element of the sphere emits to the surface the same quantity of energy that has already been absorbed by irradiation of the sphere. As soon as the elementary zone of the sphere surface dF_2 has been distinguished, it is possible to determine for heat flow that radiates through the diathermic medium and irradiates to the elementary surface dF_1:

$$dQ = \frac{E_s}{\pi} \frac{\cos^2\phi}{\ell^2} dF_1 dF_2 = \frac{E_s}{\pi} \frac{\cos^2\phi}{\ell^2} dF_1 \cdot \frac{2\pi\ell^2 \sin\phi \ d\phi}{\cos \phi} =$$

$$= 2 E_s \cos\phi \sin\phi \ d\phi \ dF_1 \tag{5}$$

If diathermic gas is substituted by a gas with the radiation-absorbing property at the same temperature as that of the sphere, the heat equilibrium of the observed parts of the sphere is not disturbed. This means that the given element of the sphere must absorb the same quantity of heat as in the preceding case, where the gas was diathermic. The difference appears only because the heat irradiated by the element of the surface is partly absorbed by a gas, which will simultaneously emit it to the sphere. However, the gas radiates energy only in a few bands of spectrum. For all other portions of the spectrum the gas has to be treated as a transparent medium, enabling the passage of irradiation from the black sphere without any loss. Then, the above equation could be expressed in the following form:

$$dQ_\lambda = 2 E_{s\lambda} \cos\phi \sin\phi \ d\phi \ dF_1 \tag{6}$$

When the gas monochromatic absorption coefficient is known, it is possible to describe the energy absorbed by the gas during monochromatic irradiation of zone dF_2, by means of Buger´s principle:

$$d(dq_{\lambda aps}) = (1 - e^{-a_\lambda \ell}) dQ_\lambda = 2 E_{s\lambda}(1 - e^{-a_\lambda \ell}) \cos\phi \sin\phi \ d\phi \ dF_1 \tag{7}$$

The expression in parentheses can be split into the series:

$$(1 - e^{-a_\lambda \ell}) = 1 - 1 + \frac{a_\lambda \ell}{1!} - \frac{(a_\lambda \ell)^2}{2!} + \frac{(a_\lambda \ell)^3}{3!} \ \ldots . \tag{8}$$

Under conditions of very low optical density of the absorbing medium, elements of higher order could be disregarded, whereby the mathematical expression is simplified:

$$(1 - e^{-a_\lambda \ell}) = a_\lambda \cdot \ell = a_\lambda \ d \cos\phi \tag{9}$$

This simplification, which results in formula (2) should not be used here because of the very high values of the optical density concerned; such simplification could provoke considerable deformation of the physical sense of the process.

In order to enable the mathematical solution of the problem, this expression (8) has to be replaced by the following:

$$(1 - e^{-a_\lambda \ell}) = k \ a_\lambda \ell + b = k \ a_\lambda \ d \cos\phi + b \tag{10}$$

which is really a straight line equation. Coefficient k is a coefficient of the direction whereas coefficient b is a segment of the ordinate. These coefficients depend on the optical density given in Figure 2.

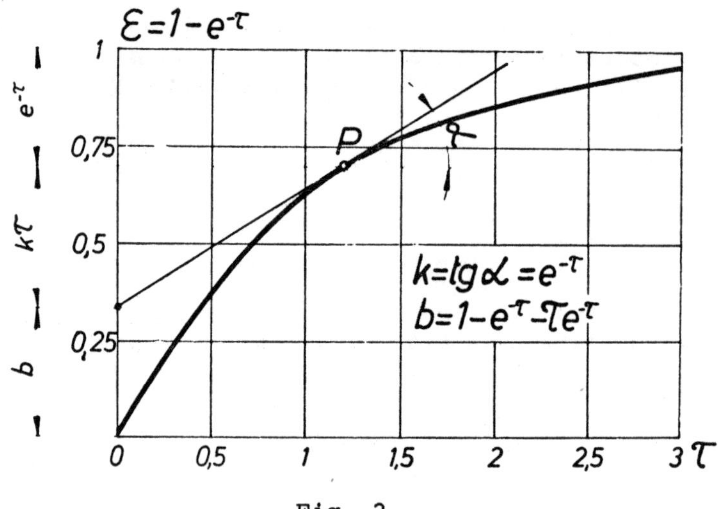

Fig. 2

With the introduction of these values into equation (7) and integration of equation within the limits of 0 to $\pi/2$, the energy absorbed in the gas as a result of irradiation from the sphere to the elementary surface dF_1 is obtained:

$$dq_{\lambda aps} = 2\, E_{s\lambda} k\, a_\lambda\, d\, dF_1 \int_0^{\pi/2} \cos^2\phi\, \sin\phi\, d\phi + 2\, E_{s\lambda} b\, dF_1 \cdot$$

$$\cdot \int_0^{\pi/2} \cos\phi\, \sin\phi\, d\phi = \frac{2}{3}\, E_{s\lambda}\, k\, a_\lambda\, d\, dF_1 + E_{s\lambda}\, b\, dF_1 \ldots . \tag{11}$$

According to the Kirchoff's law, the same quantity of energy will be irradiated from sphere volume to the elementary surface dF_1, so that the total irradiation of gas volume will be

$$q_\lambda = \left(\frac{2}{3}\, E_{s\lambda}\, k\, a_\lambda\, d + E_{s\lambda}\, b\right) \int_0^{\pi d^2} dF_1 = \left(\frac{2}{3}\, \pi\, k\, a_\lambda\, d^3 + b\, \pi\, d^2\right) E_{s\lambda} =$$

$$= (4 k a_\lambda V + bF)\, E_{s\lambda} \tag{12}$$

By application of the effective beam length it is possible to describe the heat flux to the surface of an arbitrary geometrical shape by:

$$q_\lambda = E_{s\lambda}\, a_\lambda\, \ell_o\, F \tag{13}$$

By equalizing the two last expressions (12, 13) the formula for the effective beam length is obtained

$$\ell_o = \frac{4kV}{F} + \frac{b}{a_\lambda} = \frac{4k}{f\left(1 - \frac{b}{\tau}\right)} \tag{14}$$

Coefficients k and b depend on the optical density of the gas, and their values can be easily determined. The coefficient of the direction can be expressed by the gradient of the flame emissivity:

$$k = \frac{d\varepsilon}{d\tau} = e^{-\tau} \tag{15}$$

Coefficient b is specified easily in Fig. 2

$$b = 1 - e^{-\tau} - \tau e^{-\tau} \tag{16}$$

When these values are arranged in the above equation (14), a new expression of the effective beam length is obtained:

$$\ell_o = \frac{4}{f} \cdot \frac{\tau e^{-\tau}}{\tau - 1 + e^{-\tau} + \tau e^{-\tau}} \tag{17}$$

As the effective beam length of a flame in furances of conventional geometric desing is about 10% less than the effective beam length of any other gas, the expression of the effective beam length of gas be described in a final form

$$\ell_G = \frac{3,6}{f} \cdot \frac{\tau e^{-\tau}}{\tau - 1 + e^{-\tau} + \tau e^{-\tau}} = \frac{3,6}{f} C \tag{18}$$

The coefficient C, dependent on the optical density of the absorbing medium, has been shown in Fig. 3, while the effective beam length has been represented as a function of the shape factor of the furnace and the optical density in Fig. 4.

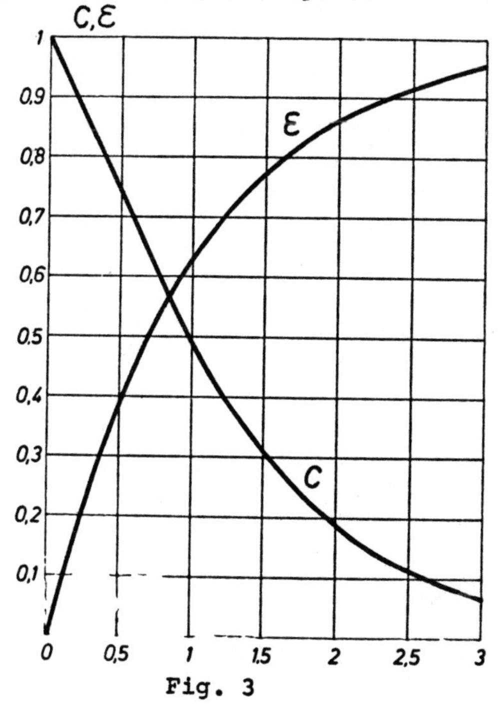

Fig. 3

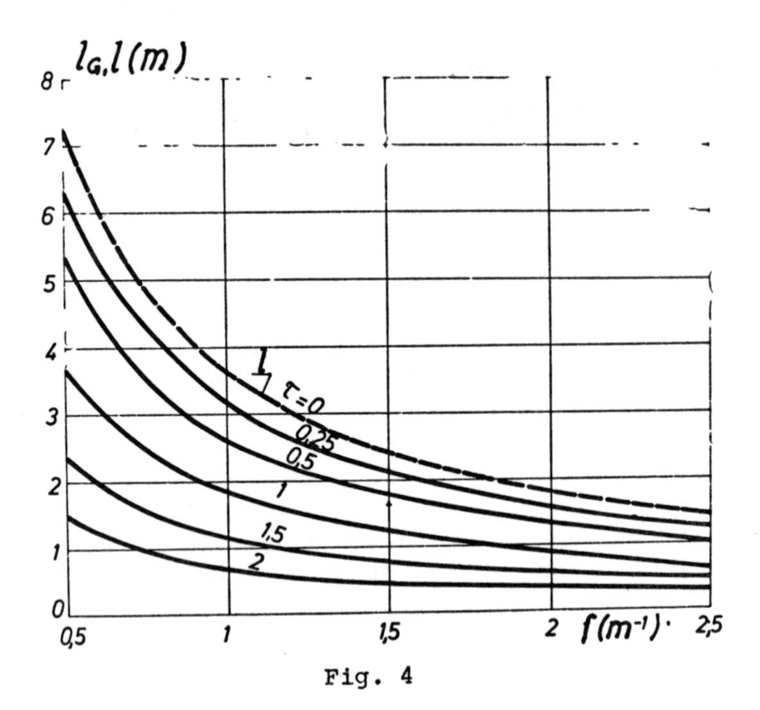

Fig. 4

A distinctive dependence of the effective beam length on the optical density can be noticed; at high values of optical density the length diminishes considerably. This is easy to understand. The increased optical density diminishes the influence of the distant points in the gas at a definite sport on the furnace surface, because of the increasing absorbing power of the gas.

This equation is more universal than the equation in present use which could be obtained with optical density approaching zero:

$$\lim_{\tau \to 0} = \frac{\tau e^{-\tau}}{\tau - 1 + e^{-\tau} + \tau e^{-\tau}} = \frac{-\tau e^{-\tau} + e^{-\tau}}{1 - e^{-\tau} - \tau e^{-\tau} + e^{-\tau}} = 1$$

for $\tau = 0$ is $\ell_G = \frac{3,6}{f}$

From the definition of the optical density it follows:

$$\ell_G = \frac{\tau}{a} \qquad (20)$$

If this value is put into formula (18) a new expression in the transcendental form is obtained which demonstrates a functional relationship among the optical density, the shape factor of the furnace, and the absorption coefficient of gas radiation.

$$\frac{f}{a} = \frac{3,6\ e^{-\tau}}{\tau - 1 + e^{-\tau} + \tau e^{-\tau}} \qquad (21)$$

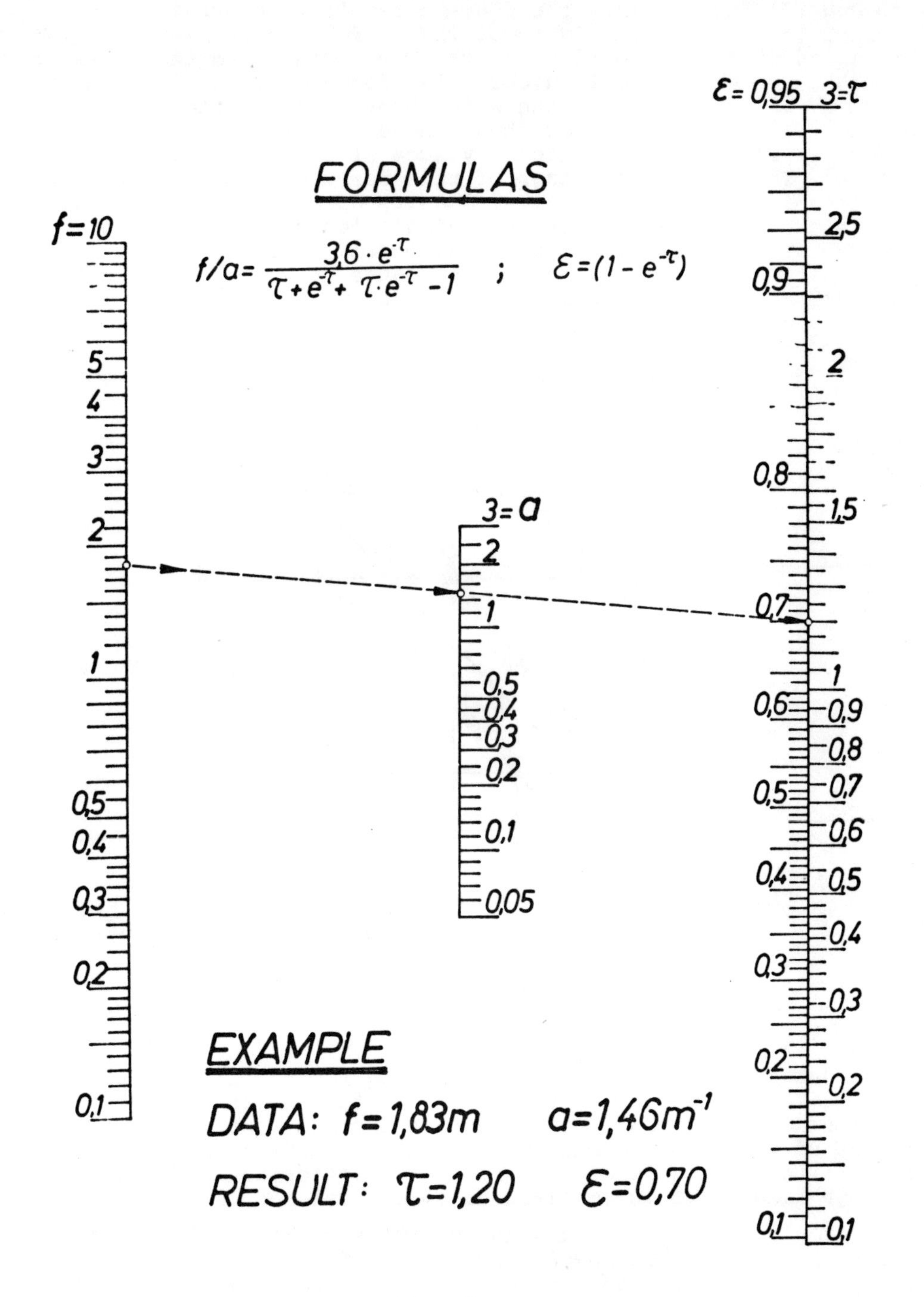
FORMULAS
f/a= 3,6 · e^-τ / (τ + e^-τ + τ·e^-τ - 1) ; ε = (1 - e^-τ)
f=10
5
4
3
2
1
0,5
0,4
0,3
0,2
0,1
3=a
2
1
0,5
0,4
0,3
0,2
0,1
0,05
ε= 0,95
3=τ
2,5
0,9
2
0,8
1,5
0,7
1
0,6
0,9
0,8
0,5
0,7
0,6
0,4
0,5
0,4
0,3
0,3
0,2
0,2
0,1
0,1
EXAMPLE
DATA: f=1,83m a=1,46m^-1
RESULT: τ=1,20 ε=0,70

Fig. 5

When the optical density determined by this formula is introduced into Buger´s formula (1), the flame emissivity is obtained by means of the diagram in Figure 5. The difference between the emissivity of flame specified by the new formula and the emissivity defined by Buger´s so-called corrected formula is represented in Figure 6. In both formulae the same absorption coefficient has been used. Obviously, and as should have been expected, scattering is more expressive at the higher values of flame emissivity (large furance). The basic advantage of this formula is that it enables the determination of the optical density of gas in the furance. This is the essential element for simulation of radiant heat exchange in a small-sized isothermal model. As is know, one of the conditions for achieving a resemblance between the furance and its model is to obtain exect optical density.

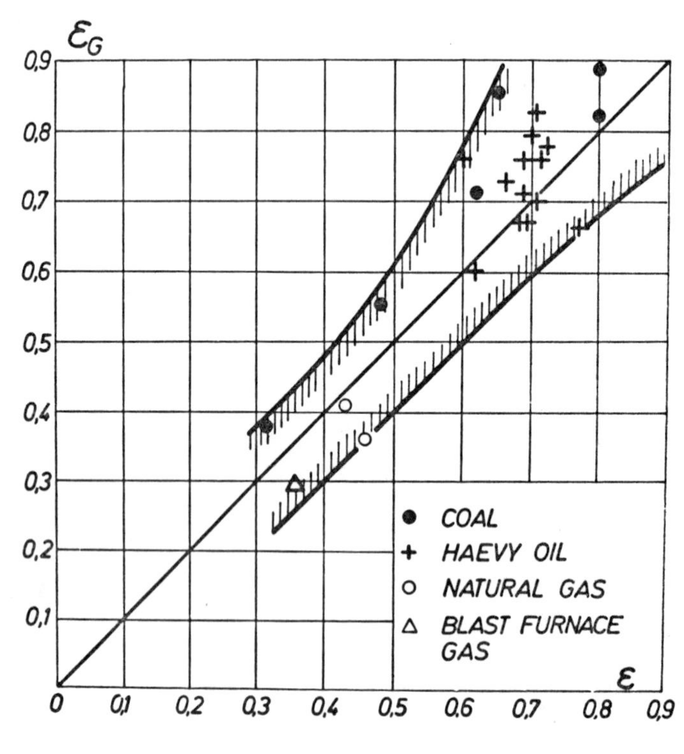

Fig. 6

If the values of both factors determining the optical density are not established correctly, the values of the absorption coefficients in the small-sized model would also be incorrect, and thus the real picture of the observed phenomenon would be distorted. The application of the formula is one of the prerequisites for forming heat exchange simulation in small-sized isothermal models.

This formula has advantages when determining flame emissivity because it enables the physically correct form of Buger´s formula (1) to be used. This involves some more critical estimation of the absorption coefficients in the thermal engineering calculations of a furance.

Chapter 13

THE INTENSIFICATION OF THE HEAT EXCHANGE PROCESS IN INDUSTRIAL FLAME FURNACES AND THE CHOICE OF RATIONAL REGIMES

A. E. Erinov

Institute of Gas, Academy of Sciences Ukrainian SSR, Kiev, USSR

Abstract

In this report an analysis is given of the possibilities of intensifying radiant and complex heat exchange in industrial flame furnaces. The improvement of the heating operation of industrial flame furnaces is possible by applying a new heat exchange regime -- an indirect radiant regime which significantly increases the role of convective heat transfer from the burning gas to the refractory walls.

NOMENCLATURE

- q : heat flow
- σ_0: radiation constant
- T : absolute temperature
- F : surface
- ϵ_g : gas emissivity
- α_{con}: convective heat transfer coefficient
- λ_a: excess air coefficient
- S_g: effective radiation layer thickness
- ϵ_{wm}: average emissivity of the metal refractory walls system
- σ_{wm}: average radiating coefficient of the system
- T'_g: gas temperature near the refractory walls surface
- T_g: volume gas temperature
- q_e : heat losses through the refractory walls
- g : gas
- w : refractory walls
- m : metal
- c : calorimeter
- rad: radiation
- con: convection

Introduction

The contemporary development of heating furnaces for the rolling industry results in their heat-output increase. Naturally, this has caused larger dimensions of furnaces and higher requirements of the heating metal quality before rolling. All this has demanded significant infensification of the heat exchange processes inside working chambers of flame furnaces as well as improvement of the uniformity of heating metal surfaces.

The study of the radiant and convective heat exchange processes and their application experience shows that increased possibilities of heat exchange intensification and specific duty in present-day installation by means of traditional methods are limited. Thus, a higher temperature at the furnace end to increase furnace efficiency results in an additional number of heating zones. This demands more complex designs of furnaces, and heating surface qualities become worse. A higher roof for higher emissivity quality of the gas layer is connected with larger dimensions and higher fuel consumption. To increase the radiation of flame by carburization is not always possible due to the necessary use of fuel gases without hydrocarbon.

Heat exchange intensification and heating operation improvement of the flame furnaces along with higher heating steel uniformity have become possible due to the new heat exchange regime, which significantly increases the role of the refractory walls as the principal radiation source. It is the indirect radiant heat exchange regime (classified by M.A. Glinkov /1/) that gives high temperature unevenness and combustion rate with the maximum near the refractory walls.

As compared with the new regime, the uniform regime of radiant heat exchange with the temperature maximum and combustion rate takes place in conventional furnaces. In spite of the fact that hundreds of heating furnaces with the new heating regime are already in operation the new regime theory has not been sufficiently described.

2. Analytic Dependences of the New Regime and its Specific Features

The resulting heat flow to the heating surface can be expressed

$$q_m = q_{rad}^{gm} + q_{rad}^{wm} + q_{con}^{gm} - q_{rad}^{m} \tag{1'}$$

or approximately

$$q_m = \sigma_o \cdot \epsilon_g \cdot T_g^4 + \sigma_o \epsilon_{wm} \cdot T_w^4 (1 - \epsilon_g) + q_{con}^{gm} - \sigma_o \epsilon_{wm} \cdot T_m^4 \tag{1}$$

Using the refractory wall temperature as the basic parameter for the flame furnace heat exchange calculation is often reasonable, for it gives the opportunity to exclude the reflected radiative heat and convective heat transfer to the refractory walls from equations.

For continuous heating furnaces of the uniform radiation regime, simple and accurate calculation methods of heat exchange are worked out which exclude refractory wall temperature and use geometric averaging of effective gas temperature /2,3,5/

$$q_m = \sigma_g (T_g^4 - T_m^4) \tag{2}$$

This is true because the principal radiation source is heating gas, and the refractory walls are a secondary source of radiation. Figure 2.2a shows the role of refractory walls at the uniform heat exchange regime depending on the combustion product emissivity.

The maximum contribution of refractory walls as the reflected heat radiation source corresponds to the average of = 0.4 - 0.6 and constitutes 30 to 50 percent of the total resulting flow. With increase or decrease of from the average value, the role of refractory walls is less significant. In its place is the role of individual constituents in the total heat flow for flame furnaces which have the indirect heat exchange regime (Fig. 2.2b). Here, at small values of the role of refractory walls becomes decisive. Therefore, the heat exchange calculation methods made for the uniform regime /2,3,4,5/ cannot be used for the new one. Since refractory walls are the principal radiation source, their temperature must be the main factor in the heat exchange calculation. Combustion temperature — is of secondary importance while calculating heat exchange but of primary importance when finding fuel consumption from the heat balance. The basic equation for the heat exchange calculation at the indirect regime will be

$$q_m = \sigma_{wm}(T_w^4 - T_m^4) \tag{3}$$

where is the average radiation coefficient considering all parameters of refractory walls, metal and gas except for the direct convection from gases to the heating surface.

Let us estimate the role and meaning of the combustion products which are between the metal and refractory walls. In Fig. 2.1 the combustion emissivity is equal to at . Let us assume that the coefficients of radiation between the heating surface and refractory walls, both in the case of absence of heat absorbing gases and in case of their presence, are equal to — and also that the direct convection from gases to the heating surface is small, i.e. the resulting heat flow as ratio of the two cases: presence and absence of radiant gas:

$$q_m = \frac{q_m^{\epsilon g \neq 0}}{q_m^{\epsilon g = 0}} = \frac{\sigma_{wm} T_w^4 (1 - \epsilon g) + \sigma_o \epsilon g T_g^4 - \sigma_{wm} T_m^4}{\sigma_{wm}(T_w^4 - T_m^4)} = 1 - \epsilon g \cdot \frac{T_w^4 - \dfrac{1}{\epsilon_{wm}} \cdot T_g^4}{T_w^4 - T_m^4} \tag{4}$$

We can say that combustion products play a negative part and serve as a screen between the radiation source and the heating surface, because in furnaces of the indirect heat exchange regime the temperatures are not equal $T_w > T_g > T_m$. It should be mentioned that this conclusion is true in theory. In practice, the presence of hot gases with temperature $T_g > T_m$ positively affects the heat and hydraulic characteristics of the system due to intensive circulation inside it. Substitution of hot semi-transparent gases for radiation would lead to a decrease of the heat exchange intensity between the refractory walls and metal.

In Fig. 2.3 the influence of ϵg is shown for the value of the relative flow $\bar{q}_m$ at different temperature relations of the system. As seen from Equation (4) and Fig. 2.3, the influence of emissivity is stronger the lower the gas medium temperature T_g is. The metal surface temperature T_m affects the gas emissivity ϵg to a lesser degree at the indirect radiation regime of heat exchange.

The system radiation coefficient from Equation 3 is approximated with 10 per cent accuracy (in interval $\epsilon g \cong 0.1 - 0.3$):

$$\sigma_{wm} = \sigma_o \cdot \epsilon_{wm} \cdot \left\{ 1 - \epsilon g \left[\sqrt[4]{\epsilon_{wm}} - \left(\frac{T_g}{T_w} \right)^4 \right] \right\} \tag{5}$$

where ϵ_{wm} is the emissivity of the refractory walls-metal system, calculated by the well-known formula:

$$\epsilon_{wm} = \frac{1}{1/\epsilon_w + (F_w/F_m)(1/\epsilon_m - 1)} \tag{6}$$

Let us consider the heat exchange conditions on the surface of the refractory walls, the principal radiation source. The heat balance equation without taking into consideration the reflected flows, can be given as

$$q_{rad}^{wm} + q_e^w = q_{rad}^{gw} + \alpha_{con}^{gw}(T_g' - T_w) \tag{7}$$

where the second constituent on the right side of the equation is the direct convection from gases with temperature of different from

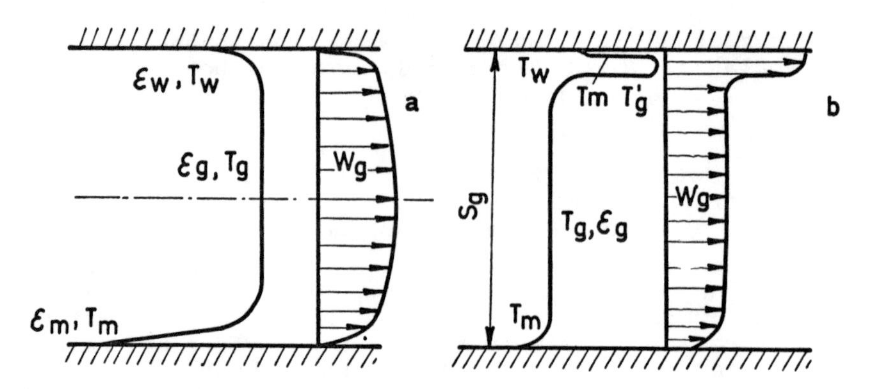

Fig. 2.1 Heat exchange pattern in flame furnaces, temperature and combustion rate distribution. a) uniform radiation regime of heat exchange; b) indirect radiation regime of heat exchange

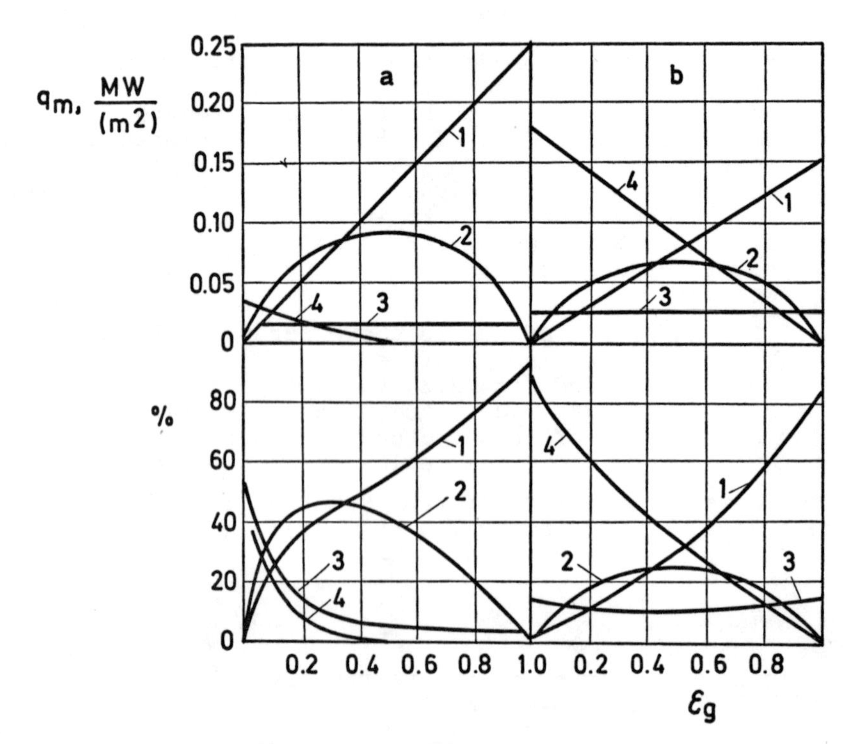

Fig. 2.2 Role of individual components of the resulting flow depending on the combustion product emissivity in furnaces.
a) unidorm radiative regime of heat exchange
b) indirect radiative regime of heat exchange
1) radiative flow from gases to heating surface
2) reflected radiation of gases by the refractory walls
3) direct convective flow from gases
4) indirect flow of convection from the refractory walls.
(T_g = 1673°K; T_w = 1573°K; T_m = 1273°K; F_m/F_w = 1)

the gas temperature inside the working chamber of the furnace, to the refractory walls with the temperature of T_W. At the uniform heat exchange regime, the direct convection from the combustion product to the refractory walls is equal to the heat losses due to the conductivity of heat and, in this case, the refractory walls are only a reflector of radiant flow from heating gases.

To form the indirect radiant heat exchange regime with temperature relation $T_W > T_g > T_m$ it is necessary to significantly intensify the convective heat exchange near the refractory walls. This condition cannot be performed with the usual volume methods of burning fuel, because it is impossible to form high temperature and rate gradients of combustion products near refractory walls under the conditions of moving flow. In such furnaces the temperature relation of basic heat exchange media and surfaces is as follows:

The high temperature and rate gradients of combustion products near the refractory walls and the indirect radiant heat exchange regime became possible using the surface fuel burning (as compared with the volume fuel burning) by means of the flat-flame (or radiant) burners. These allow fuel gas or oil to be burned near the refractory wall surface in a thin (not more than 50 mm) layer, producing large heat stresses and high temperature of combustion products. Due to the high rates of gas-air mixture output, made by high pressure and a particular burner design, and due to the higher temperature of the combustion product, the convective heat transfer from the combustion products to the refractory walls is highly intensified with its subsequent transformation into grey radiation of the refractory walls.

Estimating the value of the heat flux from the refractory walls to the heating surface under the condition of the indirect heat exchange regime from Equation (7), one can assume that the convective heat flux from burning gases to the refractory walls should be of the same order, i.e. be several times higher than in conventional furnaces with the uniform heat exchange radiation regime. To obtain such high values at reasonable meaning $(T'_g - T_W)$ is possible only if the convective heat transfer coefficient is the order of 160-200 W/(m^2deg) and higher.

A higher convective heat transfer coefficient may be obtained at the expense of both the local and the average rates of combustion products near the refractory wall surface, and at the expense of abrupt change of the heat exchange character between the unfinished burning products and the refractory walls.

3. Experimental Study

The investigation has been carried out for the indirect heat exchange radiative regime and for the gas layer role in total heat exchange in the total heat exchange in the experimental rectangular chamber (Fig. 3.1) of 0.5 x 1.5 x 1.5 m with a flat-flame burner on the vertical side wall. Fig. 6 shows the radiant heat flux distribution from the burner surface and from the opposite surface. As seen in Fig. 3.2, the heat flux in the first case decreases with an increase of the distance from the radiating wall; in the second case the heat flux increases with the distance from the refractory walls. This proves the fact that the combustion product layer is a screen and decreases the heat flux with the increase of the layer thickness S_g . The heat flux uniformity was also investigated in this chamber considering the distance in the front wall with the burner (Fig. 3.3). High uniformity (± 1%) of the heat flux is obtained at the distance of 200 mm, in spite of the fact that on the front surface there is rather

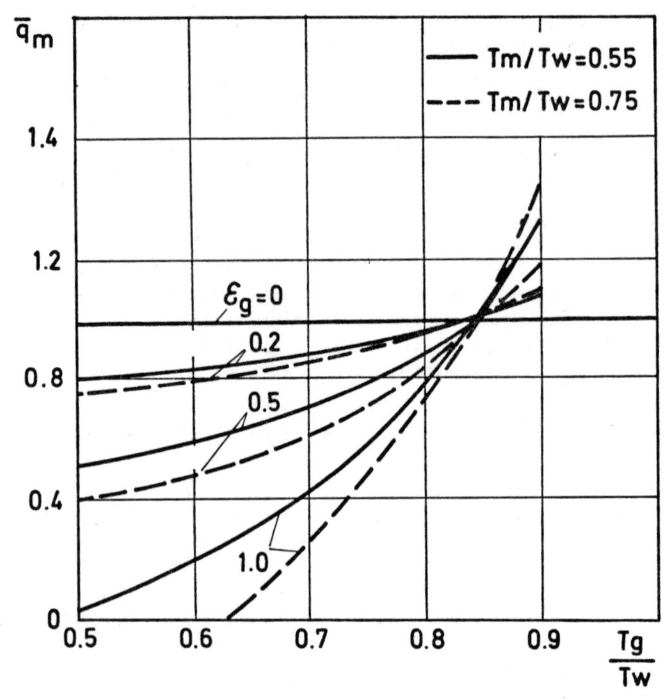

Fig. 2.3. Influence of gas temperature on the value of relative heat flux to heating surface at different emissivity of the gas layer ($\epsilon_w = 0.5$; $\epsilon_{wm} = 0.6$; $\epsilon_m = 0.9$)

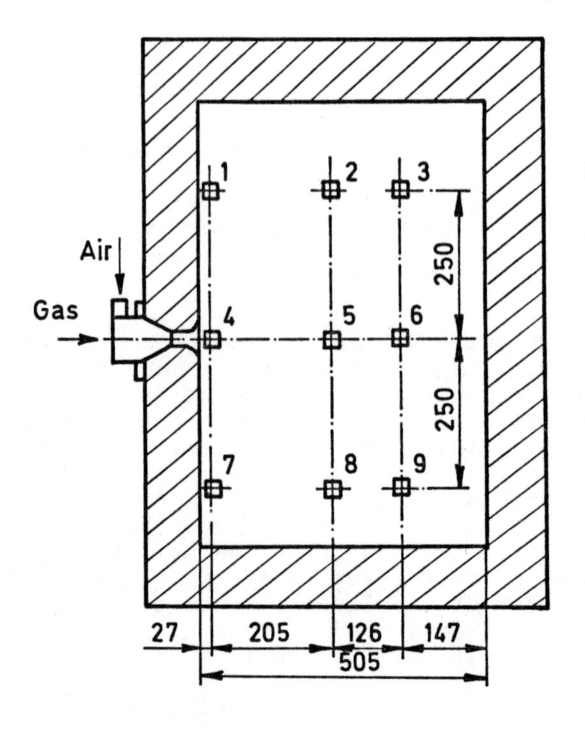

Fig. 3.1 Vertical section of experimental stand for investigation of different heat exchange regimes (1,2.... 9-points of heat flux measurement by thermosound).

high non-uniformity of the refractory wall temperature, especially in the burner zone.

The temperature field of combustion products at the distance of 200 mm from the front wall is also rather uniform ($\pm$ 2%).

The high uniformity of heat flux and temperatures in the case of the indirect heat exchange regime allows the performance of high quality heating of large surfaces with high intensity without application of the significant increase of the radiating layer thickness.

In our works /9,10/ we show the possibility of significant convective heat exchange intensification by burning fuel on a hard red-hot surface. In spite of the rather numerous works /8, 11, 12/ which showed the qualitative difference of convective heat exchange with and without burning on the hard surface, the quantitative estimation of burning effect was not carried out.

The comparative investigation of the convective heat exchange was carried out on a special stand (Fig. 3.4) which included a flat cell calorimeter with the variable temperature of the heating surface. The calorimeter was perpendicular to the burning flame, produced by an injection burner without a tunnel. The burner was supplied by a partially-aired mixture with = 0.4 + 0.8. The calorimeter surface temperature was varied by regulating the insulation layer over the calorimeter. During the experiment, the surface temperature was varied between 300 - 1280°K.

Fig. 3.5 shows the investigation results as ratio of the local convection heat exchange coefficients obtained under the variable surface temperature conditions (α_t) and the temperature constant of 320° - 330°K (α_o), depending on the absolute meaning of this temperature.

As seen in Fig. 3.5, at a certain temperature of the calorimeter surface a qualitative change of the heat transfer character takes place. This temperature during the investigation was 920° - 950°K, i.e. it approximately corresponded to the ignition temperature of the methane-air mixture used in the experiments (λ_a = 0.5+0.6). When the surface temperature exceeds this value, the ratio of —— increases to the meaning of 2.2+2.3. According to D. Spolding /6/, the same heat transfer rise is obtained when calculating the heat exchange as regards total burning-gas enthalpy changes, taking into consideration formation of heat from combustion products. As compared with our data, D. Chen /11/, performing the same investigation but with a gas-oxygen mixture, obtained a constant coefficient increase of the convection heat transfer, with a surface temperature increase of 500 to 2000°K.

Thus, one can consider that during the process of hard surface burning at the temperature over the gas-air or gas-oxygen mixture ignition, the convective heat exchange is intensified, as compared with a cold surface, by 2 - 3 times. Even V. Veron /12/ in 1948 drew attention to the possibility of the existence of "live" and "dead" convection from gases to a hard surface.

The flat-flame burners develop this specific feature in industrial furnaces. They have relatively high rates of mixture output of approximately 50-60 m/sec. and heat transfer coefficient that is twice as great. By means of such burning devices, the absolute value of α^{gc}_{con} about a 300W/(m^{2}deg) in the heat flow units gives the value 0.10 - 0.15 MW/(m^2) at the real value of the temperature gradient of 300 - 500 degrees.

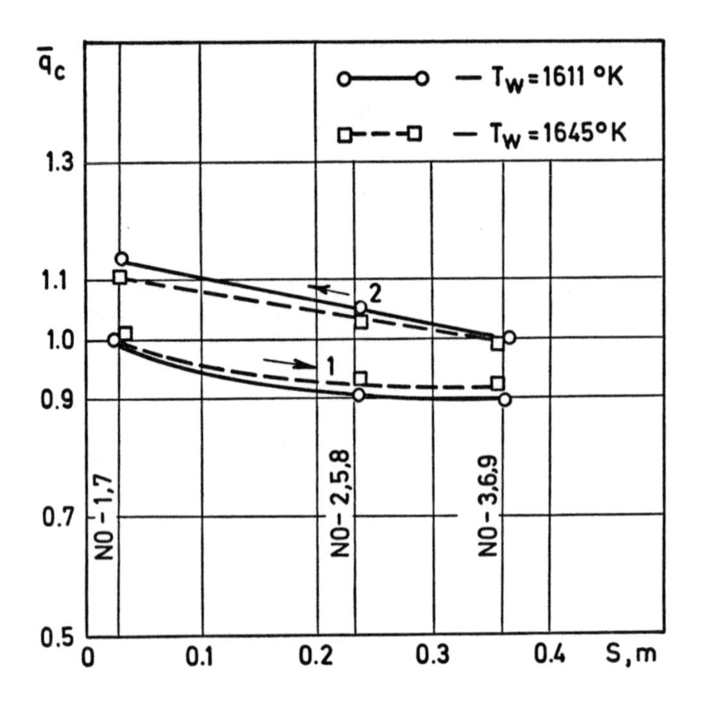

Fig. 3.2 Dependence of measured relative heat flux on the thickness of the gas layer (o-o-o - T_W = 1611°K: □-□-□ – T_W = 1645°K); 1 - by the indirect radiation regime of heat exchange; 2 - by the uniform radiation regime of heat exchange.

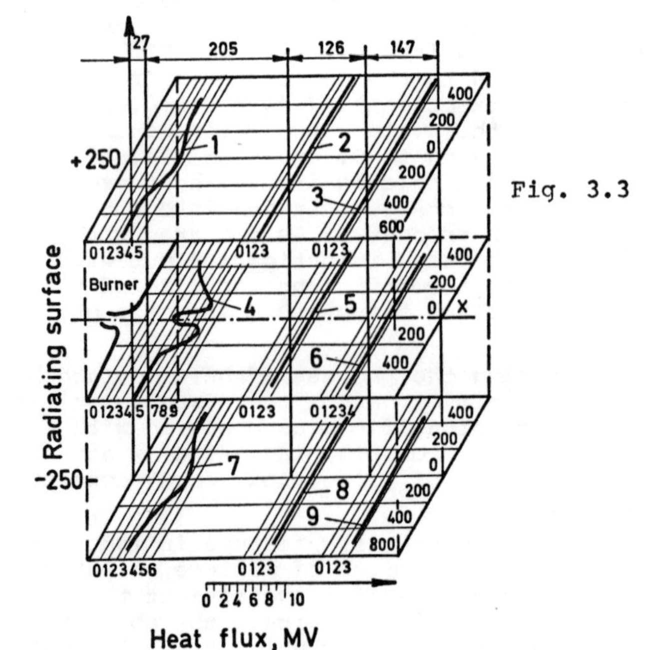

Fig. 3.3 Distribution of heat flux measured by the thermosound from the radiating front wall in different sections of the experimental chamber (measurement lines in chamber from Fig. 3.1)

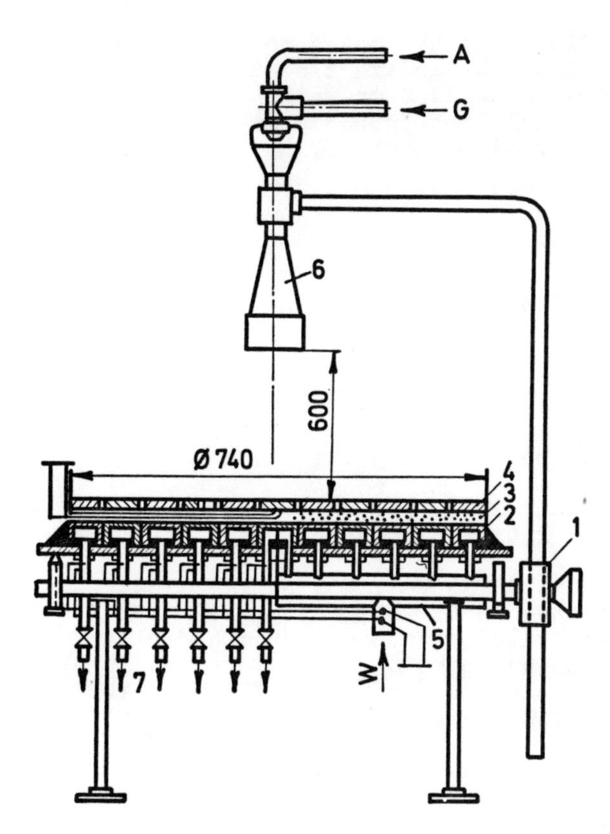

Fig. 3.4 Experimental design for investigation of the heating surface temperature effect on the convective heat transfer from burning flames;
1 - coordinate mechanism
2 - cell calorimeter
3 - insulating layer
4 - thermocouple with heating rings
5 - collector of water
6 - injection burner with two active jets
7 - water drainage
A compressed air supply
G - gas supply
W - water supply

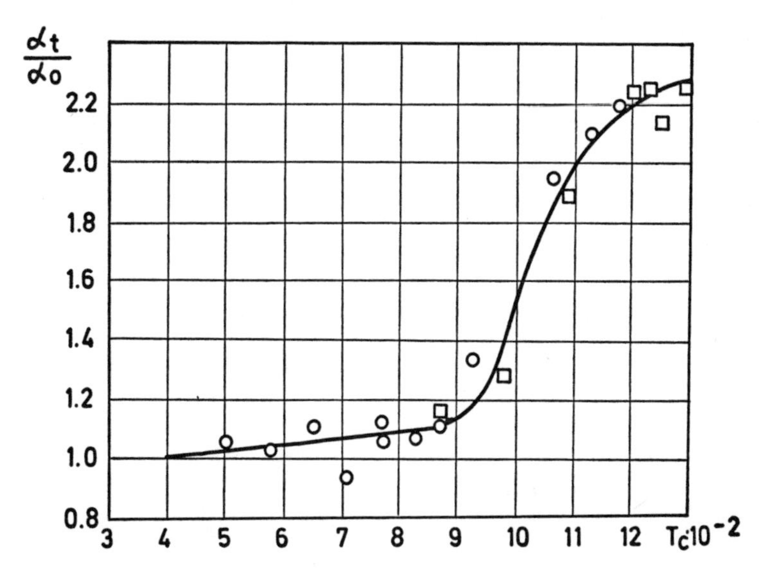

Fig. 3.5 Experimental dependence of the heating surface temperature effect on relative variation of the convective heat transfer coefficient from burning flame;
○ ○ 1 - λ_a = 0.5, w_{mi} = 35 m/s;
□ □ 2 - λ_a = 0.6, w_{mi} = 65 m/s.

Such a high value of the heat flux from the gases to the refractory walls of the heating surface allows the indirect heat-exchange regime to perform. With properly organized operation and respective design, the efficiency of industrial heat furnaces may be much higher than convectional uniform regime furnaces.

Flat-flame burners using the indirect heat exchange radiation regime are in operation in flame furnaces in many countries, including the U.S.S.R. /6,7/. Their application simplifies furnace designs and increases efficiency.

The author is obliged to his colleagues whose scientific results were used when writing this report.

REFERENCES

/1/ Glinkov, M.A.: "Osnovy obshchei teorii pechei", Moskva, Metallurgizdat (1962)
/2/ Timofeev, V.N., Sbornik Trudov UOVT V.5, Metallurgizdat, (1940)
/3/ Kavaderov, A.V.: "Teplovaya rabota plamennyh pechei", M., Metallurgizdat (1956)
/4/ Hottel, H., R. Egbert, Trans. ASME, Vol. 63, p. 297 (1941)
/5/ Eberhart, J. H. Hottel, Trans. ASME, Vol. 58, p. 185 (1936)
/6/ Erinov, A.E., B.S. Soroka: "Ratsionalnye metody szhiganiya gazovogo topliva v nagrevatelnyh pechah, Kiev, Tehnika (1970)
/7/ Soroka, B.S., A.E. Erinov, Gazovaya promyshlennost, No. 12 (1967)
/8/ Spolding, D., Sovremennye problemy teploobmena, "Energiya", M.-L. (1966)
/9/ Soroka, V.A., "Gazovaya promyshlennost", No. 11 (1972)
/10/ Soroka, B.S., A.E. Erinov, Soroka V.A., Teplofizika i teplotehnika, V. 19, p. 149 (1971)
/11/ Chen, D., Jour. of Inst. of Fuel, No. 11, p. 562 (1972)
/12/ Veron, V., La convection vive, Bull. Technique, No. 21 (1948)

Chapter 14

METHOD OF APPROXIMATE CALCULATION OF RADIANT HEAT TRANSFER BETWEEN GAS AND SURFACE

S. P. Detkov

The Ural Scientific Centre of Academy of Sciences of the USSR, Physico-Technical Problems Department, USSR, Sverdlovsk

Abstract

Formulae of bands and lines average parameters of the flame spectrum through the depth of volume are derived.Owing to them the Gas absorption is calcilated as well as in cases of constant temperature and pressure fields in the chamber's volume.Known ealier the three-parameter approximation for bands and the Curtis-Godson two-parameter narrow-scaling low are obtained as partial cases.

NOMENCLATURE

A_j, A	:	total band absorptance,cm
A_ω	:	spectral absorptivity,dimensionless
F	:	hipergeometric function
ϕ	:	degenerated hipergeometric function
Γ	:	gamma function
I_j	:	spectral intensity of incident flux,$cm^{-1}w/(m^2sr)$
P	:	gas pressure,bar
S	:	integrated band intensity,$cm^{-1}/(m\cdot bar)$
T	:	temperatire, ^oK
W_i, $\bar{w}$	:	total line absorptance,average total line absorptance for line group,cm^{-1}
b	:	line half width,cm^{-1}
c	:	constant
d	:	Line spacing,cm^{-1}
f_i, f_{i*}	:	envelopes of the bands,contoures of the lines
m_i	:	normal multiplier
P_s	:	partial pressure,bar
	:	average integrated line intensity,$cm^{-1}/(m\cdot bar)$
$u=\bar{S}x/2\pi\bar{\gamma}$	:	dimensionless thickness
$v=x_1/x$	:	dimensionless depth
x_1, x	:	depth,thickness,$m\cdot bar$
$y=\nu/\gamma; y_*=\nu/b$	:	dimensionless spectral arguments
n, n_*, a, a_*, c_*	:	indexes
α_ω	:	spectral absorption coefficient,$(m\cdot bar)^{-1}$
$\beta=2\pi b/d$	:	line structure parameter,dimensionless
γ, γ_1, γ_2	:	band width parameter,bound volume values,cm^{-1}
λ	:	dimensionless thickness
$\mu\equiv\cos\alpha$,where α =arctg y		
$\nu=\lvert\omega-\omega_0\rvert$	:	distance from band or line center,cm^{-1}
σ	:	total line intensity,$cm^{-1}/(m\cdot bar)$
τ_ω	:	optical thickness,dimensionless
ω	:	wavenumber,cm^{-1}
ω_0	:	band or line center,cm^{-1}

1. Introduction

Quantities of the total emissivity,the total absorptivity,and average absorption coefficients,form the basis of the radiative heat transfer calculation between Flame and Surface in the total spectrum.Thay are defined in terms of total absorption values in bands and lines which occupy the central place of heat engineering calculation.All values,mentioned above,correspond to the conditions of homogeneous fields of temperatures,pressures and concentrations. Any mode of the reduction of real flux path to the equivalent path with constant values of T,P and p is required while using them.This part is a weak point of the total calculation and requires a detail elaboration.In the present paper the method of the total absorption calculation in bands and lines of spectrum with quantities of T,P, , p, variable through the depth of volume is developed.Gas spectrum is represented by non-overlapping vibrational-rotational bands and separate lines.In the work /1/ the simplest methods of calculation have been compared.The Nevsky-Hottel method is of the greatest interest.More general and well-founded from the point of view of physics the method of calculation is published in /2,3/.Not depreciating the value of simplest approximations concentrate attention on investigations of the other kind.

At present,heat-engineers quickly apprehend achievements of the applied spectroscopy and themselves take part in its elaboration. The tree-parameter approximation is proposed for the total absorption calculation in spectrum's bands /4,5/.At this paper it is represented in a more general form.Formulae of the absorption are suitable for the calculation of Gas emissivity and radiative transfer between zones of any type,owing to the local thermodinamic equilibrium.Now,let us enumerate conditions,restricting the using of the three-parameter approximation:1)the considerable overlapping of spectrum's bands in the mixture of components without sure calculation of the bands overlapping, 2)the increasing of the temperature up to values,when bands run,"windows"of the spectrum disappear and the concept of a vibrational-rotational band gradually loses it meaning;this situation set in for the water vapour at temperatures higher then the 2000°C, 3)increase of the thickness up to values when edgings of the band are of essential importance and have no exact descreption for all existing wide-band models of the spectrum. For main burning products this situation sets in at $x \gtrsim 10$ m·bar.

The most fundamental method consisting in using of tables and formulae of spectral absorption coefficients,averaged in narrow intervals of spectrum /6,7,etc./is recommended in all enumerated cases.It is advisable to combine calculation methods,as in /8/,though if weak separate-situated bands frequently occur in a spectrum and it's possible to use the tree-parameter approximation for them.

The most important characteristic of a band is its envelope, descripting the dependence of the spectral absorption coefficient, averaged in a narrow range of the spectrum as function of the wave number

$$s/d = (S/\gamma) f(y), \quad y = \nu/\gamma, \quad \nu = |\omega - \omega_0| \qquad (1)$$

We consider symmetrical,two-branch bands.In this case it is easy to go on to the calculation of absorption in one branch and then in the more real two-asimmetrical branch bands /9/.There are two parameters of the band S and γ in the formula (1).It is necessary to supplement them with the rotational structure parameter $\beta = 2\pi b/\alpha$

2. The three-parameter apprtoximation

The three-parameter approximation consists in the definition of parameters $\bar{S}, \bar{\gamma}$ and β. The indistorted envelope of band is assumed, therefore the absorption is calculated by using averaged parameters so that, if fields of values T, P, p in volume were uniform. It is shown below that relationships of the approximation are rather simpl and the method as a whole has an acceptable error in case of three of envelopes

$f_1 = m_1(1+y^2)^{-(n+1)/2}$, $m_1 = \Gamma((n+1)/2)/\sqrt{\pi}\Gamma(n/2)$, $n \geq 1$

$f_2 = m_2(1+y^{n+1})^{-1}$, $m_2 = ((n+1)/2\pi)\, \mathrm{Sin}(\pi/(n+1))$, $n \geq 1$

$f_3 = m_3 y^{c_*} \exp(-y^{a})$, $c_* > 0$, $a \geq 1$

Some other versions are sufficiently simple too, for example, $f_4 = m_4(1+y)^{-(n+1)}$, $n \geq 1$. Numbers m_i are defined from the normalisation $\int_{-\infty}^{\infty} f_i dy = 1$.

The flux absorption intensity of the group of non-overlapping bands is equal to

$$\sum_j I_j A_j \ , \ A_j = \int A_\omega d\omega \tag{2}$$

The first relationship of the three-parameter approximation is obtained from the demand of exact solution in limit of small thickness: $\lim_{x \to 0} A_j = \bar{S}x$ and, hence, $\bar{S} = \int_0^1 S\, dv$ (3)

In this case the form of envelope and parameters γ, β is of no significance. The second relationship is obtained from the demand of exact aolution in case of the complete lines overlapping. The rotational structure parameter vanishes in this case

$$A_\omega = 1 - \exp(-\bar{s}x/d)$$

Here the spectral absorption coefficient is averaged not only over all lines, giving contribution to the absorption in preaent narrow interval of spectrum, but over the volume thickness too. For any thicknesses the relationship (4) is valid

$$\bar{s}/d = \int_0^1 (s/d)\, dv \tag{4}$$

or, using the formula (1), $\bar{S}\bar{f}/\bar{\gamma} = \int_0^1 (Sf/\gamma)\, dv$

The main assumption consists in that the indistorted form of envelope of an equivalent band with parameters $\bar{S}, \bar{\gamma}$ is adopted. Actually, its distortion may be considerable. It is shown in works /10,11/ where as an example the single line is examined that is fully suitable for the envelope of band. In heat engineering calculation of the envelope is no intrest, and only the total absorption is defined. In this case the assumption, drastically simplifying the calculation, does not insert the great error. For further deduction it's necessary to choose the form of envelope. When substituting the envelope f_1,

$$\bar{S}\bar{\gamma}^{\,n}/(\bar{\gamma}^2 + \nu^2)^{(n+1)/2} = \int_0^1 S\gamma^n dv \,/(\gamma^2 + \nu^2)^{(n+1)/2}$$

In the limit of great thicknesses the relationship is simplifyed. At $x \to \infty$ the absorption will be practically complete in the centre of the band and even at $\nu > \gamma$. Then the choice of value γ in the denominator under integral is of no essential significance. Moreover, it is true for $\nu \gg \gamma$, when the item γ^2 is allowed to neglect. Thus, at all ν in the denominator under integral, the sign of equality between γ and $\bar{\gamma}$ is allowed to adoptand then

$$\bar{S}\bar{\gamma}^n = \int_0^1 S\gamma^n \, dv \tag{5}$$

It is easy to conclude that using envelopes f_2 and f_4 gives the same result.The envelope f_3 requires explanation.Envelopes f_1 and f_2 may be sometimes near to the envelope f_3 when selecting the index n. The distinction in the absorption will exist only at great thicknesses of Gas.For example,

$$f_1 = 1{,}1641(1+y^2)^{-5}\ , \quad f_3 = 1{,}0998\ \exp(-3{,}8y^2)$$

Let's consider another approach now.At $x\rightarrow\infty$ the absorption is practically complete in the central part of band,and the form of the envelope is of no significance here.The edgings of the band play an essential role and the approximation is allowed for them

$$\exp(-y^{\,a}) \approx \underline{\underline{C}}y^{-a_*}$$

In this case the formula (5) is obtained too,where $n = a_* - c_* - 1 > 0$. As the thickness of Gas increases,the part of edging determing the absorption accretion goes away from the centre of band.Indexes a_* and n are increased.Thus,the first two relationships of the three-parameter approximation are based for all written down envelopes. For the latter relationship,determing the rotational structure parameter,it is necessary to choose the narrow-band model.Assume the R.Goody's statistic model

$$A_\omega = 1 - \exp(-\bar{W}/d) \tag{6}$$

$$\bar{W} = \int_0^\infty W_i \exp(-\bar{\sigma}/\bar{s})\, d(\bar{\sigma}/\bar{s}) \tag{7}$$

$$W = \int_{-\infty}^{\infty} [1 - \exp(-\tau_\omega)]\, d\nu\ , \quad \tau_\omega = \bar{\alpha}_\omega x\ , \quad \alpha_\omega = (\bar{\sigma}/\bar{b})\bar{f}_*(y_*) \tag{8}$$

Values $\bar{\sigma}, \bar{s}, \bar{b}$ and $\bar{f}_*$ are averaged through the depth of volume. The mode of averaging is obtained from the spectral optic depth formula

$$\tau_\omega = x\int_0^1 \alpha_\omega \, dv \tag{9}$$

Here α_ω -the local value of the spectral absorption coefficient, define using local values σ, b and f_* .From formulae (8) and (9) it follows that

$$\bar{\sigma}\bar{f}_*/\bar{b} = \int_0^1 (\sigma f_*/b)\, dv \tag{10}$$

The contour,calculated from the formula (10),is distorted,but we neglect the distortion for reasons,already expressed after the formula (4).When adopting contours f_{*1}, f_{*2} and f_{*3} exactly as well as envelopes f_1, f_2 and f_3 written down above and replacing $y \rightarrow y_*$ then lead to the relationship by means of proving the formula (5)

$$\bar{\sigma}\bar{b}^{n_*} = \int_0^1 \sigma b^{n_*} \, dv \tag{11}$$

The formula of the total intensity line averaging is written down on the same basis that relationship (3)

$$\bar{\sigma} = \int_0^1 \sigma \, dv \tag{12}$$

Formulae (11) and (12) represent the two-parameter approximation for a single spectral line.At $n_* = 1$ the Curtis-Godson approximation is obtained.As can be seen it is possible to consider the two-parameter approximation as a partial case in reference to the three-parameter approximation.The relationship (13) is obtained after successive substitution of formulas (9),(8) in (7) and using (10) and changing the order of integration

$$\bar{W} = \bar{b}\int_{-\infty}^{\infty} \frac{(\bar{s}/\bar{b})\bar{f}_{*}x}{1+(\bar{s}/\bar{b})\bar{f}_{*}x}\, dy_{*} \tag{13}$$

When substituting the contour $f_{*2} = m_2(1+y_*^{n+1})^{-1}$ the formula (13) takes the form

$$\frac{\bar{W}}{d} = \frac{(\bar{s}/d)\,x}{(1+\bar{s}\,m_2\,x/\bar{b})^{1-1/(n_*+1)}} \tag{14}$$

Demand the exact solution at the great thickness of Gas. From the formula (14) it is inferred

$$\frac{\bar{W}}{d} = \sqrt[n_*+1]{(\bar{s}/d)\,[\bar{b}/(m_2 d)]^{n_*}x}$$

The expression under root have a sense of the spectral optic depth, described by the formula, which is analogous to (9). The repetition of the deduction similar to transition from (9) to (11) leads to the result

$$(\bar{s}/d)\ \bar{\beta}^{n_*} = \int_0^1 (s/d)\,\beta^{n_*}\,dv$$

When substituting the formula (2), using the envelope f_1 or f_2 and repeating all reasonings, preceding the writing down of the formula (5) we obtain the last three-parameter approximation formula

$$\bar{S}\bar{\gamma}^{n}\bar{\beta}^{n_*} = \int_0^1 S\gamma^{n_*}\,dv \tag{15}$$

The dispersive contour of lines is everywhere assumed in heat-engineering calculation, therefore it's necessary to scrutinize the partial case of the formula (15) at $n_* = 1$

$$\bar{S}\bar{\gamma}^{n}\bar{\beta} = \int_0^1 S\gamma^{n}\beta\,dv \tag{16}$$

Using the formula (13) and averaging the parameter b in the group of lines according to /11/ the most wide-spread expression is obtained: $\bar{W}/d = (\bar{s}x/d)/(1+\bar{s}x/4\bar{b})^{1/2}$. At $x\to\infty, \bar{W}/d=\sqrt{4(\bar{s}/d)(\bar{b}/d)x}$,

$$\text{hence,}\ (\bar{s}/d)(\bar{b}/d) = \int_0^1 (s/d)(b/d)\,dv \tag{17}$$

The three- parameter approximation relationship (16) is possibl to obtain directly from this formula. It does replace the second formula of the two-parameter approximation in case of lines overlapping. Formulae (17) and (4) represent the Curtis-Godson approximation when using the narrow-band model.

Known formulas of the three-parameter approximation are partial to derived formulas (15),(5),(3) at $n = n_* = 1$. They were obtained for envelopes $f = 0{,}5\cdot\exp(-y)$ and $f = y\cdot\exp(-y^2)$, the dispersive contour of lines provided that the change of the parameter of width is small.

The comparison with exact results is required for the definition of limits of application of the derived formulas.

3. The control of the three-parameter approximation

Dwell on the question of check-up of first relationships (3) and (5). The rotational structure of a band is of no significance in this case. Incidentally formulae of the two-parameter approximation (12) and (11) are controled. The total absorption in the two-branch band with parameters averaged according to formulas (3) and (5) is calculated, using the formula

$A = \bar{\gamma}\int_{-\infty}^{\infty}[1-\exp(-2\pi uf)]\,dy$, $u = \bar{S}x/(2\pi\bar{\gamma})$, $y=\nu/\bar{\gamma}$. In case of the envelope f_1,

$$A = 2\pi\bar{\gamma}u\,W_n(u), \tag{18}$$

$$W_n = 2m_1(n+1)\int_0^1 \mu^{n-1}\exp(-2\pi\, m_1 u\mu^{n+1})\sqrt{1-\mu^2}\,d\mu$$

In case of the envelope f_2,

$$A = 2\pi\bar{\gamma}uM_n(u), \tag{19}$$

$$M_n = 2m_2(n+1)\int_0^1 \frac{\mu^{n-1}(1-\mu^2)^{n/2}\,d\mu}{\left[\mu^{n+1}+(1-\mu^2)^{(n+1)/2}\right]^2}\exp\left[-\frac{2\pi m_2 u\mu^{n+1}}{\mu^{n+1}+(1-\mu^2)^{(n+1)/2}}\right]$$

In case of the dispersive contour $W_1 = M_1 = L/u$, whence it appears the new expression of the Ladenburg and Raike's function

$$L = u\cdot\exp(-2u)\,\phi(3/2,2,2u) = (4/\pi)u\int_0^1\sqrt{1-\mu^2}\exp(-2u\mu^2)\,d\mu$$

It can be checked up by using the known tables of the degenerated hipergeometric function $\phi(1/2,2,2u)$ and the equation

$$\phi(3/2,2,2u) = 3\phi(-1/2,2,2u)+2(2u-1)\phi(1/2,2,2u)$$

Function W_n and M_n are tabulated. Some values are given in table 1.

Table 1. Functions W_n and M_n (only decimal figures are presented)

Function W_n.

n \ u	0,01	0,1	0,5	1	2	10	50
1	99498	95233	80132	67300	52368	24911	11254
2	98064	91427	67649	51196	35457	13133	04600
3	98783	88710	60088	42778	27971	09211	02888
4	98555	86560	54852	37514	23706	07273	02063
5	98377	84777	50932	33855	20916	06138	01728
6	98214	83229	47836	31125	18929	05446	01496
7	98082	81870	45307	28990	17426	04871	01186
9	97852	79527	41367	25834	15213	03960	01121
10	97750	78497	39784	24636	14359	03789	01099

Function M_n

n \ u	0,01	0,1	0,5	1	2	10	50
1	99503	95240	80146	67367	52378	24910	11255
2	99140	91930	68703	51902	35447	12654	04358
3	98948	90215	63281	45187	28880	08934	02687
4	98835	89213	60246	41577	25528	07246	020085
6	98710	88124	57033	37857	22196	05708	01442
8	98645	87552	55379	35979	20558	05004	01199
10	98605	87204	54380	34854	19590	04604	01068

Discrepancy of values at n=1 is equal rouph to the error of quadrature (the Gaussian quadrature with eight abscissas).

Formulas (18) and (19) are suitable for any fields of quantities T, P, p in volume. Exact solutions, following below, are obtained with limitation. It slightly reduces the value of the control: $S=c\,d\gamma/dv$. Howeve, the influence of the parameter S, is as many times lower (approximately tenfold) as that of the influence of the width parameter γ. Any field of this parameter may be assumed. It is shown in the work /10/ when the single line and the linear change of S and γ-parameters through the depth are adopted. Using formulae (3) and (5) we obtain

$\bar{S}=c(\gamma_2-\gamma_1)$, $\bar{\gamma}=\gamma_2\sqrt[n]{(1-q^{n+1})/[(1-q)(n+1)]}$, $q=\gamma_1/\gamma_2, \gamma_2>\gamma_1$

Write down exact formulae of the total absorptance. In case of envelope f_1,

$$A_1=4\lambda\gamma_2\int_0^1 (1-\mu^2+q^2\mu^2)^{\lambda}\exp(\lambda\psi)\left[\frac{1-q^2}{1-\mu^2+q^2\mu^2}-\varphi\right]\sqrt{1-\mu^2}\,d\mu$$

$$\varphi=\sum_{k=1}^{t}\mu^{2(k-1)}\left[1-q^{2k}/(1-\mu^2+q^2\mu^2)^{-(k+1)}\right]$$

$$\psi=\sum_{k=1}^{t}(\mu^{2k}/k)\left[1-q^{2k}/(1-\mu^2+q^2\mu^2)^{-k}\right]$$

$$\lambda=cm_1x/2=\pi m_1u\sqrt[n]{(1-q^{n+1})/[(n+1)(1-q)^{n+1}]}\quad,\quad n=3,5,7,9,11$$

$$t=(n-1)/2=1,2,3,4,5 \qquad u=\bar{S}x/2\pi\bar{\gamma}$$

As can be seen, four even values n are eliminated. Dimensionless total band absorptance $\bar{A}_1=A_1/\gamma_2$ are tabulated. Some values are given in Table 2.

Table 2. Values $\bar{A}_1=A_1/\gamma_2$ at envelope f_1.

q	n \ u	0,01	0,1	1	2	5	10
q=0	1	0,03114	0,29379	2,0000	3,1417	5,3333	7,7313
	3	0,03763	0,32377	1,4947	2,0135	2,8029	3,4958
	5	0,04244	0,33634	1,2658	1,6191	2,1178	2,5241
	7	0,04471	0,33523	1,1267	1,4028	1,7786	2,0737
	9	0,04622	0,33111	1,0299	1,2607	1,5683	-
q=0,1	3	0,04032	0,35087	1,5934	2,1261	2,9357	3,6480
	5	0,04374	0,35950	1,3286	1,6856	2,1898	2,6014
	7	0,04589	0,35544	1,1742	1,4514	1,8291	2,1262
q=0,5	3	0,04818	0,43054	2,0467	2,6837	3,6255	4,4549
	5	0,04935	0,42084	1,6401	2,0359	2,5835	3,0310
	7	0,05032	0,41317	1,4171	1,7122	2,1079	2,4195

In case of envelope f_2,

$$A_2=4\lambda\gamma_2(1-q^{n+1})\int_0^1\left[\frac{q^{n+1}\mu^{n+1}+(1-\mu^2)^{(n+1)/2}}{\mu^{n+1}+(1-\mu^2)^{(n+1)/2}}\right]^{[2\lambda/(n+1)]-1}\frac{(1-\mu^2)^{n/2}\mu^{n-1}d\mu}{[\mu^{n+1}+(1-\mu^2)^{(n+1)/2}]^2}\tag{20}$$

$$\lambda=cm_2x/2=\pi m_2u\sqrt[n]{(1-q^{n+1})/[(n+1)(1-q)^{n+1}]},\quad n=1,2,3,\ldots$$

Function $\bar{A}_2=A_2/\gamma_2$ is tabulated. Some values are given in Table 3.

Table 3. Values $\bar{A}_2=A_2/\gamma_2$ at envelope f_2, q=0.

n \ u	0,01	0,1	0,5	1	2	10	50
2	0,03576	0,32184	1,1409	1,7250	2,4184	4,5283	7,8837
3	0,03888	0,34060	1,1205	1,6091	2,1320	3,4851	5,3018
4	0,04118	0,35436	1,1143	1,5522	1,9892	2,9986	4,2071
6	0,04440	0,37426	1,1162	1,5029	1,8555	2,5518	3,2642
8	-	-	1,1230	1,4846	1,7966	2,3496	2,8546

Discrepancies of formulas (19) and (20) are shown in Fig.1.

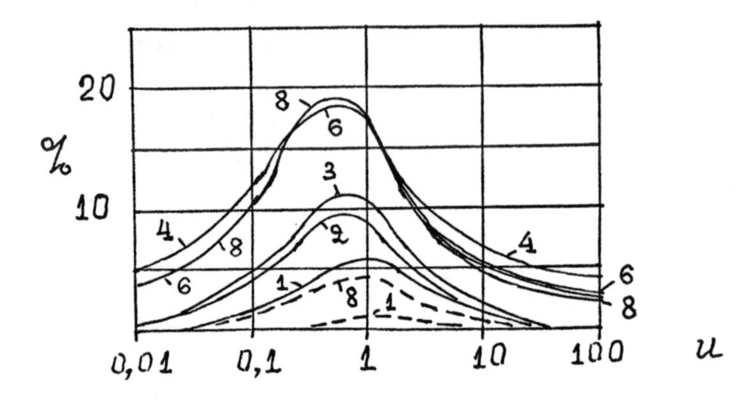

$u = \bar{S}x/(2\pi\bar{\gamma})$

Fig.1.Errors of the total band absorptance.The comparison of the results,obtained when approximation(19) and exact (20) relationship are used.Curves 1,2,...8 correspond to the indexes n in the expression for the envelope f_2.Curves at q=0 shows maxima error(-----).At q=0,5 (curves - = - -) the error is lower tenfold.The approximate values are always more then the exact one.

Curves of error of formula (18) have a little difference.

At n=1 results are available in the literature.At /12/ the expression

$$A = 2\pi\gamma_2\lambda\left[F(-\lambda, 1/2, 1, 1-q^2) - q^2 F(1-\lambda, 1/2, 1, 1-q^2)\right], \quad \lambda = cx/(2\pi)$$

is obtained.The table of the hipergeometric function F is represented ibidem.More simple result is obtained from the formula (20).

$$A = \pi\gamma_2\lambda \quad F(1-\lambda, 1/2, 2, 1-q^2)$$

Formulas agree on the basis of recurrent relationships for function F.

Four integrals,written above,both approximative and exact are obtained using transformation $\int z dy = yz - \int y dz$,where z is the primary expression under integral in front of dy.It proved to be $yz\Big|_{-\infty}^{\infty} = 0$ in all cases.Then the new variable $y = \mathrm{tg}\,\alpha$ and the designation $\mu = \cos\alpha$ are introduced.In this way:a)resulta are extremely simplifyed,b)peculiarities are liquidated in integrals,c)integrals take the form,suitable for the application of the Gauss quadrature.

As a summery,it is necessary to note that:1)the theory is corroborated;the result is exact in cases of small and great thicknesses,2) the three-parameter approximation and the two-parameter approximation always give the increased absorption in case of envelopes f_1 and f_2,3)with q increasing the error quickly decreases; at q=0,1 it decreases approximately twofold or more;at q=0,5 it decreases approximatly tenfold or more.It is believed that in heat-power-engineering chambers q> 0,4 and then the error of approximation will seldem be over 2 - 3 per cent.

REFERENCES

/1/ Detkov,S.P.; "Comparison of simplest approximations in non-isothermal radiant transfer calculations".Teplofis.Vysok. Temp.,Vol.10,N°3,(1972)

/2/ Detkov,S.P.: "Emission and absorption of non-isothermal gases". Izv.Akad.Nauk SSSR.Energetika i Transport, N°5,(1971)

/3/ Detkov,S.P.: "Absorption of heat radiation in the band spectrum medium".Teplofiz.Vysok.Temp., Vol.8, N°4,(1970)

/4/ Edwards,D.K.,S.I.Morizumi; "Scaling of vibration-rotation band parametera for nonhomogeneous gas radiations".Journ.Quant. Spectr.Rad.Transfer, Vol.10, N°3,(1970)

/5/ Chan,S.H. and C.L.Tien: "Total band absorptance of non-isothermal infrared-radiating gases".Journ.Quant.Spectr.Rad. Transfer, Vol.9, N°9,(1969)

/6/ Ludwig,C.B.: "Mesurements of Curves-of-growth of hot water vapor".Appl.Opt., Vol.10, N°5,(1971)

/7/ Ferriso,C.C.,C.B.Ludwig and A.L.Thomson: "Empirically determined infrared absorption coefficients of H_2O from 300 to 3000°K".Journ.Quant.Spectr.Rad.Transfer,Vol.6, N°3,(1966)

/8/ Leckner,B.: "Spectral and total emissivity of carbon dioxide". Combustion and Flame, Vol.17, N°1,(1971)

/9/ Malkmus,W.: "Infrared emissivity of carbon dioxide (4,3-band)".Journ.Opt.Soc.Amer., Vol.53, N°8,(1963)

/10/ Cogley,A.C.: "Radiative transport of Lorenz lines in nonisothermal gases".Journ.Quant.Spectr.Rad.Transfer, Vol.10, N°9, (1970)

/11/ Goody,R.M.: "Atmospheric Radiation".Oxford University Press, Oxford,(1964)

/12/ Yamamoto,G. and M.Aida: "Transmission in a non-homogeneous atmoaphere with an absorbing gas of constant mixing ratio". Journ.Quant.Spectr.Rad.Transfer, Vol.10, N°6,(1970)

SECTION II:

RADIATIVE PROPERTIES

Chapter 15

INFRARED GASEOUS RADIATION*

Robert D. Cess

State University of New York, Stony Brook, New York, U.S.A.

Abstract

Analytical methods for treating infrared gaseous radiation are presented, with specific emphasis upon combining techniques which have been developed in both engineering and atmospheric physics. A relatively simple procedure is illustrated for determining net radiative heating by vibration-rotation bands, and this is shown to be in excellent agreement with more detailed calculations.

1. Introduction

The object of this paper is to illustrate the incorporation of spectroscopic information into the radiative transfer equations, and to present simplified procedures for treating radiative energy transfer within gases. Restriction is made to infrared radiation resulting from permanent dipole transitions. Here the absorption and emission of thermal radiation is a consequence of coupled vibrational and rotational energy transitions, with a permanent dipole moment existing only for vibrational modes which do not possess symmetry. For unsymmetric diatomic molecules, such as carbon monoxide, the infrared spectrum consists of a fundamental vibration-rotation band occurring at the fundamental vibrational frequency of the molecule; i.e., the band arises due to an energy transition between two adjacent vibrational levels. Vibrational transitions spanning three levels produce the first overtone band located at twice the fundamental frequency of the molecule, and subsequent overtone bands occur at higher multiples of the fundamental frequency.

The picture is much the same for polyatomic molecules, except that these have more vibrational degrees of freedom. For example, carbon dioxide is a linear triatomic molecule and thus possesses four vibrational degrees of freedom. The two bending frequencies are, however, identical, while one of the stretching modes is symmetric and thus has no permanent dipole moment. Consequently, carbon dioxide has two fundamental bands. In addition to fundamental and overtone bands, the infrared spectrum of polyatomic molecules also includes combination and difference bands which occur at linear combinations or differences of the fundamental frequencies. Again choosing carbon dioxide as an example, the strongest infrared bands are the 15μ and 4.3μ fundamental bands and the 2.7μ combination band.

There are a number of important applications of infrared gaseous radiation and, of course, within the context of the present seminar the application of interest involves combustion systems. Other applications concern energy transfer within the atmospheres of planets and late-type (cool) stars. While these energy transfer systems are perhaps not of direct concern to combustion engineers, what is important is that analytical techniques developed in one field of study may have

*This work was supported by the National Science Foundation through Grant Number K036988.

direct application to other fields. It is thus the purpose of the present paper to present a rather general approach to infrared gaseous radiation, employing analytical techniques which have been developed both for combustion systems and planetary atmospheres.

2. Radiative Flux Formulation

For illustrative purposes, let us consider the energy transfer system as illustrated in Fig. 1. This consists of a gas bounded by two infinite parallel black plates, each having the same temperature T_1. With regard to applying conservation of energy locally within the gas, we first need to express the local radiative flux, q_R, and from Cess and Tiwari [1] this may be written as

$$q_R = \sum_{i=1}^{N} q_{Ri} \tag{1}$$

where the subscript i denotes the ith vibration-rotation band, N is the total number of bands, and

$$q_{Ri} = \frac{3}{2}\int_0^{u_i} \phi_i(u')A_i'\left[\frac{3}{2}(u_i - u')\right]du' - \frac{3}{2}\int_{u_i}^{u_{oi}} \phi_i(u')A_i'\left[\frac{3}{2}(u' - u_i)\right]du' \tag{2}$$

where

$$u_i = \frac{S_i P y}{A_{oi}}\ , \quad u_{oi} = \frac{S_i P L}{A_{oi}}\ , \quad \phi_i = e_{\omega i}(T) - e_{\omega i}(T_1)$$

while S_i and A_{oi} are the band intensity and bandwidth parameter for the ith band, respectively, as discussed in Section 3, P is pressure, T is temperature, and $e_{\omega i}$ is Planck's function at the wave number ω_i of the ith band. Furthermore, $A_i'(u_i)$ denotes the derivative of the band absorptance $A_i(u_i)$ with respect to u_i, where the band absorptance is defined by

$$A_i = \int_{-\infty}^{\infty} (1 - e^{-K_\omega y})\, d(\omega - \omega_i)$$

with K_ω denoting the volumetric absorption coefficient for the given band, and the integration limits simply imply integration over the entire band.

It is clear that equation (1) first requires a formulation for the total band absorptance, and this will be discussed in the following section. It will be

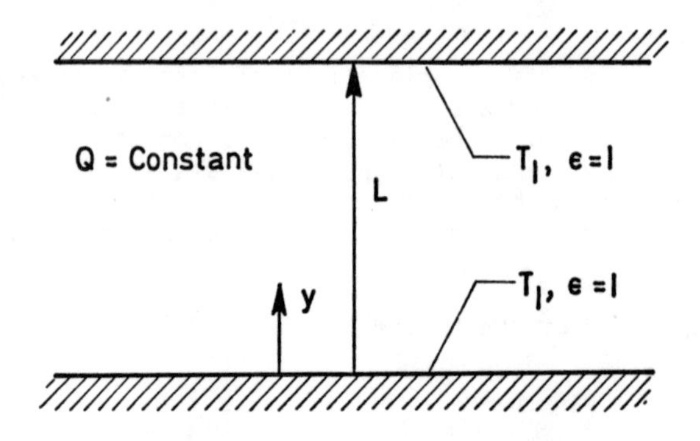

Fig. 1 Physical model and coordinate system.

useful, however, to first consider an example employing equation (1). For this purpose we shall assume a uniform heat source per unit volume, Q, within the gas, while energy is transported within the gas by both thermal conduction and radiation. Letting λ denote the thermal conductivity, local conservation of energy yields

$$\lambda \frac{d^2T}{dy^2} - \frac{dq_R}{dy} + Q = 0 \tag{3}$$

Furthermore, employing the linearization

$$e_{\omega i}(T) - e_{\omega i}(T_1) \simeq \left(\frac{de_{\omega i}}{dT}\right)_{T_1} (T - T_1)$$

then combination of equations (1) through (3), and integrating once, yields [2]

$$\frac{d\theta}{d\xi} + (\xi - \tfrac{1}{2}) = \frac{3L}{2\lambda} \sum_{i=1}^{N} \left(\frac{de_{\omega i}}{dT}\right)_{T_1} u_{oi} \left\{ \int_0^{\xi} \theta(\xi') A_i' \left[\frac{3u_{oi}}{2}(\xi - \xi')\right] d\xi' - \int_{\xi}^{1} \theta(\xi') A_i' \left[\frac{3u_{oi}}{2}(\xi' - \xi)\right] d\xi' \right\} \tag{4}$$

where $\xi = y/L$, and

$$\theta = \frac{T - T_1}{QL^2/\lambda}$$

Employing the band absorptance formulation of Tien and Lowder [3], equation (4) has been solved numerically for a variety of gases and temperatures by Cess and Tiwari [2], and center-line temperatures for CO_2 at 1000°K are illustrated in Fig. 2. The ordinate value of 0.125 corresponds to negligible radiative transfer, and Fig. 2 thus illustrates that radiation can indeed be a significant energy transport mechanism relative to thermal conduction, even for quite small physical dimensions.

It is important to note from Fig. 2 that in the limit of large pressure, pressure ceases to be a parameter. The reasons for this, as discussed in detail by Cess and Tiwari [1], is that in the $u_{oi} >> 1$ limit the band absorptance is of the form $A_i = A_{oi} \ln u_i$, such that when this asymptotic expression is employed within equation (4), the temperature profile becomes independent of pressure. Similar analytical conclusions have been reached by Edwards, et al [4], and Gierasch and Goody [5], while Schimmel, Novotny, and Olsofka [6] have experimentally illustrated this invariance with pressure for large pressure path lengths.

3. Band Absorptance Formulations

3.1. Pressure broadening

Let us now return to the question of describing an appropriate band absorptance formulation for use in equation (2), and consider first pressure levels which are sufficiently large such that the individual rotational lines of a vibration-rotation band are pressure broadened; i.e., the line shapes are described by the Lorentz line profile. A substantial body of literature exists concerning band absorptance formulations, for example references [3], [4], and [7] through [15]. Often such formulations are developed to satisfy several possible asymptotic

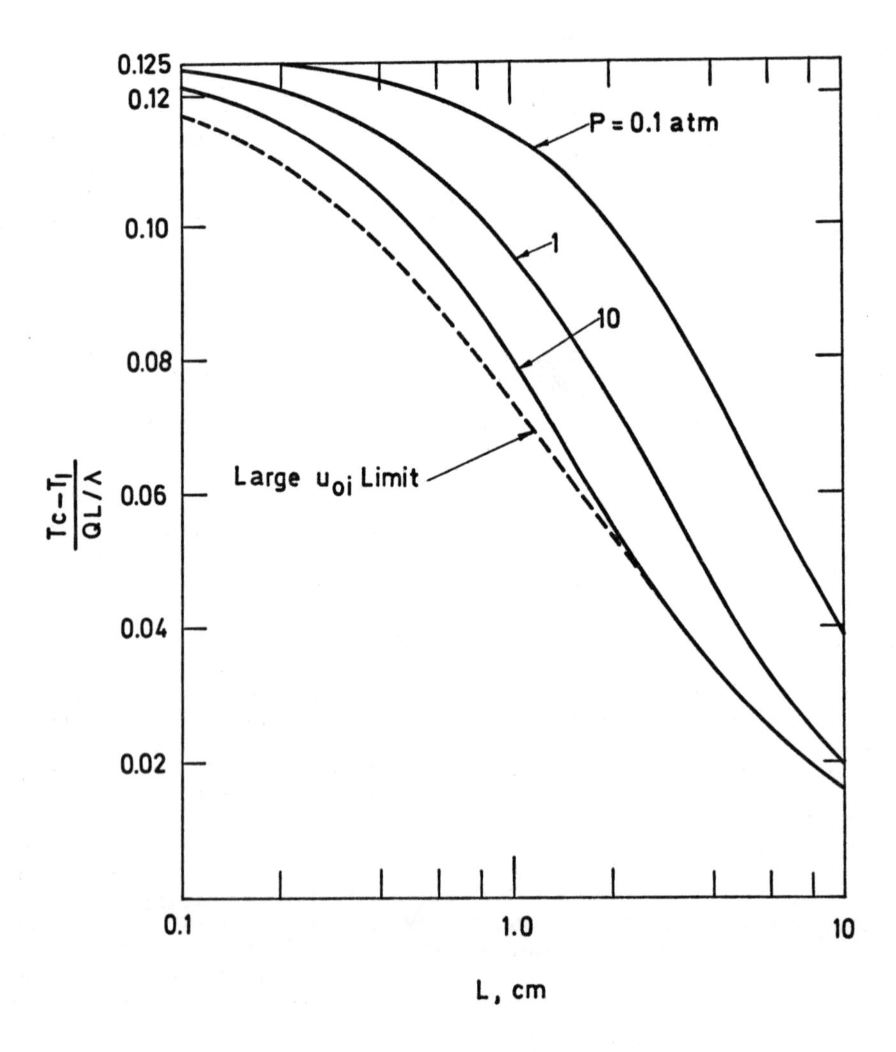

Fig. 2 Results for CO_2 with T_1 = 1000°K.

limits [3,8,11], and we presently seek as simple a formulation as possible which satisfies only the asymptotic limits which appear to be physically important with regard to realistic radiative transfer situations.

The band absorptance for pressure broadened lines depends upon two dimensionless parameters; the pressure path length, $u = SPL/A_o$, and the line structure parameter, $\beta = 4\gamma/d$, where γ is the pressure broadened line half-width and d the mean line spacing. For brevity we have dropped the subscript i which denotes a given band location. In general, the band absorptance may be expressed as [8]

$$A = A_o f(u,\beta)$$

A rather simple expression for the band absorptance has recently been proposed by Cess and Ramanathan [16]. They illustrate that for the carbon dioxide atmospheres of Venus and Mars, $u/\beta >> 1$. This condition also holds for carbon dioxide within the Earth's atmosphere as well as methane within the atmospheres of Jupiter and Saturn. The physical implication of $u/\beta >> 1$ is that the important rotational lines of a given vibration-rotation band correspond to the strong-line limit [11], and in this limit Cess and Ramanathan suggest the band absorptance formulation

$$A = 2A_o \ln (1 + \sqrt{\beta u}) \tag{5}$$

For small and large values of βu, respectively, equation (6) reduces to the limiting forms $A = 2A_o\sqrt{\beta u}$ and $A = A_o \ln (\beta u)$. The first expression represents strong nonoverlapping lines, while the second denotes strong overlapping lines. Equation (5) thus constitutes an analytic interpolation between these two asymptotic limits.

Although it has previously been shown that $u/\beta >> 1$ for atmospheric applications, it remains to illustrate this for small-scale terrestrial systems. From the definitions of u and β, and letting $\gamma = \gamma_o P$, where γ_o is the half-width per unit pressure, then

$$\frac{u}{\beta} = \frac{SLd}{4A_o\gamma_o}$$

It is important to note that u/β is independent of pressure. In the event of foreign-gas broadening, the above expression will additionally contain the ratio of partial to broadening pressures, but this will still be independent of pressure. Consider again the conditions of Fig. 2; i.e., CO_2 at 1000^oK. Choosing $L = 1$ cm, it may be shown that the 4.3μ is the main contributor to radiative transfer, and $u/\beta = O(10^2)$, clearly satisfying the strong-line limit. On the other hand, the requirement that $u/\beta >> 1$ will not be met for the 2.7μ band, but this is of no consequence, since the 2.7μ band will not be significant for such a small path length. At larger values of L for which this band is important, the condition $u/\beta >> 1$ is satisfied.

While it is not possible to conclusively prove that $u/\beta >> 1$ under all radiative transfer situations, experience indicates that this is probably a useful limit for a large number of radiative transfer problems.

Let us now turn our attention to the spectroscopic quantities A_o, γ, and S as required for use in equation (2). For illustrative purposes, particular attention will be directed to the 4.3μ fundamental band of CO_2. With respect to the bandwidth parameter A_o, Edwards and Menard [8] have shown that $A_o \sim \sqrt{T}$. Empirically determined results have been presented by Edwards and Balakrishnan [14] for a large number of gases, and for the 4.3μ CO_2 band

$$A_o = 19.4\left(\frac{T}{300}\right)^{\frac{1}{2}} \text{cm}^{-1} \tag{6}$$

Consider next the evaluation of γ for use in the line structure parameter $\beta = 4\gamma/d$. For simplicity we will limit discussion to a single-component gas, and letting j denote rotational quantum number, then γ_j represents the half-width of a given rotational line. From kinetic theory, the half-width may be shown to vary with pressure and temperature as $\gamma_j \sim P/\sqrt{T}$. More detailed quantum mechanical calculations again show the linear dependence upon pressure, but indicate that the inverse square-root variation with temperature is often true only for the band wings (large values of j). Again considering CO_2 as an example, Yamamato, et al [17] have shown that the temperature dependency of the line half-width may be described by $\gamma_j \sim T^{-n}$, and that n approaches 0.75 for small j, decreases with increasing j to approximately 0.3, and then increases with a further increase in j to the kinetic theory value of 0.5. Additional information, both theoretical and experimental, for a large number of gases under various broadening conditions has been presented by Varanasi and co-workers [18-25].

What is required in the line structure parameter β is not an individual line half-width γ_j, but rather an appropriate average over rotational quantum number. Following Cess and Tiwari [11] this average may be expressed as

$$\sqrt{\gamma} = \sum_{j=o}^{\infty} \sqrt{\gamma_j S_j} \sum_{j=o}^{\infty} \sqrt{S_j}$$

and employing the γ_j results of Yamamoto, et al [17] for CO_2, it is found that [16]

$$\gamma = 0.097P\left(\frac{300}{T}\right)^{2/3} \text{cm}^{-1}\text{atm}^{-1} \tag{7}$$

Furthermore, since the γ_j results are essentially independent of vibrational quantum number, equation (7) should apply to all CO_2 bands. Strictly speaking, equation (7) has been derived for relatively moderate temperature levels. At high temperatures, corresponding to the population of higher rotational levels, $\gamma_j \sim 1/\sqrt{T}$, and the exponent in equation (7) should approach 1/2. The net effect upon the band absorptance should, however, be small. In addition, the mean line spacing for the CO_2 bands is [16]

$$d = 1.56 \text{ cm}^{-1} \tag{8}$$

and thus the dimensionless line structure parameter may be expressed as

$$\beta = \frac{4\gamma}{d} = 0.25P\left(\frac{300}{T}\right)^{2/3} \tag{9}$$

for all CO_2 bands, with pressure in atmospheres.

The final spectroscopic quantity is the band intensity S. In reality, we are not dealing with a single vibration-rotation band at a given spectral location, but instead with a number of overlapping bands resulting from different vibrational levels. Describing the vibrational levels by $(\nu_1\nu_2\nu_3)$, where ν_1, ν_2, and ν_3 denote the symmetric stretch, bending, and asymmetric stretch modes, respectively, the individual band intensities are illustrated in Table 1 for the 4.3μ spectral region of CO_2. The bands which have been included are appropriate for the temperatures considered. The (000)-(001) transition denotes the 4.3μ ground-state band, while the remaining bands have an excited lower vibrational state and are thus hot bands. It is important to note that hot bands resulting from excited 4.3μ levels ($\nu_3 > 0$) are not significant at these temperatures.

The intensities in Table 1 have been calculated from the tabulation of molecular band intensities by Dickinson [26]. The populations of the vibrational levels

Table 1. Band intensities in the 4.3μ spectral region for CO_2.

Transition		Band Intensity, $atm^{-1}cm^{-2}$	
Lower	Upper	273°K	1000°K
000	001	2751	327
01^10	01^11	162	248
02^00	02^00	3	60
10^00	10^01	2	37
02^20	02^21	10	188
Total		2928	860

were determined in accordance with a Boltzmann distribution, except for the (10^00) level which is populated through Fermi resonance with the (02^00) and (02^20) levels. Conventionally the temperature dependence of the total band intensity for a fundamental band is $S \sim 1/T$. This is based on the assumption of harmonic wave functions, although Edwards and Balakrishnan [14] indicate that the situation might not be quite so straightforward. In any event, from Table 1 we find that $S(1000°K)/S(273°K) = 0.294$, and this is quite close to the $S \sim 1/T$ value of 0.273.

Let us now consider the incorporation of these band intensities into equation (5). A detailed approach would consist of treating each band separately, and then properly accounting for band overlap. This would, however, prove to be quite a formidable task, and alternately let us follow the approach suggested by Edwards [27].

Consider first the limit $\beta u >> 1$, such that from equation (5) $A = A_o \ln(\beta u)$. Since additionally we require $u/\beta >> 1$, this in turn reduces to $A = A_o \ln u$, which physically corresponds to a saturated band, and it may easily be shown that this expression applies to the total band (sum of individual bands) with u based upon the total band intensity. In other words, in the logarithmic limit there is no need to separately account for individual bands.

This is not the case in the opposite limit of $\beta u << 1$, for which $A = 2A_o\sqrt{\beta u}$. Here, however, the assumption of nonoverlapping lines means that we may simply add the individual band absorptances, and letting the subscript k denote an individual band at a given spectral location, then for nonoverlapping strong lines

$$A = 2A_o \sum_k \sqrt{\beta u_k} \tag{10}$$

with $u_k = S_k PL/A_o$. In turn, an equivalent line structure parameter, $\bar{\beta}$, is defined such that the square root limit for the total band absorptance may be expressed as $A = 2A_o\sqrt{\bar{\beta} u}$, and from equation (10) $\bar{\beta}$ is expressed by

$$\bar{\beta} = \left(\sum_k \sqrt{\frac{S_k}{S}}\right)^2 \beta \tag{11}$$

Although $\bar{\beta}$ is strictly applicable to the square-root limit, it is in this limit that the line structure parameter is of primary importance, and following Edwards [27] the extension of equation (5) to account for hot bands amounts to simply replacing β by $\bar{\beta}$; i.e.

$$A = 2A_o \ln(1 + \sqrt{\bar{\beta} u}) \tag{12}$$

To give an illustration, we find from Table 1 that for the 4.3μ CO_2 band

$\bar{\beta} = 1.75\ \beta\ , \quad T = 273^oK$

$\bar{\beta} = 4.38\ \beta\ , \quad T = 1000^oK$

The hot bands thus contribute to the band absorptance in two ways; first through their intensities which are incorporated in the expression for u, and secondly by their influence upon $\bar{\beta}$. Note that in the square-root limit and for 273°K, the $\bar{\beta}$ influence increases the band absorptance by the factor $\sqrt{1.75}$, or 32%, even though the hot bands contribute only 6% of the intensity. The above expression for $\bar{\beta}$ at 1000°K is probably an underestimate, since additional hot bands, although not significantly contributing to intensity, should increase $\bar{\beta}$. More detailed calculations are given by Edwards and Balakrishnan [14].
A final point concerning the band absorptance for pressure broadening refers to the pressure dependence in the square-root limit. Since both β and u are linear functions of pressure, it follows that in the square-root limit A ∿ P, which is consistent with the empirical correlation of Houghton [7]. Burch, Gryvnak, and Williams [28], on the other hand, give the dependence as $P^{0.95}$ for the 4.3μ CO_2 band, while from Edwards and Balakrishnan [14] the band absorptance varies as $P^{0.9}$. The reason for these differences is not clear, although it is possibly due different expressions which have been used to correlate data. Since there is theoretical justification for the linear dependence upon pressure, we prefer to be consistent with Houghton.

3.2. Doppler broadening

The pressure broadened half-width varies directly with pressure, and at sufficiently low pressures Doppler broadening will predominate as the line broadening mechanism. As illustrated by Cess [29], the nonoverlapping line limit is always applicable for Doppler broadening, and the band absorptance is given as

$$A = A_o u(1 - 0.18\, u/\delta)\ ;\ u/\delta \leq 1.5 \tag{13a}$$

$$A = 0.753\, A_o \delta\{[\ln(u/\delta)]^{3/2} + 1.21\}\ ;\ u/\delta \geq 1.5 \tag{13b}$$

where $\delta = \sqrt{\pi}\,\gamma_D/d$ is the line structure parameter for Doppler broadening, with

$$\gamma_D = \left(\frac{2kT\ln 2}{mc^2}\right)^{\frac{1}{2}} \omega_i$$

the Doppler half-width, and k, m, and c denote, respectively, the Boltzman constant, mass per molecule, and speed of light. For the 4.3μ CO_2 band

$$\gamma_D = 0.0022\left(\frac{T}{300}\right)^{\frac{1}{2}} \text{cm}^{-1} \tag{14}$$

A complete treatment of the combined influence of pressure and Doppler broadening requires use of the Voigt line profile, which constitutes a transition from pressure to Doppler broadening. Alternatively, however, we will employ a procedure similar to that suggested by Goody and Belton [9], which consists of Lorentz-Doppler matching. Since equation (2) employs the derivative of the band absorptance, this matching is done with respect to dA/du, such that the matching condition is given by

$$\left.\frac{dA}{du}\right|_{Lorentz} = \left.\frac{dA}{du}\right|_{Doppler} \tag{15}$$

and the transition boundary between pressure and Doppler broadening is illustrated in Fig. 3.

For Doppler broadening it is not convenient to define an equivalent line structure parameter which accounts for hot bands, as was the case for pressure broadening where $\bar{\beta}$ was introduced. Instead, since the Doppler lines do not overlap, the band absorptance is simply the sum of the individual band absorptances, as in equation (10) for pressure broadening.

In order for Doppler broadening to be of significance, one must be dealing with extremely low pressures. Considering the 4.3μ CO_2 band, and taking for example $\delta/\beta = 10$ (see Fig. 3) and $T = 1000^oK$, the corresponding pressure is only 0.004 atm. When Doppler broadening does occur, however, some very surprising things can happen, and this is best illustrated by the following example.
The most abundant isotopes of CO_2 are $C^{12}O_2^{16}$ (98.4%) and $C^{13}O_2^{16}$ (1.1%). To investigate their relative importance, consider the same band for each isotope, for example the ground-state band at 4.3μ, with $\delta/\beta = 20$ together with $u/\beta = 10^4$ for $C^{12}O_2^{16}$. Since the abundance of $C^{13}O_2^{16}$ is roughly 10^{-2} that of $C^{12}O_2^{16}$, and for a mixture the pressure in the definition for u is the partial pressure, then $u/\beta \approx 10^2$ for $C^{13}O_2^{16}$. With reference to Fig. 3, the $C^{12}O_2^{16}$ band is pressure broadened while the $C^{13}O_2^{16}$ band is Doppler broadened. From equations (5) and (13), and summing the band absorptances in accord with the assumption of nonoverlapping lines, we find that the contribution to the kernel function, dA/du, in equation (2) of the $C^{13}O_2^{16}$ band relative to the $C^{12}O_2^{16}$ band is

$$\frac{C^{13}O_2^{16}}{C^{12}O_2^{16}} \approx 0.3$$

This is quite surprising, since the relative abundance of $C^{13}O_2^{16}$ is only 0.01. The generalization of this example is that weak bands can be extremely important, relative to strong bands, when they are Doppler broadened. This feature has recently been pointed out by Dickinson [26] in his study of the atmosphere of Venus, and is discussed in the following section.

4. Illustrative Example

The purpose of this section is to present an illustrative radiative solution employing the analytical techniques which have previously been discussed. For this purpose a very convenient radiative example, which incorporates all of the features treated in the present paper, concerns the atmosphere of Venus. The composition of this atmosphere is primarily carbon dioxide, and the surface pressure of Venus is roughly 100 atm. Below altitudes corresponding approximately to a pressure of 0.1 atm (10^5 microbars), however, there are dense clouds and strong vertical convection. Thus we restrict attention to altitudes above the 10^5 microbar level. Two illustrative solutions for the atmospheric thermal structure, assuming global mean radiative equilibrium, are presented in Fig. 4. Before discussing these solutions, however, let us first consider the physical processes which govern the thermal structure.

Between roughly 10^5 and 10^4 microbars, the radiative process consists of radiative transmission of energy within the atmosphere due to the 15μ ground-state band of CO_2. This upward transmission of energy is the result of the absorption of solar energy at lower levels, with subsequent upward emission in the infrared. Thus, the radiative process at these levels is due to a single band, and the rotational lines of this band are overlapped.

Between 10^4 and 10^3 microbars, the temperature profile is modified due to solar absorption, primarily by the 4.3μ ground-state band of CO_2. Here, then, we are

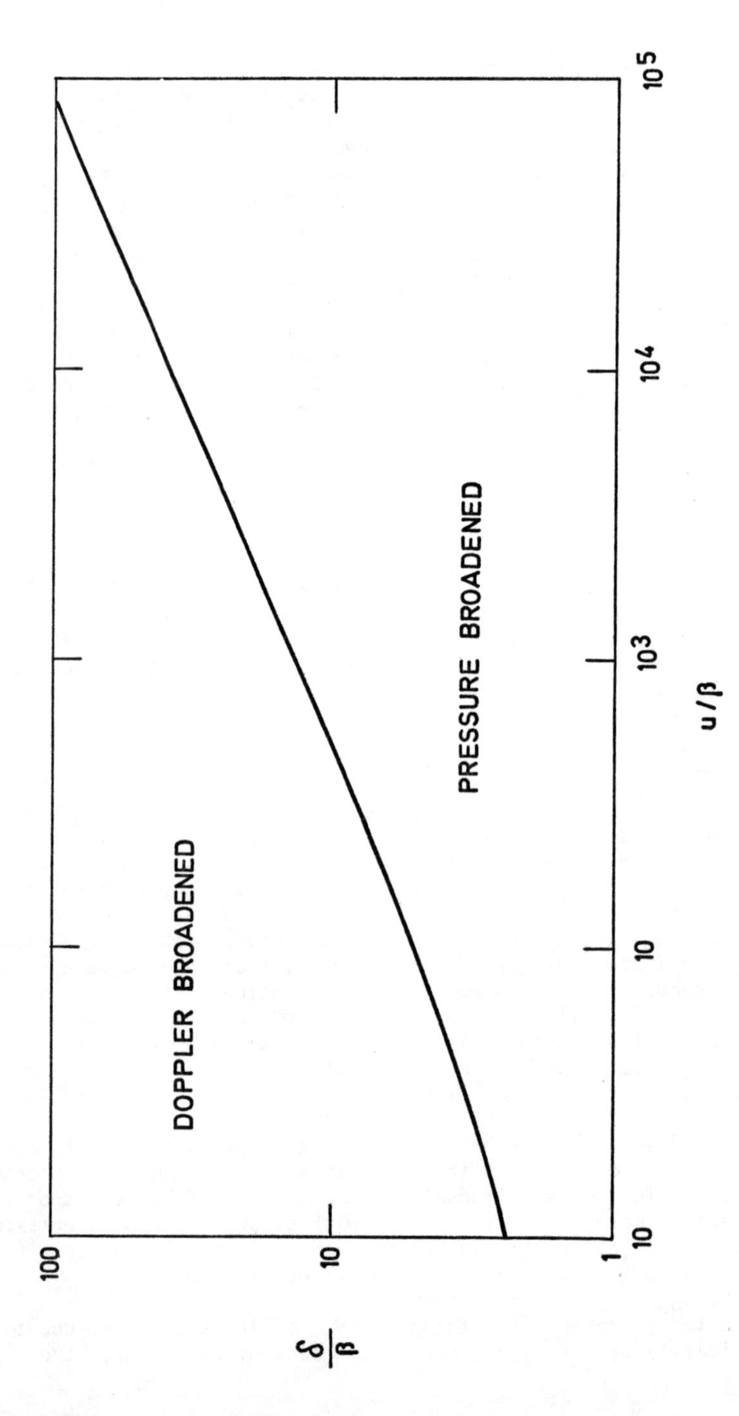

Fig. 3 Transition boundary from pressure to Doppler broadening.

concerned with two bands, and this range of pressures corresponds to a transition from overlapping to nonoverlapping lines.

Above the 10^3 microbar level, weak bands (both isotope and hot bands), which were not significant at lower altitudes, become Doppler broadened and, in accord with previous discussion, start to contribute to the radiative process. The number of such bands is quite incredible. With respect to transmission at 15μ, for example, there are six contributing hot bands, in addition to the ground-state band, as well as five isotopic species. Thus, while there is only one important 15μ band below the 10^3 microbar level, at higher altitudes the number increases to 35. The situation is even worse with regard to solar absorption, since there are weak bands located at the 2.7, 2.0, 1.6, 1.4, and 1.2μ spectral regions, with a total of 145 vibration-rotation bands which must be taken into account at the higher levels.
Doppler broadening first becomes important with respect to the short wave length solar absorption bands, and the corresponding increase in solar absorption gives rise to the increase in atmospheric temperature above roughly the 10^3 microbar level. At considerably higher altitudes, vibrational nonequilibrium becomes important, which causes a subsequent reduction in solar absorption.

Let us now consider the two solutions illustrated in Fig. 4. That due to Dickinson [26] constitutes a numerical solution of the radiative transfer equations, incorporating all of the previously mentioned vibration-rotation bands, and performing the integration on a line-by-line basis employing the Voigt line profile to correctly account for both pressure and Doppler broadening.

The solution by Ramanathan and Cess [30], on the other hand, employs the approximate band models, together with the Lorentz-Doppler matching, as discussed in the previous section. Application to a variable pressure atmosphere has been accomplished through use of the Curtis-Godson scaling approximation [31].

In view of the agreement between the two solutions, we conclude that the approximate band models discussed in the present paper certainly appear to have utility in radiative transfer analyses.

REFERENCES

[1] Cess, R. D. and S. N. Tiwari: "The Large Path Length Limit for Infrared Gaseous Radiation", Appl. Sci. Res., Vol. 19, 439 (1968).

[2] Cess, R. D. and S. N. Tiwari: "The Interaction of Thermal Conduction and Infrared Gaseous Radiation", Appl. Sci. Res., Vol. 20, 25 (1969).

[3] Tien, C. L. and J. E. Lowder: "A Correlation for the Total Band Absorptance of Radiating Gases", Int. J. Heat Mass Transfer, Vol. 9, 698 (1966).

[4] Edwards, D. K., L. K. Glassen, W. C. Hauser, and J. S. Tuchscher: "Radiation Heat Transfer in Nonisothermal Nongray Gases", J. Heat Transfer, Vol. 86, 219 (1967).

[5] Gierasch, P. and R. Goody: "An Approximate Calculation of Radiative Heating and Radiative Equilibrium in the Martian Atmosphere", Planet. Space Sci., Vol. 15, 1465 (1967).

[6] Schimmel, W. P., J. L. Novotny, and F. A. Olsofka: "Interferometric Study of Radiation-Conduction Interaction", Proc. Fourth Intern. Heat Transfer Conf., Vol. III, Paris-Versailles (1970).

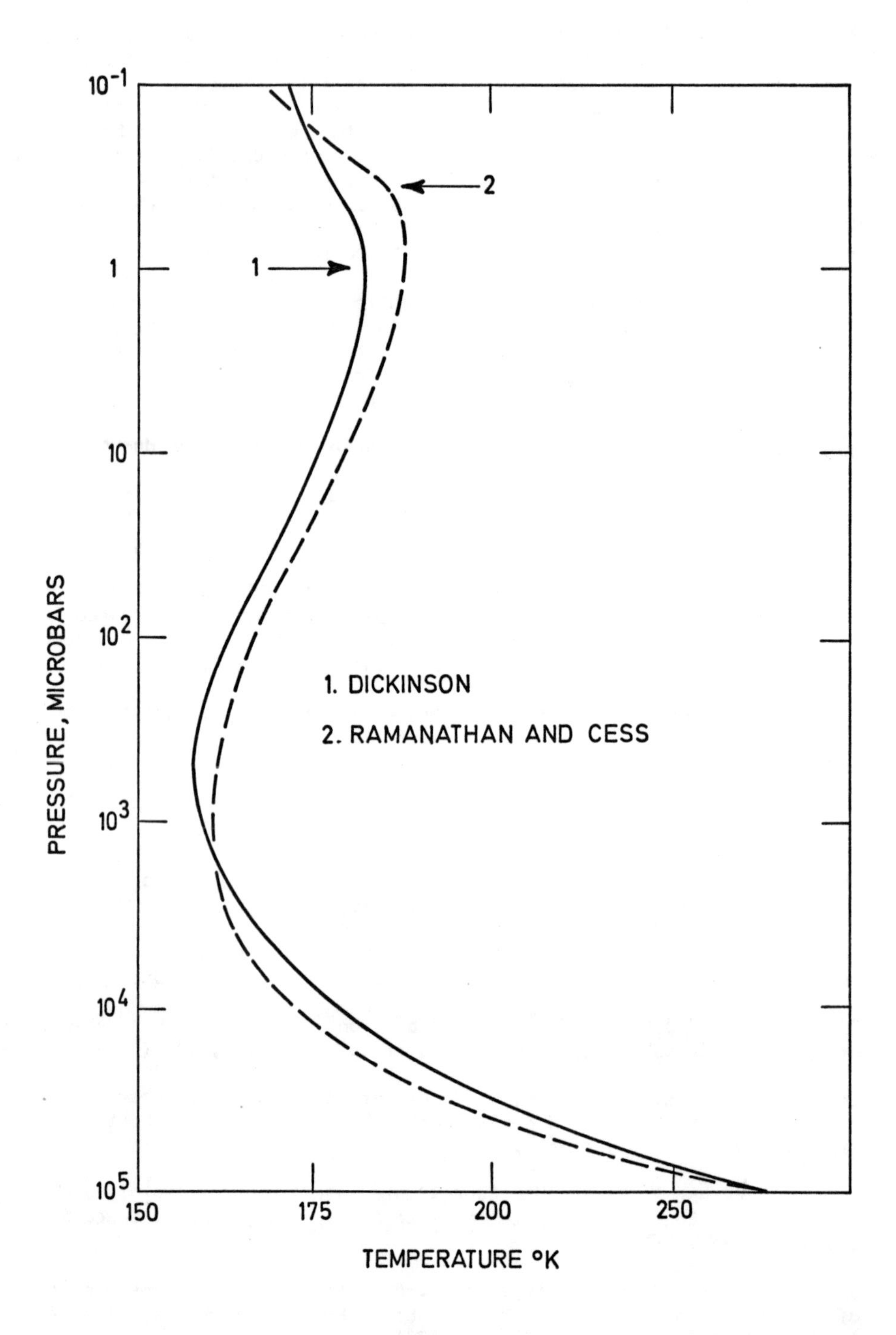

Fig. 4 Illustrative solutions for radiative equilibrium within the atmosphere of Venus.

[7] Houghton, J. T.: "The Absorption of Solar Infra-Red Radiation by the Lower Stratosphere", Quart. J. Ray. Meteor. Soc., Vol. 89, 319 (1963).

[8] Edwards, D. K. and W. A. Menard: "Comparison of Models for Correlation of Total Band Absorption", Appl. Optics. Vol. 3, 621 (1964).

[9] Goody, R. M. and M. J. S. Belton: "Radiative Relaxation Times for Mars", Planet. Space Sci., Vol. 15, 247 (1967).

[10] Weiner, M. M. and D. K. Edwards: "Theoretical Expression of Water Vapor Spectral Emissivity with Allowance for Line Structure", Int. J. Heat Mass Transfer, Vol. 11, 55 (1968).

[11] Cess, R. D. and S. N. Tiwari: "Infrared Radiative Energy Transfer in Gases", Advances in Heat Transfer, Vol. 8, Academic Press, N.Y. (1972).

[12] Edwards, D. K. and A. Balakrishnan: "Slab Band Absorptance for Molecular Gas Radiation", J. Quant. Spectrosc. Radiat. Transfer, Vol. 12, 1379 (1972).

[13] Hsieh, T. C. and R. Greif: "Theoretical Determination of the Absorption Coefficient and the Total Band Absorptance Including a Specific Application to Carbon Monoxide", Int. J. Heat Mass Transfer, Vol. 15, 1477 (1972).

[14] Edwards, D. K. and A. Balakrishnan: "Thermal Radiation by Combustion Gases", Int. J. Heat Mass Transfer, Vol. 16, 25 (1973).

[15] Felske, J. D. and C. L. Tien: "A Theoretical Closed Form Expression for the Total Band Absorptance of Infrared-Radiating Gases", Int. J. Heat Mass Transfer, submitted for publication.

[16] Cess, R. D. and V. Ramanathan: "Radiative Transfer in the Atmosphere of Mars and that of Venus above the Cloud Deck", J. Quant. Spectrosc. Radiat. Transfer, Vol. 12, 933 (1972).

[17] Yamamoto, G., M. Tanaka, and T. Aoki: "Estimation of Rotational Line Widths of Carbon Dioxide Bands", J. Quant. Spectrosc. Radiat. Transfer, Vol. 9, 371 (1969).

[18] Varanasi, P.: "Line Widths and Intensities in H_2O-CO_2 Mixtures II. High-Resolution Measurements on the ν_2-Fundamental of Water Vapor", J. Quant. Spectrosc. Radiat. Transfer, Vol. 11, 223 (1971).

[19] Varanasi, P., G. D. Tejwani, and C. R. Prasad: "Line Widths and Intensities in H_2O-CO_2 Mixtures III. Half-Width Calculations in the ν_2-Fundamental of Water Vapor", J. Quant. Spectrosc. Radiat. Transfer, Vol. 11, 231 (1971).

[20] Tejwani, G. D. T. and P. Varanasi: "Calculation of Collision-Broadened Linewidths in the Infrared Bands of Methane", J. Chem. Phys., Vol. 55, 1075 (1971).

[21] Varanasi, P.: "Collison-Broadened Half-Wdiths and Shapes of Methane Lines", J. Quant. Spectrosc. Radiat. Transfer, Vol. 11, 1711 (1971).

[22] Varanasi, P. and G. D. T. Tejwani: "Experimental and Theoretical Studies on Collision-Broadened Lines in the ν_4-Fundamental of Methane", J. Quant. Spectrosc. Radiat. Transfer, Vol. 12, 849 (1972).

[23] Varanasi, P.: "Shapes and Widths of Ammonia Lines Collision-Broadened by Hydrogen", J. Quant. Spectrosc. Radiat. Transfer, Vol. 12, 1283 (1972).

[24] Varanasi, P., S. Sarangi, and L. Pugh: "Measurements on the Infrared Lines of Planetary Gases at Low Temperatures I. ν_3-Fundamental of Methane", Astrophys. J., Vol. 179, 977 (1973).

[25] Varanasi, P.: "Addendum to 'Line-Width Measurements on CO in an Atmosphere of CO_2' ", J. Quant. Spectrosc. Radiat. Transfer (in press).

[26] Dickinson, R. E.: "Infrared Heating and Cooling in the Venusian Mesosphere. I: Global Mean Radiative Equilibrium", J. Atmos. Sci., Vol, 29, 1351 (1972).

[27] Edwards, D. K.: "Absorption of Radiation by Carbon Monoxide Gas According to the Exponential Wide-Band Model", Appl. Optics, Vol. 4, 1352 (1965).

[28] Burch, D. E., D. A. Gryvnak, and D. Williams: "Total Absorptance of Carbon Dioxide in the Infrared", Appl. Optics, Vol. 6, 759 (1962).

[29] Cess, R. D.: "A Band Absorptance Formulation for Doppler Broadening", J. Quant. Spectrosc. Radiat. Transfer (in press).

[30] Ramanathan, V. and R. D. Cess: "Radiative Transfer within the Mesospheres of Venus and Mars", submitted for publication to Astrophys. J.

[31] Goody, R. M.: "Atmospheric Radiation", Oxford University Press (1964).

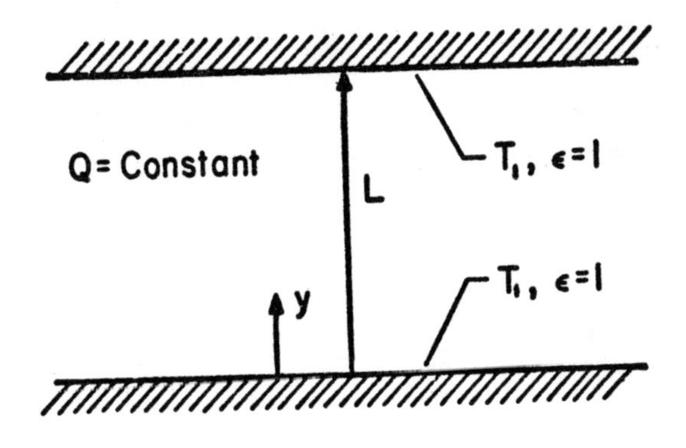

Fig. 1 Physical model and coordinate system.

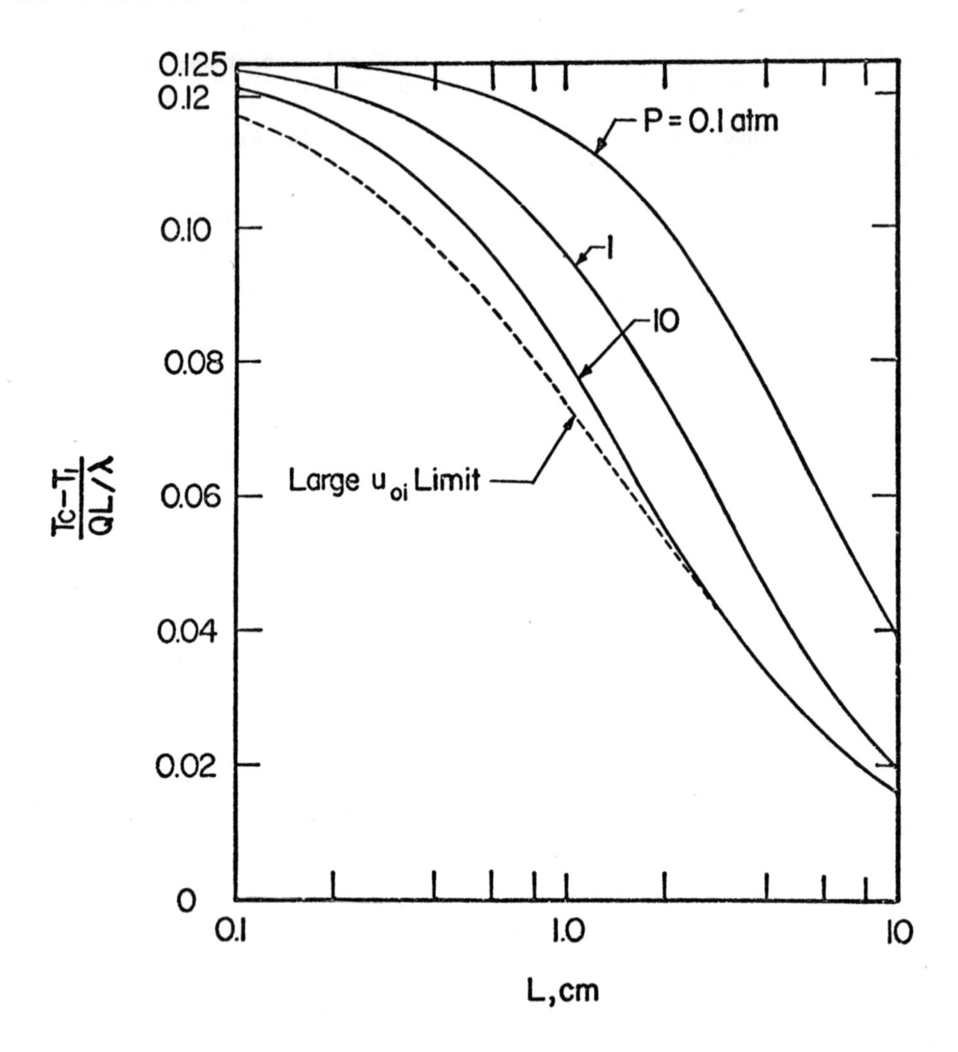

Fig. 2 Results for CO_2 with T_1 = 1000°K.

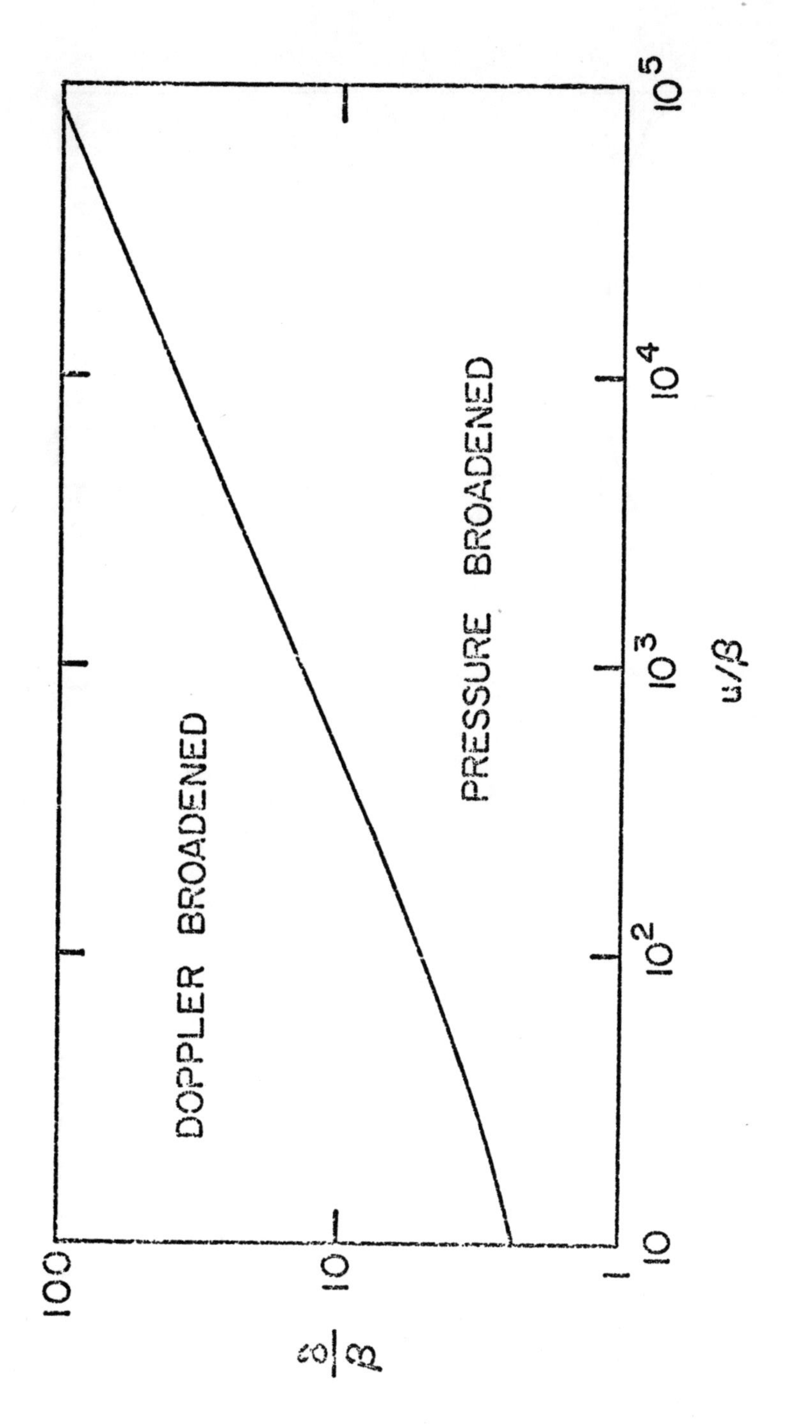

Fig. 3 Transition boundary from pressure to Doppler broadening.

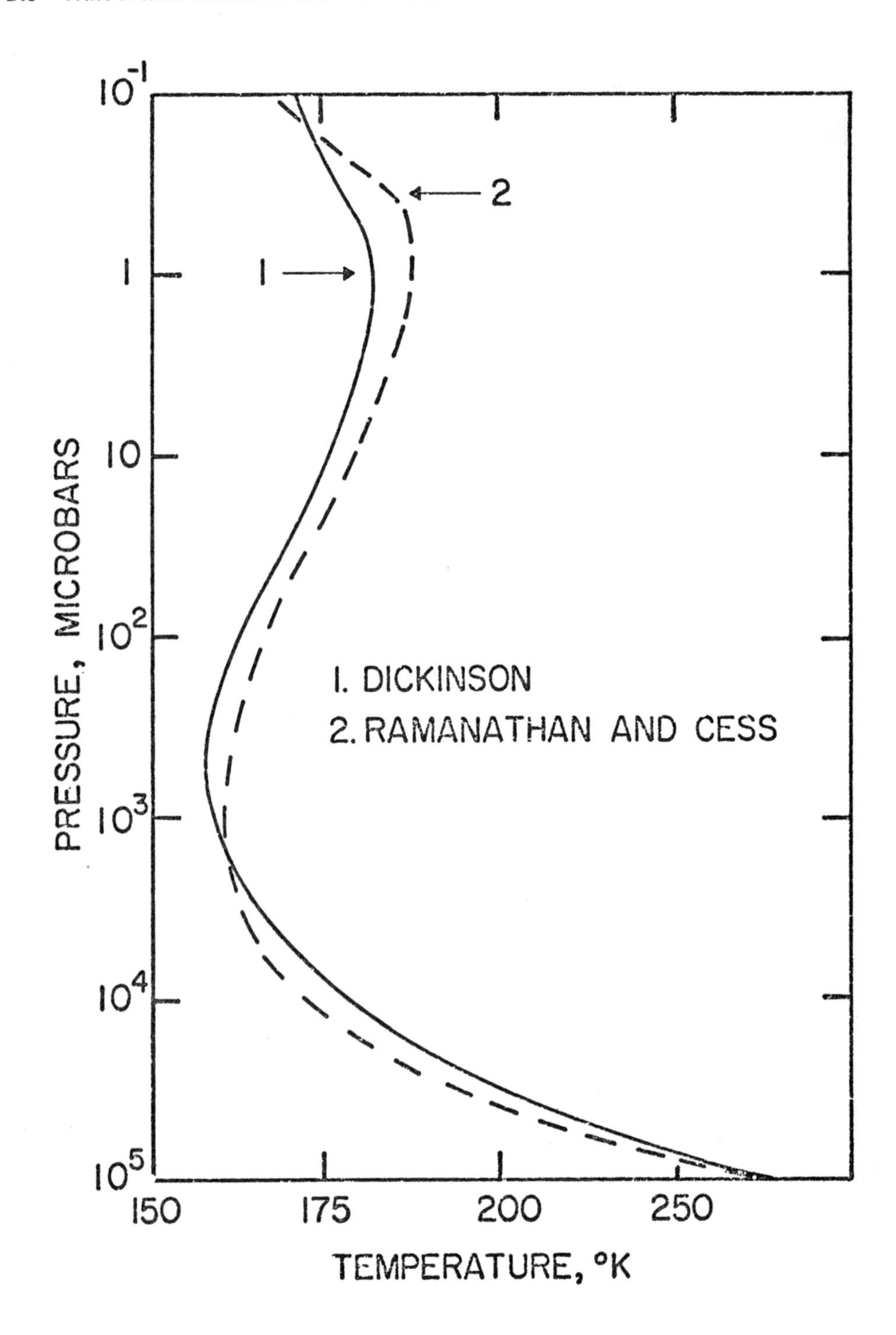

Fig. 4 Illustrative solutions for radiative equilibrium within the atmosphere of Venus.

Chapter 16

EXPERIMENTAL AND THEORETICAL RESULTS WITH INFRARED RADIATING GASES

Ralph Greif

Department of Mechanical Engineering
University of California, Berkeley, California, USA 94720

Abstract

Experimental and theoretical results are presented for the turbulent flow of air, carbon dioxide and steam. The effects of thermal radiation are clearly presented and discussed. Theoretical results for the basic radiative properties of gases have also been obtained and the usefulness of these results is noted.

NOMENCLATURE

Nu : Nusselt Number
Re : Reynolds Number
β : Heating Parameter
T_b : Bulk Temperature
T_w : Wall Temperature

ν : frequency
$\Delta\nu$: band width
k_ν : spectral absorption coefficient
x : pressure pathlength
A : total band absorptance
J : rotational quantum number
α : integrated band intensity

The determination of the energy transport in many systems requires the inclusion of the contribution from thermal radiation. Unfortunately, the determination of the radiative transport is made difficult due to the variation of the properties with respect to frequency and temperature, as well as the complexity of geometry effects [1-4]. This has frequently caused investigators to simplify or at times neglect this contribution. To clearly demonstrate the importance of radiation, as well as our ability to calculate its effects under various conditions, experiments have been carried out with carbon dioxide and with steam in a heated, vertical stainless steel tube. The tube is 22 feet long with an outside diameter of 2.0 inches and a tube wall thickness of 0.049 inches. The tube is heated by employing it as the resistance element of an electric circuit with electrodes attached to each end of the tube for this purpose. A stainless steel bellows is used to permit expansion of the tube.

Thirty thermocouples were spot welded to the outer surface of the tube. In addition, a temperature probe was placed at an axial location 108 tube diameters downstream to measure the hot gas temperature. The centerline gas inlet and outlet temperatures were also measured with thermocouples. Additional information

Table 1. Summary of results

Gas	Pressure (atm)	$Re_{0.4}$	β	$Nu_{0.4}$* (expt., non-rad.)	$Nu_{0.4}$* (theor., non-rad.)
Air	1.1	12,200	0.0091	37.1	36.5
Air	1.1	14,700	0.0066	42.7	42.0
Gas	**Pressure (atm)**	$Re_{0.4}$*	**β**	$Nu_{0.4}$* (expt., rad.)	$Nu_{0.4}$* (hypoth., non-rad.)
CO_2	1.1	12,200	0.0092	39.8	36.7
CO_2	1.1	16,200	0.0092	48.5	45.9
CO_2	1.1	19,000	0.0071	52.8	48.2
H_2O	1.0	22,100	0.0065	71.4	56.4
H_2O	1.0	15,300	0.0059	56.2	44.1
H_2O	1.0	15,300	0.0061	54.8	43.9

* Based on $T_{0.4} = T_b + 0.4(T_w - T_b)$.

concerning the experimental system is available in reference [5].

A check on the system was first carried out with air, a non-absorbing gas, at the temperatures of interest. The experimental data for the temperature profiles and the Nusselt numbers were in very good agreement with previous results for fully developed turbulent flow of air [6]. The results for air are summarized in part in Table 1.

Subsequent experiments were carried out with carbon dioxide [5] and more recently with steam [7] and the results are also summarized in Table 1. The column denoted by hypothetical non-radiating refers to calculations that were made which omit the radiation contribution. Thus, a comparison between the two Nusselt numbers directly provides a basis for an assessment of the overall effects of radiation. For example, for the three steam runs the contribution of radiation results in a 20% increase in the Nusselt number. It should be noted that the wall temperature for the three steam runs is approximately 600°K. Values of various parameters are given in Table 1 and in references [5] and [7].

An informative comparison can also be made between the experimental temperature profile and a hypothetical non-radiating profile which corresponds to the same wall flux and mass flow rate. Thus, the bulk temperature is the same for both conditions. The effect of radiation is to increase the energy transport thereby decreasing the wall temperature and resulting in a flatter temperature profile [5, 7].

It should also be noted that calculations were made which include the radiation contribution. The radiative properties of carbon dioxide and steam were taken from the work reported in references [8] -[12] and good agreement between the experimental and theoretical temperature profiles was obtained. Subsequent experiments in laminar flow should provide a more definitive comparison between theory and experiment.

Over the range of temperatures and pressures tested experimental data was available for the evaluation of the absorption properties. Unfortunately, this information is often not available and the determination of the absorption properties is then frequently made in an approximate, if not arbitrary, manner. The objective of some recent studies has been the determination of the absorption properties of infrared radiating gases from the fundamental spectroscopic parameters, such as line width, line spacing, etc. The procedure used is to begin with the basic expression for the spectral absorption coefficient and then obtain analytic expressions from function theory. These results are then substituted directly into the relation for the total band absorptance, A, which is defined by

$$A = \int_{\Delta\nu} [1 - e^{-k_\nu x}] \, d\,(\nu - \nu_o)$$

where k_ν is the absorption coefficient, x is the pressure pathlength, ν_o is the frequency of the band center and $\Delta\nu$ is the band width. Analytic results have been obtained for k_ν and A based on an approximate specification of the intensity; namely, $S_J = S_0 \exp[-(\nu_{J,R} - \nu_0)/F_R]$ for the R branch with a corresponding relation for the P branch [13,14].Here, J is the rotational quantum number associated with the lower rotational energy level in the transition which results in the spectral line $\nu_{J,R}$, F_R is a characteristic band (branch) width and S_J is the intensity of the Jth line. In the intensity relation used there are two constants S_0 and F_R. However, the requirement that the integrated band intensity, α, equals $\int S_J dJ$

leaves only one constant which then appears in the subsequent expressions for the absorption coefficient and the total band absorptance. For the fundamental vibrational-rotational band of carbon monoxide, which has the band center at 4.7 microns, a value of unity for the constant gave very good agreement with the experimental data [14]. This value of the constant also yields good agreement with experimental data for the 2.7 micron bands of carbon dioxide and of water vapor [15].

At high temperatures the contribution resulting from transitions between higher energy levels, that is, the "hot bands" must be considered. This effect was calculated a priori for carbon monoxide and clearly demonstrated the importance of the hot bands [16].

Furthermore, in solving for the absorption under non-isothermal conditions the analytical relations are especially useful. Calculations made were in excellent agreement with the experimental data of Edwards et al. [17]. Previous studies were made by Chan and Tien [18], Edwards and Morizumi [19], and Cess and Wang [20].

One important task that remains, among others, is the determination of the absorption directly from the fundamental spectroscopic variables, without recourse to an arbitrary constant. This would involve the specification of the more accurate and more complex relation for the line intensity [ ] and would provide a critical comparison between theoretical and experimental results. If this approach proves to be successful it should be possible to predict the basic radiative properties of absorbing gases. This would be of special importance when data is either lacking or incomplete.

It is a pleasure to acknowledge the support of the National Science Foundation.

REFERENCES

[1] Sparrow, E.M. and R.D. Cess: "Radiation Heat Transfer," Wadsworth Publishing Co., California (1966).

[2] Hottel, H.C. and A.F. Sarofim: Radiative Transfer, McGraw-Hill Book Co., New York (1967).

[3] Siegel, R. and J. Howell: Thermal Radiation Heat Transfer, McGraw-Hill Book Co., New York (1972).

[4] Ozisik, M.N.: Radiative Transfer and Interactions with Conduction and Conversion, John Wiley & Sons, Inc., New York (1973).

[5] Habib, I.S. and R. Greif: "Heat Transfer to a Flowing Nongray Radiating Gas: An Experimental and Theoretical Study," International Journal of Heat and Mass Transfer, 13, p. 1571 (1970).

[6] Deissler, R.G. and C.S. Eian: Analytic and experimental investigation of fully developed turbulent flow of air in a smooth tube with heat transfer with variable fluid properties, NACA Tech. Note 2629 (February 1952).

[7] Chiba, Z. and R. Greif: "Heat Transfer to Stream Flowing Turbulently in a Pipe," International Journal of Heat and Mass Transfer, 1973, accepted for publication.

[8] Tien, C.L.: Thermal radiation properties of gases, Advances in Heat Transfer, Vol. V. Academic Press, New York (1968).

[9] Edwards, D.K., B.J. Flornes, L.K. Glassen, and W. Sun: "Correlation of Absorption by Water Vapor at Temperatures from 300°K to 1100°K,"Appl. Optics 4, 715-721 (1965).

[10] Ferriso, C.C. and C.B. Ludwig: "Spectral Emissivities and Integrated Intensities of the 2.7μ H_2O band between 530 and 2200 K. J. Quant. Spectros. Radiat. Transfer 4, 215-227 (1964).

[11] Ludwig, C.B., C.C. Ferriso, and C.N. Abeyta: "Spectral Emissivities and Integrated Intensities of the 6.3μ Fundamental Band of H_2O," J. Quant. Spectros. Radiat. Transfer 5, 281-290 (1965).
[12] Ludwig, C.B., C.C. Ferriso, W. Malkmus, and F.P. Boynton: "High-temperature Spectra of the Pure Rotational Band of H_2O," J. Quant. Spectros. Radiat. Transfer 5, 697-714 (1965).
[13] Edwards, D.K. and W.A. Menard: "Comparison of Models for Correlation of Total Band Absorption," Applied Optics, Vol. 3, p. 621 (1964).
[14] Hsieh, T.C. and R. Greif: "Theoretical Determination of the Absorption Coefficient and the Total Band Absorptance Including a Specific Application to Carbon Monoxide," International Journal of Heat and Mass Transfer, p. 1477 (1972).
[15] Lin, J.C. and R. Greif: "Total Band Absorptance of Carbon Dioxide and Water Vapor Including the Effects of Overlapping," submitted for publication.
[16] Lin, J.C. and R. Greif: "Theoretical Determination of Absorption with an Emphasis on High Temperatures and a Specific Application to Carbon Monoxide," J. of Heat Transfer, accepted for publication.
[17] Edwards, D.K., L.K. Glassen, W.C. Hauser and J.S. Tuchscher: "Radiation Heat Transfer in Nonisothermal Nongray Gases," J. Heat Transfer, 89, 219-229 (1967).
[18] Chan, S.H. and C.L. Tien: "Total Band Absorptance of Non-Isothermal Infrared-Radiating Gases," J. Quant. Spectros. Radiat. Transfer, 9, 1261-1271 (1969).
[19] Edwards, D.K. and S.J. Morizumi: "Scaling of Vibration-Rotation Band Parameters for Nonhomogeneous Gas Radiation," J. Quant. Spectros. Radiat. Transfer, 10, 175-188 (1970).
[20] Cess, R.D. and L.S. Wang: "A Band Absorptance Formulation for Nonisothermal Gaseous Radiation," Int. J. Heat Mass Transfer, 13, 547-555 (1970).

Chapter 17

THE EFFECT OF PRESSURE ON HEAT TRANSFER IN RADIATING GASES

J. L. Novotny

University of Notre Dame, Notre Dame, Indiana, U.S.A.

Abstract

This paper presents a summary of a research program directed to the determination of the pressure behavior of molecular gas radiation in heat transfer situations. The ability of analytically predicting the interaction of gas radiation with conduction and convection is examined for a horizontal heated-from-above gas layer and a free convection boundary layer. Experimental data gathered with a Mach-Zehnder interferometer are compared to analyses based on the gray-gas model, the box model and the exponential wide-band model.

1. Introduction

The role of molecular gas radiation in heat transfer processes has been extensively investigated in recent years. The determination of the influence of molecular gas radiation on heat transfer rates is of particular interest to designers of furnaces, combustion chambers, boilers and chemical reactors to name a few users of such information. Because of the wide range of pressures encountered in combustion situations, it is extremely important to have at hand a full understanding of the effect of pressure, partial as well as total, on heat transfer rates where gas radiation plays a role. This is not only important from the standpoint of being able to analytically predict with confidence the interaction of radiative transfer with the other heat transfer modes but also from the standpoint of knowing when molecular gas radiation is negligible compared to the other modes.

This paper concentrates on a review of a modest research program conducted for the purpose of examining the pressure behavior of radiating gases in heat transfer situations as well as the ability to analytically predict this behavior. The program starts with an examination of the role of molecular gas radiation in conduction situations; this, of course, allows one to concentrate on simple geometries. Specifically, the radiation-conduction interaction problem is a heated-from-above horizontal gas layer. Analyses generated using a gray gas model, a box (top hat) model and the more sophisticated exponential wide-band model are compared to experimental heat flux data. These comparisons allows one to judge the reliability of the models. It should be noted that the gray and box models have been widely used in heat transfer calculations whereas the wide-band model has not yet been accepted for wide-spread use. Heat flux data gathered with a Mach-Zehnder interferometer exist for pure CO_2, CH_4, N_2O and NH_3, mixtures of CO_2-CH_4 and CO_2-N_2O and nitrogen broadened CO_2, N_2O and NH_3 in the pressure range of 1/4 to 3 bar. Although H_2O vapor which is an important product of combustion is not included in this list, this does not hamper the overall objectives of this work. The limited pressure range (1/4 to 3 bar) is also sufficient for the purposes of this study.

Because of the significant role of pressure in convection processes, it also is necessary to examine the role of molecular gas radiation in a convection interaction problem. As will be discussed, the role of pressure in the conduction-radiation case is significantly different from that encountered in a convection-radiation situation. To simplify the experimental apparatus, this study focuses on a free-convection boundary layer; the basic convection mechanisms are present in such a study. Due to the information gained in the horizontal layer study, the analysis here will be limited to the exponential wide-band model; the general pressure behavior of the gray gas and the box model will be discussed. The experiments were limited to the boundary layer regime of a vertical enclosure containing pure NH_3*. The choice of ammonia and the experimental configuration was dictated by the desire to keep the entire apparatus near room temperature.

It is not the purpose of this paper to present analyses and/or experimental data for extremely complicated physical situations. Rather, it concentrates on some simple problems that can be analytically formulated without appeal to the volume interchange approach. Thus, the magnitude and behavior of radiation interaction can be studied with a relative lack of complicating factors. Although the gas radiation models are intorduced into integro-differential conservation equations, the knowledge gained from this approach can be extended to more complicated physical problems that have to rely on the volume interchange approach. Also, this paper does not contain a comprehensive review of the literature; only selected portions of the literature will be cited.

2. Analysis

The brief analytical formulation presented here will be limited to one-dimensional radiative transport. This is a severe approximation but it is felt that the one-dimensional approach does not introduce significant errors into the problems studied here. The presentation will begin with the one-dimensional horizontal gas layer; Figure 1a is a sketch of the geometry.

The radiative flux gradient, dq_r/dy, for two gray diffuse bounding surfaces is given by [1, 2]

$$\frac{dq_r}{dy} = -\frac{3}{2}\left\{\int_0^L \frac{dT}{dt}\sum_{n=0}^{\infty}(\rho_1\rho_2)^{2n}\left[\rho_1\Omega\left\{\frac{3}{2}(2nL+t+y\right\}\right.\right.$$
$$-\rho_2\Omega\frac{3}{2}[2(n+1)-t-y]\Big\}+\rho_1\rho_2\Omega\left\{\frac{3}{2}[2(n+1)+t-y]\right\}$$
$$\left.-\rho_1\rho_2\Omega\left\{\frac{3}{2}[2(n+1)-t+y]\right\}\right]dt-\int_0^y\frac{dT}{dt}\Omega\left[\frac{3}{2}(y-t)\right]dt$$
$$\left.+\int_y^L\frac{dT}{dt}\Omega\left[\frac{3}{2}(t-y)\right]dt\right\} \qquad (1)$$

where n is a summation index and t is a dummy variable for y. The quantity

$$\Omega(y) = \int_0^{\infty}\frac{\partial e_{\omega}}{\partial T}A'_{\omega}(y)\,d\omega \qquad (2)$$

where

$$A'_{\omega}(y) = \varkappa_{\omega}\exp(-\varkappa_{\omega}y) \qquad (3)$$

contains the spectral description of the absorption-emission process. It is obvious that the radiative flux gradient, which appears in the energy conservation equation, could be written in other forms [3, 4]; this particular formulation has certain advantages [1, 2]. Although it is necessary only in the wide-band model to invoke the exponential kernel approximation [1, 2, 3], equation (1), which contains the kernel approximation, will be used for all models for purposes of consistancy.

The only existing analytical boundary-layer results for the wide-band model are for a perfectly-reflecting surface with a black surface located outside the boundary layer. The radiative flux gradient for this situation takes the form [5, 6]

*As will be mentioned later, N_2 broadened NH_3 data exist; these will not be discussed here.

$$\frac{\partial q_r}{\partial y} = \frac{3}{2}\left\{\int_0^y \frac{\partial T}{\partial t}\Omega\left[\frac{3}{2}(y-t)\right]dt - \int_0^\delta \frac{\partial T}{\partial t}\Omega\left[\frac{3}{2}(t+y)\right]dt - \int_y^\delta \frac{\partial T}{\partial t}\Omega\left[\frac{3}{2}(t-y)\right]dt\right\} \quad (4)$$

where δ refers to some point in the free stream.

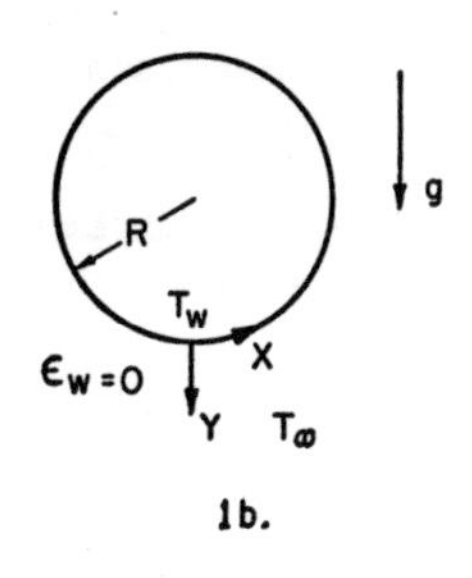

Fig. 1a & 1b. Sketch of vertical gas layer and free-convection stagnation flow geometries.

Equations (1) and (4) constitute the radiative term in the energy conservation equation for radiation-conduction interaction in a horizontal gas layer and radiation-convection interaction in a boundary layer where one-dimensional radiative transport is a valid approximation. The remaining portion of the analytical formulation consists of dictating a model for the Ω functions and substitution into the appropriate energy equation.

For the gray gas model, which has been extensively used in heat transfer analyses (e.g. see [3,7]), Ω takes a rather simple form

$$\Omega(y) = [\bar{\kappa}\exp(-\bar{\kappa}L)]\, 4\sigma T_r^3 \quad (5)$$

where $\bar{\kappa}$ is an average absorption coefficient and T_r is a suitable reference temperature. Linearized radiation* has been assumed in the analysis.

In the box model [8], each vibrational-rotational band is replaced by a box of height $\bar{\kappa}_i$ and width $\Delta\omega_i$. In this approximation, the Ω function is

$$\Omega(y) = \sum_i [\bar{\kappa}_i\, \Delta\omega_i \exp(-\bar{\kappa}_i\, y)\,(de_{\omega i}/dT)_r] \quad (6)$$

where it has been assumed that the gradient of the Planck function† does not vary across the band width $\Delta\omega_i$; the integral has been replaced by the summation over the i bands in the spectrum.

The wide-band model approach is based on correlations for the total band absorptance developed by Edwards and Menard [9]. One of the first examples of a comparison between the wide-band model and the other two models is given by Cess, Mighdoll and Tiwari [8] which points out that there can be significant differences between the three models. The wide-band correlation [9] is based on the spectral absorptivity as given by Goody [10], which is a narrow-band statistical model with an exponential distribution for the mean line intensity. The wide-band model accounts for the line structure within a vibrational-rotational band and the wings of a band; the box model and gray gas model do not account for these effects.

Instead of using the original limits given by Edwards and Menard [9], the continuous correlation proposed by Tien and Lowder [11] will be used in the analyses. This relationship for the total band absorptance is

$$\bar{A} = \frac{A_i}{A_{oi}} = \ell n\left\{u_i\, f(\beta_i)\left[\frac{u_i+2}{u_i+2f(\beta_i)} + 1\right]\right\} \quad (7)$$

*The majority of the results presented here will be for linearized radiation; in a few instances, results will be presented for $T_1/T_2=1.1$ which is close to the linearized case.

†For linearized radiation, $de_{\omega i}/dT$ is introduced at a suitable reference temperature.

where

$f(\beta_i) = 2.94\,[1-\exp(-2.6\,\beta_i)]$

$u_i = C_{oi}^2\, P_a y$

$\beta_i = B_i^2\, P_e$

$P_e = (b\,P_a + F_k P_k)^n$

The parameters A_o, B^2, C_o^2, b and n are correlation parameters. The pressure P_e is the effective broadening pressure where P_a is the absorbing-gas partial pressure, P_k is the partial pressure of the broadening gas and F_k is broadening coefficient for the kth constituent. Since equation (7) is rather cumbersome to use for discussion purposes, the various limiting regimes proposed by Edwards and Menard [9] are given in Table 1. Three basic regimes are indicated. When the absorbing gas is not broadened ($\beta<1$), the total band absorptance consists of a linear, square root and logarithmic region. When the gas is broadened ($\beta>1$), the square root region ceases to exist. A detailed discussion of the various absorption limits can be found in [4, 9]. There are more recent closed form expressions for the total band absorptance [12]; however, equation (7) has been used for the work presented here.

To return to the radiative flux formulation, the wide-band Ω function is given by

$$\Omega(y) = \sum_i \left[A_i'(y) \left(\frac{de_{\omega i}}{dT} \right)_r \right] \tag{8}$$

where

$$A'(y) = \int_{\Delta\omega_i} \varkappa_\omega\, e^{-\varkappa_\omega y}\, d\omega$$

In equation (8), it has been assumed that the gradient of the Planck function is constant across the width of the band; the summation is over all bands in the spectrum. The derivative of the total band absorption, $A'(y)$, is obtained by differentiating equation (7) with respect to the physical length.

To serve as a basis for discussion, the final dimensionless conservation equations will be presented; a detailed development and solution methods used for these equations appear elsewhere [1, 2, 5, 6]. Treating the one-dimensional conduction problem first, the final governing energy equation takes the general form

$$\begin{aligned}\frac{d^2\theta}{d\eta^2} = -\frac{3}{2}\Bigg\{ \int_0^1 \frac{d\theta}{dt} \sum_{n=o}^{\infty} (\rho_1\rho_2)^{2n}\, [\rho_1 \overline{\Omega}\,\{\tau_i(2n+\eta+t)\} \\ -\rho_2 \overline{\Omega}\,\{\tau_i[2(n+1)-t-\eta]\} + \rho_1\rho_2\, \overline{\Omega}\,\{\tau_i[2(n+1)+t-\eta]\} \\ -\rho_1\rho_2\, \overline{\Omega}\,\{\tau_i[2(n+1)-t+\eta]\}]\, dt - \int_0^{\eta} \frac{d\theta}{dt}\, \overline{\Omega}\,[\tau_i(\eta-t)]\, dt \\ + \int_{\eta}^{1.0} \frac{d\theta}{dt}\, \overline{\Omega}\,[\tau_i(t-\eta)]\, dt \Bigg\}\end{aligned} \tag{9}$$

with the boundary conditions

$\theta(0) = 1.0, \quad \theta(1.0) = 0$

where $\theta = (T-T_2)/(T_1-T_2)$, and $\eta = y/L$. The dimensionless $\overline{\Omega}$ functions are dependent on the gas radiation model under consideration.

For the gray-gas model, $\overline{\Omega}$ is given by

$$\overline{\Omega}\,(\tau_i\eta) = N\, \exp(-\tau\,\eta) \tag{10}$$

$$N = \frac{L^2}{\lambda} \bar{\kappa} \; 4\sigma T_r^3 \tag{11}$$

$$\tau = \frac{3}{2} \bar{\kappa} \; L \tag{12}$$

In the gray gas case, the summation index i has no meaning. The box model is characterized by

$$\bar{\Omega} (\tau_i \eta) = \sum_i N_i \; \exp(-\tau_i \eta) \tag{13}$$

$$N_i = \frac{L^2}{\lambda} \bar{\kappa}_i \, \Delta\omega_i \left(\frac{de_{\omega i}}{dT}\right)_r \tag{14}$$

$$\tau_i = \frac{3}{2} \bar{\kappa}_i \; L \tag{15}$$

where the integrated band intensity is $S_i = (\bar{\kappa}_i \Delta\omega_i)/P_a$. The wide model expression for $\bar{\Omega}$ is

$$\bar{\Omega}(\tau_i \eta) = \sum_i N_i \; \bar{A}_i'(\tau_i \eta) \tag{16}$$

$$N_i = \frac{L^2}{\lambda} S_i \, P_a \left(\frac{de_{\omega i}}{dT}\right)_r \tag{17}$$

$$\tau_i = \frac{3}{2} S_i \, P_a L / A_{oi} \tag{18}$$

where $S_i = A_{oi} \, C_{oi}^2$ is the integrated band intensity. Except for total band absorptance term, the box and wide-band models are quite similar. It should be noted that two interaction parameters appear in each model. The parameter N denotes the relative role of radiation to conduction and τ denotes the relative optical thickness of the layer. A complete description of the pressure behavior of equation (9) will be presented in the next section of this paper.

Analytical results will be given for two boundary layer situations; the pertinent conservation equations will be presented for free-convection at the stagnation point of a horizontal cylinder and a free-convection boundary layer on a flat plate. Schematic description of the two problems are presented in Figures 1b and 1c. The conservation equations for the two problems will be presented for the wide-band model with a perfectly reflecting plate.

The free-convection stagnation flow situation is governed by the following set of equations

$$f_{\eta\eta\eta} + f f_\eta - (f_\eta)^2 + \theta = 0 \tag{19}$$

$$\theta_{\eta\eta} + Pr\, f\, \theta_\eta = -\frac{3}{2} \int_0^\infty \theta_t \bar{\Omega}_i \left[\tau_i(\eta + t)\right] dt - \int_0^\infty \theta_t \bar{\Omega}_i \left[\tau_i(\eta - t)\right] dt + \int_\eta^\infty \theta_t \bar{\Omega}_i \left[\tau_t(t - \eta)\right] dt \tag{20}$$

with boundary conditions

$$\theta(0) = 1.0, \quad f(0) = f_\eta(0) = 0$$

$$\theta(\infty) = 0, \quad f_\eta(\infty) = 0$$

The $\bar{\Omega}$ functions are of the form

$$\bar{\Omega} (\tau_i \eta) = \sum_i \xi_i \bar{A}_i' (\tau_i \eta) \tag{21}$$

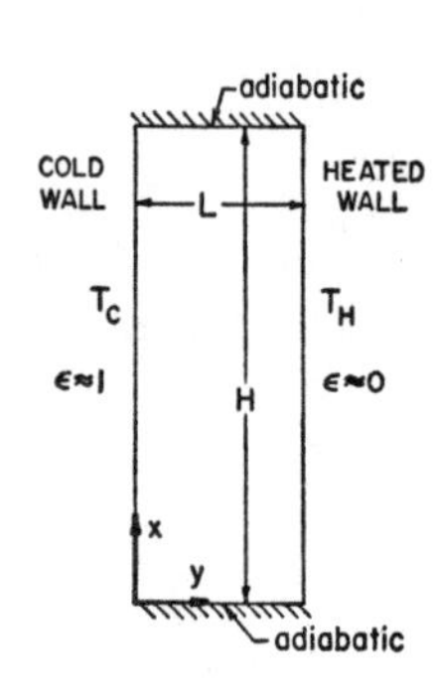

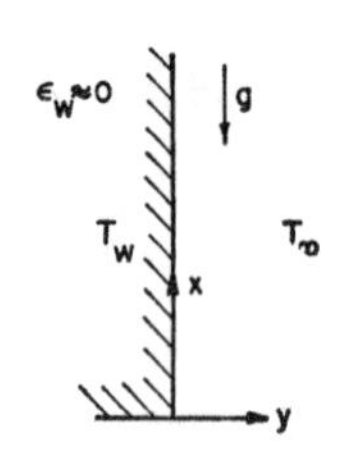

Fig. 1c. Sketch of vertical enclosure and two-dimensional free-convective boundary layer geometries.

$$\xi_i = \frac{R^2 S_i P_a}{k\, Gr^{1/2}} \left(\frac{de_{\omega i}}{dT}\right)_r \tag{22}$$

$$\tau_i = \frac{3}{2} S_i\, P_a\, R/A_{oi}\, Gr^{1/4} \tag{23}$$

where $\eta = y(Gr)^{1/4}/R$ and $Gr = g\beta(T_w - T_\infty)R^3/\nu^2$. A detailed discussion of the formulation and the solution of the above equations is given by Novotny [5].

As in the conduction problem, there are two governing parameters. The parameter ξ_i denotes the relative role of radiation to convection where as the τ_i denotes the relative optical thickness of the boundary layer. In regard to the pressure, it should be noted that Gr α P_T where P_T is the total pressure; P_a denotes partial pressure of the absorbing gas. Although, the convection results will be limited to a pure absorbing gas, it is advantageous to discuss the pressure dependency [6]. The parameter ξ_i is independent of pressure for a pure gas and τ_i varies as $(P_a)^{1/2}$ for a pure gas. In a mixture, ξ_i varies as the mole fraction of the absorbing gas and τ_i is proportional to $P_a/(P_T)^{1/2}$. The pressure also plays a role through the line broadening parameter β_i (see equation (7)). The detailed behavior of $\theta_\eta(0)$ with respect to pressure for a pure absorbing gas will be discussed in a later section.

The free-convection vertical-plate boundary layer problem is governed by

$$f_{\eta\eta\eta} + 3 f f_{\eta\eta} - 2(f_\eta)^2 - 2\xi \{f_\eta f_{\xi\eta} - f_\xi f_{\eta\eta}\} + \theta = 0 \tag{24}$$

$$\frac{1}{Pr}\theta_{\eta\eta} + 3f\theta_\eta - 2\xi(f_\eta\theta_\xi - f_\xi\theta_\eta) + \frac{3}{Pr}\left\{\int_0^\infty \frac{\partial\theta}{\partial t}\,\overline{\Omega}_i\,[\tau_i(\eta+t)]\,dt - \int_0^\eta \frac{\partial\theta}{\partial t}\,\overline{\Omega}_i\,[\tau_i(\eta-t)]\,dt + \int_\eta^\infty \frac{\partial\theta}{\partial t}\,\overline{\Omega}_i\,[\tau_i(t-\eta)]\,dt\right\} = 0 \tag{25}$$

with the boundary conditions*

$\theta(0,\xi) = 1.0, \qquad f(0,\xi) = f_\eta(0,\xi) = 0$

$\theta(\infty,\xi) = 0, \qquad f_\eta(\infty,\xi) = 0$

where $\eta = y\left(\frac{Gr}{4}\right)^{1/4}/x$, $\theta = (T - T_\infty)/(T_w - T_\infty)$ and

$$\overline{\Omega}_i(\tau_i\eta) = \sum_i \xi_i\, \overline{A}_i'(\tau_i\eta) \tag{26}$$

$$\xi_i = \frac{P_a\, x^2 S_i}{k\, Gr^{1/2}}\left(\frac{de_{\omega i}}{dT}\right)_r, \quad \xi = \sum_i \xi_i \tag{27}$$

$$\tau_i = \frac{3}{2}\,\frac{S_i P_a\, x}{A_{oi}(Gr/4)^{1/4}} \tag{28}$$

*These boundary conditions are sufficient for the purposes of the solution given in [5].

The Grashof number is defined as $Gr = g\beta(T_W - T_\infty)\, x^3/\nu^2$.

As before, there are two parameters (ξ_i, τ_i) connected to the role of radiation; their relationship to the partial and total pressures is the same as previously discussed in regard to the stagnation problem. Since the stream function $f(\xi, \eta)$ and the dimensionless temperature $\theta(\xi, \eta)$ are functions of two variables, it was necessary to solve equations (24) and (25) by expansions of the form

$$f(\xi, \eta) = f_o(\eta) + f_1(\eta)\, \xi^m + f_2(\eta)\, \xi^{2m} + \text{-------}$$

$$\theta(\xi, \eta) = \theta_o(\eta) + \theta_1(\eta)\, \xi^m\, \theta_2(\eta)\, \xi^{2m} + \text{--------}$$

The reader is referred to Bratis and Novotny [6] to sort out the details of the solution.

To logically discuss the pressure behavior of radiation interaction and the theoretical-experimental comparisons, it was necessary to present, at least in a brief manner, the previous equations. It is hoped that the clarity of the previous presentation is sufficient for the following discussion.

3. Radiation-Conduction Interaction

Although there exist two purely analytical investigations [1, 13] using equation (9), this discussion will concentrate on the papers by Schimmel, Novotny and Olsofka [2] and Novotny and Olsofka [14]. Both investigations contain experimental data for comparison purposes.

Since a rather detailed description of the experiments exists in [2] and [14], this presentation will be brief. The experimental apparatus consisted of two horizontal highly polished aluminum* plates (50.8 cm square) with a spacing of 2.55 cm contained in a pressure cell. The temperature profile in the gas layer was obtained with a Mach-Zehnder interferometer. A sample of the fringe profiles obtained from the interferograms is given in Figure 2; the fringe profile is rotated such that center portion is vertical in the figure. Wall heat flux data were then obtained by differentiating the temperature profile at each wall. Pure N_2 was used in the apparatus in a series of test runs to estimate the convection in the cell; at 2 bar, the convection in the cell increased the pure-conduction heat flux by no more than 4%.

Figure 3 presents a comparison between the experimental data and the gray gas, box and wide-band models evaluated for the experimental conditions [2].† The figure presents the conductive flux (actually the ratio of the temperature gradient at the wall to that if gaseous radiation were not present) as a function of pressure for pure CO_2 and N_2O. Since the wall temperature ratios correspond very closely to linearized radiation, the two wall heat fluxes should be very nearly equal; the slight differences for the data are attributed to the spurious convection in the gas layer. The gray gas model uses the mean Planck coefficient [2] as an average absorption coefficient ($\overline{\kappa} = \kappa_P$) whereas the box model parameters are determined from band widths listed in Penner [15] and $\overline{\kappa}$'s such that the integrated band intensity is conserved $S_i = \overline{\kappa}_i\, \Delta\omega_i$. The details of the input data for the wide-band model as well as the other models are given in [2].

It is clear that the gray-gas and box models do not properly represent the

*The reflectance of the plates varied from approximately 0.8 to 0.96 in the wavelength range of 4 to 18 μm; within experimental error, both plates exhibited the same reflectance.

†The bands included in the calculations are CO_2 (4.3 μm, 15 μm) and N_2O (4.5 μm, 7.8 μm and 17 μm). The gray-gas interaction in Figure 3 tends toward zero at P> 3 bar in a manner similar to the box model results.

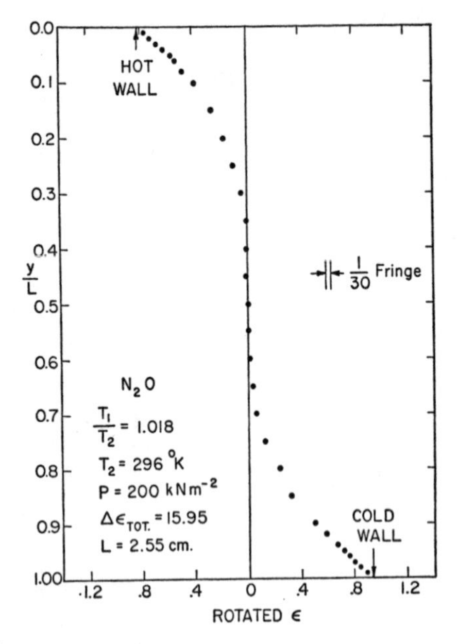

Fig. 2. Sample of interferogram fringe profile.

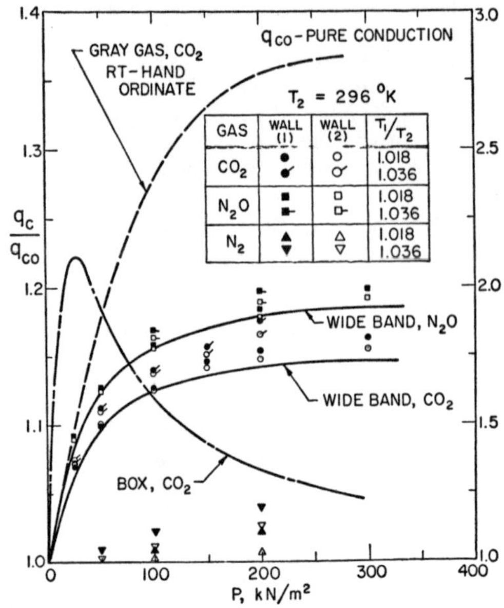

Fig. 3. Radiation-conduction interaction; pure absorbing gases.

pressure trends of the data. The exponential wide-band model is a better representation of the actual interaction problem. The reason that the interaction is independent of pressure at the higher pressures can be determined from equations (7, 9, 16-18). As the absorbing-gas pressure increases, the total band absorptance becomes logarithmic in behavior or $\overline{A}_i'(\tau_i\eta)\ \alpha\ (1/\tau_i\eta)$. Thus, the ratio $\sum_i N_i/\tau_i$ is the parameter governing the interaction; $\sum_i N_i/\tau_i$ is independent of the pressure. In other words, the wings of the vibrational-rotational band, which decay exponentially, govern the interaction at higher pressures; the band never becomes opaque due to these wings. It is interesting to note that for CO_2 and N_2O the interaction becomes pressure independent close to 1 bar. Examining equations (9, 10-12, 13-15), the gray-gas model and the box model indicate (as shown in Figure 3) that the magnitude of the interaction should approach zero (opaque limit) as P_a increases.* It is also interesting to note that the gray gas overestimates the magnitude of the interaction. This magnitude could be adjusted by a better choice of the average absorption coefficient (other than $\overline{\kappa} = \kappa_p$). This choice, of course is dependent on the problem and difficult to determine without a solution to the problem. It is clear from Figure 3 that the wings of an absorption band can not be ignored, at least in radiation-conduction interaction, if a description of the proper pressure behavior is required.

Figures 4 and 5 present results for CO_2-N_2O and CO_2-CH_4† mixtures [2]. Again, it is clear that the wide-band model is a superior representation of the interaction. The analytical gray-gas and box model results shown in Figure 5 could be adjusted in level by a proper choice of input parameters; this adjustment would be a function of total pressure. The rapid increase in interaction with a slight addition of another absorbing gas is due to filling the windows in the spectrum. The flatness of the interaction curve at intermediate mole fractions for P = 3 bar is due to both gases being in the pressure saturated (logarithmic) regime.

* Note that κ_p/P_a is independent of the pressure.

† For CH_4, the 7.6 μm band was the only band included in the calculations.

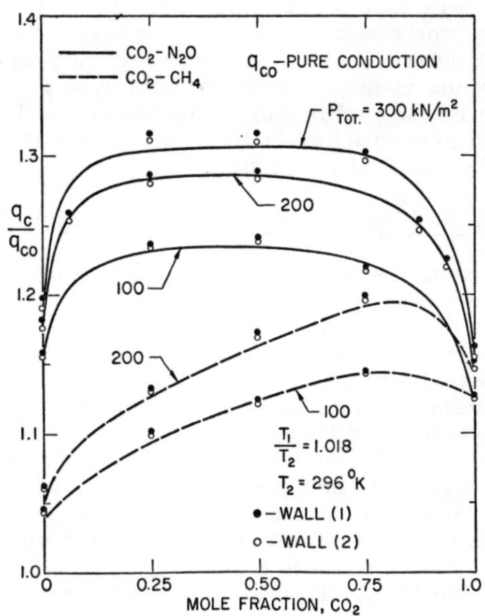

Fig. 4. Radiation-conduction interaction; mixtures of absorbing gases.

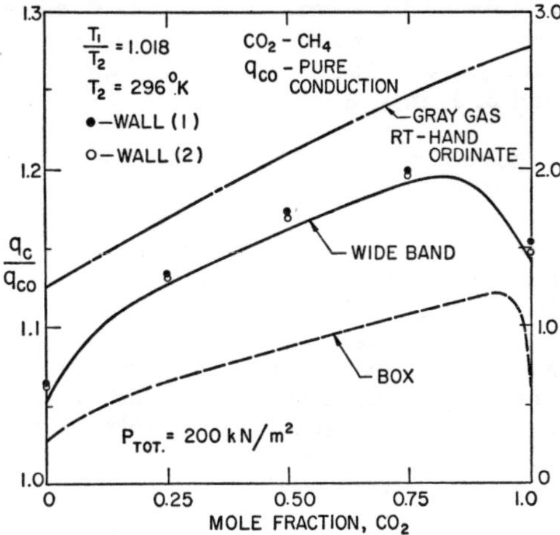

Fig. 5. Radiation-conduction interaction; mixtures of absorbing gases.

Although the exponential wide-band model correctly describes the interaction pressure trends for pure absorbing gases and mixtures of absorbing gases, the effect of a non-absorbing broadening gas such as nitrogen and the ability of the wide-band approach to describe the effect of line broadening has to be determined. Figures 6 and 7 present results for N_2 broadened CO_2 and N_2O [14]. Because of the failure of the gray-gas and box model in the pure absorbing-gas case, the analytical results in Figures 6 and 7 are limited to the wide-band model. The experimental configuration [14] is the same as that used for the results given in [2].

Before discussing the comparison between the theoretical and experimental pressure-broadened results, it is of interest to discuss the expected interaction behavior. Table 1 indicates that for $\beta < 1.0$ (unbroadened), the total band absorptance has three distinct regions (linear, square root, logarithmic) depending on the optical path length. For $\beta > 1.0$ (broadened), the square root region disappears and there is a direct transition from the linear to the logarithmic region. Thus at a fixed partial pressure of the absorbing gas, as the partial pressure of the nitrogen increases, the total band absorptance approaches the situation of a direct transition from the linear to the logarithmic regions. Although the derivative of the total band absorptance in equation (9) will generally encompass all three regions, the radiation interaction should increase at a more rapid rate with respect to P_a in a nitrogen broadened mixture than for a pure absorbing gas. However, the interaction parameter N_i (equation (17)) is inversely proportional to the thermal conductivity of the mixture. For CO_2-N_2 and N_2O-N_2 mixtures, the mixture thermal conductivity decreases with an increase in absorbing gas partial pressure; N_i decreases with respect to increasing P_a. This behavior of N_i and A_i could possibly cause a crossover in the interaction curves for a pure and nitrogen broadened absorbing gas.

This is exactly the behavior observed in Figures 6 and 7; the interaction curves for a N_2 broadened gas increase at a more rapid rate with respect to P_a at low values of P_a and crossover the pure absorbing gas curve intersecting again at $P_a = P_T$. Although, the effects are not large, the experimental and theoretical results are in good agreement. The β_i's for the important bands of N_2O are larger than those of CO_2. Thus, one would expect, at least at low P_a, that the N_2O-N_2 curves would be closer to the pure N_2O curves than the equivalent curves for CO_2; this behavior is observed in Figures 6 and 7. The reader is referred to [14] for a more detailed discussion of these results.

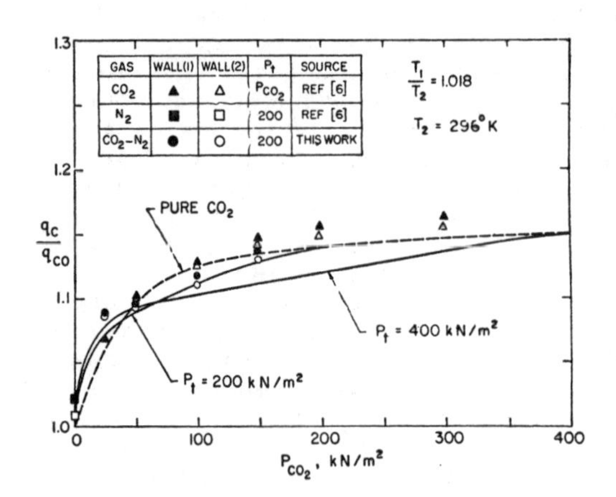

Fig. 6. Radiation-conduction interaction; N_2 broadened CO_2.

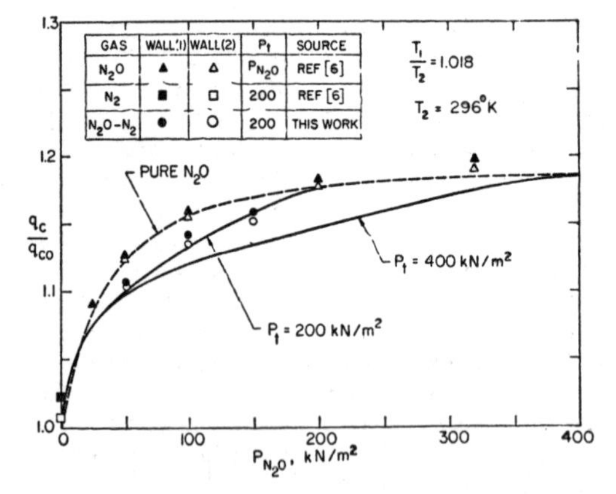

Fig. 7. Radiation-conduction interaction; N_2 broadened N_2O.

There exists one set of NH_3 data [16] for the conduction-radiation interaction situation. This set, however, corresponds to the conduction regime of the vertical enclosure; these results will be presented after the description of the vertical enclosure.

4. Radiation-Convection Interaction

The material presented in this section concerns two specific free-convection problems, the stagnation region of a horizontal cylinder and a two-dimensional boundary layer on a vertical surface. The stagnation problem, which is limited to an analytical study, will be discussed first. In an effort to determine the pressure behavior of gaseous radiation in a convection situation, the exponential wide-band model was applied to a very specific situation [5]. Equations (19-23) were solved for pure CO_2, CH_4 and H_2O vapor at a temperature ratio of T_w/T_ω = 1.1 and in the pressure range of 0.01 to 10 bar* for a fixed cylinder radius of 30.5 cm. The bands used in the calculations are H_2O (6.3 μm), CO_2 (4.3 μm and 15 μm) and CH_4 (7.6 μm and 3.3 μm).

Figure 8 presents the results of the calculations. The results are presented as a ratio of the temperature gradient at the wall with radiation to that for pure free convection versus the pressure of the absorbing gas. It should be noted that the radiative flux at the wall is zero for the perfectly-reflecting wall condition. The most interesting aspect of the results is the fact that the interaction approaches zero for low and high absorbing gas pressures with a peak in the range of 0.1 to 1 bar. Referring to equations (19-23) and the band absorptance behavior given in Table 1, one can gain some insight to this pressure behavior. At high pressures, the derivative of the band absorptance is proportional to $1/\tau_i$. Therefore, the ratio $\sum_i \xi_i/\tau_i$ determines the relative effect of the interaction.† This ratio is inversely proportional to the square root of P_a; the interaction decreases with increasing pressure at high pressures.

*At 0.01 bar, the Grashof number is approximately 10^4; thus, equations (19) and (20) are of questionable validity. Also, at low pressures, phenomena such as a change in line shape, slip flow, and a breakdown in local thermodynamic equilibrium enter the problem.
† The parameter $\Sigma(\xi_i/\tau_i)(Gr^{1/4}/R)$ where $R/Gr^{1/4}$ is a measure of boundary layer thickness, can be used for measuring the relative magnitude of interaction for a particular gas. This dimensional parameter is independent of the flow situation and pressure. As noted by Cess and Tiwari [17], it governs optically-thick interaction in conduction-radiation problems.

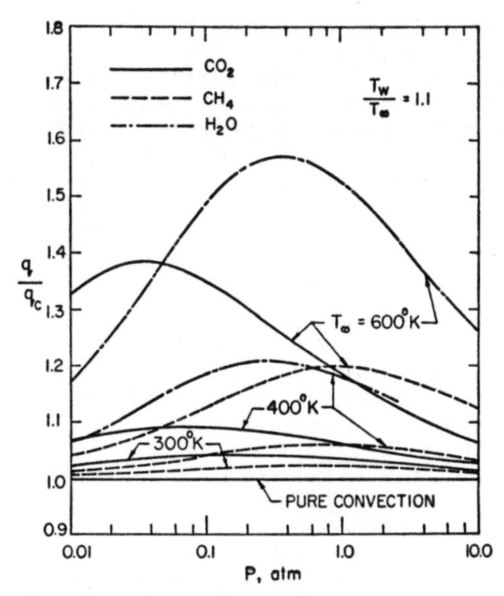

Fig. 8. Radiation-convection interaction; free-convection stagnation flow.

As the pressure decreases, the line structure plays an increasing larger role in the results. At low pressures ($\beta_i < 1.0$) if the linear region (Table 1) dominates, the equations would be independent of pressure. Since the logarithmic regime corresponds to the large optical path limit, it is also discarded. If the square-root absorption region dominates, the parameter appearing in equation (20) is $\sum_i (\xi_i/\tau_i)(\beta_i)^{1/2}$ which is proportional to P_a raised to the power $(2n-1)/4$. This assumes that the broadening exponent n is constant for all bands; this is not quite correct but sufficient for this discussion. Since $(2n-1)/4>0$, the magnitude of interaction decreases as P_a decreases. The square-root absorption regime therefore must dominate at low pressures. The observations concerning the high and low pressure behavior of the interaction suggest approximations that will be used in the two-dimensional boundary-layer problem.

Since the magnitude of radiation interaction is relatively small except at higher temperatures (see Figure 8) and since it was desired to use a Mach-Zehnder interferometer in the experimental program, the vertical enclosure [16] (see Figure 1c) was chosen for the experimental configuration rather than the stagnation problem. This choice, of course, introduces other difficulties. For example, there exist three basic flow regimes [18, 19] in the enclosure depending on the Rayleigh number, Prandtl number and the aspect ratio. At small values of Ra, the heat is transferred mainly by conduction with a small unicellular motion in the fluid. At large Ra, there exist two layers on the vertical surfaces resembling the boundary layer formed on a vertical flat plate. At moderate values of Ra, there is a transition regime. Additionally, thermal stratification exists in the midplane of the enclosure for the boundary layer regime; this has to be accounted for in the analytical-experimental comparisons. There are also stability difficulties with the enclosure [20]. For a detailed discussion of the various aspects of enclosure problem the reader is referred to [16, 21]. The experiments were run for two values of the aspect ratio such that results could be obtained for the stable conduction and boundary layer regimes.

The basic experimental apparatus is the same as that reported in [2, 14] except for the vertical rather than the horizontal geometry and the fact that the radiative surface conditions were modified for the boundary-layer problem. To enhance the interaction and also to simulate a free-convection boundary layer in an infinite fluid, the cool plate was blackened such that the surface reflectance was less than 0.03.

To check out the apparatus and also to serve as another confirmation of the utility of the wide-band model, experiments were run for the conduction regime of the enclosure. Figure 9, which is arranged in the same manner as Figures 3, 6 and 7, presents the results for pure NH_3. The analytical results based on equation (9) take into account the 950 cm^{-1} and 1627 cm^{-1} bands of NH_3. It is important to note that although the surface reflectance conditions are different than those previously presented, the agreement between theory and experiment is excellent; convection in the cell is evident at P_a = 2 bar.

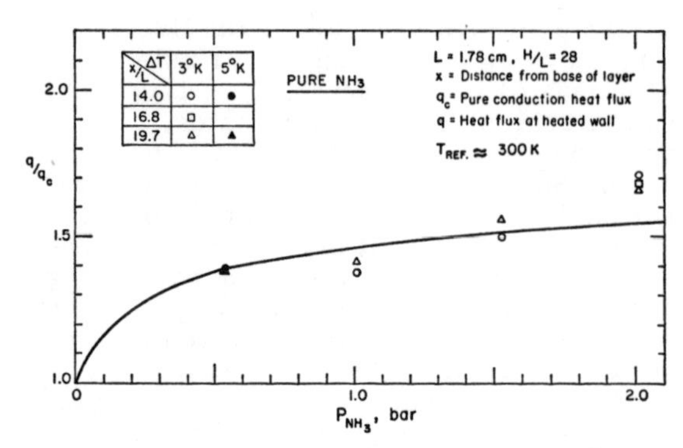

Fig. 9. Radiation-conduction interaction; pure NH_3.

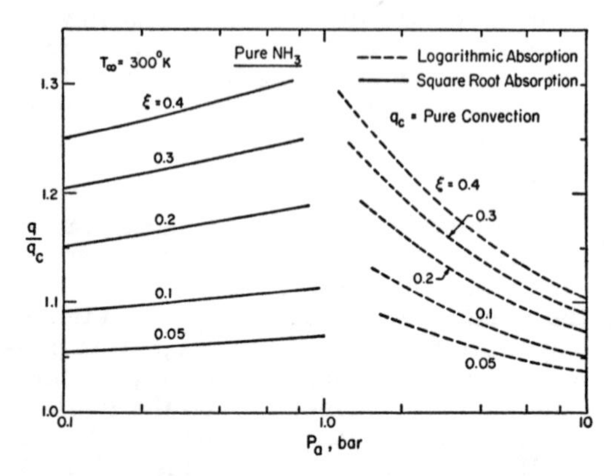

Fig. 10. Radiation-convection interaction; free-convective flow on a vertical flat plate.

For the analytical-experimental comparisons in the boundary layer case, the boundary layer along the heated surface in the enclosure is theoretically approximated by a boundary layer analysis for a perfectly-reflecting flat plate in an infinite fluid [6] (see equations (24) and (25)). Based on the results for the stagnation problem, the derivative of the total band absorptance, equation (26), is approximated by expressions representing the square root limit and the logarithmic limit for low and high absorbing gas pressures, respectively [6, 16]. Figure 10 presents an example of the analytical NH_3* results for a range of ξ values (equation (27)) and absorbing gas pressures for 0.1 to 10 bar. As in the stagnation case, the results indicate a peak in the interaction. Although the two limiting curves do not allow the exact specification of the pressure at the peak, it appears to be in the range of 0.5 to 2 bar depending on ξ. Of course, this peak will shift for different gases (see Figure 8).

Figures 11 and 12 present comparisons between the boundary layer analysis and enclosure experiments. The figures present the local Nusselt number

$$Nu = q\,x/(T_W - T_\infty)\,\lambda$$

versus the local Grashof number

$$Gr = g\,\beta\,(T_W - T_\infty)\,x^3/\nu^2$$

where x denotes the distance from the starting corner and T_∞ denotes the local midplane temperature in the enclosure. Nitrogen data is also shown in the figures for comparison purposes. At the higher absorbing gas pressures (logarithmic approximation), the agreement is excellent.† The agreement at the lower pressures (square-root limit) is not quite as good; the analysis has a tendency of over-predicting the data at the higher x values along the surface. It is strongly suspected that this failure is due to the square-root approximation. This approximation not only ignores the linear absorption region near the surface of the plate but also ignores the outer region of the boundary layer which should be logarithmic. The logarithmic approximation does not have this tendency; the logarithmic approximation improves with the optical thickening of the boundary layer.

*Results for N_2-NH_3 mixtures are also given in [6, 16]; the N_2 broadened results will not be discussed here.

†The analysis is based on the assumption that if the strongest band is logarthmic all bands in the spectrum are logarithmic [16]. The actual results should fall between all-band logarithmic results and the analysis based solely on the strongest band (dashed curve).

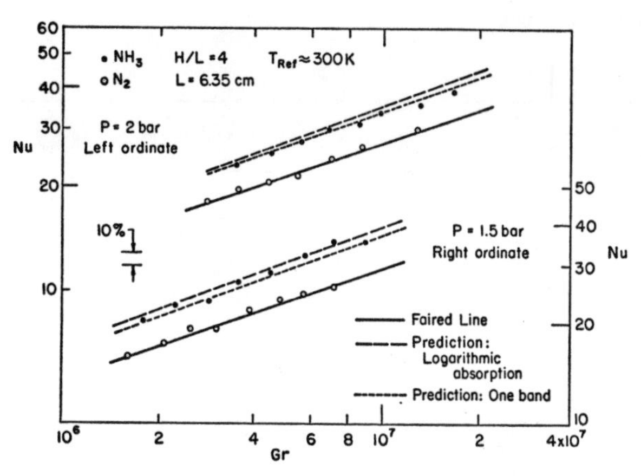

Fig. 11. Radiation-convection interaction; comparison of enclosure data to boundary layer analysis.

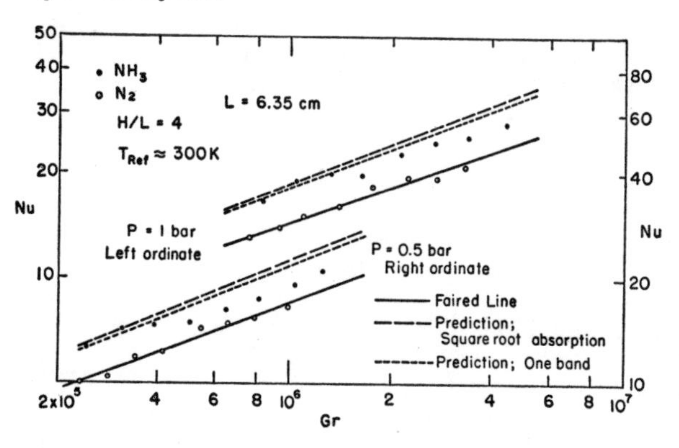

Fig. 12. Radiation-convection interaction; comparison of enclosure data to boundary layer analysis.

Due to the failure of the gray-gas and box model approximations in the conduction situation, these approximations were not used for comparative purposes in the convection problems. However, an examination of equation (20), with equations (21), (22) and (23) written in terms of the gray-gas model or the box model, reveals a different pressure behavior. Although the magnitude of interaction decreases with pressure at high pressures as in the case of the wide-band model, the gray-gas and box models predict that the interaction is independent of pressure with the maximum interaction occurring at low pressures. This is entirely different from the behavior observed in Figures 8 and 10. It is fully realized that this convection study concentrates on the situation where the wall is perfectly reflecting. This, of course, is not the situation generally encountered in actual practice; a nearly black wall is more common in combustion problems. However, the limitation on the surface reflectance does not detract from the fact that it is necessary to take into account the wings of a band for the proper description of pressure behavior.

5. Conclusion

Although this investigation does not concentrate on heat transfer in combustors or other combustion problems, it is clearly evident that the findings of this investigation are important in terms of heat transfer calculations in combustion situations where molecular gas radiation is an important mode of transfer. When band radiation is important, an accurate prediction of radiation interaction must include a description of the wings of a band. It is shown here, for two specific heat transfer problems, that the exponential wide-band model of Edwards and Menard [9] contains enough physics of the absorption-emission process to give an accurate description of the effect of gas radiation. It has been found that the gray-gas model and the box model do not properly describe the magnitude or effect of absorbing gas pressure. This does not imply that these models are useless. For example, the interaction magnitude could be adjusted with the proper input of parameters particular to the gray-gas or box models.

In terms of additional research, it is necessary to extend the wide-band approach to more complicated problems. This, of course, dictates the use of a volume interchange approach such as that proposed by Edwards and Balakrishnan [22]. The extension of the wide-band model to the volume interchange approach still needs additional research including experimental verification before it could be applied with confidence.

Table 1: Total Band Absorption Regimes [9].

$\beta_i \le 1$	$\beta_i > 1$
Linear $(\bar{A}_i \le \beta_i)$	Linear $(\bar{A}_i \le 1)$
$\bar{A}_i = u_i, \quad \bar{A}_i' = 1$	$\bar{A}_i = u_i, \quad \bar{A}' = 1$
Square root $(\beta_i \le \bar{A}_i \le 2-\beta_i)$	Logarithmic $(1 \le \bar{A}_i)$
$\bar{A}_i = 2(u_i\beta_i)^{\frac{1}{2}} - \beta_i, \quad \bar{A}_i' = (\beta_i/u_i)^{\frac{1}{2}}$	$\bar{A}_i = \ell n(u_i)+1, \quad \bar{A}_i' = 1/u_i$
Logarithmic $(2-\beta_i \le \bar{A}_i)$	
$\bar{A}_i = \ell n\,(u_i\beta_i) + 2-\beta_i, \quad \bar{A}_i' = 1/u_i$	

ACKNOWLEDGMENTS

The author gratefully acknowledges the support of the National Science Foundation under Grants GK 672, 2093, and 20382. This author is grateful for the significant contributions of J. C. Bratis, F. A. Olsofka and W. P. Schimmel. Without their efforts, this work would have been impossible.

REFERENCES

[1] Schimmel, W. P., J. L. Novotny and S. Kast: "Effect of Surface Emittance and Approximate Kernels in Radiation-Conduction Interaction," Wärme-und Stoffübertragung, 3, 1, (1970).

[2] Schimmel, W. P., J. L. Novotny, and F. H. Olsofka: "Interferometric Study of Radiation-Conduction Interaction," Proceedings of the Fourth Int. Heat Trans. Conf., paper R2. 1, Elsevier, Amsterdam (1970).

[3] Sparrow, E. M. and R. D. Cess: Radiation Heat Transfer, Wadsworth, Belmont, California (1966).

[4] Cess, R. D. and S. N. Tiwari: "Infrared Radiative Energy Transfer in Gases," Advances in Heat Transfer, 8, Academic Press, New York, (1972).

[5] Novotny, J. L.: "Radiation Interaction in Nongray Boundary Layers," Int. J. Heat Mass Transfer, 11, 1823, (1968).

[6] Bratis, J. C. and J. L. Novotny: "Radiation Interaction in Real Gases," AIAA Progress in Astronautics and Aeronautics, 31, (1973).

[7] Hottel, H. C. and A. F. Sarofim: Radiative Transfer, McGraw-Hill New York, (1967).

[8] Cess, R. D., P. Mighdoll and S. N. Tiwari: "Infrared Radiative Heat Transfer in Nongray Gases," Int. J. Heat Mass Transfer, 10, 1521 (1967).

[9] Edwards, D. K. and W. A. Menard: "Comparisons of Models for Correlation of Total Band Absorption," Appl. Optics, 3, 621 (1964).

[10] Goody, R. M.: Atmospheric Radiation, Oxford, London, (1964).

[11] Tien, C. L. and J. E. Lowder: "A Correlation of Total Band Absorptance of Radiating Gases," Int. J. Heat Mass Transfer, 9, 698 (1966).

[12] Felske, J. D. and C. L. Tien: "A Theoretical Closed Form Expression for the Total Band Absorptance of Infrared-Radiating Gases," submitted to the Int. J. Heat Mass Transfer, (1973).

[13] Novotny, J. L. and M. D. Kelleher: "Conduction in Nongray Radiating Gases" Int. J. Heat Mass Transfer, 11, 365 (1968).

[14] Novotny, J. L. and F. A. Olsofka: "The Influence of a Non-absorbing Gas in Radiation-Conduction Interaction," AIAA Progress in Astronautics and Aeronautics, 24, 410 (1971).

[15] Penner, S. S.: Quantitative Molecular Spectroscopy and Gas Emissivities, Addison-Wesley, Reading, Mass. (1959).
[16] Bratis, J. C. and J. L. Novotny: "Radiation-Convection Interaction in the Boundary-Layer Regime of an Enclosure," submitted to the Int. J. Heat Mass Transfer, (1973).
[17] Cess, R. D. and S. N. Tiwari: "The Interaction of Thermal Conduction and Infrared Gaseous Radiation," Appl. Scient. Res. 19, 439, (1968).
[18] Eckert, E. R. G. and W. O. Carlson: "Natural Convection in an Air Layer Enclosed Between Two Vertical Plates with Different Temperatures," Int. J. Heat Mass Transfer, 2, 106 (1961).
[19] Elder, J. W.: "Laminar Free Convection in a Vertical Slot," J. of Fluid Mech., 23, 77 (1965).
[20] Gill, A. E. and A. Davey: "Instabilities of a Bouyancy Driven System," J. Fluid Mech., 35, 775 (1969).
[21] Bratis, J. C.: "Interaction of Gaseous Radiation and Natural Convection Heat Transfer in an Enclosed Layer Between Two Vertical Parallel Plates," Ph. D. Thesis, University of Notre Dame (1972).
[22] Edwards, D. K. and A. Balakrishnan: "Volume Interchange Factors for Non-homogeneous Gases," J. Heat Transfer, 94, 181, (1972).

Chapter 18

LUMINOUS FLAME EMISSION UNDER PRESSURE UP TO 20 ATM

Takeshi Kunitomo

Department of Mechanical Engineering, Kyoto University, Kyoto, Japan

Abstract

In order to predict radiation from a luminous flame of liquid fuel, the influence of the air excess factor, temperature, flame scale, pressure and other factors upon the emission of nonluminous gas and that of soot particle clouds are examined. A method is developed to calculate the luminous flame emission under pressure up to 20 atm. The maximum, average, and endzone absorption coefficients can be easily predicted.

NOMENCLATURE

A : band absorption [cm^{-1}]
K : absorption coefficient
l : optical path [ft] and relative distance
p : partial pressure and total pressure [atm]
T : temperature [K]
γ : specific gravity of fuel
φ : excess air factor
b : self-broadening coefficient
L : flame breadth and luminosity factor
m : pressure exponent
R : C/H
r : overall fuel/air ratio
T_b : boiling temperature [°C]
ε : emissivity
ω : wavenumber [m^{-1}]
a_1, c_1, c_2, c_3, c_4, d, e, n, α, β : constant

Subscripts
A : infrared active gas
av : average
end : endzone of flame
max : maximum
B : broadening gas
e : effective
m : average
min : minimum
CO, CO_2, H_2O, N_2 : carbon monoxide, carbon dioxide, water vapor and nitrogen
D, M, W : CO_2, CO and H_2O
f, g, s : flame, nonluminous gas and soot particle cloud
o : at an atmospheric pressure and at band center

1. Introduction

This author previously reported experimental and analytical studies/1/∿/5/ on soot formation, the emission characteristics and the effect of several factors on radiative heat transfer in the luminous flame for spray combustion of liquid fuel. The heat transfer in a combustion furnace can be estimated by various methods of calculation after the scale and the structures of the furnace and of the luminous flame, the emissivity of the luminous flame, the absorptivity of the furnace wall and so on are known. However, at the present time, it is impossible to estimate the emissivity of the luminous flame. For example, Holliday and Thring/6/ presented the following expression from the results of a small scale furnace,

$$\varepsilon_{av} = 0.282 \ln\{(R-5.0)/4\} + 0.002(T-200) + 0.484. \quad (1)$$

However, they did not succeed in obtaining the general expression which can be applied to the flame of a different scale. Furthermore, the important excess air factor is not included. Marsland, Odgers and Winter/7/ proposed the expression,

$$\varepsilon_f=1-\exp[-1.6\times10^4(r1)^{0.5}pT_f^{-1.5}L] \quad (2)$$

where $L[=7.53(C/H-5.5)^{0.84}]$ is the luminosity factor. In their experiment, the fuel/air ratio in the primary zone is not equal to r/8/. This author supposes that the large scattering of the data around the eq.(2) may be caused by the different fuel/air ratios in the primary zone. The primary zone fuel/air ratio must be considered instead of the overall ratio. Storojyk and Antonofskii/9/ presented the following expression for the absorption coefficient of the soot particle cloud in the luminous flame.

$$K_s=a_1f_1(T)f_2(\varphi)f_3(p)f_4(C/H), \quad (3)$$

$$\text{where}\quad f_1(T)=C_1(0.34\times10^{-3}T+0.14)/T,\quad f_2(\varphi)=0.1+(1.85/\varphi),$$
$$f_3(p)=C_3p^n \quad\text{and}\quad f_4(C/H)=C_4(C/H)^2. \quad (4)$$

The functions, f_1, f_2 and f_3 are obtained only for one liquid fuel and f_4 is obtained from the relation between one gas fuel and one liquid fuel. The applicability of the above formula is doubtful.

In the present study, the method of estimating the emissivity or the absorption coefficient of luminous flames is examined under various conditions of combustion which include flame scale, fuel nature, excess air factor, pressure, temperature and so on.

2. Emission from soot particle cloud and from gas under atmospheric pressure

In order to apply the result of flame study to the usual technological case, the flame scale must approach as closely as possible the actual scale. So this author used the furnace of 2.7 m length and of 0.5 m width square section.

If the main part of the flame, where the soot formation becomes prominent and the flame emits luminously, is considered and the extreme top region where the emission is not so large is excluded, the average partial pressures of infrared-active gases may be determined and the non-luminous gas emissivity or its absorption coefficient may be calculated. So, if the relationship between the emissivity of the soot particle cloud ε_s and that of gas ε_g, or the relationship between the absorption coefficient of the former K_s and that of the latter K_g, is experimentally clarified for the various conditions of combustion, the emissivity ε_f or the absorption coefficient K_f may finally be estimated. In the earlier papers/1/∿/5/, it was found that ε_s was affected by the soot concentration and the particle size and that the soot formation was influenced greatly by the excess air factor, the kinds of burner and the types of fuel, and was also found that the effect of the quantity of atomizing air was smaller than those of the other factors. With reference to the results, in the case of inadequate atomization, the flame contained highly concentrated and large soot particles. However, the soot particles of large size show weak radiancy. On the contrary, in the case of proper atomization, the flame contains a low concentration of small particles. Small soot particles show strong radiancy. It is therefore assumed that the two effects of the soot concentration and radiancy of the particles on the absorption coefficient may cancel each other out, that is, the effect of atomization or that of the kind of burner may be reduced.

On the effect of the degree of swirl and the spray angle of burner, Storojyk and Antonofskii/10/ and IFRF/11/ performed the experiment and the remarkable change of the flame emissivity was not observed. The decrease of soot concentration usually corresponds to the increase of flame breadth. Thus the decrease of the absorption coefficient of the soot particle cloud coincides with that of nonluminous gas.

The effects of the heat release rate and of the kinds of atomizing fluid are also found to be small, by IFRF/11/ and Kunitomo/2/.

On C/H in soot, this author previously presented the result shown in Fig.1/2/. Fig.1(a) shows the longitudinal variation of the absorption coefficient K_s in each flame and Fig.1(b) shows that of C/H of soot. It is confirmed by electron microscopy that in each flame shown in the figure, the size distributions of soot do not change along the flame, although each flame has its own size distribution of soot. The effect of C/H in soot is much smaller than that of particle size. Therefore, it is conceivable that the C/H of fuel affects $K_s[m^{-1}]$ only through the soot concentration.

At constant optical depth (partial pressure x length), the gas emissivity decreases as the temperature increases. In the case of the soot particle cloud, its emissivity may become smaller as the temperature becomes higher, since the effective soot concentration may decrease because of the increased volume of combustion gas and because of the acceleration of the combustion of soot particles. Thus it is supposed that the temperature effect becomes comparatively small if the data is rearranged by calculating the ratio $\varepsilon_s/\varepsilon_g$.

Next, the influence of the flame breadth is considered as follows. The average absorption coefficient of gas K_g, defined by the following expression,

$$K_g=-[\log_e(1-\varepsilon_g)]/p_gL \quad (5)$$

decreases as the flame scale increases at constant partial pressures of infrared-active gases, because of the characteristics of gaseous radiation. For the soot particle, as the flame scale becomes larger, the formation becomes weaker since the eddy diffusivity which dominates the combustion of diffusion flame is proportional either to the product of the mixing length and the intensity of turbulence, or to the product of the mean velocity of combustion gas and the burner diameter. Consequently, the larger the flame scale, the smaller the K_s. Thus, it is supposed that the effect of the flame scale on ε_g and ε_s is similar.

On the basis of the above discussion, about 90 kinds of experiments on luminous flames were carried out at atmospheric pressure, in which three kinds of liquid fuels and two kinds of burners, high pressure air atomizing and low pressure air atomizing, were used and the conditions of combustion were changed respectively. The average values of ε_{so} and ε_{go} for each luminous flame were obtained and the ratios of the two were calculated. The results are shown in Fig.2. The specific gravity γ was chosen as the parameter to represent the fuel nature, because the boiling point, viscosity and other properties of fuel are reflected well in the specific gravity. The correspondence of C/H, which is often used in reports, to γ of the fuels that are mentioned in this study is shown in Fig.3. The results in Fig.2 show that the ratios depend mainly on the nature of the fuel and the excess air factor φ and scarcely on the other combustion conditions, although the wall temperature and the flow rates of fuel and primary and secondary air are varied. The temperature of the points in Fig.2 ranges from 900°K to 1600°K and the flame breadth from 0.3 m to 0.5 m. The temperature and the flame breadth at each point are not shown in the figure since their effect has not been detected. The left side group of the data of the fuel of specific gravity 0.85 corresponds to the case of the lumps of soot particles of 100 μ ∿ 500 μ floating in the flame. Therefore, these data are excluded from the following discussion.

It would be very convenient for the estimation of the emissivity of the luminous flame if the results obtained above are generally accepted. The results of the IFRF/11/ are referred to in this investigation. The IJmuiden furnaces have scales of about 4 ∿ 6 times that of the present study furnace and the fuels used are also different from those in the present study. The comparison between the present study and the IFRF study may clarify the effect of the flame scale and of the fuel nature. At IJmuiden, detailed experiments, containing many tests of burner, fuel nature, heat release rate, atomizing air and steam quantity, excess air factor and so on, were carried out. Though most IFRF reports give only the total emissivities ε_{fo}, ε_{go} can be estimated from the diagrams. Then the emissivities of the soot particle cloud in the main part of each flame can be obtained by the following equation,

$$\varepsilon_{so}=1-\{(1-\varepsilon_{fo})/(1-\varepsilon_{go})\} \quad (6)$$

The results are also shown in Fig.2. The effects of many other factors, except the fuel nature and can also be considered small in the IFRF work as in the present work. The effects of the fuel nature in both studies show the same definite trend and coincide with each other very well. The effect of the flame scale can also be neglected.

In order to apply the results in Fig.2 to the estimation of the emissivity of luminous flame, $\varepsilon_{so}/\varepsilon_{go}$ is expressed by the simple function of γ and φ as follows

$$\varepsilon_{so}/\varepsilon_{go}=0.09/(\varphi-\gamma^2+0.35\gamma-0.38)+6.8\gamma-5.95. \quad (7)$$

The value of $\varepsilon_{so}/\varepsilon_{go}$ is set equal to zero if a negative value is obtained. The values calculated by Eq.(7) are shown by solid lines in Fig.2.

3. Partial pressure of infrared-active gases under atmospheric pressure

The partial pressure of gases can not be calculated easily by the assumption of thermal equilibrium because of soot formation and must therefore be determined experimentally. The present experimental data are used and extrapolated to regions beyond the experimental conditions. The partial pressures of infrared-active gases may vary slightly if the flame scale is enlarged. However, as the final nonluminous gas emissivity is only slightly influenced by such small variations of partial pressures, the effect of the scale can be neglected. The average partial pressures of H_2O, CO_2 and CO are obtained in the main part of the flame except in the extreme top region. p_{CO} are shown in Fig.4 by using the same parameter as in Fig.2. The larger the γ, the smaller the p_{CO}. This is caused by the fact that the evaporation velocity from the fuel droplet becomes larger as γ decreases. At the first step of spray combustion of the fuel of small γ in a diffusion flame the quantity of fuel vapor is large in comparison with that of the oxygen and a large portion of the fuel vapor burns to CO. After the above reaction, the variation occurs rarely because of insufficiency of oxygen in the case of a small φ and because of low temperature in the case of a large φ. p_{CO} are expressed by the function of γ and φ as follows.

$$p_{CO}=2.2(1.0-\gamma)(\ -1.2)^2, \qquad \varphi \leqq 1.2, \quad \gamma \leqq 1.0$$
$$=0 \qquad 1.2<\varphi<1.7 \text{ or } \gamma>1.0 \quad (8)$$

Solid lines in Fig.4 show Eq.(8). p_{CO_2} are shown in Fig.5. It is found that the larger the γ, the larger the p_{CO_2}. The reason is the same as in the above discussion on p_{CO}. The emprirical equation for p_{CO_2} is as follows.

$$p_{CO_2}=0.0348\gamma-2.619(1.01-\gamma)(\varphi-1.05)^2+0.092, \quad \varphi<1.1$$
$$=-0.075\varphi+0.033\gamma+0.1754 \qquad 1.1<\varphi<1.7 \quad (9)$$

In the region of the large φ, the calculated values are adopted. p_{H_2O} are shown in Fig.6. In the region of a large φ, the partial pressure is small for the fuel of large γ, because of low concentration of hydrogen in fuel. In the region of the small φ, the higher the γ, the higher the p_{H_2O}. The influence of γ on p_{H_2O} is the same as that on p_{CO_2}, but the effect is smaller because of the lower concentration of hydrogen in the fuel of larger γ. The relationship obtained is

$$p_{H_2O}=0.126-0.72\{\varphi+0.887\gamma-1.724-1.42(\gamma-0.805)^2\}^2 \qquad \varphi<1.05$$
$$=-0.0783\gamma-0.068\varphi-0.259 \qquad 1.05\leqq\varphi<1.7 \quad (10)$$

In the region of the large φ, the calculated values are adopted.

4. Effect of pressure

In the case of industrial furnaces or combustion chambers where pressures often increase over atmospheric pressure, the above results can not be applied. So the pressure dependences of soot particle cloud radiation and of non-luminous gas radiation are examined. To this problem, only the deductive inference based on the empirical fact has possiblility of success, and the analytical method can not be adopted at present.

4. 1 Soot particle concentration

On referring to this author's experiment at atmospheric pressure, the main fraction of carbon particles at low excess air factor is the soot formed from evaporated fuel, and that at high excess air factor is the coke of large diameter formed directly by pyrolysis from fuel spray. Accordingly, it is appropriate to examine the pressure dependence of carbon particle formation separately in the region of low φ and in the region of high φ. (In this paper, soot and coke particles are often lumped together and called soot particles.)

In the region of the low φ, the experimental results by Macfarlane et al./13/, who investivated the soot formation at pressures up to 20 atm, are suggestive. At constant pressure, the soot concentration increases at first as φ decreases and it comes to the maximum at φ 0.3 ∿ 0.4. When φ decreases further, the soot concentration decreases. Under constant φ, the soot concentration increases in proportion to p^m. The exponent m is about 2.5 at φ 0.5. The threshold of φ at which the soot formation begins is independent of pressure. With reference to these results and this author's earlier investigation, the region where the curve of $\varepsilon_s/\varepsilon_g$ begins to rise, i.e., the soot formation becomes strong, may be independent of pressure.

Secondly, the pressure dependence of coke particle formation is examined. In the case of spray combustion, the temperature of the fuel droplet is raised after ejection and the pyrolysis-carbonization process begins to proceed in some part of a fuel droplet. This process proceeds rapidly since the contribution of radiation becomes great because of high absorptivity of coke. The longer the period of time after the coke formation begins, the more the heat is absorbed to form the coke. So, the concentration of coke particles is assumed to be proportionate to the combustion lifetime. Hall and Diederichsen/14/ obtained the empirical relation $t \propto p^{-0.25}$ between the combustion lifetime t of a fuel droplet and the pressure p. The exponent m is then assumed to be -0.25 at the infinite φ. Thus the pressure dependence of soot particle formation is expressed as p^m both in the low and high φ. In the middle region, the two kinds of soot particles exist in the certain ratio according to φ. The exponent m is assumed to decrease gradually from about 2.5 around at φ of 0.5 to -0.25 at the infinite excess air factor. The expression of the exponent m is assumed as follows:

$$m=e/(\varphi+d)-0.25. \tag{11}$$

Although m may be supposed to vary with fuel nature, d and e are actually regarded as constant, by comparing the calculated values with the empirical values of the emissivity, as described in Section 5. This may be caused by the fact that, the main effect of fuel nature appears already at atmospheric pressure and that the pressure dependence of each fuel is similar.

4. 2 Partial pressures of infrared-active gases

Macfarlane et al./13/ showed the partial pressures of various gases for the combustion of n-hexane and benzene at 8 atm and 15 atm. For example, when the pressure increased from 8 atm to 15 atm at the excess air factor of 0.65, the percentages of CO and H_2O in all gases at 15 atm showed the decrease of 0.16 times and 0.04 times the percentages at 8 atm respectively and the percentage of CO_2 at 15 atm showed the increase of 0.16 times that at 8 atm. The empirical formulas (8)∿(10) can be utilized to estimate the partial pressures at high pressure. The partial pressures at a high pressure are obtained by multipling the above values by the pressure p. The differences of the percentages of gases at different pressures above mentioned may have little influence on the overall non-luminous gas radiation since their effects fairly cancel each other.

4. 3 Non-luminous gas radiation at high pressure

Edwards and Menard/15/ and Edwards et al./16/ proposed the exponential wide band model and presented the comparatively simple relations for CO_2, CO and H_2O, which can be applied for high pressures. In the present work, their results are used. The band absorption $A(cm^{-1})$ can easily be calcualted for each band of CO_2, CO and H_2O. The effective pressure p_e is defined as

$$p_e=(p_B+bp_A)^n. \tag{12}$$

The values of constants and the expressions of A were adopted from the work of Edwards et al./16/.

5. Calculation

5. 1 Procedure

The procedure of computing the non-luminous gas emission or emissivity is described firstly. The combustion gas is composed of N_2, O_2, CO_2, CO, H_2O, H_2 and etc. The broadening effect of each gas on the gas whose emission is to be calculated is different, However, the values of b and n for N_2 given by Edwards et al. are used for all broadening gases, by assuming the other co-existing gases to be N_2. Then

$$p_B\equiv p_{N_2}=p-p_A \tag{13}$$

As the representative wavenumber, the band center ω_o is chosen when it is given, and

$$\omega=\omega_{max}-0.5A \qquad \text{or} \qquad \omega=\omega_{min}+0.5A \tag{14}$$

is chosen respectively when ω_{max} or ω_{min} is given.

In calculating the non-luminous gas emissivity, φ and γ (or C/H) of fuel are given first and then the partial pressures of infrared active gases at atmospheric pressure are calculated by eqs. (8)∿(10). The partial pressures at a high pressure are obtained by multipling the above values by the pressure p. Next, the temperature T_f and the breadth L are set and then the mass path lengths of the infrared-active gases are determined. The band absorptions are calculated by the procedure of Edwards et al. The emissivities of the infrared-active gases are obtained. The average overall non-luminous gas emissivity ε_g at the pressure is given by the following expression,

$$\varepsilon_g=1-(1-\varepsilon_D)(1-\varepsilon_M)(1-\varepsilon_W). \tag{15}$$

The average absorption coefficient K_g is then obtained by

$$K_g=-\{\log_e(1-\varepsilon_g)\}/L. \tag{16}$$

In calculating the emissivity of the carbon particle cloud, it is supposed that the absorption coefficient at a certain pressure and a certain combustion condition can be obtained by multipling the absorption coefficient at atmospheric pressure by the ratio of the carbon particle concentration at the pressure to the concentration at atmospheric pressure. The above supposition is equivalent to the assumption that the size distribution of the carbon particle cloud at atmospheric pressure is nearly equal to that at any high pressure if φ and the fuel are unchanged.

The average non-luminous gas emissivity ε_{go} at atmospheric pressure is determined at first by the above procedure and the average emissivity of the soot particle cloud ε_{so} is calculated by eq.(7). Then the average absorption coefficient of the soot particle cloud K_{so} is given by

$$K_{so}=-\{\log_e(1-\varepsilon_{so})\}/L \tag{17}$$

The average absorption coefficient K_s at pressure p is obtained by multipling K_{so} by the concentration ratio p^m as follows.

$$K_s=K_{so}\cdot p^m \tag{18}$$

The average emissivity of soot particle cloud ε_s, the average emissivity of the luminous flame ε_{fm} and the average absorption coefficient K_{fm} are obtained as follows;

$$\varepsilon_s = 1 - \exp(-K_s L) \tag{19}$$

$$\varepsilon_{fm} = 1 - (1-\varepsilon_g)(1-\varepsilon_s) \quad \text{and} \tag{20}$$

$$K_{fm} = -\{\log_e(1-\varepsilon_{fm})\}/L. \tag{21}$$

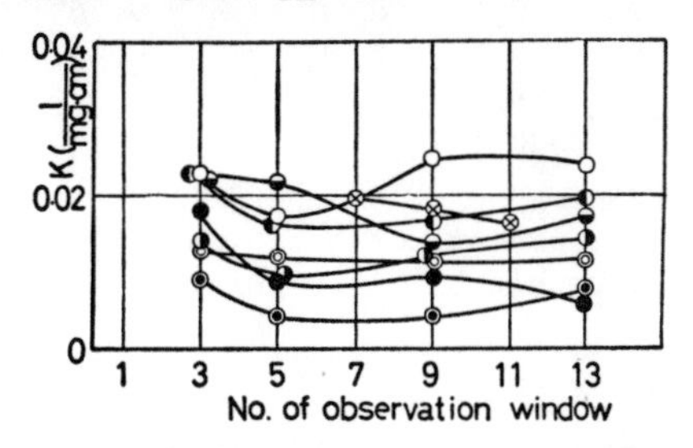

Fig.1(a) Absorption coefficient of soot particle cloud.

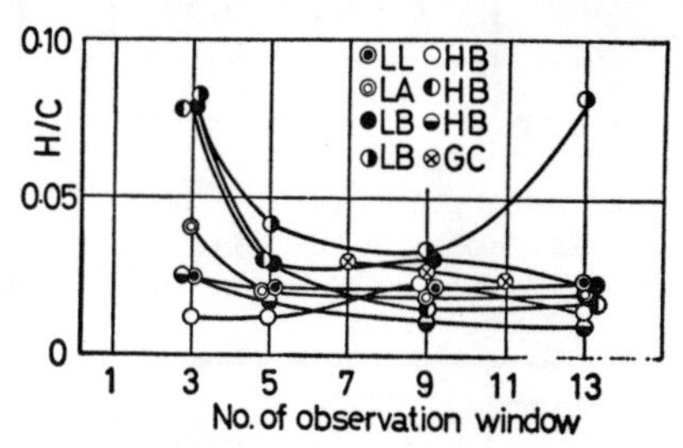

Fig.1(b) Mass ratio of hydrogen and carbon in soot particle.

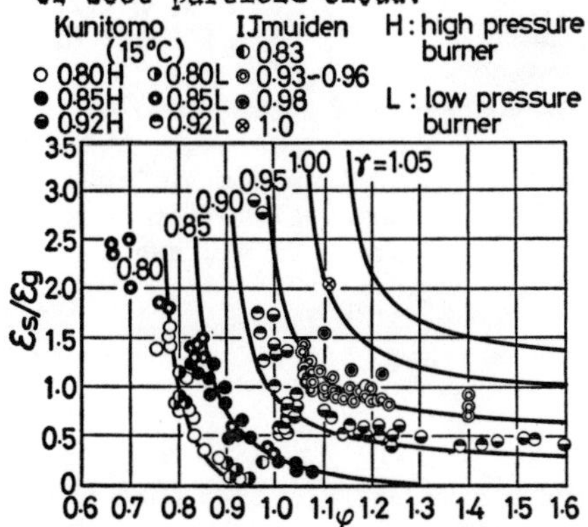

Fig. 2 Ratio of the emissivity of the soot particle cloud and the nonluminous gas emissivity

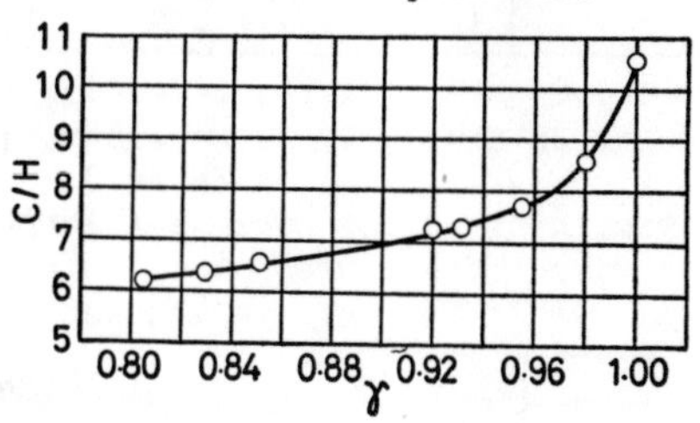

Fig.3 Relationship between the specific gravity and carbon-hydrogen ratio of fuel

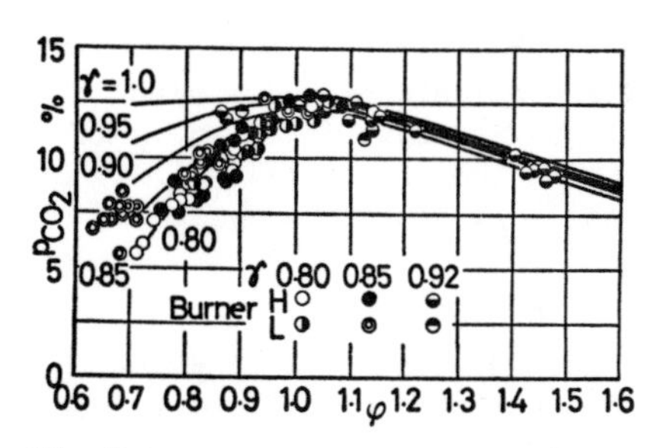

Fig.5 Partial pressure of CO_2.

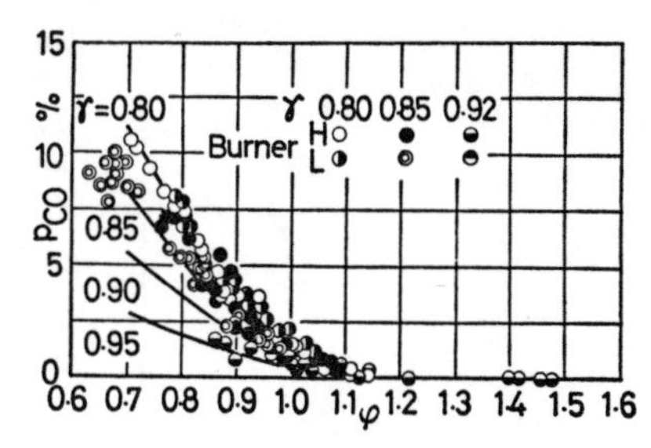

Fig.4 Partial pressure of CO.

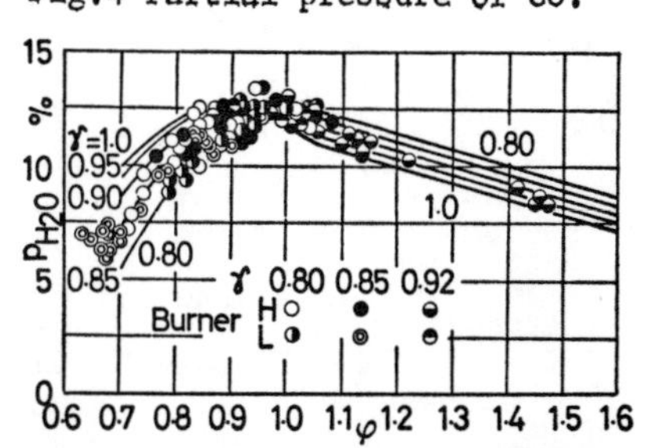

Fig.6 Partial pressure of H_2O.

5. 2 Results

Macfarlane and Holderness/17/, Macfarlane/18/, Storojyk and Antonofskii /10/, Storojyk/19/, Marsland et al./7/, Hill and Dibelius/20/ and Schirmer et al. /21/ carried out experiments on the radiation of luminous flames at high pressures. Among the above experiments, the last two can not be applied to make the estimation formula since the air quantity is not clearly indicated. The unknown constants in eq.(11) are determind on referring to the experiments by Storojyk who presented the

Pressure (atm)	K_{fm} (m^{-1}) Present	Storojyk
1.5	0.37	0.375
2.0	0.49	0.50
2.5	0.62	0.625
3.0	0.75	0.75

Table 1
Average absorption coefficient (Comparison of the present calculation with the experimental result by Storojyk)

Pressure (atm)	K_{fmax} (m^{-1}) Present	Storojyk	K_{fend} (m^{-1}) Present	Storojyk
1.5	0.49	0.48	0.25	0.27
2.0	0.67	0.64	0.31	0.36
2.5	0.88	0.80	0.36	0.45
3.0	1.09	0.96	0.41	0.54

Table 2
Maximum and endzone absorption coefficients (Comparison of the present calculation with the experimental result by Storojyk)

Combustion condition*	Present K_{fmax}	K_{fm}	K_{fend}	Storojyk & Antonofskii K_{fmax}	K_{fm}	K_{fend}
1	0.61	0.47	0.33	0.80	0.46	0.32
2	1.21	0.87	0.53	1.25	0.90	0.40

*1 : p=1.05 atm, =1.21, L=0.364m, γ=0.89, T_f=1300∿1400°K
2 : p=2.23 atm, =1.33, L=0.364m, γ=0.89, T_f=1300∿1400°K

Table 3
Maximum, average and endzone absorption coefficients (Comparison of the present calculation with the experimental result by Storojyk and Antonofskii)

longitudinal distribution of the absorption coefficient under the condition that γ is 0.89, p 1.3∿3.3 atm, φ 1.22∿1.26, the diameter of combustion chamber 0.4 m and T_f 1600∿1900°C. In his experiment, the maximum and minimum absorption coefficients are obtained as

$$K_{fmax}=0.32p \quad \text{and} \quad K_{fend}=0.18p \ , \tag{22}$$

respectively, so the average absorption coefficient is regarded as

$$K_{fm}=0.25p. \tag{23}$$

The constants d and e are determined so that the average absorption coefficients calculated by the above procedure under the condition of φ 1.24, T_f 1900∿1950°K and p 1.5, 2.0, 2.5 and 3.0 atm agree well with the results of eq.(23). These are

$$d=1.5 \quad \text{and} \quad e=4.95. \tag{24}$$

The comparison of the calculated value with the experimental result by Storojyk is shown in Table 1.

In order to check further the applicability of the constants of eq.(24), K_{fend} and K_{fmax} are approximated by

$$K_{fend}\simeq K_g \quad \text{and} \quad K_{fmax}\simeq 2K_{fm}-K_g. \tag{25}$$

The comparison of eq.(25) with eq.(22) is shown in Table 2. It is seen that the calculated values hold fairly good approximations of the actual values. Further, the comparison of the present calculation with the experimental result by Storojyk and Antonofskii/10/ is shown in Table 3. Close agreement between observed and calculated values is also obtained. In the next place, the experimental result by Macfarlane and Holderness in the model gas turbine of diameter 0.076 m using the aviation kerosine of specific gravity 0.788 under pressures 6, 11, 16 and 21 atm is used to examine the present procedure. As the position where the data are obtianed is the maximum radiation zone, the comparison must be carried out on

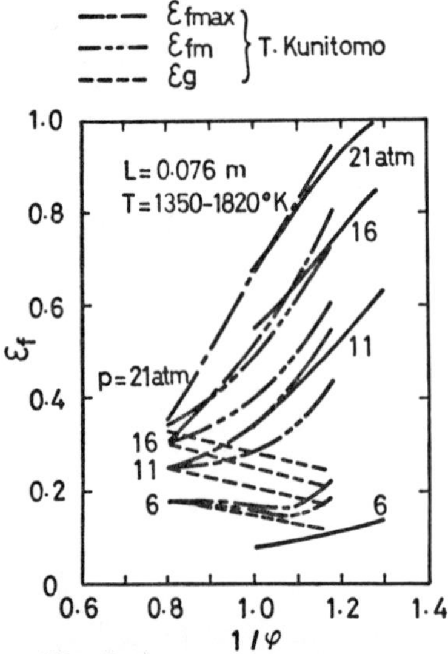

Fig.7 Emissivity of luminous flame (Comparison of the present calculation with the experiment by Macfarlane and Holderness)

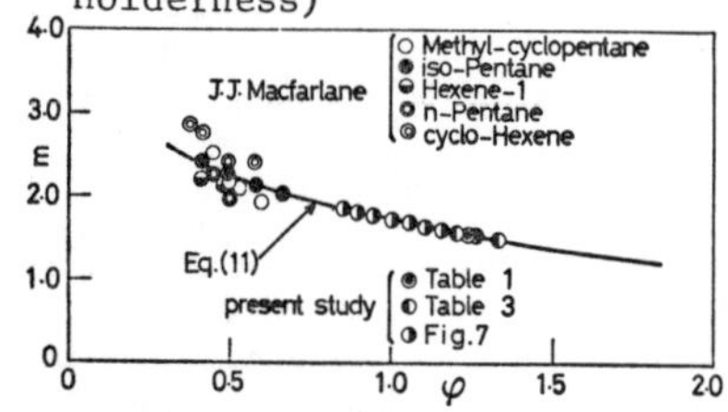

Fig.8 Pressure exponent m

ε_{fmax}. The results are shown in Fig.7. The values of ε_g and ε_{fm} are also shown. In the cases of the pressures 11, 16 and 21 atm, the calculated values agree well with the observed values. The large difference at 6 atm originates in the large experimental error due to the small output of radiometer as Macfarlane and Holderness described. From the above results, it is estimated that the constants d and e of eq.(24) can be used for other kinds of fuel. In the experiment of Marsland et al., φ for each flame in the primary zone was not precisely measured but ranged from 0.64 to 1.22/8/. γ is 0.791. The emissivities calculated by the present procedure sufficiently cover the experimental results for the pressures near 4, 12 and 20 atm.

For reference, the adopted values of the exponent m in the above comparisons are shown in Fig.8. From the close agreement between the observed and the calculated values in the above, it may be concluded that the standard values of the average, maximum and endzone absorption coefficients, accordingly the respective emissivities, can be estimated by the present procedure. If the effect of the special factor is considered to be large, the perturbation from the standard value must be estimated with reference to the reports of IFRF or other authors.

The average absorption coefficients and the average non-luminous gas absorption coefficients are calculated and shown in Fig.9 for the convenience of engineers. The absorption coefficients for the flame breadth 2.0 m are $0.6\sim0.7$ times those for 1.0 m and the absorption coefficients for flame breadth 0.1 m are $1.6\sim1.9$ times those for 0.3 m. The present procedure has now a little doubt about the applicability at the pressure much larger than 20 atm.

On the calculation of the heat transfer in a combustion chamber, K_{fm} can directly be used for the well-stirred furnace model. When the long furnace model is adopted, K_{fmax}, K_{fm} and K_{fend} enable one to predict the heat transfer. As the function which describes the longitudinal change of the absorption coefficient or the emissivity, $a(e^{-\alpha l} - e^{-\beta l})$/22/ or $ale^{-\alpha l}$ may be adopted, where l is a relative distance to the position where the combustion completes 99%. The constants in these relations may be determined easily from K_{fmax}, K_{fm} and K_{fend}.

References

/1/ Sato, T., T. Kunitomo, F. Nakashima & H. Fujii, Bull. of JSME, Vol. 9 (1966) 768.

/2/ Sato, T. & T. Kunitomo, Mem. of Fac. of Engng., Kyoto Univ., Vol. 31-1 (1969) 47.

/3/ Sato, T., T. Kunitomo, S. Yoshii & T. Hashimoto, Bull. of JSME, Vol. 12 (1969) 1135.

/4/ Kunitomo, T. & T. Sato, Bull. of JSME, Vol. 14 (1971) 58.

/5/ Kunitomo, T. & T. Sato, Heat Transfer 1970, Vol. 3 (Elsevier Publishing Co., Amsterdam, 1970) R. 1. 6.

/6/ Holliday, D. K. & M. W. Thring, J. Inst. of Fuel, Vol. 30 (1957) 127.

/7/ Marsland, J., J. Odgers & J. Winter, 12th Symp. on Comb. (The Comb. Inst., Pittsburgh, 1969) 1265.

/8/ Marsland, J., Private Communication.

/9/ Storojyk, T. P. & V. I. Antonofskii, Rep. Central Boiler Turbine Inst., Vol. 75 (1967) 148.

/10/ Storojyk, T. P. & V. I. Antonofskii, Teploenergetika, Vol. 11 (1964) 39.

/11/ The International Flame Research Foundation, J. Inst. of Fuel, Vol. 26 (1953) 189, Vol. 29 (1956) 23, 27, Vol. 30 (1957) 553, 556, 561, 564, Vol. 32 (1959) 328, 338.

/12/ Kunitomo, T., Heat Transfer - Japanese Research, Vol. 1 (1972) 57.

/13/ Macfarlane, J. J., F. H. Holderness & F. S. E. Whitcher, Comb. and Flame, Vol. 8 (1964) 215.

/14/ Hall, A. R. & J. Diederichsen, 4th Symp. on Comb. (Williams and Wilkins, Baltimore, 1953) 837.

/15/ Edwards, D. K. & W. A. Menard, Applied Optics, Vol. 3 (1964) 847.

/16/ Edwards, D. K., L. K. Glassen, W. C. Hauser & J. S. Tuchscher, Trans. ASME, Ser. C, Vol. 89 (1967) 219.

/17/ Macfarlane, J. J. & F. H. Holderness, Inst. Mech. Engineers, 184 (1969) 43.
/18/ Macfarlane, J. J., 12th Symp. on Comb. (The Comb. Inst., Pittsburgh, 1969) 1255.
/19/ Storojyk, T. P., Energomachinostroenie, Vol. 8 (1962) 1.
/20/ Hill, W. E. & N. R. Dibelius, Trans. ASME, Ser. A. Vol. 92 (1970) 310.
/21/ Schirmer, R. M., L. A. McReynolds & J. A. Dayley, SAE Trans., Vol. 68 (1960) 554.
/22/ Beer, J. M. & C. R. Howarth, 12th Symp. on Comb. (The Comb. Inst., Pittsburgh, 1969) 1205.

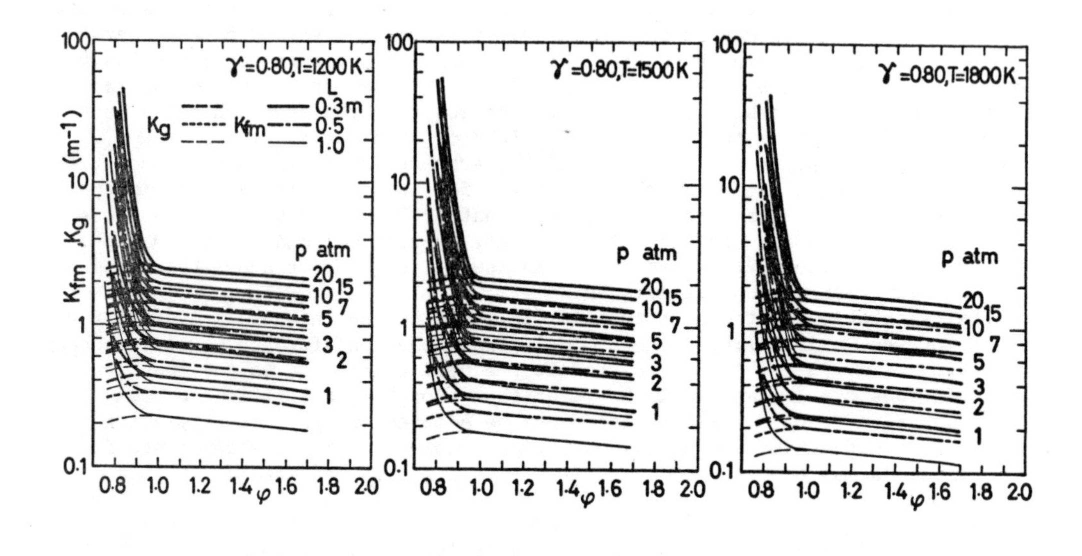

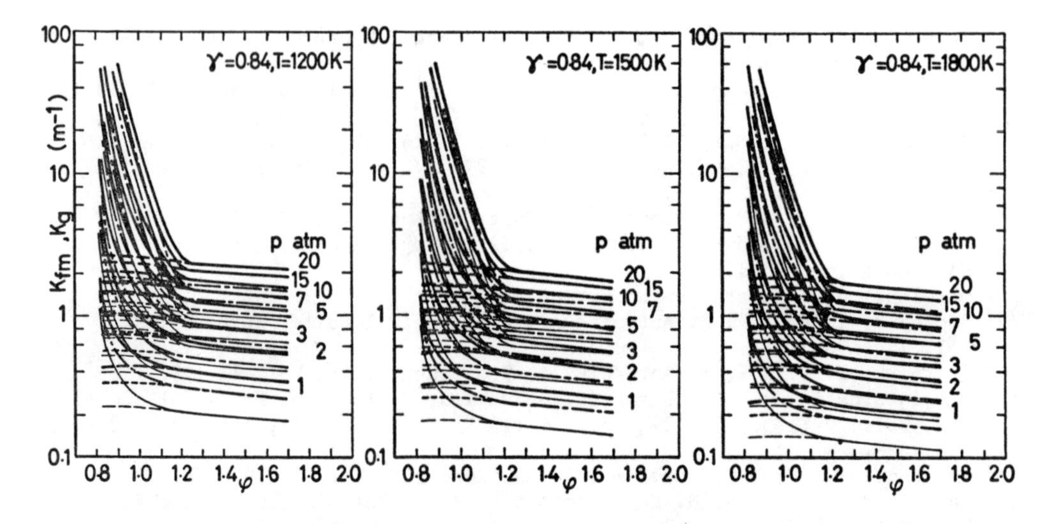

Fig.9 Calculated average absorption coefficient and endzone absorption coefficient of luminous flame

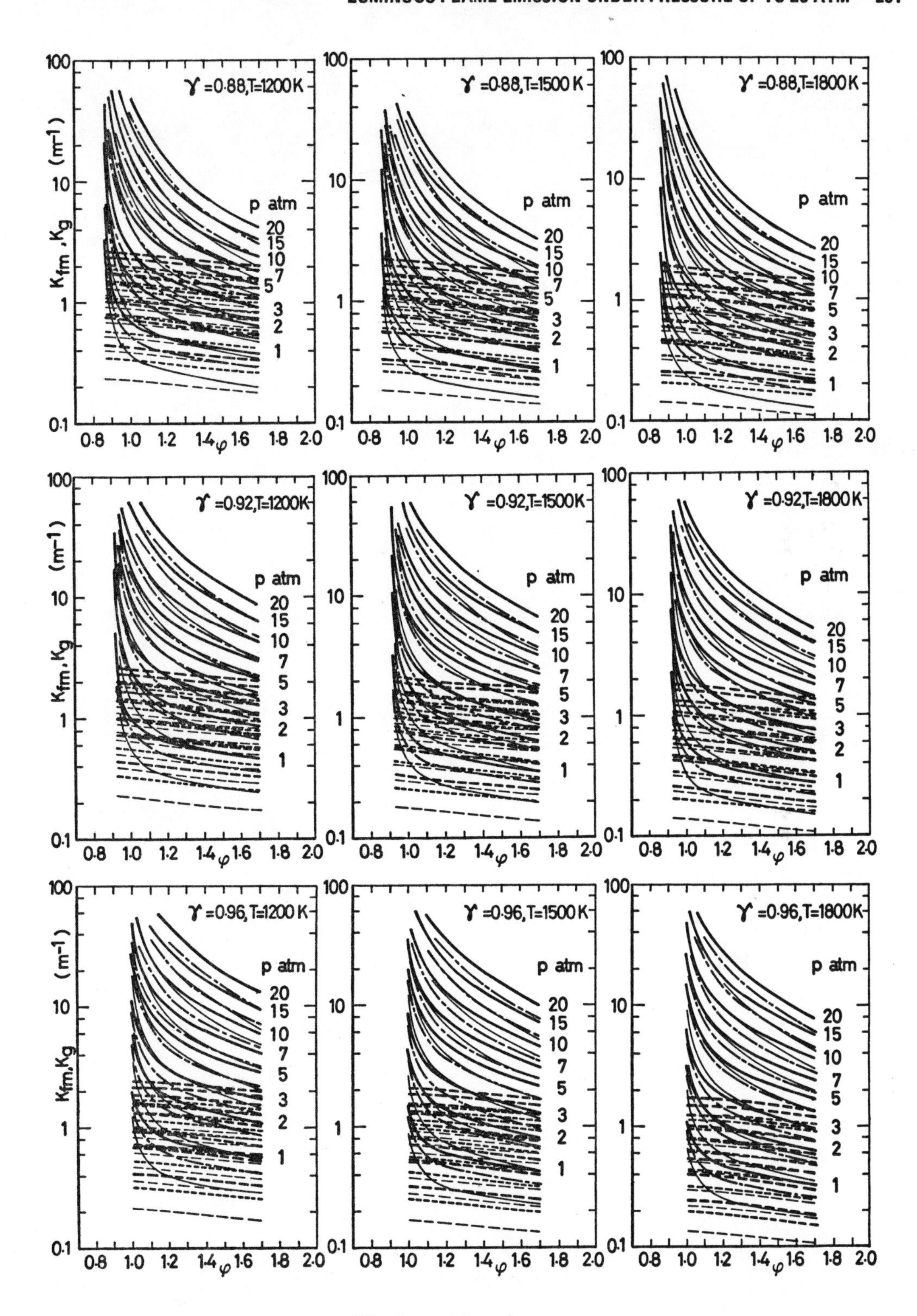

Fig.9 continued

Chapter 19

SPATIAL DISTRIBUTION OF SPECTRAL RADIANT ENERGY IN A PRESSURE JET OIL FLAME

E. G. Hammond* and J. M. Beér**

**Ministry of Defence, Ship Department, Bath*
***Professor, Department of Chemcial Engineering, University of Sheffield, England++*
Superintendent of Research, International Flame Research Foundation, Ijmuiden, Holland

Abstract

Spectroradiometric measurements have been made of slices of variable thickness of flame and distributions of concentrations of soot and those of temperature along the same optical paths were determined also experimentally. Spectral attenuation coefficients of soot bearing flames ($K_\lambda = k_\lambda \bar{c} L$) were calculated for various flame thicknesses together with values of the specific attenuation coefficient k_λ along the flame and for various C/H ratios of the soots in the flame.

NOMENCLATURE

A	:	constant in the equation $k_\lambda = A\lambda^{-\alpha}$
B	:	constant in the equation $k_\lambda = B\lambda^{-\beta}$
c	:	soot concentration g cm^{-3}; $\bar{c}$ mean value across the flame
d	:	particle size cm
E	:	extinction efficiency factor. The ratio of energy absorbed and scattered to the energy incident on the projected area of particle per unit time dimensionless.
I	:	radiant intensity watts cm^{-3} steradian^{-1}
K_m	:	mean spectral attenuation coeff. $K = k_1 \bar{c} L$
k_λ	:	specific spectral attenuation coeff. mg^{-1} cm 10^3
k	:	real absorption index of particle
L	:	flame width cm
m	:	complex refractive index $m = n(1 - ik)$
n	:	real refractive index of particle
N	:	number of particles per unit volume cm^{-3}
T	:	temperature ^oK
W	:	black body emission watts cm^{-3} steradian^{-1}
x	:	linear dimension along traverse across the flame cm
α	:	constant
β	:	constant
ε	:	emissivity dimensionless
λ	:	wavelength μm
ρ	:	density of soot particles g cm^{-3}

Subscripts

o	:	initial value
F	:	flame
S	:	source
x	:	at a distance x within the flame
λ	:	monochromatic

++The experimental work was carried out at the C.E.G.B. research laboratories at Marchwood as part of one of the author's (EGH) thesis work for a Ph.D. degree at the University of Sheffield.

1. Introduction

In recent years there has been a growing interest in the study of thermal radiation for industrial applications. This is partly due to the improvements in the ability to calculate the detailed heat flux distributions from flames to their bounding surfaces (1) and also due to the steadily increasing combustion intensities in industrial applications which in turn make more stringent requirements on design. Volumetric heat release rates in modern oil fired power station boilers can be some 50% higher than those in coal fired units, and as shown by a recent study (2) peak values of radiative heat flux can exceed mean heat flux values in boilers by more than 100%. In gas turbines the higher operating pressures in modern plant have increased the relative contribution of radiation from the flame to the walls of the combustor so as to make it commensurable with convective heat transfer (3). Also, a recent review of radiative heat transfer research in rocket combustors is indicative of the significance of radiation in the design of rocket motors (4).

In flames of gaseous liquid, or solid fuel the principal contributors to the radiation from the flame and from the fully burned combustion products are CO_2, H_2O, soots and fly ash particles. In hydrocarbon flames, as the C/H ratio in the fuel increases, there is an increasing tendency for soot formation which makes the flame luminous. In luminous flames the continuous radiation due to soot particles dominates (5). For a theoretical treatment of radiative transfer from such flames knowledge of radiative properties of the particles is necessary. In particular there is lack of detailed experimental data determined in situ in flames. In the present study spectral absorption coefficients were determined from measurements in a pressure jet oil flame and were correlated with soot concentrations measured along the same optical paths in the flame.

2. The Emissivity due to soot particles in luminous flames

In a cloud of particles with varying sizes the attenuation of a monochromatic parallel beam after penetrating a distance x in the cloud can be given as

$$I_x = I_o \exp\left[- c\, x \sum_z E_z N_z \, \pi d_z^2/4 \right] \tag{1}$$

For particles small compared with the wavelength of the incident radiation the extinction efficiency factor E can be given as a simple function of material properties of particles expressed in terms of the complex refractive index $m = n(1 - ik)$ and hence the spectral emissivity due to soot particles can be given as

$$\varepsilon_\lambda = \frac{I_o - I_x}{I_o} = 1 - \exp\left[- c\, x\, 36\, \pi/\rho \; f(n,k)/\lambda \right] \tag{2}$$

or in the form $\varepsilon_\lambda = 1 - \exp\left[- c\, x\, k_\lambda \right]$ (2a)

and we can give the spectral attenuation coefficient as

$$k_\lambda = A\lambda^{-\alpha} \tag{3}$$

with A a constant incorporating material properties and α a constant with a value close to unity (6). If $m \neq f(\lambda)$ it can be shown, Hammond (15) that $\alpha = 1$.

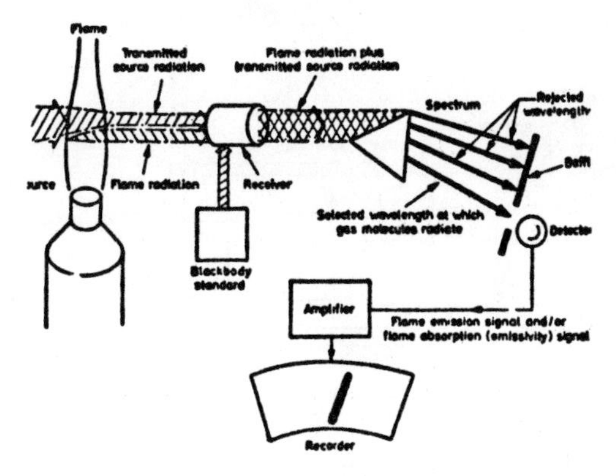

Fig.1. Schematic diagram illustrating radiometric methods of gas pyrometry.

The experimental technique adopted in the present study is a combination of the spectral emission - absorption method first reported by Schmidt (7) and the traversing method developed by Beér and Claus for total radiation measurements (8). The schematic diagram in Fig.1. illustrates the principle of the spectral emission - absorption method. In modern practice, as described by Silverman (9) the radiation is first chopped at some frequency, between the flame and receiver, so that the amplified signal corresponds to the emission from the flame and the emission from the source that is transmitted from the flame also,

$$I_{SF}(\lambda, T_S, T_F) = I_F(\lambda, T_F) + I_S(\lambda, T_S)[1 - \alpha_F(\lambda, T_F)] \tag{4}$$

Similarly, by chopping the radiation between the source and flame the only emission that is detected and amplified is the source emission that is transmitted by the flame, i.e.

$$T_{SF}(\lambda, T_S T_F)_{I_F=0} = I_S(\lambda, T_S)[1 - \alpha_F(\lambda, T_F)] \tag{5}$$

The emission from the flame alone is simply the difference between Equations 4 and 5, and the flame absorptivity is

$$\alpha_F(\lambda, T_F) = 1 - \frac{I_{SF}(\lambda, T_S, T_F)_{I_F=0}}{(I_S(\lambda, T_S))} \tag{6}$$

By introducing probes into the flame it is possible to make measurements of the contribution of slices of flame to the spectral radiation along an optical beam (8). When the soot concentration and temperature distributions are known along the same line of sight, the radiation traverse can be predicted and compared with measurements.

3. Apparatus

The furnace and associated apparatus is shown diagrammatically in Fig.2. Basically the apparatus consisted of a vertical furnace of 10" internal diameter. A single pressure jet burner was mounted in the furnace and could be adjusted some 30", so that radiation could be studied after different degrees of combustion. On the left hand side of the furnace was a spectrometer (a Grubb Parsons Model M2 monochromator and a Redeer thermocouple detector) and tuned amplifier and on the right hand side a reference furnace of known spectral emissivity. The spectrometer was sighted, through the flame, onto the reference furnace through ceramic sight tubes. By introducing the reference furnace sight tube onto the furnace by different amounts it was possible to measure the radiation from within the flame of varying thicknesses. Measurements being taken at 1" increments up to the full flame thickness of 10"

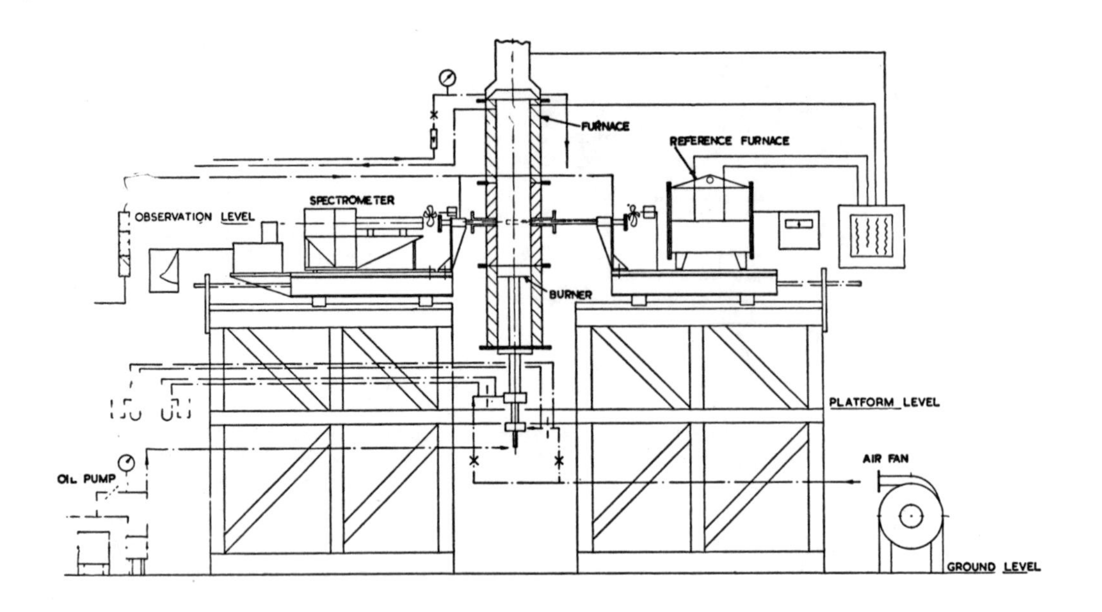

Fig. 2. Line Diagram of Apparatus.

In order to prevent flame and combustion products from entering the sight tubes Ca F_2 windows were provided and a small quantity of N_2 used as a purge.

Soot concentrations and temperature profiles were also measured along the same line-of-sight as the spectrometer measurements. Also, a number of soot samples were taken for C/H ratio analysis and for studying under an electron microscope.

The fuel was diesel oil burning at a rate of 35.5 lb/hr to 5% O_2 excess at the furnace exit.

4. Experimental results

Typical transmission measurements shown in Fig.3. illustrate the continuous absorption due to soot superimposed by the banded absorption of products of combustion such as CO_2 and H_2O. Also there is evidence of the presence of unburned fuel vapour in the flame particularly in the wavelength region beyond 3 μm. as indicated by Weeks and Saunders (10) and Tourin (11). As can be seen from Fig.3. soot particles are the main contributors to the radiation from the flame up to about 2.5 μm. Beyond this wavelength the relative significance of banded radiation increases.

Fig.4. is a plot of log (τ) vs wavelength calculated for various flame thicknesses from the transmission measurement data in Fig.3. As can be seen from Fig.4. the inverse relationship of attenuation coefficient with wavelength is only appropriate at shorter wavelength up to about 2.5 μm. For longer wavelengths banded absorption becomes predominant and this simple relationship does not hold. The full line in figure 4. is a least square fit of the equation $K_\lambda = B\lambda^{-\beta}$ to the experimental points. (K_λ is the mean spectral attenuation coefficient: $K_\lambda = k_\lambda \bar{c}L$). Thring et al (6) and Beér (12) have shown that the specific total attenuation coefficient across the full flame thickness can be calculated

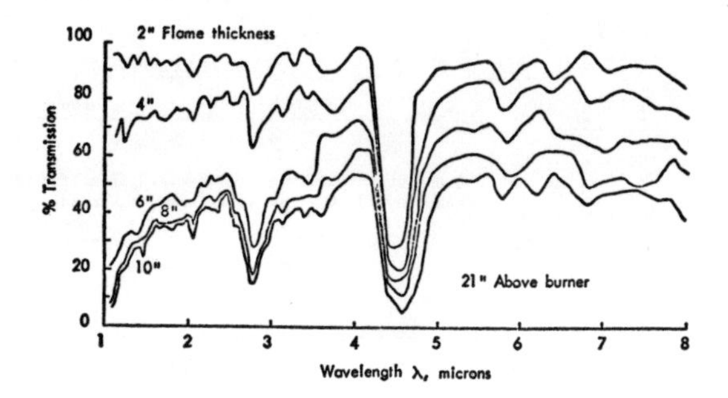

Fig.3. Variation of transmission with wavelength for increasing flame thickness.

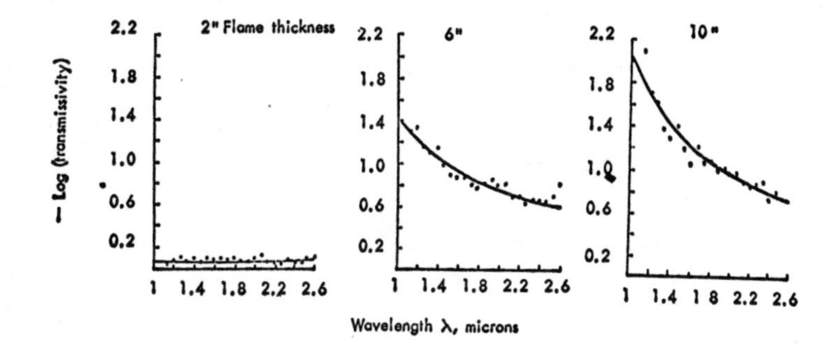

Fig.4. Variation of −Log (transmissivity) with wavelength for increasing flame thicknesses 21" above burner.

from a plot of $-\log\tau$ versus $\int c\,dx$ where τ is the transmissivity of the flame, c is the mass concentration of soot and x is distance along the optical beam. The slope of the above plot gives k_m the specific total attenuation coefficient. The same technique has been used in the present study to determine the mean spectral attenuation coefficient across the flame. From measurements of soot concentrations and of spectral transmissions of various thicknesses of flame

$$-\log(\tau_{\lambda,x}) \text{ is plotted against } \int_0^L c_x\,dx \text{ in Fig.5.}$$

The slope of the resulting straight line gives the attenuation coefficient at the particular wavelength. This procedure was then repeated for a number of wavelengths to obtain the spectral variation of the attenuation coefficient.
A comparison of the parameters A and α derived from transmission measurements over the full flame thickness and internal transmission measurements respectively is given in Table 1. As can be seen from table 1. the values of A and α derived from external transmission measurements are, for the same axial station in the flame, in good agreement with those derived from internal traversing measurements, the difference being less than 13%. This tends to confirm theoretical predictions by Howarth et al (16) showing that attenuation coefficients determined experimentally along non isothermal paths can stand without temperature correction.

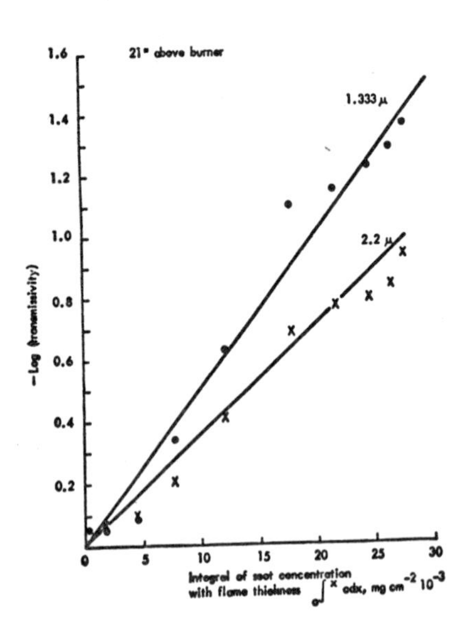

Fig.5. Typical variation of -log (transmissivity) plotted against the integral of soot concentration with increasing flame thickness.

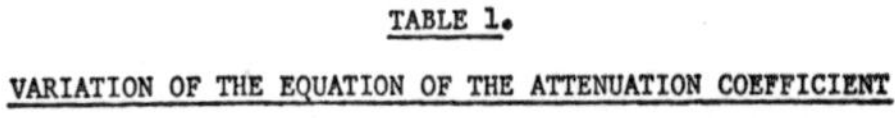

TABLE 1.

VARIATION OF THE EQUATION OF THE ATTENUATION COEFFICIENT

$k_\lambda = A\lambda^{-\alpha}$ AT VARIOUS HEIGHTS ABOVE THE BURNER

In method 1 the equations have been derived from transmission measurements over the full flame thickness.

In method 2 the equations have been derived from internal transmission measurements.

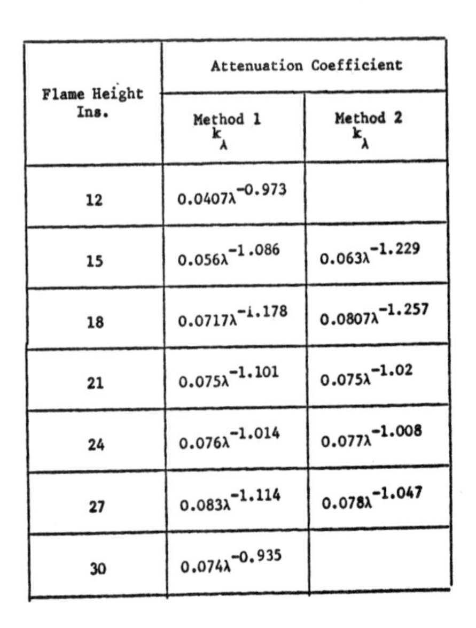

Flame Height Ins.	Attenuation Coefficient	
	Method 1 k_λ	Method 2 k_λ
12	$0.0407\lambda^{-0.973}$	
15	$0.056\lambda^{-1.086}$	$0.063\lambda^{-1.229}$
18	$0.0717\lambda^{-1.178}$	$0.0807\lambda^{-1.257}$
21	$0.075\lambda^{-1.101}$	$0.075\lambda^{-1.02}$
24	$0.076\lambda^{-1.014}$	$0.077\lambda^{-1.008}$
27	$0.083\lambda^{-1.114}$	$0.078\lambda^{-1.047}$
30	$0.074\lambda^{-0.935}$	

Mean attenuation coefficients can be estimated from

$$K_m = \frac{1}{\bar{c}L} \log \left[\frac{\int_0^\infty W_\lambda \, d\lambda}{\int_0^\infty \exp(-A\lambda^{-\alpha}\bar{c}L) \, W_\lambda \, d\lambda} \right] \qquad (7)$$

Where $\bar{c}L = \int_0^L C_x dx$ and W_λ is Planck's function evaluated at the experimentally determined mean radiant temperature (15). Values of K_m varied from 0.019 and 0.036 at 12" and 30" from the burner respectively. These results are in good agreement with results of flame studies at Sheffield and Ijmuiden (6,12) in which mean absorption coefficients were determined from total radiation measurements, (K_m = 0.030 at λ_m = 2.45 μm and K_m = 0.032 at λ_m = 2.79 μm), and with results of more recent studies in which radiation traverses predicted from species concentration and temperature measurement data were compared with radiation traverses measured in an oil flame at Ijmuiden (13). Dalzell and Sarofim (14) have shown from qualitative theoretical considerations that the parameter A should increase

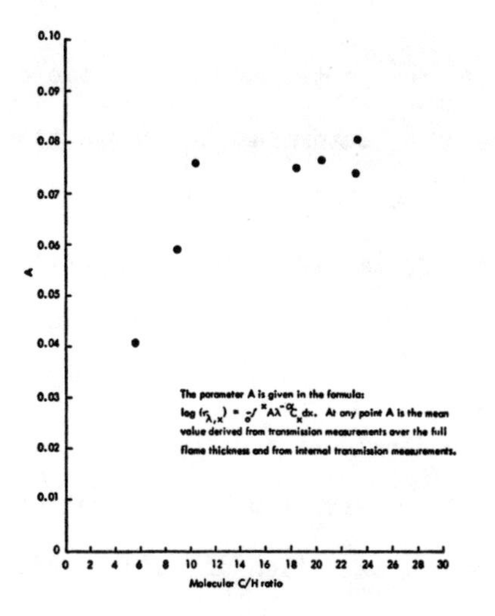

Fig.6. Variation of the parameter A with mean C/H ratio of the soot on the flame centre line.

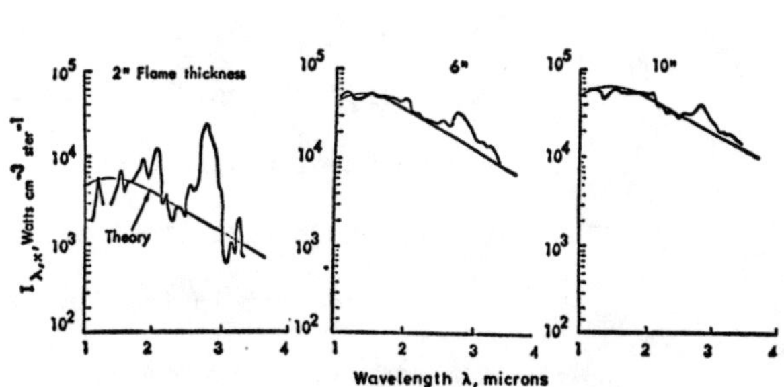

Fig.7. 21" above burner, variation of radiation heat flux with wavelength for increasing flame thickness, compared with theory.

with C/H ratio. This is confirmed in Fig.6. which shows the variation of the parameter A with the C/H ratio of the soot on the centre line of the flame. Once the spectral variation of the attenuation coefficient had been obtained it became possible with the aid of soct concentration and temperature profiles to predict the spectral distribution of radiant flux within the flame. This was achieved by a numerical solution cf the transport equation by Hammond (15). The transport equation can be derived by considering the change in spectral intensity of a pencil beam of radiation in an emitting absorbing medium from point x to x + dx and is given by

$$d\, I_{\lambda,x} = -\, k_{\lambda,x}\, c_x I_{\lambda,x}\, dx + k_{\lambda,x}\, c_x\, W_{\lambda,x}\, dx \qquad (8)$$

after rearranging and integrating we have

$$I_{\lambda,x} = \exp\left(-\int k_{\lambda,x}\, c_x\, d_x\right)\left[\int \exp\left(\int k_{\lambda,x} c_x dx\right) k_{\lambda,x} c_x\, W_{\lambda,x}\, dx + C_1\right] \qquad (9)$$

where C_1 is a constant depending upon the boundary conditions.

When $I_{\lambda,0} = 0 \qquad C_1 = 0$

In Fig.7. comparison is made of calculated and measured values of radiant heat flux as a function of wavelength and for various flame thicknesses. As the flame thickness increases the agreement between calculation and measurement improves and is good at full flame thickness.

Conclusions

1) The spectral attenuation coefficient of soot particles in the flame can be represented by $k_{\lambda} = A\lambda^{-\alpha}$. A value of K_m determined by integration of Equation 7 and that measurements at 30" from the burner when $k_{\lambda} = 0.074\ \lambda^{-0.935}$ and $\int_0 Cdx = 12.5\ mg/\ell\ cm$ is in good agreement with those values of luminous flames from total radiation measurements at emission mean wavelengths.

2) The values of the parameters A and α are dependent upon the material properties of the soot in the flame. They vary also with the degree of combustion possibly as a consequence of the variation of the C/H ratio of the soot along the flame.

3) Spectral attenuation coefficients derived from measurements along non isothermal paths can be used without temperature correction.

4) Temperature and soot concentration profiles together with appropriate expression of the attenuation coefficient can be used to predict spatial distributions of the spectral radiation within luminous flames with good approximation.

Acknowledgement

The authors are grateful to the C.E.G.B. Research Laboratories at Marchwood for providing the experimental facilities and for their permission to publish this paper.

REFERENCES

(1) Hottel, H.C. and Sarofim, A. Radiant Heat Transfer, McGraw Hill, 1967.
(2) Hammond, E.G. and Godridge, A.M. CEGB Report RD/M/R88, 1967.
(3) Lefebvre, A.H., Twelfth Symposium (Int.) on Combustion, 1247 (1969).
(4) Rochelle, W.C., NASA Tech. Memo TM-X5379 (1967).
(5) Beér, J.M. and Howarth, C.R.: 12th Symposium on Combustion, Combustion Inst. 1969.
(6) Thring, M.W., Foster, P.J., McGrath, I.A. and Ashton, J.S. Int. Development of Heat Transfer, 96, 796, 1961.
(7) Schmidt, H. Ann. der Physik, 29, 1-27, 1909.
(8) Beér, J.M., and Claus, Ir. J. J.Inst.Fuel, 35, 437, (1962).
(9) Silverman, S.J. Opt.Soc.Am., 39, 275 (1949).
(10) Weeks, D.J. and Saunders, D.A., J.Inst.Fuel, 31, 247 (1958).
(11) Tourin, R.H. "Temperature, its measurement and control in science and industry", 3, 2, Reindhold (1962).
(12) Beér, J.M., J.Inst.F. Vol.XXXV. No. 2, Jan.1962, p.3.
(13) Johnson, T.R., and Beér, J.M., Fourth Symposium on Flames and Industry, Paper No. 4, Inst. of Fuel, London 19/20 Sept. 1972.

(14) Dalzell, H.C. and Sarofim,A.F. Trans.ASME, J.Heat Transfer, 1969, 91, p.100.

(15) Hammond, E.G., Ph.D. Thesis, University of Sheffield, 1971.

(16) Howarth, C.R., Foster, P.J., and Thring, M.W. Proc.Third Heat Transfer Conference, Chicago, 1966, Vol. V, p.122.

SECTION III:

EXPERIMENTAL METHODS

Chapter 20

NONLINEAR INVERSION TECHNIQUES IN FLAME TEMPERATURE MEASUREMENTS

C. M. Chao and R. Goulard

Purdue University, School of Aeronautics and Astronautics
West Lafayette, Indiana 47907

ABSTRACT

Flame temperature profiles can be retrieved on the basis of a single-line-of-sight multifrequency set of radiance measurments. Nonlinear inversion techniques, developed mostly for satellite meterology, are used and compared in this highly nonlinear application. It is found that accuracy and rapid convergence depend critically on the availability of transmittance kernels featuring sharp and well spaced maxima. A systematic use of pivotal techniques is also necessary. Several typical flame configurations are tested, including the practical case where the unknown concentration profile is also to be retrieved. A closed form model for the wings of molecular band absorption is also developed.

I. INTRODUCTION

Temperature profiles inside combustion flames and plasmas are often difficult to measure with thermocouples and other immersed pyrometers, because such temperature probes may have to be immersed in an inaccessible and possibly destructive medium and they are also apt to distort the field to be measured. Determination of gas temperatures from radiation measurements is an alternative to the use of immersion probes. The concept of temperature determination in hot gases from their radiation is an old one and many spectroscopic methods of this kind have been developed [1].

This study concerns itself with the measurement of temperature profiles in hot gases by spectral scanning of the radiance of a flame along a particular line of sight. The radiance thus measured at each frequency can be expressed as a Fredholm integral of the first kind [2]. The problem consists to extract the temperature profile from this set of integral equations. This "inversion" problem has been the subject of numerous papers in recent years. A recent survey report [3] discusses recent developments in this field.

Briefly, there exist three classes of inversion methods: linearized, nonlinear (iterative) and statistical. In all three, the accuracy of the solution profiles depends on the transmittance properties, on temperature profile and – of course – on the various sources of errors introduced in both measurements and calculations. It is a well known fact of integral calculus that inversion procedures tend to be non-unique and unstable. This feature is worsened by the non-linear character of high temperature radiation properties (absorption, Planck function). Hence the poor performance of linearized techniques when applied to flames. This aspect has been discussed extensively in a recent report [4].

In this report, nonlinear (iterative) techniques will be considered. Although quite effective for low temperature media (see Chahine's pioneering paper [5]) they are inherently suited for the nonlinear features of high temperature media. We shall be especially concerned with the role played by the kernel in the accuracy of the solution. An optimum iterating strategy (the pivotal condensation technique) will be discussed extensively.

II. DEVELOPMENT OF THE MATHEMATICAL MODEL

2.1. The Radiative Transfer Equations

For a one-dimensional, nonscattering plane flame in local thermodynamic equilibrium, the spectral radiance $N^+(\omega, b)$ emerging from a flame at a point [b] along a given line of sight and at a frequency ω is given (4) by:

$$N^+(\omega;b) = N^+(\omega;a) \exp\left\{-\int_a^b \rho(s)k[\omega,T(s)]\,ds\right\} + \int_a^b B[\omega,T(s)] \exp\left\{-\int_s^b \rho(s')k[\omega,T(s')]\,ds'\right\} \rho(s)k[\omega,T(s)]\,ds \quad (1)$$

$N^+(\omega;a)$ is the spectral radiance emitted from the flame boundary [a] furthest from the observer [see Fig. (1)]. The first term of Eq. 1 will be omitted in subsequent equations by assuming that $N^+(\omega;a)$ is negligible, known or that it can be calibrated out. In the second term, $\rho(s)$ is the specific mass of the absorbing gas at the distance s along the line of sight, $k[\omega,T(s)]$ is the mass spectral absorption coefficient and $B[\omega,T(s)]$ is the Planck blackbody radiance function at wavenumber ω, and the temperature T(s).
The spectral Planck blackbody radiance function is given by [6]

$$B[\omega,T(s)] = \frac{(C_1/\pi)\omega^3}{\exp(C_2\omega/T) - 1} \quad (\text{watt/cm}^2\ \text{ster}^{-1}\ \text{cm}^{-1}) \quad (2)$$

Where $C_1 = 3.7405 \times 10^{-12}$ watt cm^2 and $C_2 = 1.43879$ cm °K. In practice, the measurement is not made "at one wave number" but rather over a finite spectral interval $\Delta\omega$ center on the wave number of interest ω. $\Delta\omega$ istypically a value of 10 cm^{-1} which is determined by the slit size and the spectrometer used in the measurement [7].

2.2. Average Measurements over Small Frequency Intervals

Let us define $\bar{B}[\omega,T]$ as the spectral radiance at wavenumber ω which would be received from a blackbody at a temperature T, by a sensor with an instrument function $g(\omega,\omega')$ and a spectral slit width $\Delta\omega$. Since the blackbody radiance is effectively constant over $\Delta\omega$, $\bar{B}[\omega,T]$ can then be separated into two factors, one of which is independent of temperature and the other is independent of instrument function:

$$\bar{B}[\omega,T] = B[\omega,T] \int_{\Delta\omega} g(\omega,\omega')\,d\omega' \quad (3)$$

If we define a thermally inhomogeneous flame into a series of m imaginary zones, each of which is isothermal within the precision of measurement, the spectral radiance of the detector, $N^+(\omega;b)$ is given approximately by

$$N^+(\omega;b) = \sum_{i=1}^{m} K[\omega,T(s_i)]\,\bar{B}[\omega,T(s_i)]\,\bar{W}(s_i) \quad (4)$$

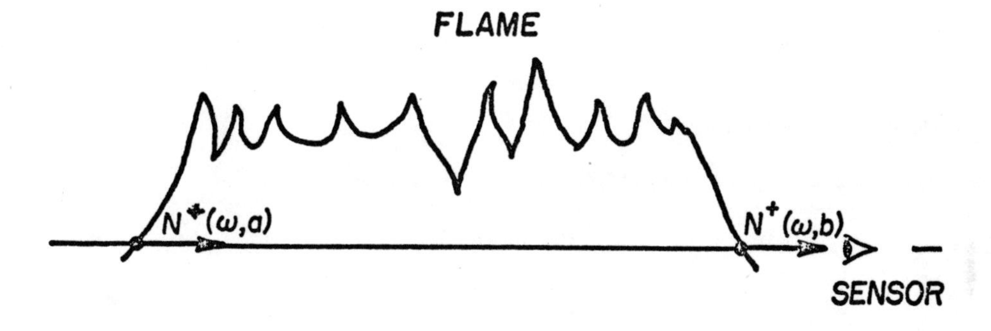

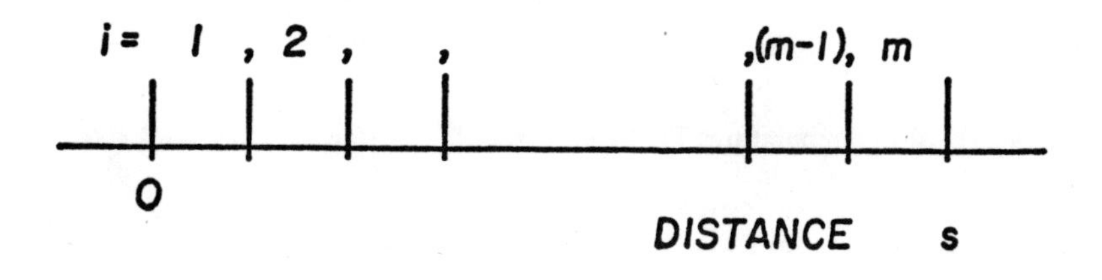

Fig. 1 Line of Sight Radiance Measurement

where $k[\omega,T(s_i)]$ is the observed derivative of the transmission function with respect to the spectral width $\Delta\omega$ and the instrument function $g(\omega,\omega')$, which can be expressed as follows:

$$K[\omega,T(s_i)] = \int_{\Delta\omega} \rho(s)k[\omega,T(s_i)] \exp\left\{-\int_{s_i}^{b} \rho(s')k[\omega,T(s')]\,ds'\right\} g(\omega,\omega')\,d\omega' \Big/ \int_{\Delta\omega} g(\omega,\omega')\,d\omega \quad (5)$$

and $\overline{W}(s_i)$ is the quadrative weight factor [8] used in the approximation of the integration of Eq.(1). Since $B[\omega,T(s)]$ is, given by Eq.(2), the measurement of apparent radiance $\overline{B}[\omega,T(s)]$ of a blackbody at any convenient temperature T(s) can be used to evaluate $\int_{\Delta\omega} g(\omega,\omega')d\omega'$ from Eq.(3). Thus, once the values of $\overline{B}[\omega,T(s)]$ are solved from Eq. (4), the temperature profile T(s) can then be obtained from the corresponding values of $B[\omega,T(s)]$, and the results will be independent of the instrument function.

2.3 The Inversion Problem

Eq. (4) can be expressed in a compact "matrix" equation form as:

$$N^+(\omega_j) = \overline{K}[\omega_j,T(s_i)] \otimes \overline{B}[\omega_j,T(s_i)] \qquad i = 1, 2, \ldots, m \qquad j = 1, 2, \ldots, n \quad (6)$$

where $\otimes$ is defined as a selective multiplication matrix operator (see Appendix A). $\overline{K}$ is a matrix of order nxm of elements $K[\omega_j,T(s_i)]\overline{W}(s_i)$, and $\overline{B}$ is a matrix of order mxn of elements $\overline{B}\omega_j,T(s_i)]$. N^+ is a vector of order n.

The determination of a set of m temperatures $T(s_i)$ from a set of n measurements $N^+(\omega_j)$ is the problem at hand. In the following chapter, we shall review briefly linearized and nonlinear (iterative) techniques usually applied to the solution of this inversion problem. An evaluation of the third large class of possible approaches, the statistical methods [9,10], will not be included in this report.

III. INVERSION TECHNIQUES

3.1. Linearized Techniques

If the kernel matrix $\overline{K}$ which appears in Eq. (6) is either temperature independent or directly measurable, then the total number of independent variables in the system of n equations would still be nxm becaise of $\overline{B}$'s dependence on frequency ω and temperature T. This is an underdeterminate problem.

However, when the problem at hand involves a small temperature range only (as in meteorology, for instance [11]), the Planck function $B(\omega,T)$ is often approximated by a separated functional relation:

$$\overline{B}[\omega,T(s)] = B_T(\omega)B_\omega(T(s)) \quad (7)$$

Thus Eq. (6) can be reduced to an ordinary matrix equation:

$$N^+(\omega_j) = K_T[\omega_j,\rho(s_i)]\,B_\omega[T(s_i)] \quad (8)$$

where $K_T[\omega_j, \rho(s_i)] = B_T(\omega_j)\bar{K}[\omega_j, \rho(s_i)]$ is a density functional only. The solution of Eq. (8) can be divided into three cases:

a) Case 1: $n < m$ (less measurements than unknown variables.

This is an underdeterminate problem. By definition K_T has a right hand inverse [12] [13] and the general solution of $B_\omega[T(s_i)]$ is

$$B_\omega[T(s_i)] = K_T'(K_T K_T')^{-1} N^+(\omega_j) + [1 - K_T'(K_T K_T')^{-1} K_T] Q(\omega_j) \tag{9}$$

where K_T' is the transport matrix of K_T, 1 is the identity matrix, and $Q(\omega_j)$ is an arbitrary vector

b) Case 2: $n = m$ (as many measurements as unknown variables)

This is an exact problem, the solution of $B_\omega(T(s_i)$ is

$$B_\omega[T(s_i)] = K_T^{-1} N^+(\omega_j) \tag{10}$$

where K_T^{-1} is the inverse matrix of K_T.

c) Case 3: $n > m$ (more measurements than unknown variables.

This is an overdeterminate problem and no exact solution for $B_\omega(T(s_i))$ exists. However it is recognized that in practice $N^+(\omega_j)$ is measured with modest accuracy only [11]. Therefore, only this case provides the information necessary to estimate a best approximation solution.

For a Gaussian-type random radiance measurement errors, the best approximation solution can be written [11] as:

$$B_\omega[T(s_i)] = K_T'[KK_T' + \gamma I]^{-1} N^+(\omega_j) \tag{11}$$

where γ is the noise-to-signal-power ratio which must be determined by statistical expectation or estimation.

This method has been presented recently with great clarity in [13]. It is used extensively in meteorology and also in the one previous attempt at flame inversion [7]. Its application to high temperature cases is discussed extensively in [4].

3.2. Nonlinear (iterative) Techniques

The transmittance kernel matrix $\bar{K}[\omega_j, T(s_i)]$ is, a function of the specific mass of the absorbing gas ρ and of the mass spectral absorption coefficient k which are both temperature dependent. Whether this dependence is strong or weak depends on the properties of the absorbing gas, on the selected frequencies and on the thermodynamic property distribution (ρ,T) inside the flame. Most high temperature cases display such strong dependence. Therefore both $\bar{K}$ and $\bar{B}$ are nonlinear and a non-linear iterative sheme should be introduced to effect the inversion of Eq. (6).

An interative method is, in principle, a rule for operating on a previous approximate solution to obtain an improved solution. For almost diagonal or dominate-diagonal kernel matrix systems, convergence is rapid. Otherwise a possibility of slow or irregular convergence could occur.

Let us write, after [10], a first-degree non-linear iteration as follows:

$$\bar{B}[\omega_j, T^{(k+1)}(s_i)] = \bar{B}[\omega_j, T^{(k)}(s_i)] + C^{(k)}\Lambda[N^{+}(\omega_j) - N^{(k)}(\omega_j)]\,S^{(k)} \tag{12}$$

where $T^{(k)}(s)$ is the temperature profile vector from the kth iteration, $C^{(k)}$ is an error-correcting matrix, $S^{(k)}$ is a weight spreading matrix and $N^{(k)}(\omega_j)$ is the calculated spectral radiance from the kth matrix and $N^{(k)}(\omega_j)$ is the calculated spectral radiance from the kth iteration [Eq. (6)]:

$$N^{(k)}(\omega_j) = \bar{K}[\omega_j, T^{(k)}(s_i)] \otimes \bar{B}[\omega_j, T^{(k)}(s_i)] \tag{13}$$

The choices for the error-correcting matrix $C^{(k)}$ and the weight-spreading matrix $S^{(k)}$ (see reference [8] and [15]) depend on the particulars of the inversion problem at hand.

If the transmittance kernel matrix $\bar{K}$ is almost diagonal or "dominate-diagonal" matrix, then $C^{(k)}$ and $S^{(k)}$ can be chosen both as diagonal matrices Λ of the form:

$$C^{(k)} = \Lambda\left\{\frac{1}{\bar{K}[\omega_j, T^{(k)}(s_j)]}\right\} \qquad j = 1, 2, \ldots, m \tag{14}$$

and

$$S^{(k)} = \Lambda[\bar{W}(s_j)] \qquad j = 1, 2, \ldots, m \tag{15}$$

Thus the iterative scheme is similar to Jacobi's method [15], and is given by

$$\bar{B}[\omega_j, T^{(k+1)}(s_j)] = \bar{B}[\omega_j, T^{(k)}(s_j)] + \frac{\bar{W}(s_j)}{\bar{K}[\omega_j, T^{(k)}(s_j)]}[N^{+}(\omega_j) - N^{(k)}(\omega_j)] \tag{16}$$

This process must be started, of course, with some initial estimate $T^{(0)}(s_i)$.

The success of this iteration method will depend greatly on whether of the kernel $\bar{K}$ is dominate diagonal or not.

a) If $K[\omega_j, T(s_i)]$ is dominate diagonal, the following approximation can be applied to Eq. (13):

$$N^{(k)}(\omega_j) = \bar{K}[\omega_j, T^{(k)}(s_j)]\,\bar{B}[\omega_j, T^{(k)}(s_j)]$$

or

$$\frac{1}{\bar{K}[\omega_j, T^{(k)}(s_j)]} = \frac{\bar{B}[\omega_j, T^{(k)}(s_j)]}{N^{(k)}(\omega_j)} \tag{17}$$

After substitution of Eq. (17) into Eq. (16), and assuming an uniform weight-spreading $S^{(k)} = 1$, Eq. (16) becomes

$$\bar{B}[\omega_j, T^{(k+1)}(s_j)] = \bar{B}[\omega_j, T^{(k)}(s_j)] \frac{N^+(\omega_j)}{N^{(k)}(\omega_j)} \tag{18}$$

which is precisely the method offered by Chahine in 1970[5].

The error-reducing matrix can be approximated[15] as Jacobi's iteration matrix:

$$R^{(k)} = (1 - D^{-I(k)}\bar{K}^{(k)}) \tag{19}$$

where $D^{(k)}$ is the diagonal of $\bar{K}^{(k)}$, at the kth iteration. The corresponding iteration converges if, and only if

$$\lim (M^{(k+1)})_{k\to\infty} \to 0 \tag{20}$$

with

$$M^{(k+1)} = \prod_{k=0}^{k} R^{(k)}$$

If all the eigenvalues of $M^{(k+1)}$ are less than one in absolute value, then the iteration converge. Let $\lambda(M^{(k)}$ be the largest magnitude of the eigenvalue of $M^{(k)}$. Then the average rate of convergence of the iteration for a fixed k is given [15] by:

$$G(M_K) = -\frac{1}{k} \log [\lambda(M^{(k)})] \tag{21}$$

b) In practice, the kernel matrix $\bar{K}$ is not a dominate-diagonal matrix.

Therefore, a certain selection strategy of the largest pivotal elements in the kernel matrix must be used in setting up the error-correcting matrix $C^{(k)}$ and weight-spreading matrix $S^{(k)}$.

This optimum pivot strategy [8] (or pivotal condensation [15]) consists in constructing a smaller kernel matrix made only of the elements of significant magnitude (pivots) of the original matrix K. Hence one chooses in succession magnitude in the transmittance sub-kernel matrix which is still available for pivoting (not in a row or column already containing a pivot) as the pivot for the ℓth pass.

Let r_ℓ or C_ℓ be, respectively the row and column subscript of the ℓth pivot element $\bar{K}[\omega_{r_\ell}, T^{(k)}(s_{C_\ell})]$, then let

$$C^{(k)} = \Lambda\left\{\frac{1}{\bar{K}[\omega_{r_l}, T^{(k)}(s_{C_l})]}\right\} \tag{22}$$

and

$$S^{(k)} = \Lambda[\bar{W}(s_{C_l})] \tag{23}$$

The object of pivoting is to prevent the loss of accuracy due to the rapid decrease in the size of elements of successive $\bar{K}[\omega_j,T^{(k)}(s_i)]$. This "ill-conditioning" is of course due to the physical nature of the transmittance and it cannot be avoided in most real life problem. Hence the frequent need for a pivot strategy.

IV. TYPICAL APPLICATIONS AND THE CORRESPONDING KERNELS

It was shown in the previous chapter that the accuracy of the inversion procedure depended a great deal on the set of kernels used in the radiance integral equation. Given a specific situation, there exists a particular set of observation frequencies which yield kernels which optimize in turn the inversion procedure. The purpose of this chapter is to consider several physical applications and to investigate the corresponding kernels for their suitability to the inversion process.

In Eq. (1), the kernel $\bar{K}[\omega_j,T(s_i)]$ is given by:

$$\bar{K}[\omega_j,T(s_i)] = \rho(s_i)k[\omega_j,T(s_i)] \exp\left\{-\int_{s_i}^{b} \rho(s)k[\omega_j,T(s)]\, ds\right\} \tag{24}$$

where $\rho(s)$ is the concentration profile of the absorbing gases and $k[\omega_j,T(s)]$ is the mass absorption coefficient in the flame. It is clear that $\bar{K}$ is also the derivative of the transmittance $\bar{T}$:

$$\bar{T}_{\omega_j}(s_i) \equiv \exp\left\{-\int_{s_i}^{b} \rho(s)k[\omega_j,T(s)]\, ds\right\} \tag{25}$$

with respect to the optical length

$$\tau_{\omega_j}(s_i) \equiv \int_{s_i}^{b} \rho(s)k[\omega_j,T(s)]\, ds \tag{26}$$

Hence the importance of a realistic representation of the transmittance. This implies first a good knowledge of the absorption coefficient k as a function of frequency and temperature. Such information has been collected for a typical set of combustion gases and is now available in handbook form[6]. In Appendix B, current transmittance theories are briefly discussed. Also a reasonably accurate model of the absorptivity of the 2.7μ band of CO_2 is derived for later application in this report.

It can also be surmised from Eq. 24 that the role of the density profile $\rho(s)$ is prominent in the make up of a good kernel. In the following sections we shall investigate four different applications and the nature of their kernels.

4.1. Exponential Concentration Profiles

This kernel corresponds to those situations where the density ρ of the medium under observation increases exponentially with distance away from the boundary facing the sensor. This of course is typical of the atmospheres of both planets and stars when observed from above (satellites or earth observatories respectively).

Let us assume a theoretical atmosphere, for the sake of discussion, where the density would have exactly the exponential form:

$$\rho = \rho_0 \exp\left[-\frac{gs}{R_m T(s)}\right] \tag{27}$$

where ρ_o is the absorbing gas concentration on the ground level. T(s) is the temperature profile of the atmosphere, and R_m is the specific gas constant of the absorbing gas.

The optical thickness $\tau(\omega_j,s_i)$ can be integrated from s_i to infinity:

$$\tau(\omega_j,s_i) = \rho_o k(\omega_j) \int_{s_i}^{\infty} \exp\left[-\frac{gs}{R_m T(s)}\right] ds \tag{28}$$

If an isothermal atmosphere T_o is assumed, the optical thickness becomes

$$\tau(\omega_j,s_i) = \rho_o k(\omega_j) \frac{R_m T_o}{g} \exp\left(-\frac{gs_i}{R_m T_o}\right) \tag{29}$$

The transmittance kernel matrix then can be written as

$$K(\omega_j,s_i) = \rho_o k(\omega_j) \exp\left[-\frac{gs_i}{R_m T_o} - \rho_o k(\omega_j) \frac{R_m T_o}{g} \exp\left(-\frac{gs_i}{R_m T_o}\right)\right] \tag{30}$$

The height of the maximum $K(\omega_j,s_i)$ can be found by differentiating the function in the bracket of the exponential term:

$$s_m = \frac{R_m T_o}{g} \ln\left[\frac{\rho_o k(\omega_j) R_m T_o}{g}\right] \tag{31}$$

If a scale height H is defined as

$$H = \frac{R_m T_o}{g} \ , \tag{32}$$

the following transmittance kernel matrix K can be expressed:

$$K(\omega_j,s_i) = \frac{g}{R_m T_o} \exp\left[-\frac{s_i - s_m}{H} - \exp\left(-\frac{s_i - s_m}{H}\right)\right] \tag{33}$$

The maximum value of the kernel K_m corresponding to s_m can be found to be:

$$K_m = \frac{g}{R_m T_o e} = \frac{1}{He} \tag{34}$$

Thus, Eq. 33 can be rewritten:

$$K(\omega_j,s_i) = K_m \exp\left[1 - \frac{s_i - s_m}{H} - \exp\left(-\frac{s_i - s_m}{H}\right)\right] \quad (35$$

This is a well separated set of kernels (Fig. 2). The wide spacing between maxima (s_m) will be always obtainable, provided the absorbing gas allows for a suitable range of absorption coefficients $k(\omega_j)$ for the available measurement frequencies ω_j (Eq. 31).

In real atmospheres, the temperature T is not constant and therefore, strictly speaking, such good kernels as those shown in Fig. 2 are not available. However, the temperature dependence of most molecular specie (see Fig. B-F2 in Appendix, for instance) is not very sensitive to the relatively small range of atmospheric temperatures. Hence it is no surprise that real earth atmospheric kernels, such as the 6.3μ band of water vapor, are also well behaved (Fig. 3). This explains also Chahine's success in using nonlinear method in satellite meteorology [14].

In the following discussions the exponential concentration profile [Eq. (35)] will be used as an "ideal" point of comparison for other kernels. It should be kept in mind that it does not deviate too much from the reality of satellite meteorology applications (although ground based meteorology is quite another matter: see Appendix C).

4.2. Uniform Mixing Ratio Profiles

A configuration rather akin to heated jets would be that of a uniform mass fraction of an absorbing substance in a heated mixture. In such a case the density variation across the gas sample would be due to temperature distribution on a constant pressure path for instance (perfect gas law):

$$\rho(s) = \rho_o \frac{T_o}{T(s)} \tag{36}$$

where ρ mean here the concentration of the absorbing species in the mixture.

The optical thickness of the flame can be expressed as

$$\tau(\omega_j, s_i) = \rho_o T_o \int_{s_i}^{b} \frac{k[\omega_j, T(s)]}{T(s)} \, ds \tag{37}$$

If one considers again the simplified case where the temperature variations do not affect measurably the transmittance (a less realistic assumption than in the previous case, since $1/T$ appears in the denominator of Eq. 37), then Eq. (37) can be simplified:

$$\tau(\omega_j, s_i) = \rho_o T_o \frac{\bar{k}(\omega_j)}{T_{av}} (b - s_i) \tag{38}$$

where T_{av} and $\bar{k}(\omega_j)$ are the average values of $T(s)$ and $k(\omega_j, T(s))$ across the flame. The transmittance kernel matrix in this particular case yields

$$\bar{K}(\omega_j, s_i) = \rho_o \left(\frac{T_o}{T_{av}}\right) \bar{k}(\omega_j) \exp\left[-\rho_o \left(\frac{T_o}{T_{av}}\right) \bar{k}(\omega_j)(b - s_i)\right] \tag{39}$$

Unlike the transmittance kernel matrix for the atmospheric sounding problem [Eq. (30)], this kernel has no maximum along the line of sight. In fact, the kernels for different frequencies are nearly proportional to each other [Eq. (39)] a prime reason for ill-conditioning in matrix inversion.

Actually, an exact representation of the temperature profile in Eq. 37 introduces deviations from this proportionality (Fig. 4) and therefore the ill-conditioning of this kind of kernel could be removed.

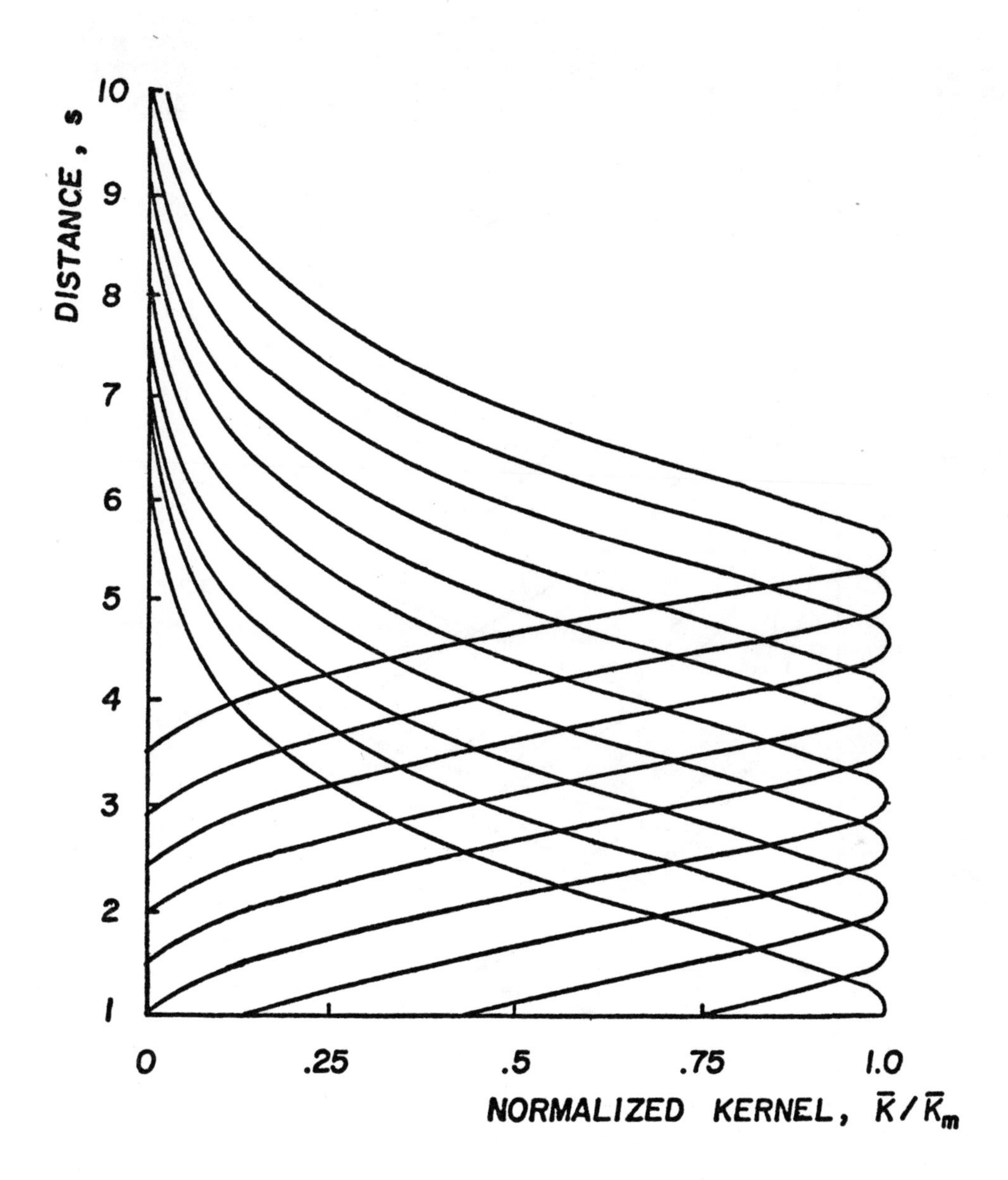

Fig. 2 Reduced Exponential Kernels [Eq. (35)]

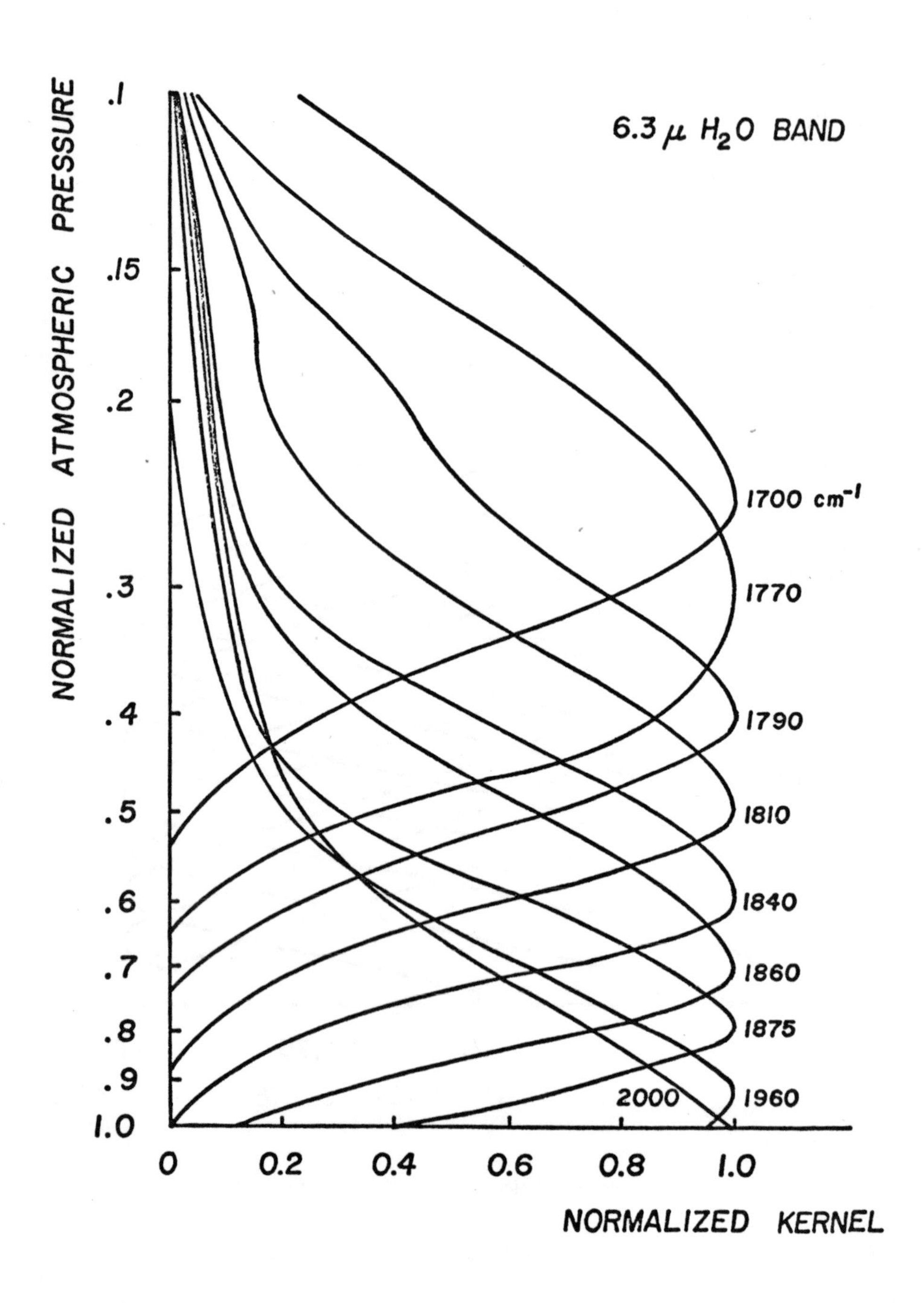

Fig. 3 A Kernel for Water Vapor in the Earth Atmosphere [14].

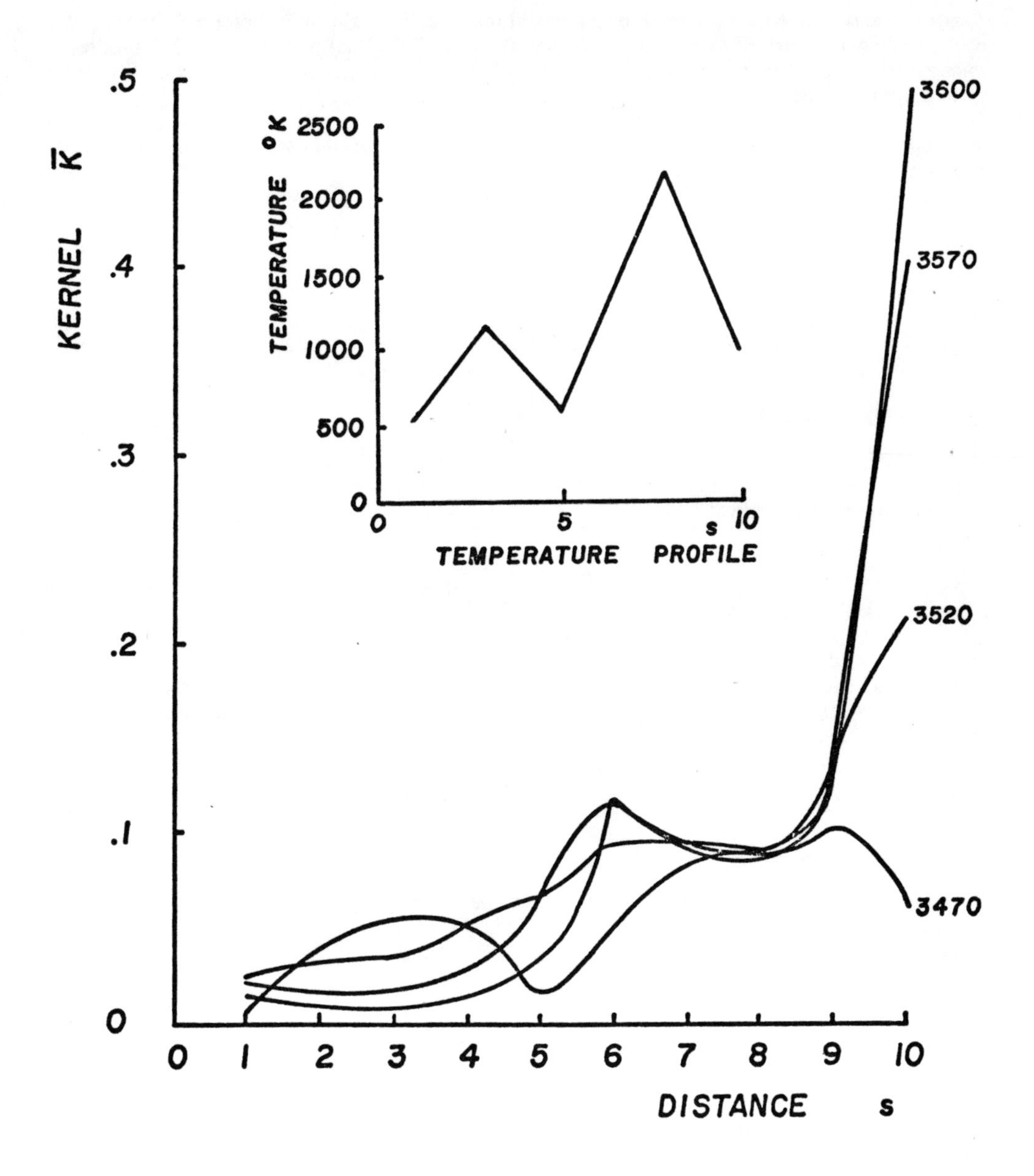

Fig. 4 A Kernel for a Heated Mixture [par. 4.2.]

4.3. Diffusion Concentration Profiles

When a specie is injected or generated in an ambient medium (such as combustion products for instance), their spreading is controlled by diffusion processes (Fig. 5). This situation leads to density profiles which are affected both by temperature (perfect gas law) and the exponential character of diffusion concentration profiles.

Let us consider the case of a symmetrical single peak of temperature and concentration such as they exist downstream of a combustion zone (Fig. 5). Typically:

$$\rho = \rho_o \frac{T_o}{T(s)} \exp\left[-\frac{|s - s_o|}{hT(s)}\right] \tag{40}$$

where h is a scale factor related to the thermal diffusion rate of the absorbing gas and the distance from the burner.

The optical thickness in this model can be expressed in following

$$\begin{aligned} \tau(\omega_j, s_i) &= \rho_o \bar{k}(\omega_j)\left(\frac{T_o}{T_{av}}\right) hT^* \exp\left(-\frac{b - s_o}{hT^*}\right)\left[\exp\left(-\frac{s_i - b}{hT^*}\right) - 1\right], \quad \text{for } s_i > s_o \\ &= \rho_o \bar{k}(\omega_j)\left(\frac{T_o}{T_{av}}\right) hT^*\left[2 - \exp\left(-\frac{b - s_o}{hT^*}\right) - \exp\left(-\frac{s_o - s_i}{hT^*}\right)\right], \quad \text{for } s_i < s_o \end{aligned} \tag{41}$$

where T* is a different average temperature with respect to exponential weight distribution. For the purpose of this analysis, a further simplification is assumed: $T_o = T_{av} = T^*$. The transmittance kernel matrix is given then by:

$$\bar{K}(\omega_j, s_i) = \rho_o \bar{k}(\omega_j) \exp\left[-\frac{|s_i - s_o|}{hT_o} - \tau(\omega_j, s_i)\right] \tag{42}$$

The maximum of the kernel for each frequency j can be found by differentiating the function terms in the bracket of the exponential in Eq. (42). An important result (Fig. 6) is that the maximum peaks of the kernel are always found for values of s_m larger than s_o:

$$\frac{s_m - s_o}{hT_o} = \ln\left[\rho_o \bar{k}(\omega_j) hT_o\right] \tag{43}$$

In other words there are no satisfactory kernels to observe beyond the center of symmetry of the flame (i.e. for $s_m < s_o$).

As we proceed with more transformations, we obtain at $s_i = s_m$, the peak values of the kernel

$$\begin{aligned} \bar{K}_m(\omega_j) &= \bar{K}(\omega_j, s_m) \\ &= \frac{1}{hT_o} \exp\left[\exp\left(\frac{s_m - b}{hT_o}\right)\right] \end{aligned} \tag{44}$$

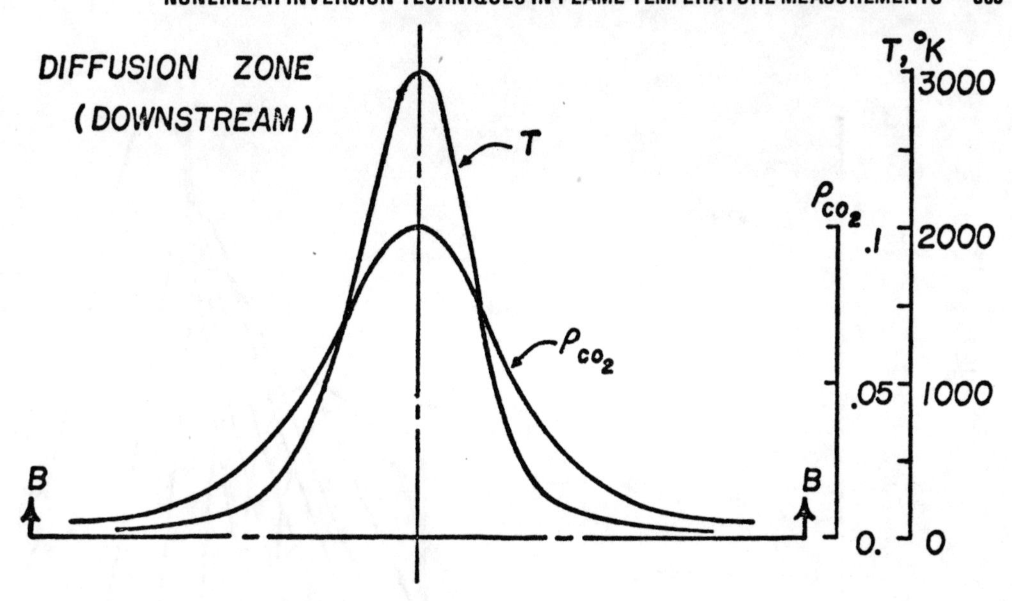

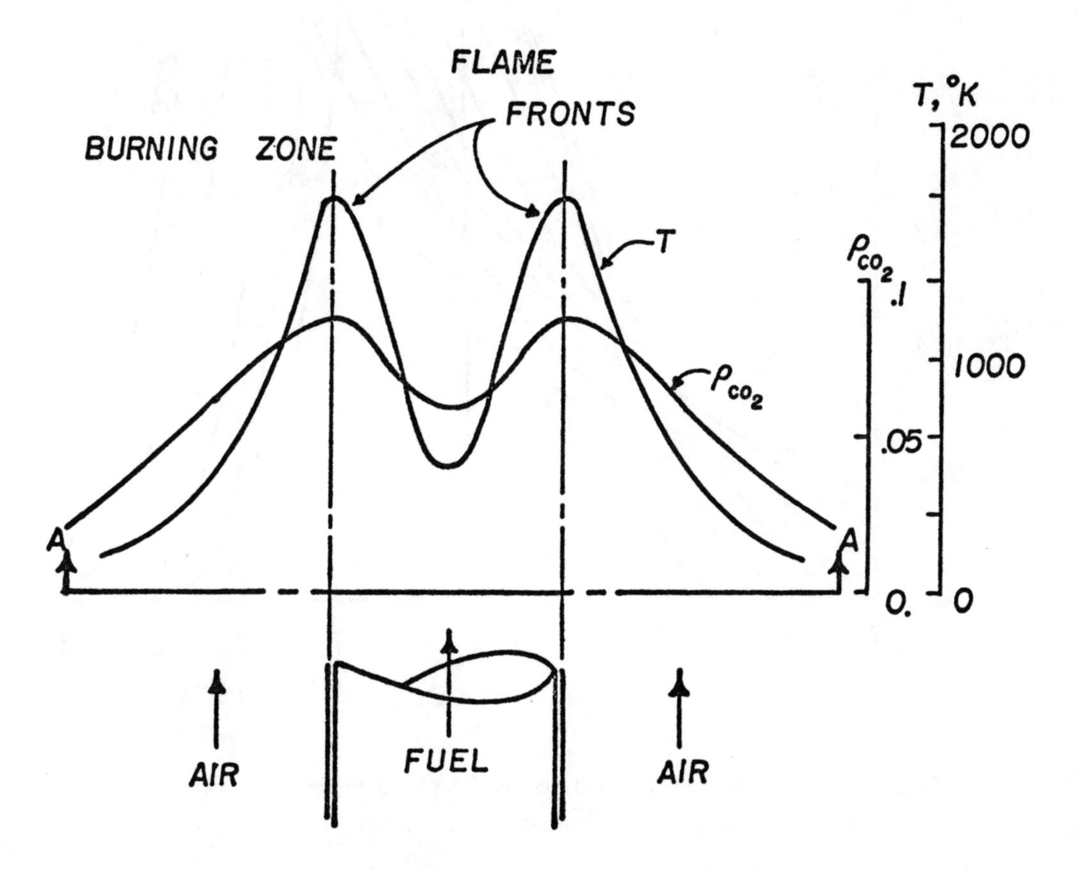

Fig. 5 Combustion and Diffusion Zones in a Flame

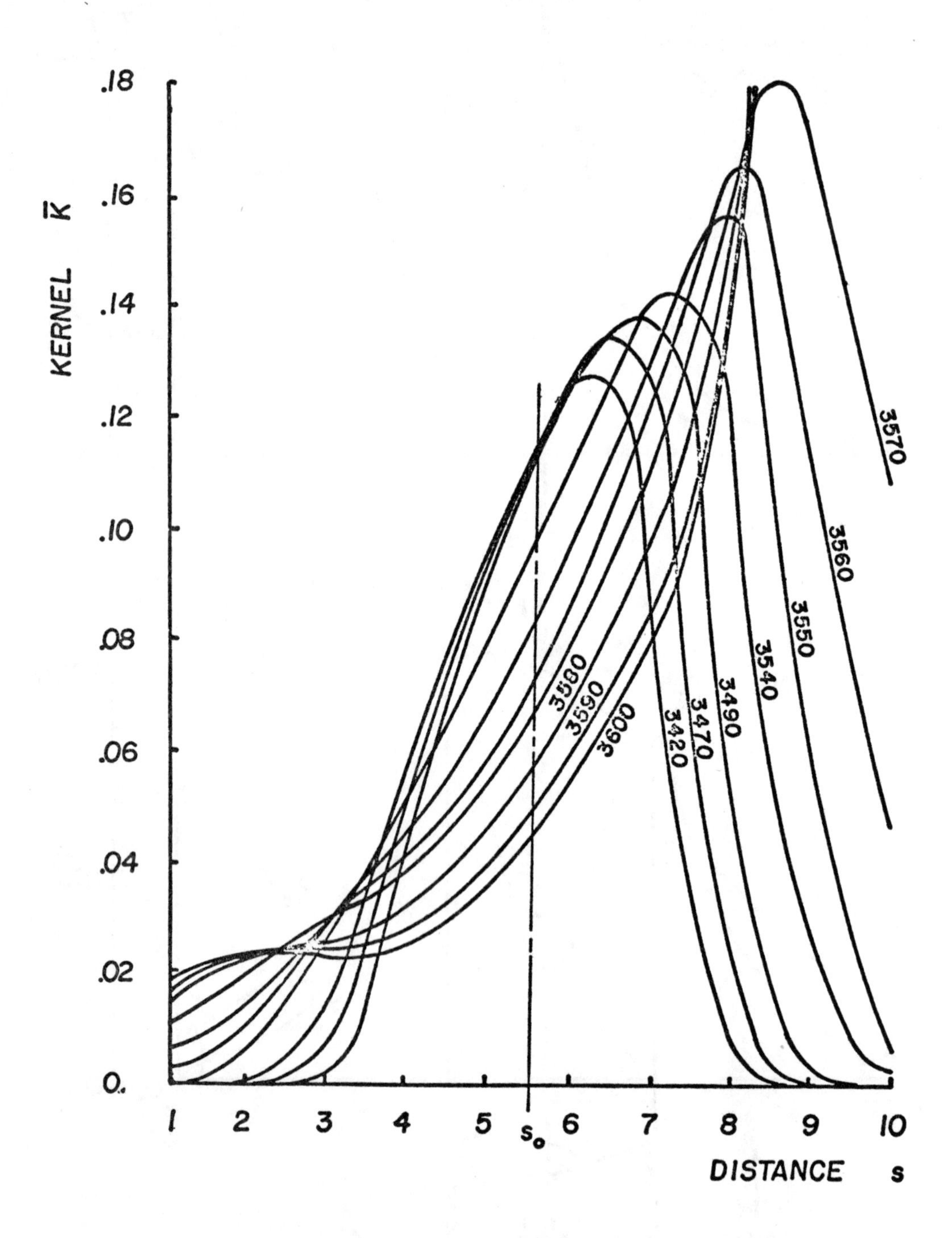

Fig. 6 A Kernel for Diffusion CO_2 in a Flame [Eqs. (40) and (B-8)], with Axis of Symmetry for T and ρ on $s = s_o$.

while at the center of the flame, the values of the kernel are

$$\begin{aligned} K_o(\omega_j) &= K(\omega_j, s_o) \\ &= K_m(\omega_j) \exp\left[\frac{s_m - s_o}{hT_o} - \exp \frac{s_m - s_o}{hT_o}\right] \end{aligned} \tag{45}$$

The transmittance kernel matrix can then be expressed in terms of K_m and s_m as:

for $s_i > s_o$:

$$K(\omega_j, s_i) = K_m(\omega_j) \exp\left[\frac{s_m - s_i}{hT_o} - \exp\left(\frac{s_m - s_i}{hT_o}\right)\right] \tag{46-a}$$

for $s_i < s_o$:

$$K(\omega_j, s_i) = K_m(\omega_j) \exp\left\{\frac{s_i - s_o}{hT_o} - \frac{K_m(\omega_j)}{hT_o}\left[2 - \exp\left(\frac{s_i - s_o}{hT_o}\right)\right]\right\} \tag{46-b}$$

As stated earlier, these kernels, while better behaved than the constant mixing profile ones (Parag. 4-2), are satisfactory only for probing the side of the flame facing the sensor. If axisymmetry cannot be assumed, they are inadequate.

For profiles in the combustion zone, several maxima can occur (Fig. 5). The following concentration profile with two peaks occurred at the flame fronts could be proposed, again for kernel behavior comparisons:

$$\rho = \frac{\rho_o}{r + l} \frac{T_o}{T(s)} \left\{ r \exp\left[-\frac{|s - s_o - r|}{hT(s)}\right] + l \exp\left[-\frac{|s - s_o + l|}{hT(s)}\right]\right\} \tag{47}$$

where r and ℓ are the location distances of the right and left flame fronts with respect to the center of the flame. Complex kernels can be derived, with less sharply defined peaks as before (Fig. 7) and with no satisfactory way to measure the properties between the areas of maximum concentration.

4.4. Unknown Concentration Profiles: Auxiliary Variables

The 3 cases considered above were illustrative of the effect of given density profiles on inversion conditioning. In practice, however, the concentration profiles may be just as unknown as the temperature profile. Hence the need to retrieve both profiles.

One line of approach dominates the literature: It consists in seeking a temperature profile T(x) not in terms of the physical distance s, but rather in terms of a new auxiliary variable x which "absorbs" in Eq. 1 all functions of the species density ρ as well as of s. If additional relationships between specie densities and temperature can be found from chemistry, fluid dynamics, etc. . . . , a second step is then to eliminate x and to express eventually all property profiles, including ρ, in terms of the physical distance s.

An obvious candidate for this role is the optical thickness [Eq. (26)] at some reference frequency ω_o, since the species density ρ enters the problem only through the product $\rho k_\omega ds$ [Eq. (1)]. The astrophysical literature presents a number of such solutions, although the dependence of k_ω on T in the hot stellar layers is a troublesome complication [16,17].

A similar approach is used in meteorology with the additional advantage that k_ω depends only weakly on temperature and pressure, and it is therefore more legitimate to use as an auxiliary variable $x \equiv \int \rho_A ds$, where A refers to the particular molecule under observation. As it turns out, the atmospheric CO_2 has, conveniently, a constant mixing ratio throughout the atmosphere. Hence the possibility to use directly the atmospheric density ρ
in the auxiliary variable, and then to proceed with the final step, i.e. to bring in the perfect gas law and hydrostatic equations, and obtain T directly in terms of the altitude s.

In fact, it is common practice in meteorology to express directly the kernel of Eq. (1) in terms of the derivative of the transmittance with respect to the atmospheric pressure p [18,19]:

$$N^+(\omega) = \int_0^{p_0} B[\omega, T(p)] \frac{\partial \bar{T}(\omega,p)}{\partial p} dp \tag{48}$$

where $\bar{T}$ is the transmittance [Eq. (25)]. The solution is of the form T(p), which can be further reduced to T(s) by the additional relationships mentioned above, or by pressure probe measurements. This procedure, too, ignores the small dependence of the kernel on temperature and there might be limitations to the accuracy of this method.

In any case, the solution of Eq. (1) in two consecutive steps involving a well chosen auxiliary variable is an attractive method which should be applicable to other inversion problems.

V. ACCURACY OF NONLINEAR TECHNIQUES

In this chapter, two nonlinear iterative techniques will be tested against the four configurations discussed in the previous chapter (paragraphs 4.1. to 4.3.). The case of an unknown concentration profile will be discussed also.

The two methods to be compared are the general iteration method [Eq. (16) and the nonlinear relaxation method [Eq. (18). In both cases, the pivotal technique has been included when necessary. Statistical methods were also tested. but with no success: they will not be discussed in this report. Transmittances were calculated on the basis of the absorption model of the 2.7μ CO_2 band, as presented in Appendix B.3., Eq. (B-8).

5.1. Well Behaved Kernels

The meteorological literature [5, 10, 11, etc.] presents a variety of atmospheric kernels which all feature well spaced, fairly distinct maxima (Fig. 3). This character is due to the near-exponential decrease of atmospheric density with altitude. The idealized model based on a standard atmosphere profile (paragraph 4.1.) brings out the key feature of this kind of kernel [Eq. (45) and Fig. (2)]. The spacing of the peaks [Eq. (31)] is optimized by a choice of frequencies which spaces out evenly the magnitudes of the absorption coefficients $k(\omega_j)$. The thickness of each peak is [Eq. (31)] and the density profile itself [Eq. (27)]: H reflects simply the scale of the observed layers; it does not help or hinder their measurement. The choice of the proper set of frequencies is the one critical design decision. The properties of such kernels are further discussed in [20].

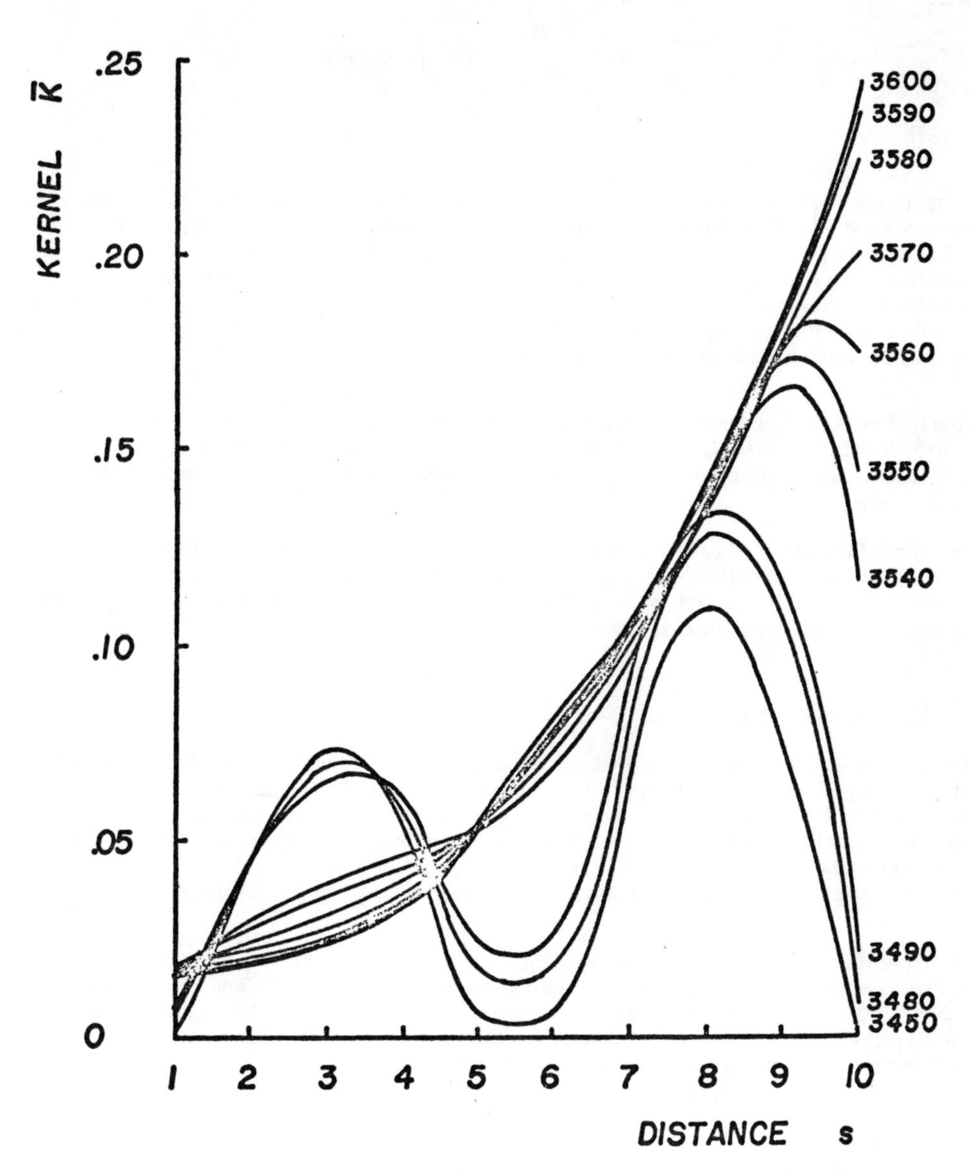

Fig. 7 Typical Kernel in a Flame Combustion Zone [Eqs. (47) and (B-8)]

In this section we shall discuss several solutions based on a kernel $\bar{K}$ of this general class, which we chose for computational convenience:

$$\bar{K}[\omega_j, s_i] = \bar{K}_m \exp\left[-\frac{(s_i - s_{mj})^2}{H}\right] \tag{55}$$

10 frequencies ω_j where chosen, corresponding to 10 kernels $\bar{K}$ peaking at values of s_{mj} equally spaced between 1 and 10. The value of H was unity. The temperature profile to recover is shown in a full line on Fig. 8, and the initial guess (T = 1500° K) in a small dashed line. It can be seen that except for the over-relaxation method, all procedures converge quickly to the exact solution,* due to the exceptional suitability of this idealized kernel [Eq. (55)]. Fig 9 shows that the rate of convergence obtained for the general iteration method tends to be slowest, whereas the nonlinear relaxation method converges monotonically to an error less than 10% in 10 iterations.

Note that most of these errors take place at both ends of the flame (low T) where the source term B_i is small and the kernels K_j are truncated. If these 2 values were ignored (or presumed to be room temperature) in the average error calculation, $\bar{\sigma}_T$ would converge to about 3% instead of 10%.

The point of this idealized case was to illustrate the fact that the kind of success met in atmospheric measurements could be extended to higher and irregular temperature profiles if similar kernels (i.e. favorable density distributions) could be found in applications such as flames, plasmas, etc. . . Unfortunately, this is not often the case.

5.2. Uniform Mixing Ratio Kernels

It was shown in Section 4.2, that kernels corresponding to heated uniform mixtures present their maximum at the point of minimum temperature, since they vary inversely with T [Eq. (39)]. Hende the kernels of Fig. 4 offer no spacing in their maxima, since they all tend to occur near the same low temperature point of the flame. To alleviate this poor conditioning, 20 frequencies were selected for the CO_2 band model defined in Appendix B.3., and the pivotal method was used to select the 10 best kernels out of this set. (We shall follow this procedure in all subsequent runs of this report).

In spite of the use of the pivotal technique, Fig. 11 shows that the minimum average temperature error on Fig. 10 is more than 10% for the general iteration method ($\bar{\sigma}_T^2 > 10^{-2}$), and about equal to 10% ($\bar{\sigma}_T^2 \simeq 10^{-2}$) at the optimum number of iterations for the nonlinear relaxation method. Notice on Fig. 10 that even though the general iteration method is not, on the average, as effective as the nonlinear relaxation method, it gives better results for the temperature peaks. There is no priori criterion to help in choosing a method which would answer all needs best.

5.3. Diffusion Flow Kernels

The density profiles characteristic of the flame region downstream of the combustion zone [Fig. 5] lead to the interesting situation where the kernel maxima exist on the side of the flame facing the observer only [Fig. 6]. As a result, the two inversion procedures used in this case show less accuracy as they did in the idealized case discussed in paragraph 5.1, even though the temperature profile is a simpler one. On Fig. 12 in particular, it is seen that the general iteration method does not give satisfactory convergence, in spite of a rather favorable initial guess. Fig. 13 emphasizes this lack of convergence for the general iteration method, regardless of the number of iterations performed. [Note that the profile plotted for general iteration on Fig. 12 corresponds to the minimum variance $\bar{\sigma}_T^2$ displayed on Fig. 13: N = 5].

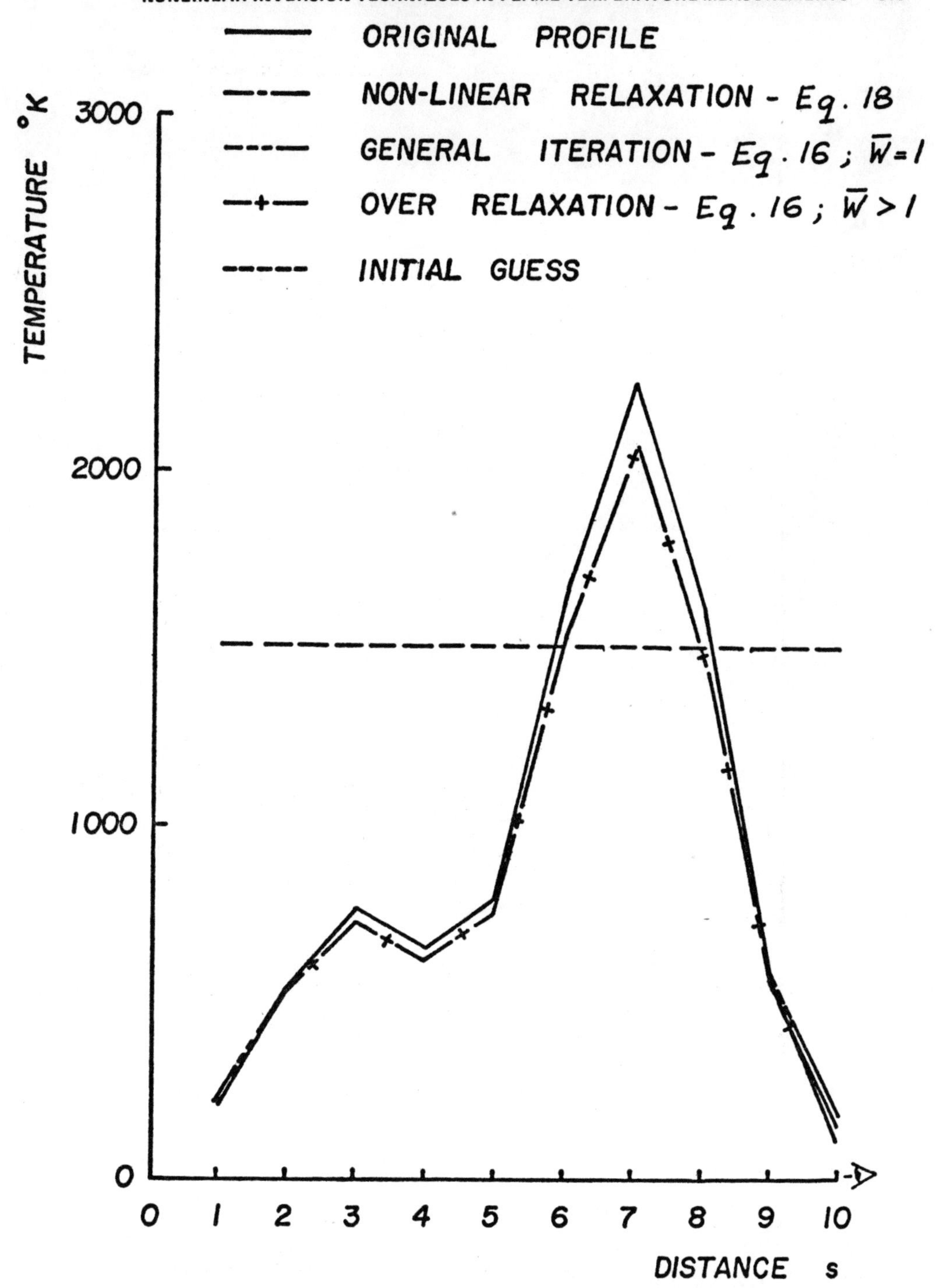

Fig. 8 Temperature Profile Retrieval [Exponential Kernel, par. 5.1.]

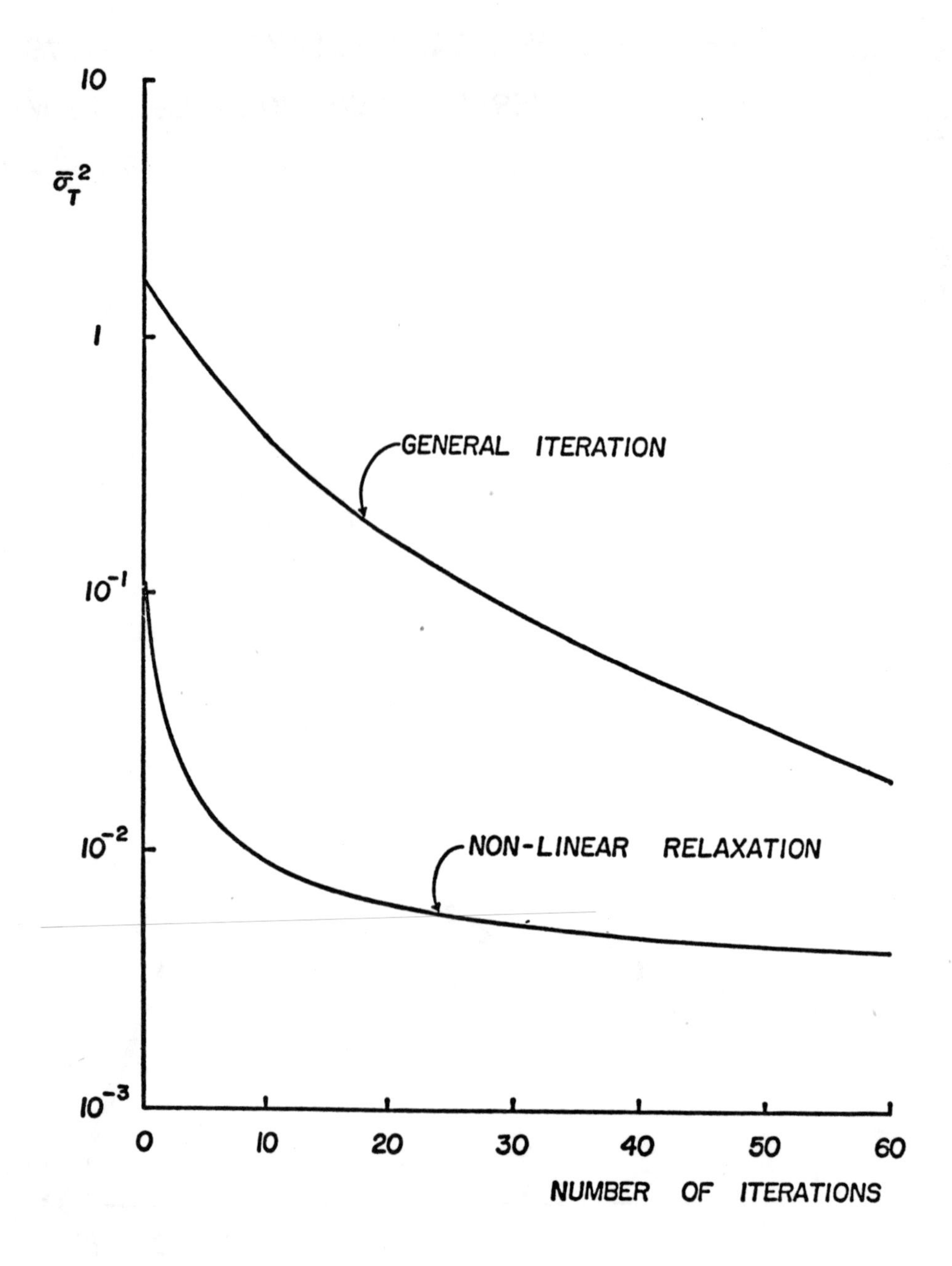

Fig. 9 Profile-Averaged Squared Error for Successive Iterations [Exponential Kernel, par. 5.1.]

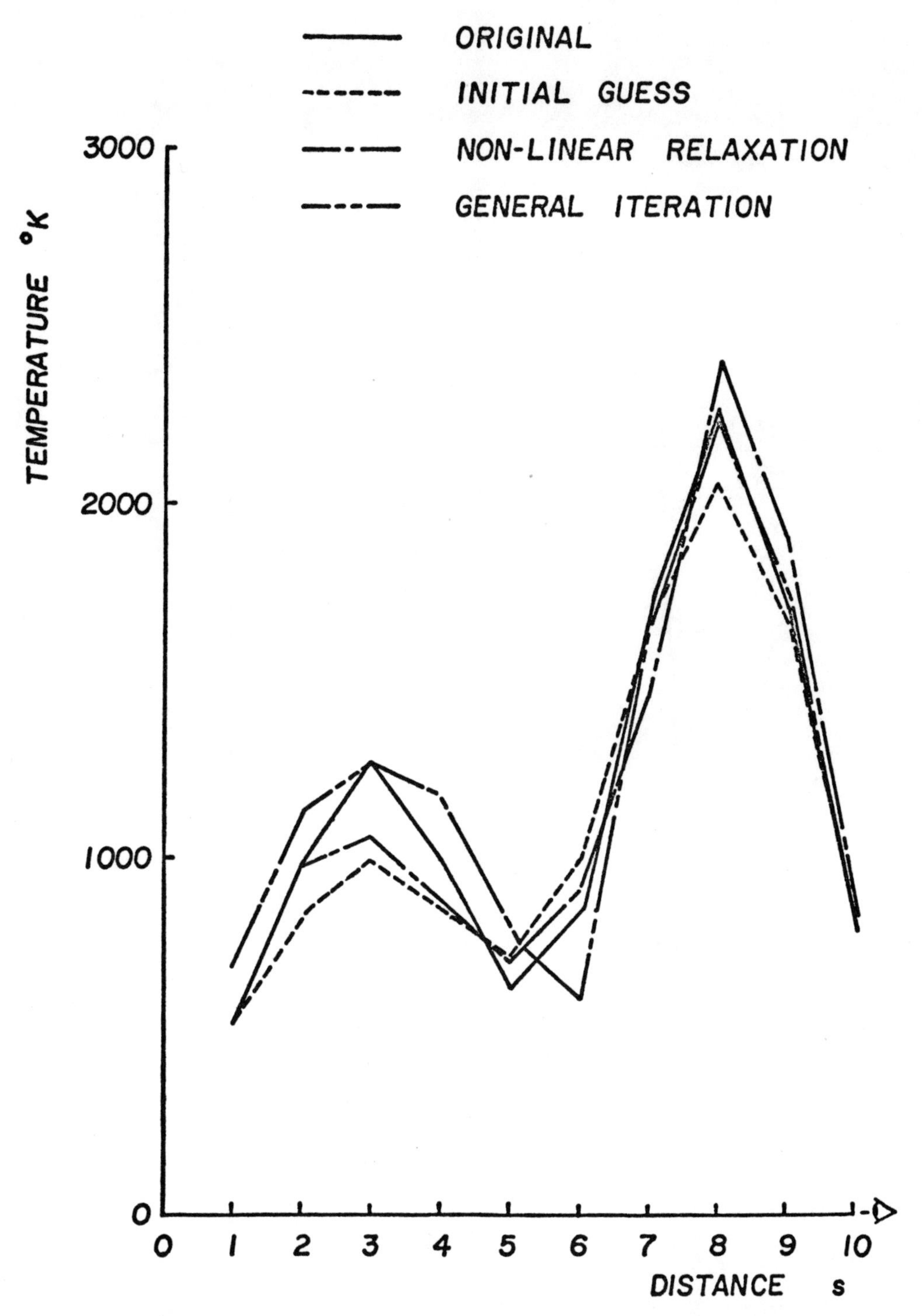

Fig. 10 Temperature Profile Retrieval [Uniform Mixing Ratio Kernel, par. 5.2.]

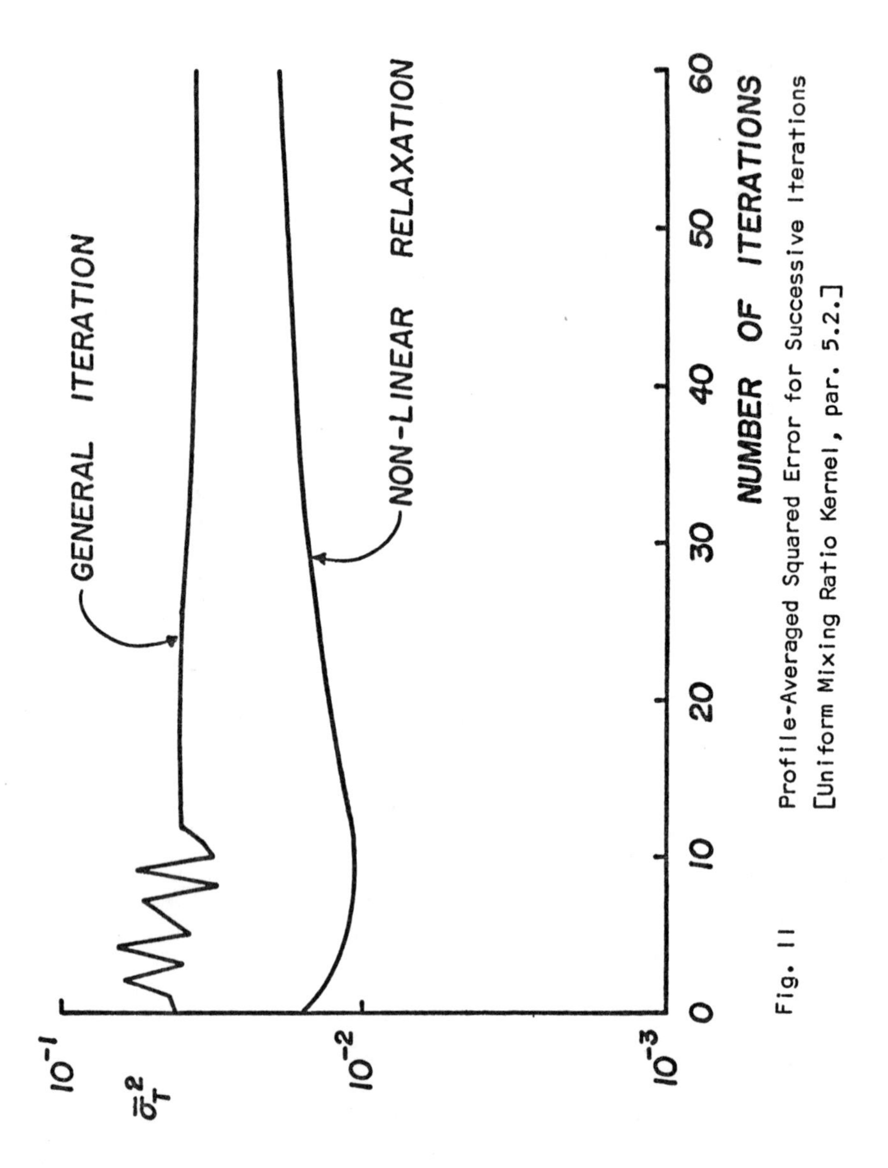

Fig. 11 Profile-Averaged Squared Error for Successive Iterations [Uniform Mixing Ratio Kernel, par. 5.2.]

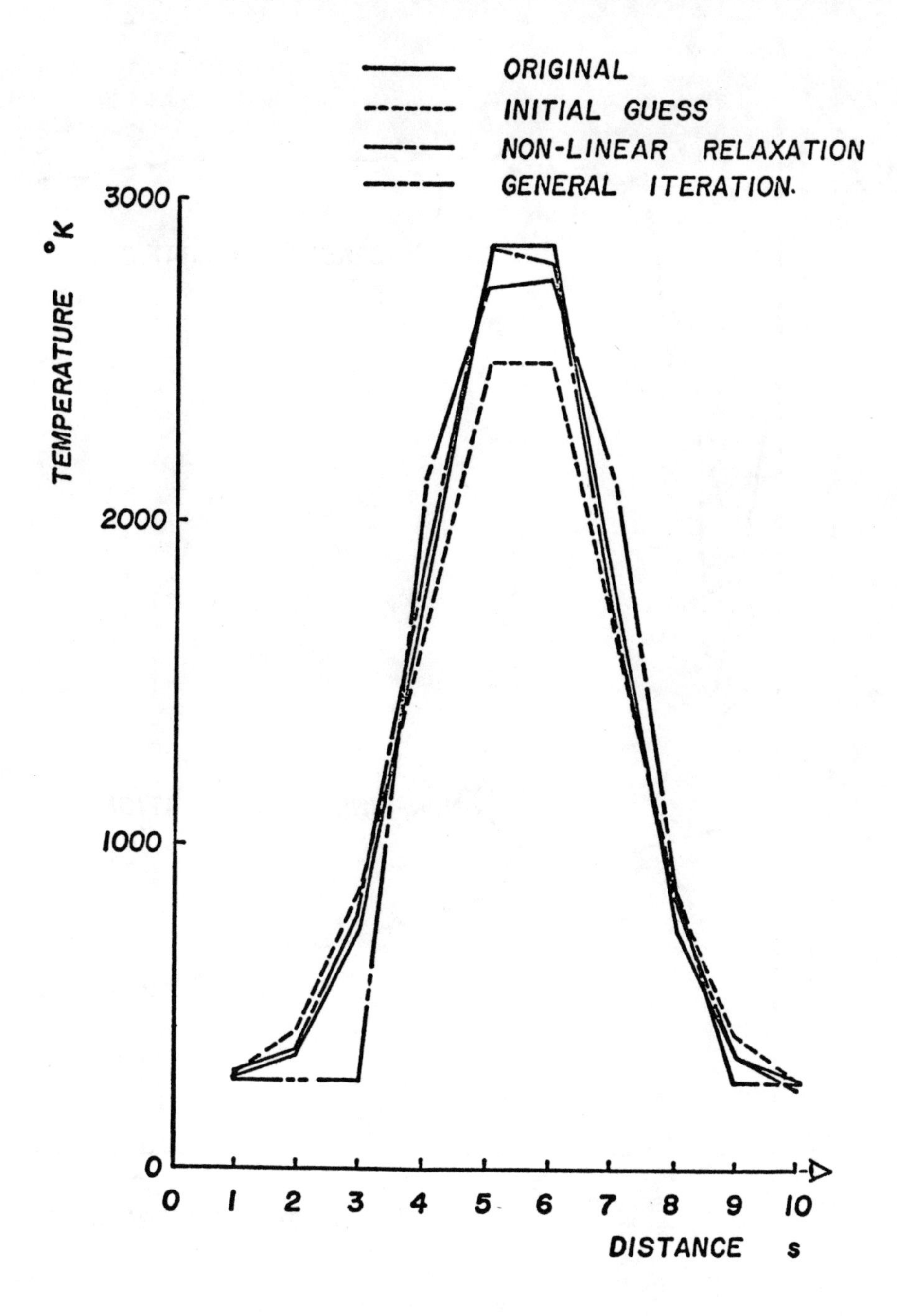

Fig. 12 Temperature Profile Retrieval Diffusion Flow Kernel, par. 5.3.]

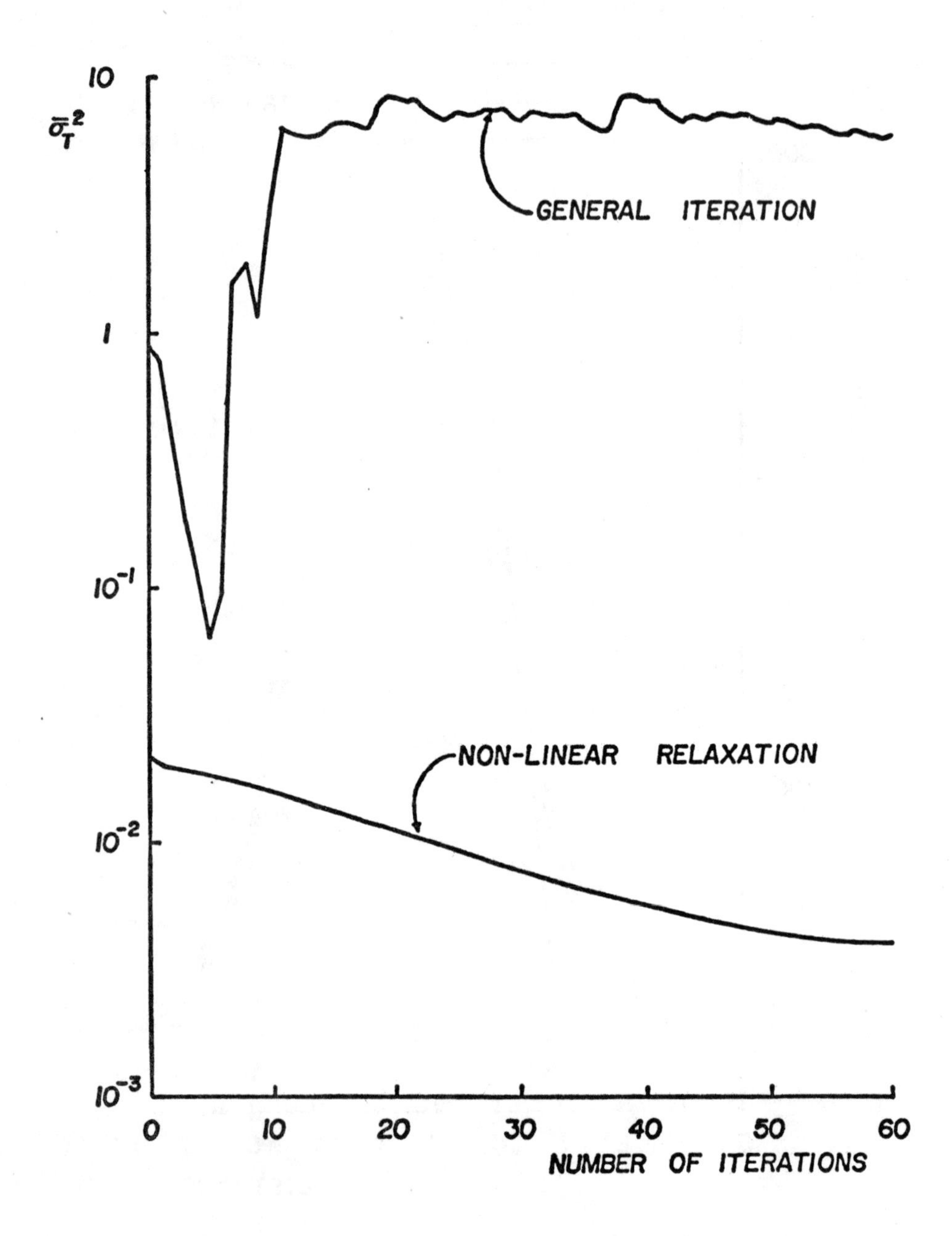

Fig. 13 Profile-Averaged Squared Error for Successive Iterations [Diffusion Flow Kernel, par. 5.3.]

As expected, accuracy is especially poor on the left side, where the kernels are weakest [Fig. 6]. Fig. 14 shows that if a less favorable initial temperature profile had been assumed, the general iteration method would have diverged altogether. Hence the apparent superiority of the nonlinear relaxation approach when kernels are not too well behaved.

5.4. Combustion Zone Kernels

The other type of flame profile likely to be encountered is that which presents several (typically two) flame fronts [Fig. 5]. In such a case, one can expect a relative "blindness" of the method for those parts of the profile which lie beyond the first maximum concentration zone as seen from the sensor. This situation is indeed illustrated on Figs. 15 and 16 where the trends and conclusions discussed in the preceding paragraph 5.3. are in evidence again.

REFERENCES

1. Tourin, R.H., "Spectroscopic Gas Temperature Measurement – Pyrometry of Hot Gases and Plasmas," Elsevier Publishing Co., New York, 1966.
2. Kondratyev, K.Ya., Timofeev, Yu.M., Pokrovsky, O.M., Tikhnov, A.N., and Glasko, V.B., "On the Solution of Some Inverse Problems of Atmospheric Optics," COSPAR Proceedings, pp.357-360, Leningrad, 1971.
3. Wang, J.Y., "Theory and Application of Inverse Techniques: A Review," Purdue University Report AA & ES 70-6, September 1970.
4. Goulard, R. and Wang, J.Y., "Linearized Inversion Techniques in the Temperature Measurement of Radiating Flows," Lectures at the International Center for Mechanical Sciences, Udine, Italy (Oct. 1972); monograph to appear. See also Purdue University Reports AA & ES 72-1-1 and 72-9-4.
5. Chahine, M.T., "Inverse Problems in Radiative Transfer: Determination of Atmospheric Parameters," J.Atmospheric Sci., Vol. 27, No. 6, pp. 960-967, September 1970.
6. Ludwig, C.B., Malkmus, W., Reardon, J.E., and Thomson, J.A.L., Handbook of Infrared Radiation from Combustion Gases, (Goulard, R., and Thomson, J.A.L., Editors) NASA SP-3080, 1973.
7. Krakow, B., "Spectroscopic Temperature Profile Measurements in Inhomogeneous Hot Gases," Applied Optics, Vol.5, No.2, p.201, February 1966.
8. Carnahan, B., Luther, H.A., and Wilkes, J.O., "Applied Numerical Methods," John Wiley, New York, 1964.
9. Backus, G.E., and Gilbert, J.F., "Uniqueness in the Inversion of Inaccurate Gross Growth Data," Phil. Trans. R. Soc. London, A266, 1970, pp.123-192.
10. Conrath, B.J. "Vertical Resolution of Temperature Profile Obtained from Remote Radiation Measurement," Goddard Space Flight Center, X-622-71-51, November 1971.
11. Smith, W.L., Woolf, H.M., and Fleming, H.E., "Retrieval of Atmospheric Temperature Profiles from Satellite Measurements for Dynamical Forecasting," J.Appl. Meteorology, Vol. II, Feb. 1962, pp.113-122.
12. Lewis, T.O., and Odell, P.L., Estimation In Linear Models, Prentice Hall, New Jersey, 1971.
13. Rust, B.W., and Burrus, W.R. "Mathematical Programming and the Numerical Solutions of Linear Equations," American Elsevier Publishing Company, New York, 1972.
14. Chahine, M.T., "A General Relaxation Method for Inverse Solution of the Full Radiative Transfer Equations," J. Atmospheric Sci., Vol. 29, May 1972, pp. 741-747.
15. Westlake, J.R., A Handbook of Numerical Matrix Inversion and Solutions of Linear Equations, John Wiley & Sons, Inc., New York, 1968, pp.37.
16. Goldberg, L. and Pierce, A.K., "The Photosphere of the Sun" from the "Handbuch der Physik", Flugge Ed., Vol. 52, Leipzig.
17. Jeffries, J.T., Spectral Line Formation, Blaisdell Publishing Co., 1968.
18. Wark, D.Q., and Fleming, H.E., "Indirect Measurements of Atmospheric Temperature Profile from Satellites: 1. Introduction," Monthly Weather Review, Vol.94, No. 6, June 1966.

19. Smith, W.L., Woolf, H.M., and Fleming, H.E., "Retrieval of Atmospheric Temperature Profiles from Satellite Measurements for Dynamical Forecasting," Journal of Applied Meteorology, Vol. II, pp.113-122, February 1972.
20. Chao, C.M., "Optimal Selection of the Design Kernel for the Inverse Problem of Remote Sounding in the Earth's Atmosphere," (an A & A School Report)
21. Goody, R.M., Atmospheric Radiation – Theoretical Basis, Oxford Press, 1964.
22. Kondratyev, K.Ya., Radiative Heat Exchange in the Atmosphere, Pergamon Press, 1965.
23. Ambartsumyan, V.A., Theoretical Astrophysics, Pergamon Press, 1958.

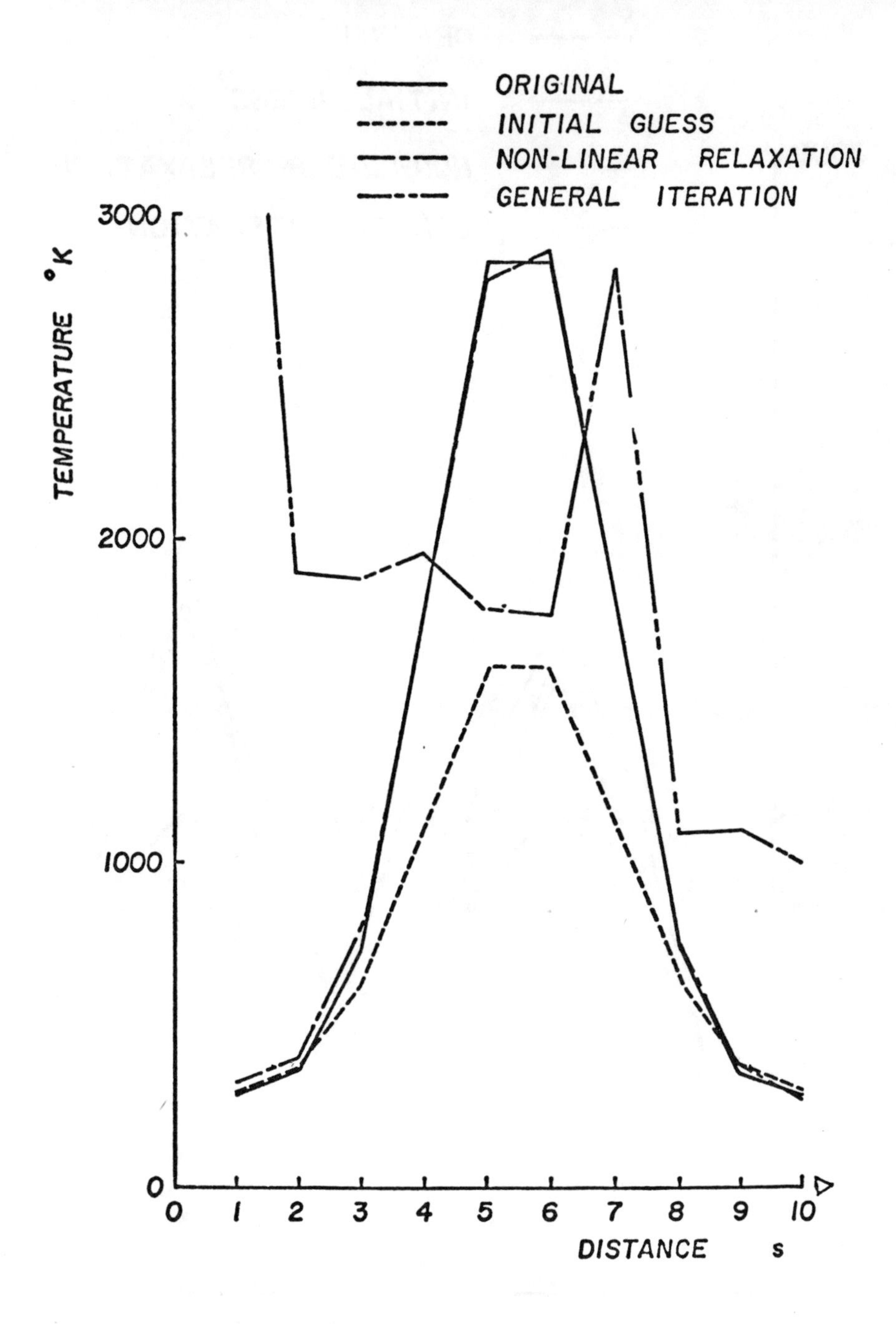

Fig. 14 Temperature Profile Retrieval, with Poor Initial Guess [Diffusion Flow Kernel, par. 5.3.]

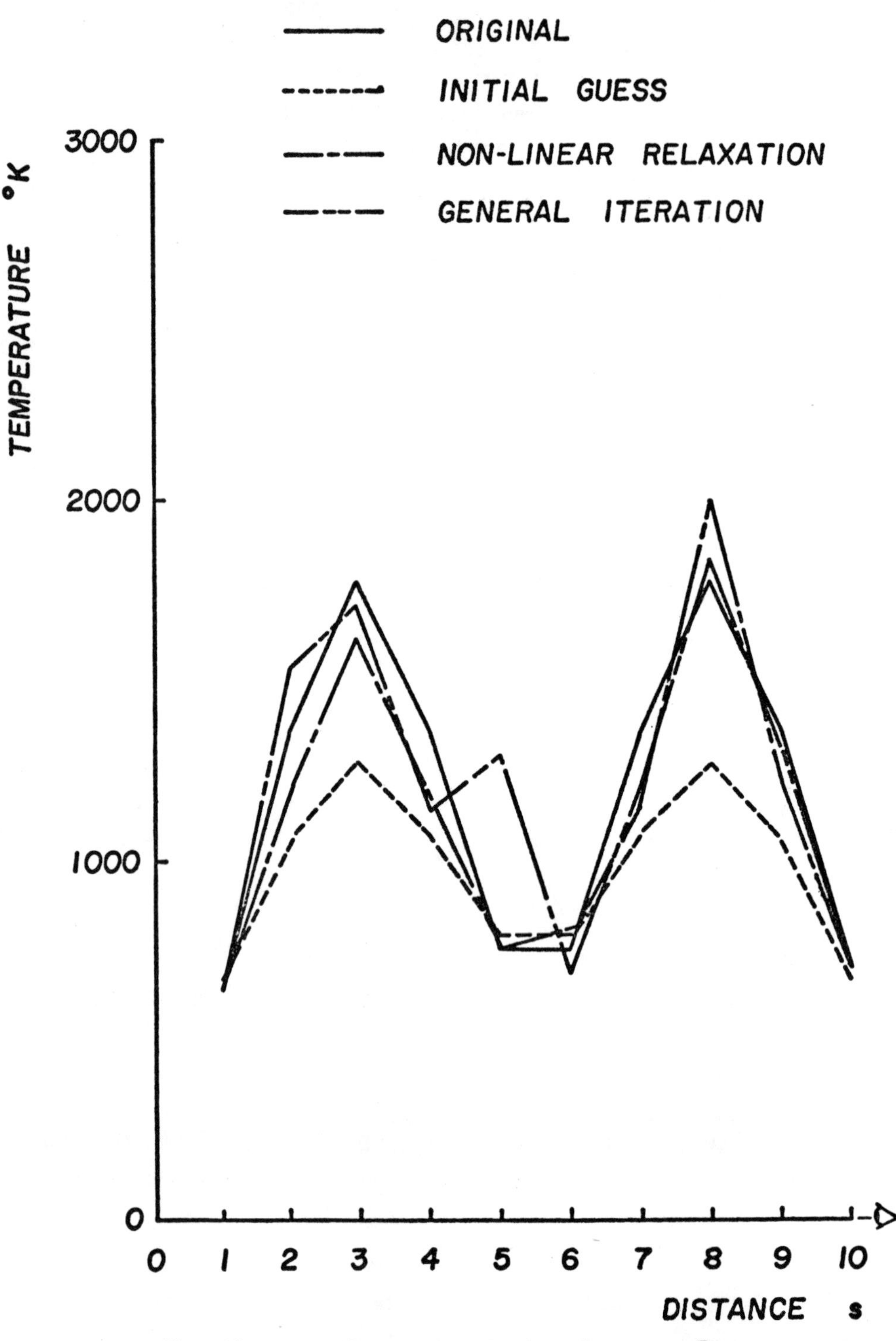

Fig. 15 Temperature Profile Retrieval [Combustion Zone Kernel, par. 5.4.]

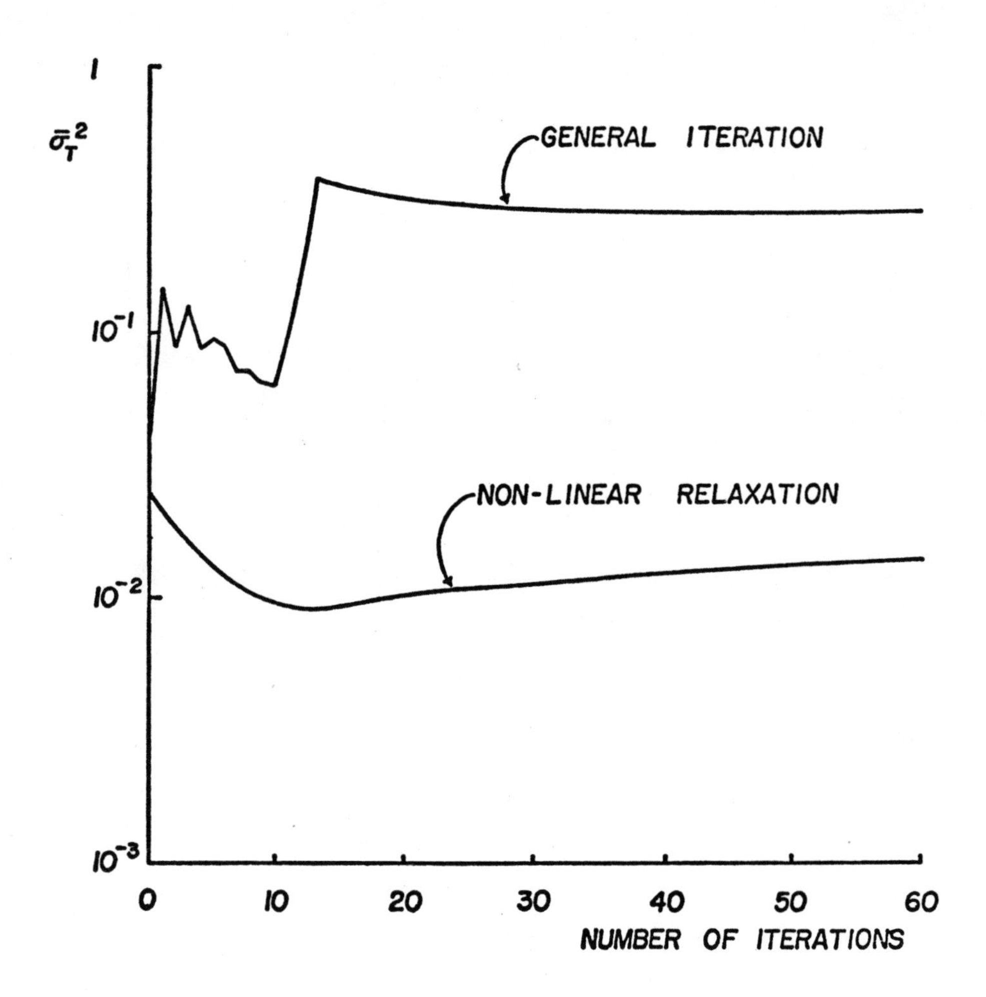

Fig. 16 Profile-Averaged Squared Error for Successive Iterations [Combustion Zone Kernel, par. 5.4.]

APPENDIX A

A.1. The Selective Mutiplication Matrix Operator ⊗

Eq. 4 introduces a new symbol ⊗

$$N^+(\omega_j) = \bar{K}[\omega_j, T(s_i)] \otimes \bar{B}[\omega_j, T(s_i)] \quad \text{(A-1)}$$

This selective multiplication matrix operator ⊗ defines the following extraordinary matrix operation:

$$\begin{bmatrix} \bar{K}(\omega_1, s_1), \bar{K}(\omega_1, s_2) \ldots \bar{K}(\omega_1, s_m) \\ \bar{K}(\omega_2, s_1), \bar{K}(\omega_2, s_2) \ldots \bar{K}(\omega_2, s_m) \\ \ldots\ldots\ldots\ldots\ldots \\ \bar{K}(\omega_n, s_1), \bar{K}(\omega_n, s_2) \ldots \bar{K}(\omega_n, s_m) \end{bmatrix} \otimes \begin{bmatrix} \bar{B}(\omega_1, s_1)\bar{B}(\omega_2, s_1) \,.\, \bar{B}(\omega_n, s_1) \\ \bar{B}(\omega_1, s_2)\bar{B}(\omega_2, s_2) \,.\, \bar{B}(\omega_n, s_2) \\ \vdots \qquad\qquad \vdots \\ \bar{B}(\omega_1, s_m)\bar{B}(\omega_2, s_m)\, \bar{B}(\omega_n, s_m) \end{bmatrix}$$

$$\equiv \begin{bmatrix} \sum_{i=1}^{m} \bar{K}(\omega_1, s_i)\bar{B}(\omega_1, s_i) \\ \sum_{i=1}^{m} \bar{K}(\omega_2, s_i)\bar{B}(\omega_2, s_i) \\ \ldots\ldots\ldots \\ \sum_{i=1}^{m} \bar{K}(\omega_n, s_i)\bar{B}(\omega_n, s_i) \end{bmatrix} \quad \text{(A-2)}$$

The distribution law can be applied to this particular matrix operation. If

$$N = (\bar{K}_1 + \bar{K}_2) \otimes \bar{B}$$

then:

$$N = \bar{K}_1 \otimes \bar{B} + \bar{K}_2 \otimes \bar{B} \quad \text{(A-3)}$$

The transpose operations is valid also:

$$N = \bar{B}' \otimes \bar{K}' \quad \text{(A-4)}$$

where the superscript' indicates the transpose matrix [Eq. (49)].

APPENDIX B

In inversion procedures, the dependence on temperature of the absorption coefficient k_ω presents a serious obstacle to inversion procedures. The next two paragraphs (borrowed mostly from [4]) will review briefly k_ω for lines and molecular bands. A model for the 2.7μ band of CO_2 will also be presented. Continuum absorption does not provide sharp enough variations of k_ω across reasonable spectral intervals to be useful in inversion techniques.

B.1. Line Absorption

There exists three fundamental processes which control the shape of spectral lines: damping, Doppler and collision. (At very high temperatures, magnetic and electrical fields introduce further distortions: we shall not consider them here.) Eqs. (B-1) and (B-2) state the analytical form of this dependence:

Doppler Broadening:

$$\rho k_\nu = AN \frac{1}{\Delta\nu_D} \exp\left(-\frac{(\nu-\nu_o)^2}{\Delta\nu_D^2}\right) \tag{B-1}$$

with N = number of radiating particles per unit volume ν_o is the resonance frequency (center) of the line $\Delta\nu_D = \nu_o/c\sqrt{2RT}$, and A is a physical constant

Natural and Collision Broadening:

$$\rho k_\nu = AN \frac{1}{\sqrt{\pi}} \frac{\frac{\gamma}{4\pi}}{(\nu-\nu_o)+\left(\frac{\gamma}{4\pi}\right)^2} \tag{B-2}$$

with $\gamma \equiv \gamma_n + \gamma_c$
$\gamma_n \equiv$ natural damping "half width", a constant
$\gamma_c \equiv$ collision broadening "half width" $\simeq Bp\,(1/\sqrt{T}$

We observe that both $\Delta\nu_D$ and γ_c are temperature dependent. Also, although these three effects are additive (see Fig. B-F1), practical applications tend to bring out only one effect at a time. Low temperature applications (e.g. meteorology, flames) tend to bring out collision broadening [21,22] while high temperature and low density ones (e.g. stellar atmospheres) tend to bring out Doppler and natural broadening [i7,23].

The difficulties inherent to these nonlinearities have resulted in the frequent recourse to angle scanning approaches (Abel inversion, solar disk scanning, etc...). Also, "synthetic" temperature profiles are commonly used, where the unknown function T(s) is given a series expansion form with only two or three unknown coefficients to be determined [17].

B.2. Molecular Band Absorption

At ambient or flame temperatures, remote sensing is usually based on the infrared radiation of molecules. Because of the relative weakness of the signal emitted by single lines, it is customary to measure the emission of all the lines within a frequency interval of the order of 10 or 50 cm^{-1} (wavenumbers). The formulation of the measured signal under these conditions leads to the averaging of the transmittance T_ν over a small interval $\Delta\nu$ determined by the instrument.
Since all the lines included in each of these measurements are affected by temperature and pressure in the way discussed in the preceding paragraph, the averaged transmittance $\bar{T}_{\nu_o}$ depends also on these properties. More important even than the relative spreading of individual lines, is the appearance, even at moderately high temperatures, of a significant increase of rotationally excited states populations. These yield in turn a number of radiative transitions at frequencies further removed from the center of

the band. Hence the spreading of molecular bands with increased temperatures, as observed for instance of Fig. B-F2 (2.7μ CO_2 band) and Fig. B-F3 (15μ CO_2 band).

The infrared literature provides procedures and tables to calculate the transmittance of any part of most molecular bands [6]. Typically it uses for molecules like H_2O, CO_2, a "random" model [21] where the optical thickness τ_ν of Eq. (26) is replaced by an equivalent thickness τ_ν^*:

$$\tau_\nu^* \equiv \frac{\int_0^s \rho \bar{k}\, ds'}{\left(1 + \dfrac{\int_0^s \rho \bar{k}\, ds'}{4\dfrac{\gamma_c}{d}}\right)^{1/2}} \tag{B-3}$$

where $\bar{k}$ is an average absorption coefficient (to be used only as part of this particular model), a function of both ν and T (see Figs. B-F2 and B-F3).
d is a characteristic spacing of lines for a particular frequency interval and temperature [6].
γ_c is an average line thickness (we considered collision broadening only); it is also a function of frequency and temperature (Eq. B-2).

Eq. (B-3) could be generalized easily [6] to include other collision mechanisms and other specie. What Eq. (B-3) illustrates is that the averaging of a group of lines into a single optical thickness τ_ν^* can be accomplished by the single expedient of the denominator of this equation. This correction factor brings about the "gappiness" of the band through the line spacing ratio γ_c/d.

B.3. A Closed Form Model for Molecular Bands

It is often convenient, even in numerical computations, to have a closed form expression of a physical property, rather than to use handbook tables directly. In this paragraph, a model of the mass absorption coefficient $k(\omega, T)$ on the 2.7μ band of CO_2 (Fig. B-F2) is derived. It aims at curve fitting closely the low frequency wing of this band with the half branches of a set of hyperbolas centered on a reference frequency $\omega_0 = 3600\ cm^{-1}$.

The fit is of the form:

$$x = -a\left[1 + \left(\frac{y}{b}\right)^2\right]^{1/2} \tag{B-4}$$

where

$$x = \ln k(\omega, T) - \ln k(\omega_0, T)$$
$$y = \omega_0 - \omega$$

For each temperature there is a different set of [a,b]. A linear regression procedure was carried out and the resulting values of a and b are given in Table B-T1. Such results can be correlated in closed form by the following expressions:

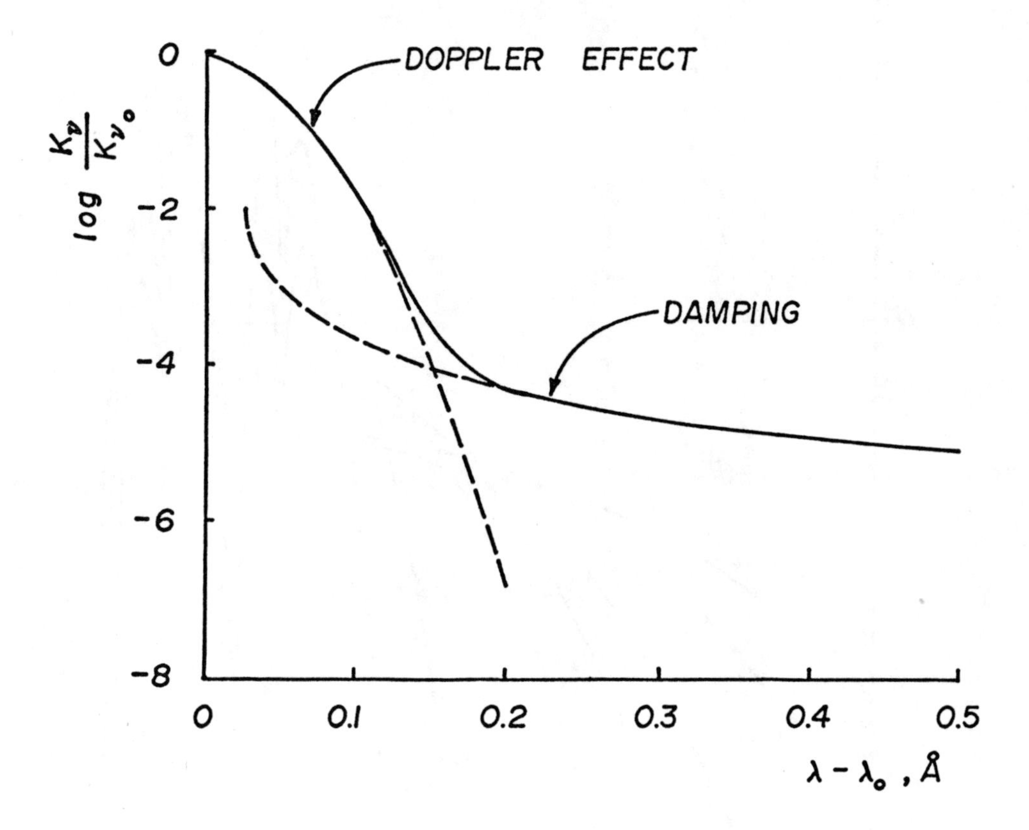

Fig. 6-F1 Line Broadening for a Classical Oscillator, After Ambartsumyan [23] (T = 5700 K, $\Delta\lambda_D$ = 0.05 Å, $\gamma/4\pi\Delta\nu_D = 3 \times 10^{-3}$)

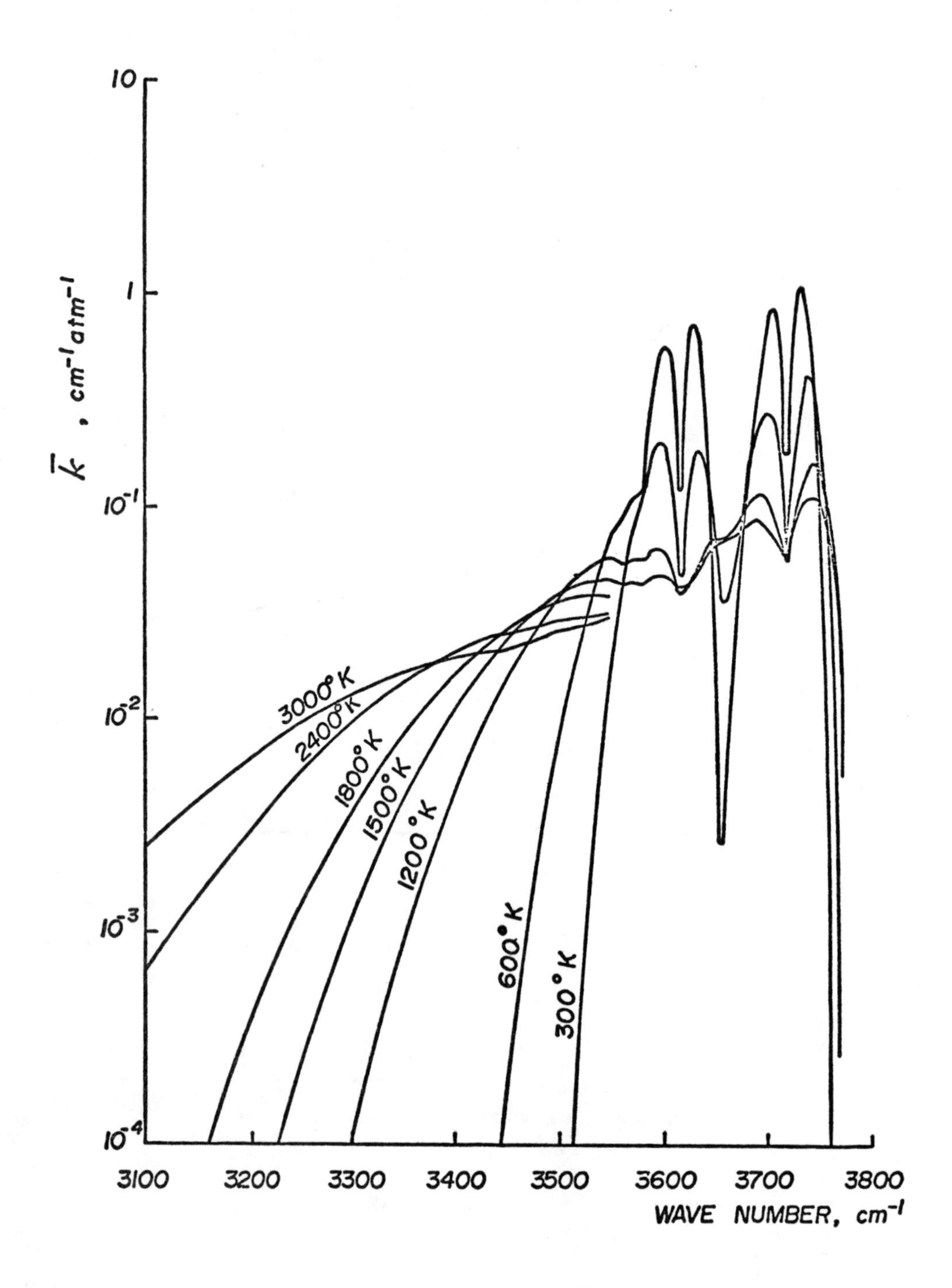

Fig. B-F2 Average Absorption Coefficient $\overline{k}$ for the 2.7μ Band of CO_2 [6]

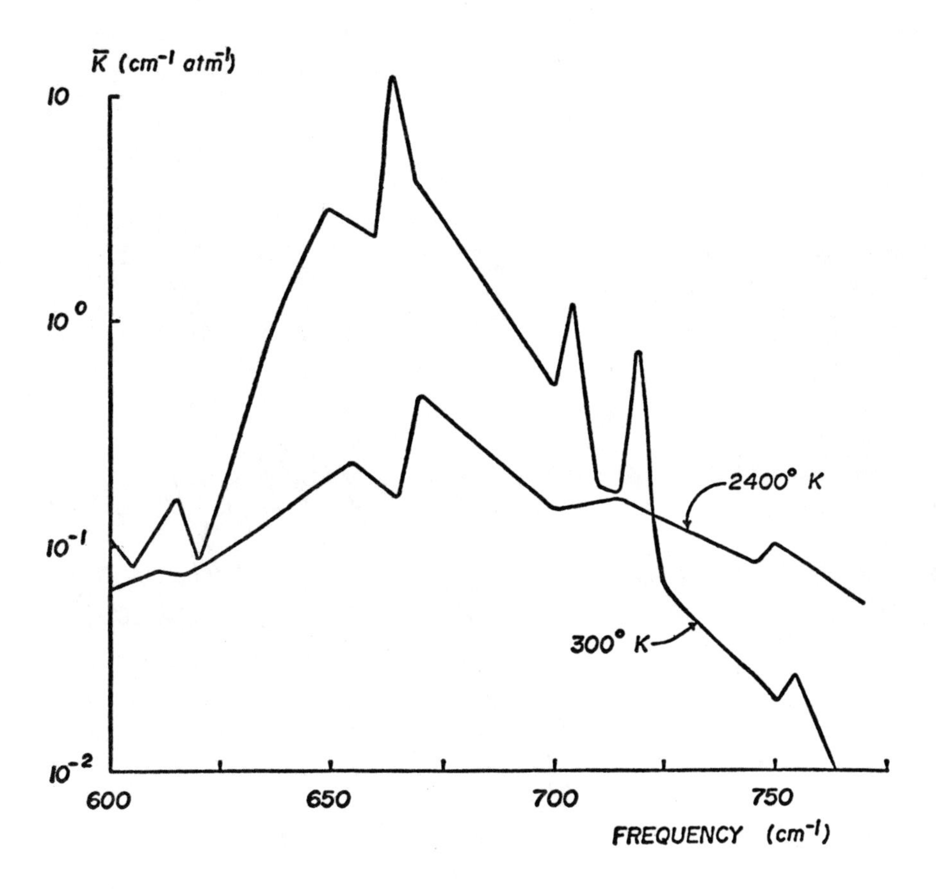

Fig. B-F3 Average Absorption Coefficient $\overline{k}$ for the 15μ Band of CO_2 [6]

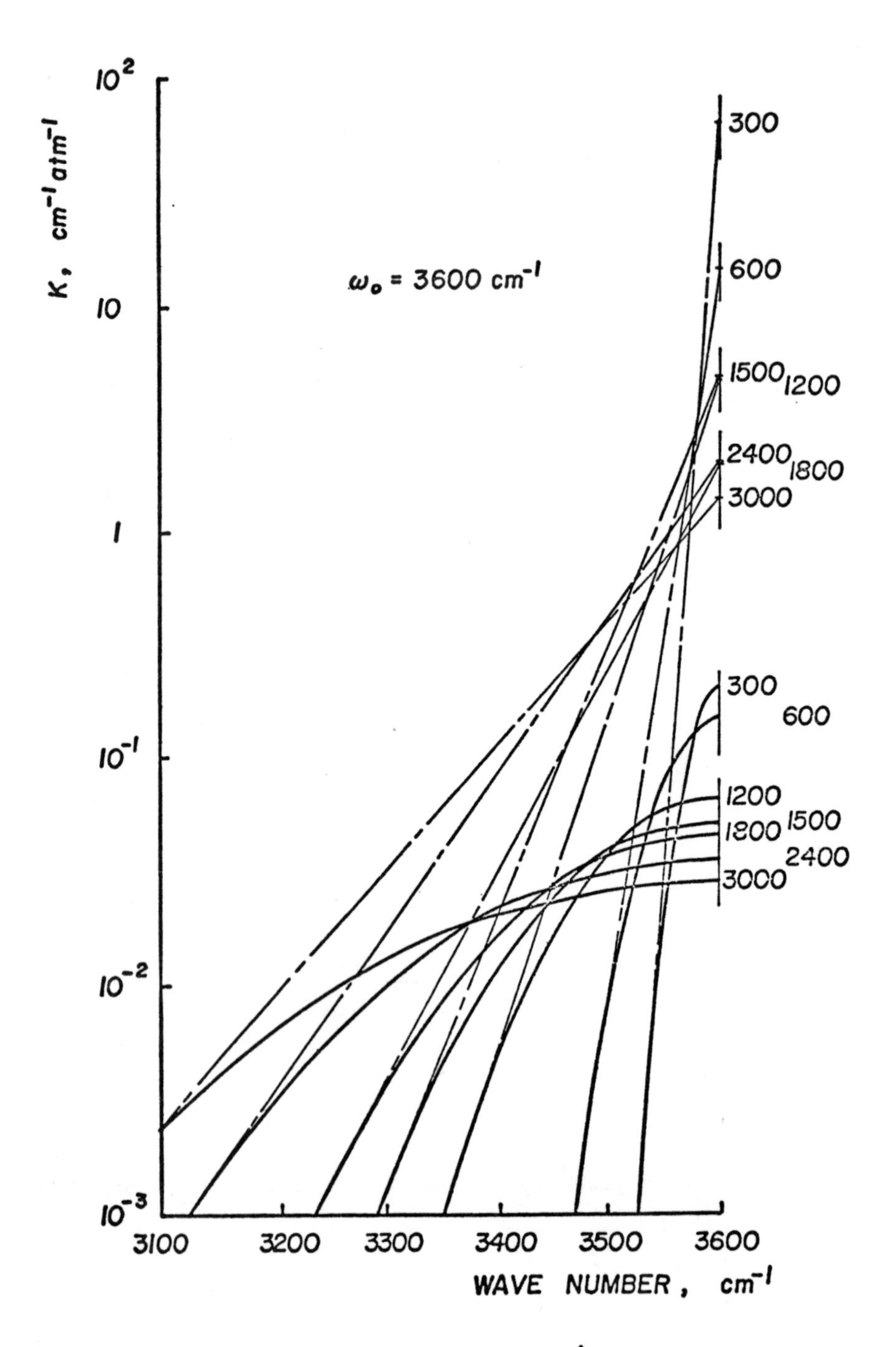

Fig. B-F4 Graphical search for a, b and $\ell n\ k(\omega_o, T)$ for Each Temperature [Eq. (B-4)]

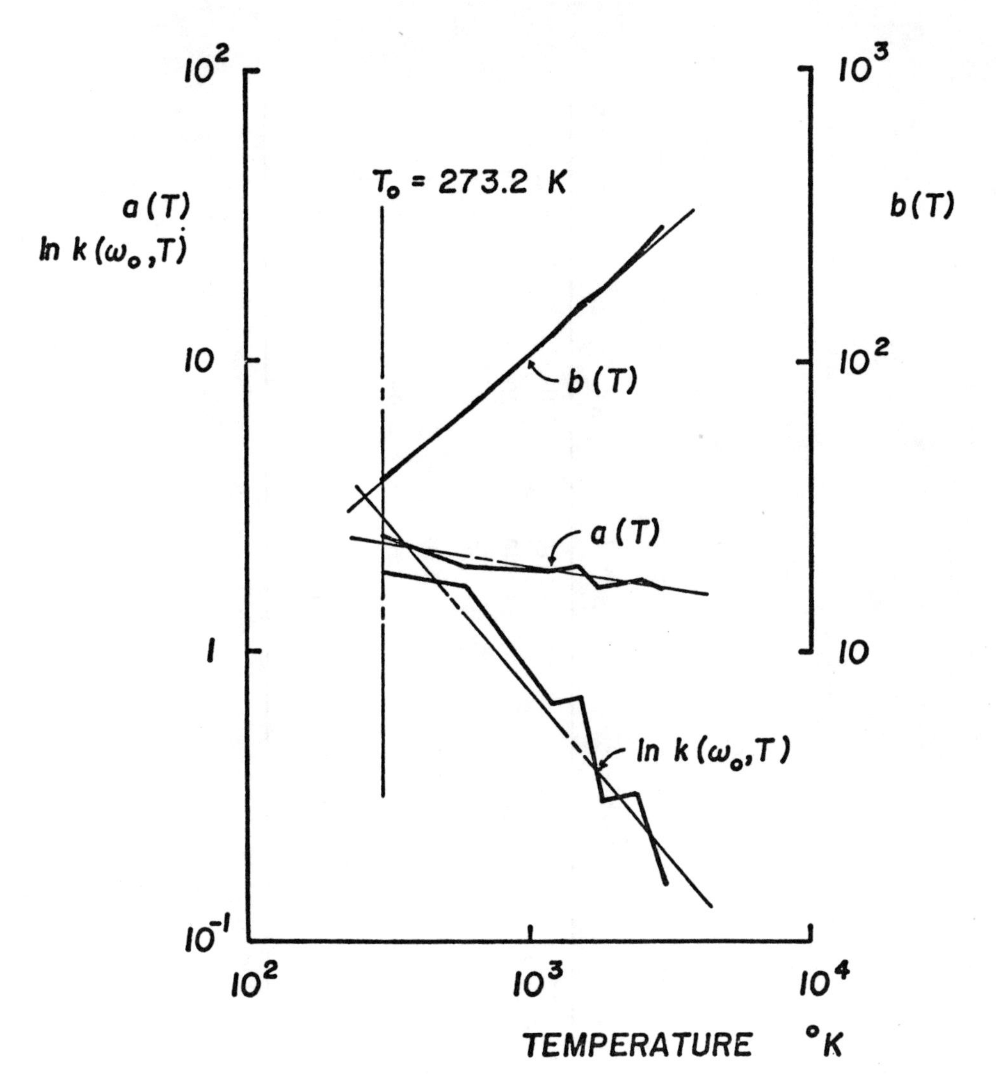

Fig. B-F5 Graphical Search for the Temperature Dependence of a, b and $\ln k(\omega_o, T)$ [Eqs. (B-5, 6 and 7)]

Table B-TI

T °K	$\ln k(\omega_o, T)$	a(T)	$\frac{b}{a} = \frac{y}{x}$	cm^{-1} b(T)
300	1.815	2.470	16.6	39.2
600	1.168	1.940	30.6	59.5
1200	0.675	1.860	65.6	122.5
1500	0.703	1.960	81.3	160.0
1800	0.308	1.630	109.2	178.5
2400	0.324	1.740	140.0	242.0
3000	0.157	1.680	173.0	292.0

$$\ell n\ k(\omega_o, T) = c_1 \left(\frac{T_o}{T_E}\right)^{n_1} \tag{B-5}$$

$$a = c_2 \left(\frac{T_o}{T}\right)^{n_2} \tag{B-6}$$

and

$$b = c_3 \left(\frac{T_o}{T}\right)^{n_3} \tag{B-7}$$

where T_o is a reference temperature. Table B-T2 gives the values of T_o, c_1, c_2, c_3, n_1, n_2, and n_3. The closed form expression obtained for k is then:

$$k(\omega, T) = \exp\left\{c_1 \left(\frac{T_o}{T}\right)^{n_1} - c_2 \left(\frac{T_o}{T}\right)^{n_2} \left[1 + \left(\frac{\omega_o - \omega}{c_3}\right)^2 \left(\frac{T_o}{T}\right)^{2n_3}\right]^{1/2}\right\} \tag{B-8}$$

Fig. B-F4 illustrates the graphic procedure used for obtaining for each temperature the parameters a, b, and $\ell n\ k(\omega_o, T)$ for the 2.7μ CO_2 band. Fig. B-F5 illustrates the linear regression procedure which yielded the temperature dependence of these coefficients (i.e. the c's and the n's). Finally Fig. B-F6 compares the original data of Fig. B-F2 with the model given by Eq. (B-8).
It is recognized, incidentally, that Eq. B-8 represents in reality the parameter $\bar{k}$ of Eq. (B-3), rather than the absorption parameter k which is used throughout the report. Only for zero "gappiness" ($d \to 0$) would the two quantities be equivalent. However, the model is representative of the frequency and temperature dependence of the absorption coefficient; thus it is a convenient tool to analyze kernel behavior. It is always possible to include in practice the accurate tabulated values of k, but at considerable computational cost.

APPENDIX C

C.1. Ground Based Atmospheric Observations

It has been sometimes commented that nonlinear techniques are not effective when used in connection with ground based observations, in contrast with their repeated successes in satellite observations [(5)(10)(11), (14)]. One simple explanation has to do with the poor kernels of such inversions.
If one assumes the same theoretical atmosphere as in paragraph 4,1., the optical thickness differs from Eq. (28) by its integral limits only:

$$\tau(\omega_j, s_i) = \rho_o k(\omega_j) \int_o^{s_i} \exp\left(-\frac{gs}{R_m T(s)}\right) ds \tag{C-1}$$

Hence the new integrated expressions for the optical thickness and the transmittance kernel matrix:

$$\tau(\omega_j, s_i) = \rho_o k(\omega_j) \frac{R_m T_o}{g} \left[1 - \exp\left(-\frac{gs_i}{R_m T_o}\right)\right] \tag{C-2}$$

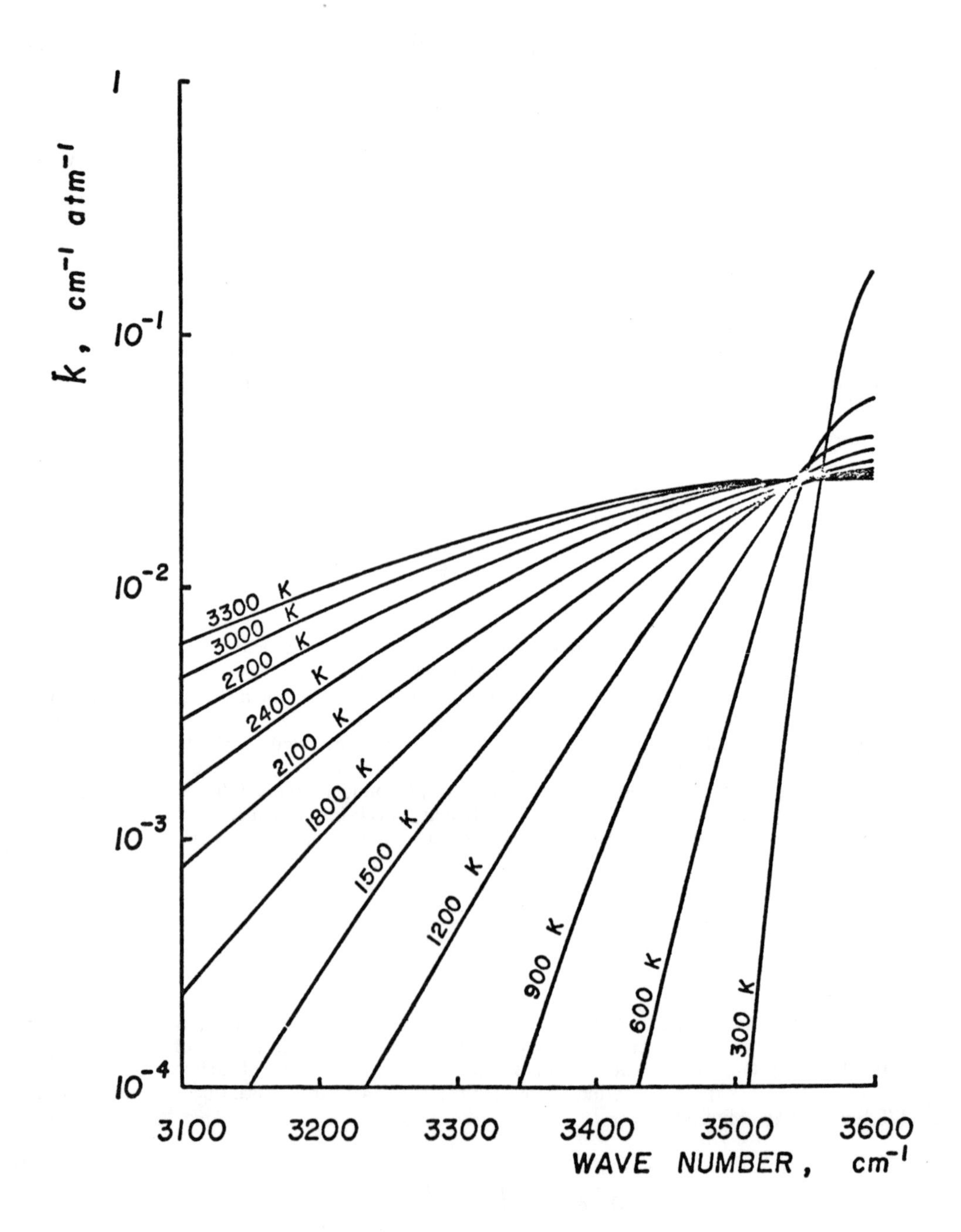

Fig. B-F6 Model of the Average Absorption Coefficient k for the 2.7μ Band of CO_2 [Eq. (B-8)]

and, from Eq. (25):

$$\bar{K}[\omega_j, s_i] = \rho_o k(\omega_j) \exp \left\{ -\frac{s_i}{H} - \rho k(\omega_j) H \left[1 - \exp\left(-\frac{s_i}{H} \right) \right] \right\} \tag{C-3}$$

where the scale height $H \equiv \dfrac{R_m T_o}{g}$.

The maximum of this kernel is obtained by setting the derivative of Eq. (C-3) equal to zero:

$$-\frac{1}{H} - \rho k \exp\left(-\frac{s_i}{H} \right) = 0 \tag{C-4}$$

However, it is clear that both terms on the left side of Eq. (C-4) are always negative. Hence the kernel $\bar{K}[\omega_j, s_i]$ has no maximum for any value of s_i or of the frequency ω_j. The conditions for a good dominate-diagonal kernel matrix are never met.

Chapter 21

STEADY AND UNSTEADY RADIANT HEAT FLUX MEASUREMENT ON THE SCREEN TUBE OF A POWER BOILER FURNACE

P. Pavlović, L. Jović. Lj. Jovanović, N. Afgan

Boris Kidrich Institute, Vinca, Beograd Yugoslavia

Abstract

This paper presents the results of the development of a resistant type radiant heat fluxmeter. It is shown that with adequate selection of material and fluxmeter dimensions a constant sensitivity is obtained. In this paper is also presented a method for the determination of static and dynamic characteristics of the fluxmeter. The developed fluxmeters have been used in a 200 MW thermal power station for the measurement of spatial distribution within the boiler furnace. Following the differences between the readings of a fixed fluxmeter welded onto screen tubes and a movable fluxmeter with a known absoptivity, the scaling problem of the ash deposit was studied.

1. Introduction

A different type of insert tube instrumented for surface temperature measurement has been extensively used recently for investigating heat transfer on boiler screen tubes. However, in many cases in order to define an undesirable temperature and flow regime within the boiler furnace, it is necessary to determine some other parameters which characterize working conditions at different points on screen tubes. In order to verify operating conditions on boiler surfaces of large power stations under different regimes, it has been necessary to develop methods for the local measurement of total heat flux (primary heat flux) and the local heat transfer coefficient. The purpose of this paper is to describe the development of radiant heat fluxmeter and the results of the measurements performed in a 200 MW power station.

For local heat flux measurement at the boiler screen tubes the following three methods are used: the method based on the enthalpy measurement /1,2,3/; the method based on the measurement of thermal resistance /2,4,5/; and the inertial method /6,7/. The first two methods are normally used for stationary measurement and can be used for continual measurement. The third method is time dependent and can be used only as a discontinuous device. It should also be mentioned that the first and third methods need a special cooling system and would be impractical for large scale measurement if used as fixed instruments.

The determination of radiant flame characteristics and the emissivity of boiler screen tubes are certainly special problems. However, in our case, we have focused attention on the investigation of heat transfer on surfaces under the assumption that emissivity of the surface is known or will be determined by an adequate method and that the spectral characteristics of the flame will be constant during the measurement. Since our aim was to concern ourselves with the processes at the surfaces, these assumptions have been made.

The fluxmeter which we have developed and used for the determination of radiative heat flux distribution at a boiler screen tube is based on the thermal resistence of a circular disc. One side of the disc is exposed to thermal radiation and the other side is isolated by a gas gap.

The heat received with the disc is radially conducted to the periphery of the disc which is welded onto the ring-form of the main body of the fluxmeter. The main body of the fluxmeter is welded onto the screen tube so that its temperature corresponds to the screen tube surface temperature (Fig. 1). The obtained temperature

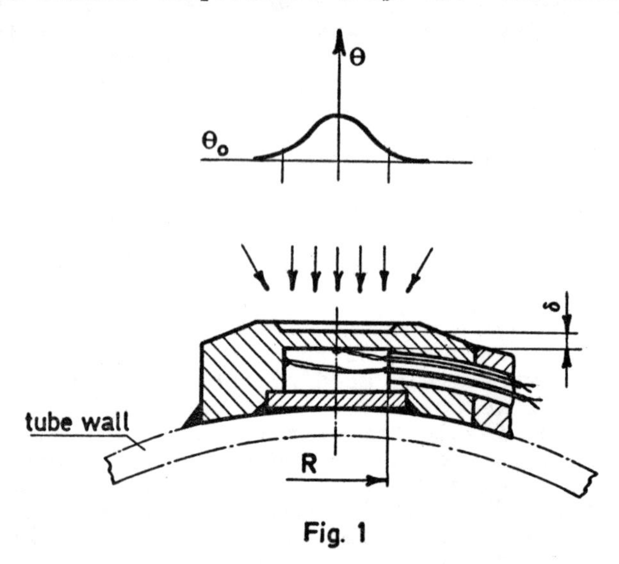

Fig. 1

difference between the center of the disc and its periphery is proportional to the heat flux at the surface of the fluxmeter disc. With adequate selection of disc material and its dimensions it is possible to obtain the desired sensitivity and dynamic characteristics of the fluxmeter. The simplicity of the design of this type of fluxmeter makes it very attractive for stationary and non-stationary measurement within the boiler. The temperature measurement on the fluxmeter can be used also as an indication of the outside surface temperature of the screen tubes. It can be expected that the emissivity of the disc surface will be changed during the operation due to the scaling of ashes on the screen tube surface. If the change in reading of the fluxmeter during the same period of time is also followed for other relevant parameters of the boiler, this type of measurement can be used also as an indication of scaling deposits on the surface. As will be shown in this paper, this is one of the possible methods for studying the problem of deposits and their effect on the heat transfer performance of the boiler

2. Static Characteristic of the Fluxmeter

If it is assumed that convective and radiative heat transfer from the inner surface of the disc can be neglected and that conduction losses through the thermocouple wire are negligible and also that the heat sink (main part of the fluxmeter) is at a constant temperature, the relation between the primary heat flux and temperature difference between the center and periphery of the disc can be obtained by solving the nonlinear differential equation for heat conduction in a circular disc.

The heat conduction differential equation for a circular disc is

$$\nabla^2 V = \frac{\partial^2 V}{\partial r^2} + \frac{1}{r}\frac{\partial V}{\partial r} + \frac{\partial V}{\partial z^2} = 0 \tag{1}$$

where

$$V = \frac{1}{\lambda_o}\int_{\theta_o}^{\theta} \lambda(\theta)d\theta = \int_o \frac{\lambda(\vartheta)}{\lambda_o}\, d\vartheta \tag{1a}$$

The boundary condition for this case can be written as follows:

$$\lambda_o \frac{\partial V}{\partial z} = q \qquad (z = 0, 0 \leqslant r < R) \tag{2}$$

$$\lambda_o \frac{\partial V}{\partial r} = HV \quad (r = R) \tag{3}$$

$$\lambda_o \frac{\partial V}{\partial z} = 0 \qquad (z = \delta, 0 \leqslant r < R) \tag{4}$$

The general solution for the stationary temperature field in the disc is:

$$V(r, z) = \frac{2R^2}{4\lambda_o\delta}\sum_{n=1}^{\infty} \frac{8 I_o\ \beta_n \frac{r}{R}\ I_1(\beta_n)\beta_n \frac{\delta}{R} \mathrm{ch}\,\frac{\delta - z}{R}\beta_n}{\beta_n^3[I_o^2(\beta_n) + I_1^2(\beta_n)]\mathrm{sh}\beta_n \frac{\delta}{R}}$$

If it is assumed that the linear dependence between the conductivity of the material and temperature is in the form:

$$\lambda(\vartheta) = \lambda_o(1 + \alpha\vartheta)$$

After integration, it can be obtained as

$$V = \vartheta\left(1 + \frac{\alpha}{2}\vartheta\right)$$

In this case electromotive forces of the thermocouples, which measure temperature difference ϑ, can be expressed in the following form:

$$E = A\vartheta\left(1 + \frac{\alpha'}{2}\vartheta\right)$$

With this in mind, the expression for sensitivity of the fluxmeter has the following form:

$$K_o = \frac{E_m}{2} \approx \frac{AR^2}{4\lambda_o\delta}\left(1 + \frac{\alpha' - \alpha}{2}\vartheta_m\right)\sum_{n=1}^{\infty} \frac{8\beta_n \frac{\delta}{R} I_1(\beta_n)/\mathrm{sh}\beta_n \frac{\delta}{R}}{\beta_n^3[I_o^2(\beta_n) + I_1^2(\beta_n)]} \tag{7}$$

If the maximum temperature difference is not large and if α' and α are approximately equal, the term in brackets can be neglected, so that the expression for sensitivity has a constant value for the chosen geometry of the disc:

$$K_o = \frac{AR^2}{4\lambda_o\delta}\sum_{n=1}^{\infty} \frac{8\beta_n \frac{\delta}{R} I_1(\beta_n)/\mathrm{sh}\beta_n \frac{\delta}{R}}{\beta_n^3[I_o^2(\beta_n) + I_1^2(\beta_n)]} \tag{8}$$

As is obvious from the expression (8), the sensitivity of the fluxmeter can be increased by increasing the radius of the disc or by decreasing conductivity of the disc material. A fluxmeter with a large diameter will have a large temperature gradient along the radius. This might cause an increase in error due to heat losses from the disc. This will happen also if disc material of low conductivity is selected. Thus, in order to obtain a good resolution of the measurement and at the same time prevent the secondary effects which will affect the accuracy of the measured heat flux, it is necessary to theoretically and experimentally optimize parameters with the aim of satisfying the working conditions and design requirements.

For our case, we have made an extensive analysis of the parameters which affect selection of the dimensions and material of the fluxmeter and which correspond to temperature conditions in the boiler with special consideration of the expected heat flux, type of boiler furnace and fuel used in the power station. From this analysis /7/ stainless steel - 25% Cr, 20% Ni has been selected as the material for the fluxmeter. The diameter of the disc is D = 8 mm, thickness δ = 1 mm, and other dimensions are adjusted in accordance with the design requirements.

2.1 Experimental determination of the statis characteristic of the fluxmeter

Determination of the static characteristic of the fluxmeter is performed on the apparatus shown in Fig. 2. The radiation source was

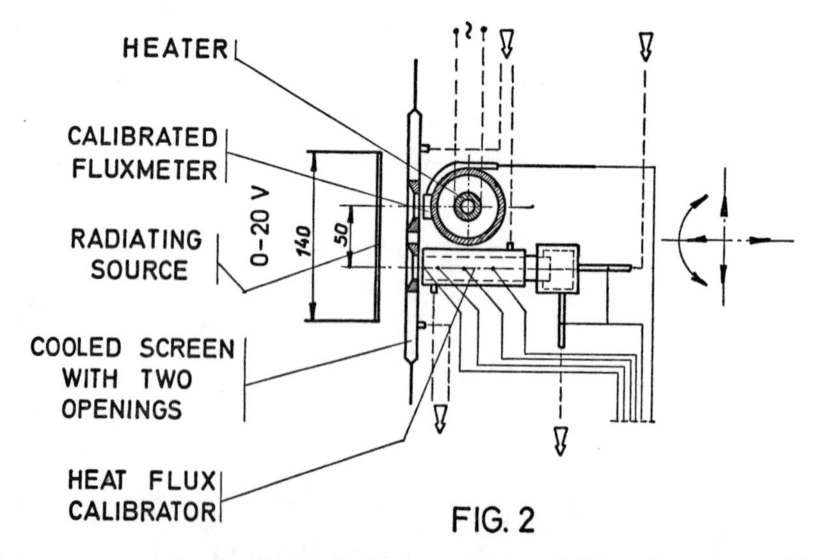

FIG. 2

a stainless steel electrical heater 300 x 140 x 1 cm supplied by D.C. current. The surface temperature of the heater was up to 1100°C. Calibration of the fluxmeter was made by comparing the temperature difference obtained at the fluxmeter disc with the reading of one caloric type fluxmeter under the same conditions. The referent fluxmeter was based on the thermal resistance measurement along copper specimen. A parallel measurement was made of the enthalpy difference of cooling water. The front side of the referent fluxmeter was exposed to the radiation of the heater and the back side of the copper specimen was cooled by water. From the obtained temperature gradient the heat flux eas calculated at the referent fluxmeter. In order to control this measurement, the enthalpy difference of cooling water was measured and compared with the heat flux obtained from temperature gradient measurement.

In the design of referent fluxmeter special attention was given to the radial heat losses from the copper specimen by a proper thermal shield.

In order to obtain the same emissivity on the front surface exposed to the radiation, both fluxmeters, referent and calibrated, were coated with a "pyromark" color. Special consideration was given to the surface temperature differences between the referent and calibrated fluxmeter by measuring the surface temperature on both fluxmeters and controlling them by adequate heating and cooling system.

The static characteristic of the referent and calibrated fluxmeters is shown in Fig. 3. It can be seen that the latter is in good agreement with the linear relation which corresponds to the sensitivity of the fluxmeter

$$K_o = E/2 = 1.22 \times 10^{-4}\ \mathrm{mV/w/m^2}$$

and it is lower than the theoretical sensitivity which should be

$$K_o = 1.60 \times 10^{-4}\ \mathrm{mV/w/m^2}$$

In the same figure is presented the theoretical relation which corresponds to the case when measured temperature difference is

$$\theta_m(0,\delta) - \theta(R,\delta)$$

and also corresponds to the sensitivity S = 0.83.

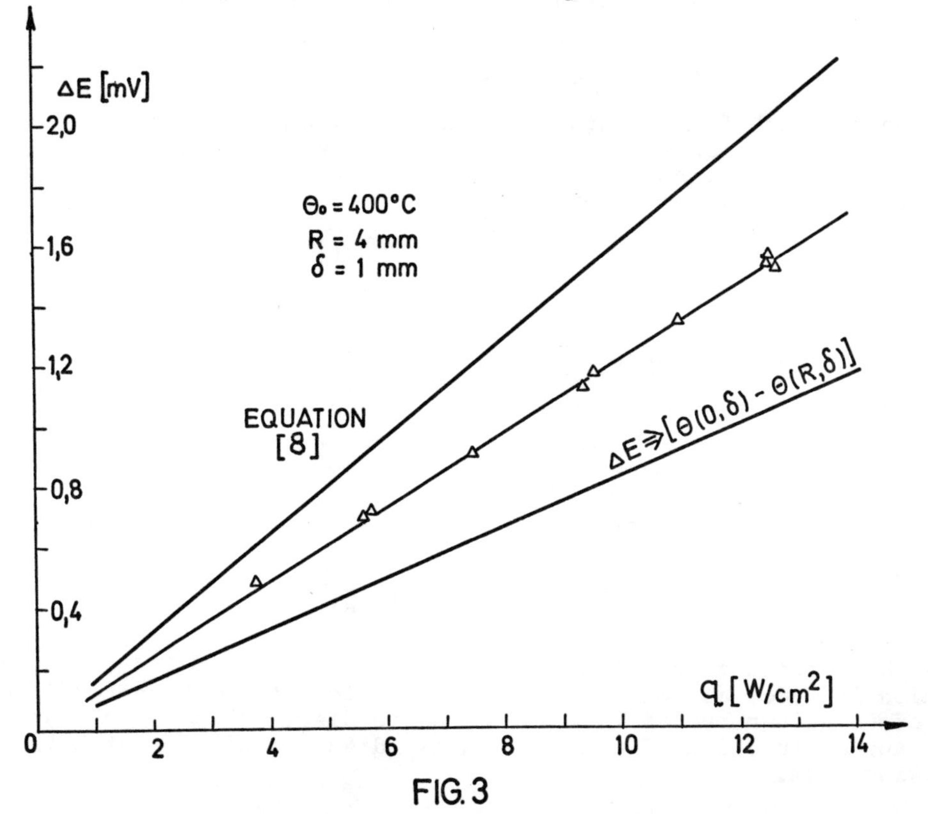

FIG. 3

This difference between the theoretical and experimental results is due to the geometrical and temperature effects which have not been taken into consideration in this analysis. It should be mentioned that besides the temperature effect on the conductivity of disc material /15/ there is also some influence resulting from technical uncertainly which is very difficult to control in the production of the fluxmeter. The accuracy of the applied calibrated method is within the limits ± 5%.

3. Dynamic Characteristic of the Fluxmeter

Non-stationary heat conduction in a metallic disc, neglecting axial heat conduction, can be described by the following differential equation

$$\delta \frac{\partial}{\partial r}\left(\lambda(\theta)\frac{\partial \theta}{\partial r} + \frac{\delta}{r}\lambda\theta\frac{\partial \theta}{\partial r}\right) = \rho c \frac{\partial \theta}{\partial t} \tag{9}$$

In order to make an adequate model which corresponds to the disc type fluxmeter, the following boundary and initial conditions are accepted:

Boundary conditions

$$\theta = \theta_o \quad \text{at} \quad r = R \quad \text{for} \quad t > 0$$

$$\frac{\partial \theta}{\partial r} = 0 \quad \text{at} \quad r = 0 \quad \text{for} \quad t > 0$$

Initial conditions

$$\theta = \theta_o \quad \text{for} \quad t = 0$$

$$q = 0 \quad \text{for} \quad t \leqslant 0 \tag{10}$$

$$q = q(t) \quad \text{for} \quad t \geqslant 0$$

By using the Kirhof temperature function (1a) and applying Laplace transformation to equation (9) with the elementary transformation, the following solution, which satisfied the mentioned boundary and initial conditions, can be obtained

$$\bar{V}(r,s) = \frac{A\bar{q}(s)}{\delta\lambda_o(s/k)}\left[1 - \frac{I_o(\sqrt{s/k}\, r)}{I_o(\sqrt{s/k}\, R)}\right] \tag{11}$$

The transfer function of the fluxmeter is defined as the ratio between the temperature difference at the center and periphery of the disc and energy absorbed at the front surface of the fluxmeter.

$$W(s) = \frac{\bar{\theta}(o,s) - \bar{\theta}_o}{\bar{q}(s)} = \frac{(\lambda_o/\lambda_m)\bar{V}(o,s)}{\bar{q}(s)}$$

Using (1a), it can be written

$$W(s) = \frac{A}{\delta\lambda_m(s/k)}\left[1 - \frac{1}{I_o(\sqrt{s/k}\, R)}\right] \tag{12}$$

It can be shown that the linear approximation of the modified Bessel function includes all physical phenomena relevant to the determination of transient behavior of the metallic disc. This means that the transfer function of a thin metallic disc can be presented in the following form:

$$W(s) = \frac{K_o}{1 + Ts} \tag{13}$$

where

$$K_o = \frac{AR^2}{4\delta\lambda_m} \quad \text{- sensitivity}$$

$$T = \frac{R^2}{4K} \quad \text{- time constant}$$

It follows that in our case this transfer function should describe the dynamic behavior of the fluxmeter as an initial block of the first order.

3.1 Experimental determination of the fluxmeter transfer function

The experimental determination of the fluxmeter transfer function was performed by the measurement response signal of temperature differences $\theta(0,t) - \theta_o$ to the incident impulse of radiative heat flux. The impulse rubin laser was used as the source for the radiative heat flux. The investigated fluxmeter was placed in a furnace whose temperature could be controlled. The outlet energy of the laser impulse was about 6.5 and 13 Joules in the time interval 2.10^{-3} sec. The diameter of the laser beam, about 8 mm, was focused on the center of the fluxmeter. In this way the normal working regime of the fluxmeter was simulated. The temperature difference signal was amplified about 1300 times and fed to the storage oscilloscope.

The measurements were made at several different temperatures (184°C, 324°C, 417°C, 484°C, and 500°C). In Fig. 4 is shown the typical response signal, obtained at ambient temperature θ_o = 500°C with incident laser energy of 13 Joules.

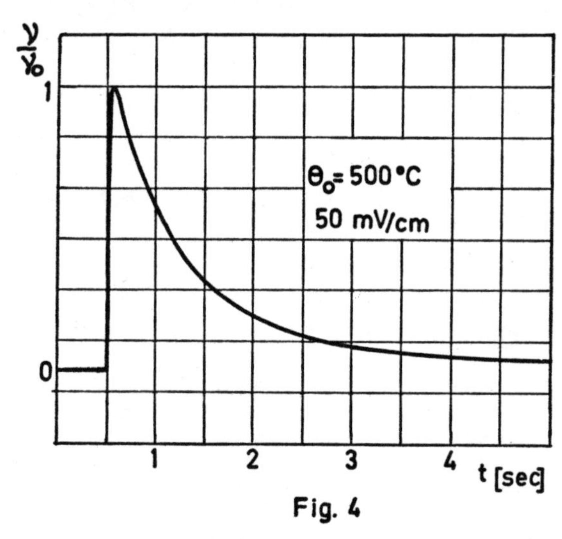

Fig. 4

The obtained signal was analized as a response to the Dirac δ function and characteristic parameters were determined. By the complex transformation of the obtained results, an experimental transfer function in good agreement with expression (13) was obtained.

In Fig. 5 are given the amplitude-frequency and phase-frequency characteristics of a fluxmeter for an ambient temperature θ_o = 200°C, 400°C and 700°C. It can be seen that the frequency of the fluxmeter is between 0.15 - 0.2 Hz depending on the ambient temperature θ_o.

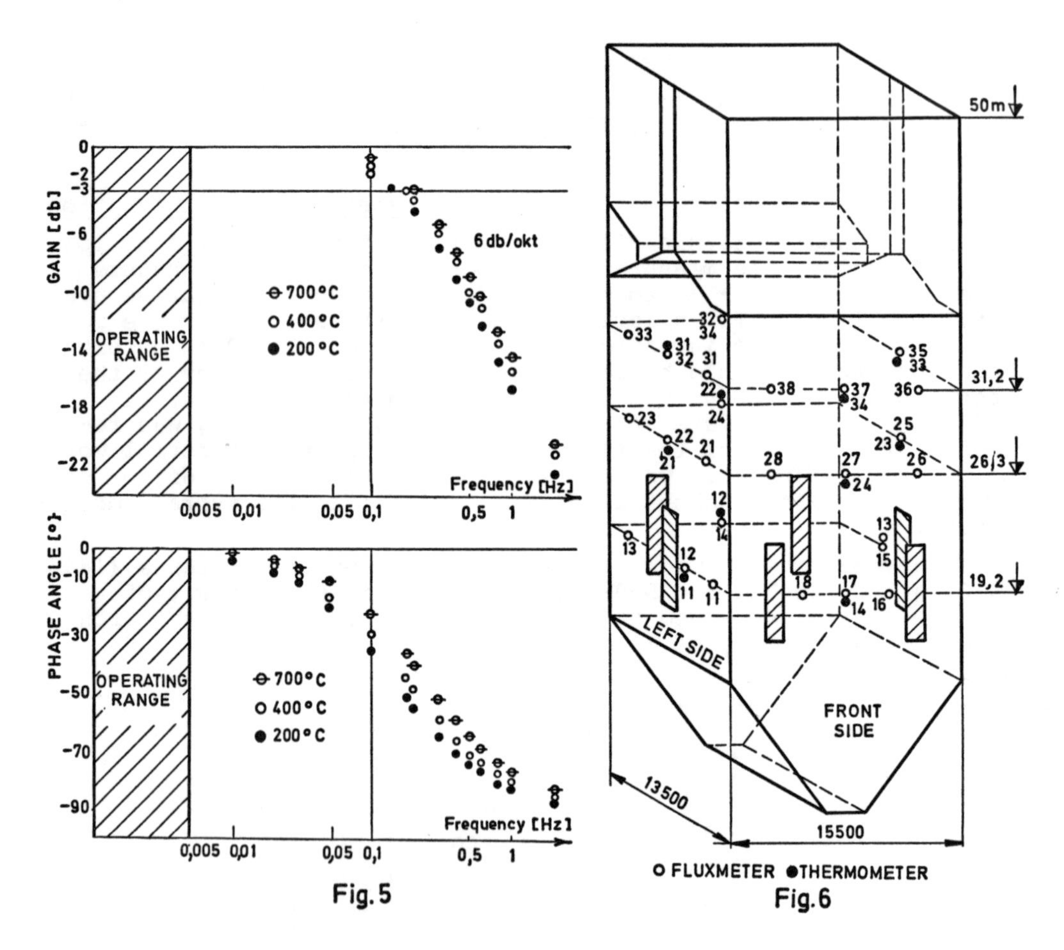

Fig. 5

Fig. 6

4. Experimental Results

4.1 Stationary measurement

The developed radiant heat fluxmeter has been used for the measurement of total radiant heat flux on the boiler screen tube in a 210 MW power station. The investigated boiler furnace is designed with six pulverised coal fuel burners. During this investigation liquid fuel with combustion heat of 6300 kj/kg and up to 70% moisture and mineral water (noncombustible ingredients) was used. The fluxmeters and thermocouples were fixed on three levels on the furnace (Fig. 6). The front and left sides of the furnace surface were equipped with nine fluxmeters on each side in order to obtain special temperature and flux distribution. Measurement was made at the different regimes of the power station. In Figs. 7a and 7b distribution of radiant heat flux on the screen tube and surface temperature are presented. These measurements correspond to the two different steam production and other constant parameters. In order to control the measured

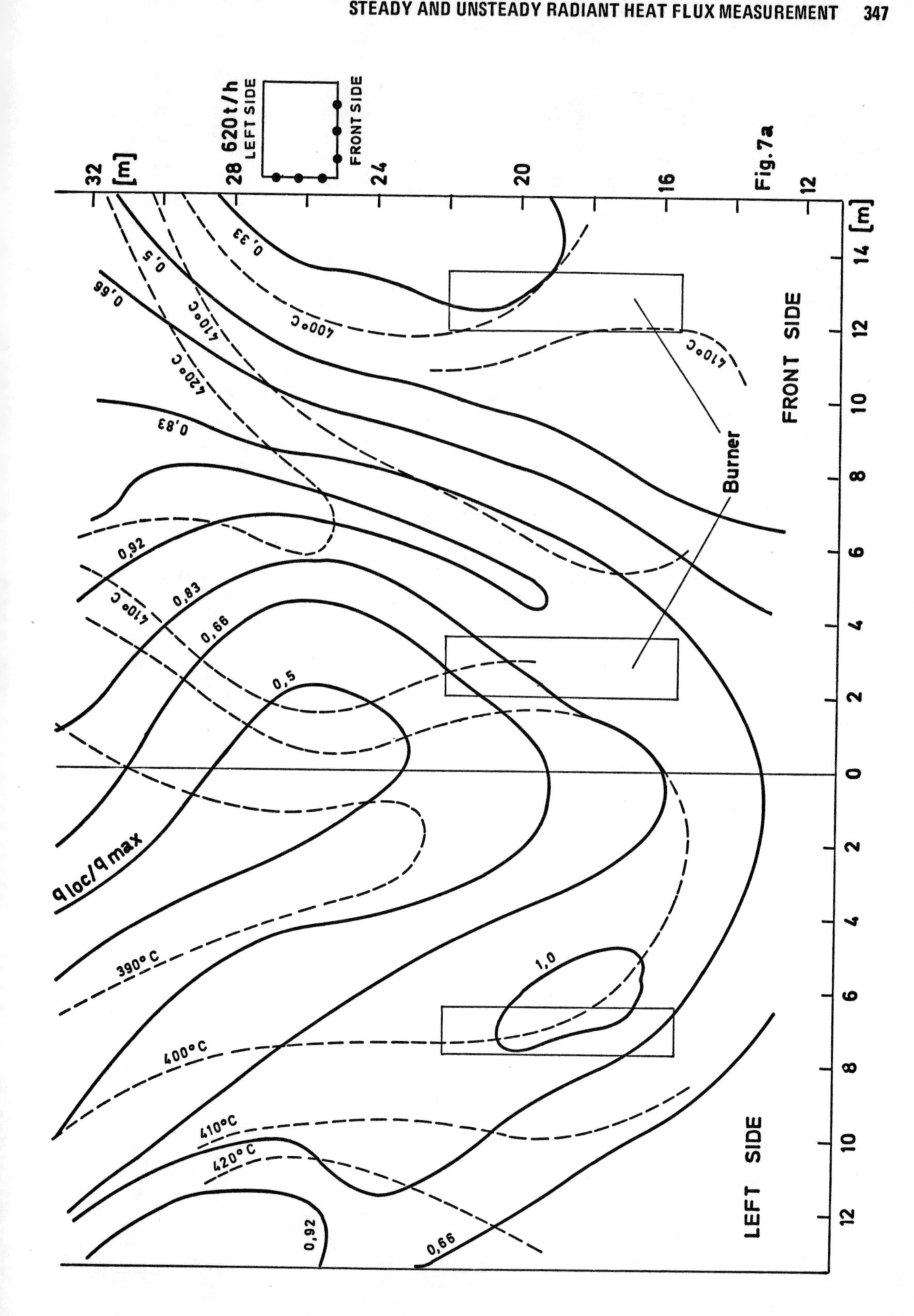
620 t/h
LEFT SIDE
FRONT SIDE
Burner
q loc/q max
FRONT SIDE
LEFT SIDE
[m]
[m]
0,33
0,5
0,66
0,83
0,92
1,0
390° C
400°C
410°C
420°C

Fig. 7a

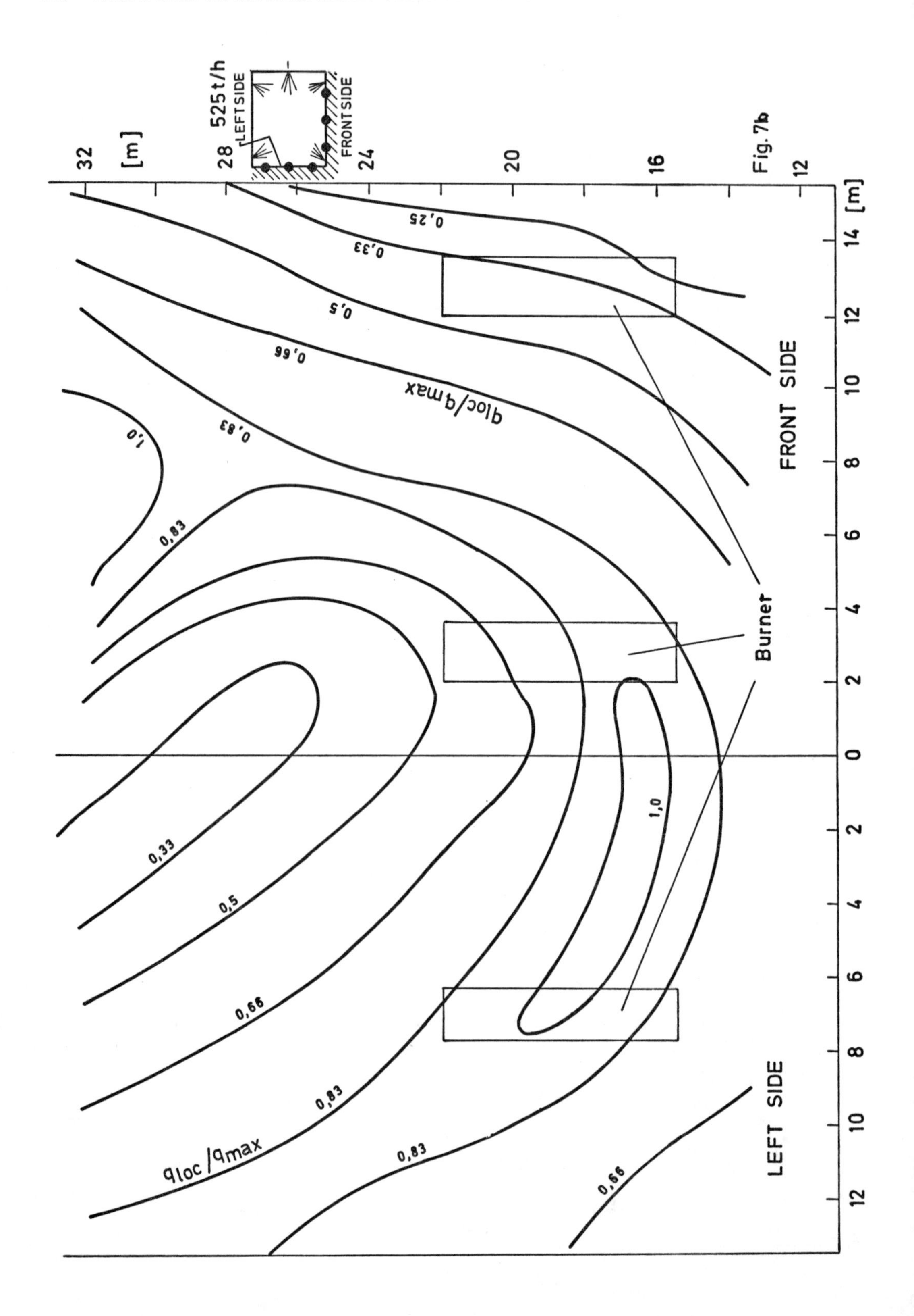
525 t/h
LEFT SIDE
FRONT SIDE
[m]
32
28
24
20
16
12
14 [m]
12
10
8
6
4
2
0
2
4
6
8
10
12
0,25
0,33
0,5
0,66
0,83
1,0
q_{loc}/q_{max}
0,83
0,33
0,5
0,66
0,83
1,0
q_{loc}/q_{max}
0,83
0,66
Burner
FRONT SIDE
LEFT SIDE

Fig. 7b

radiant heat flux, a movable fluxmeter with a known absorptivity of the reserving surface was used, and measurement was made through the existing openings at the furnace walls. By comparing the heat flux measured by the fixed fluxmeter and the heat flux obtained by movable sonde, good agreement was observed. From these comparisons it was found that there exists an effect of the surface deposits on the ratio between these two measurements. In Figure 8 the time dependence of the ratio between the reading of the fixed fluxmeter and the referent fluxmeter is shown. The lines in Fig. 8 correspond to the four different fluxmeters placed at a different position on the surface of the boiler furnace. It can be seen that in a few days a thin layer of dust diminishes absoptivity of fluxmeters F_{21} and F_{25} and after that it remains constant. The fluxmeters F_{24} and F_{27} have shown a steady decrease in the ratio $\frac{q_{loc}}{q_{ref}}$ as a result of deposit changes during the operation of the power station.

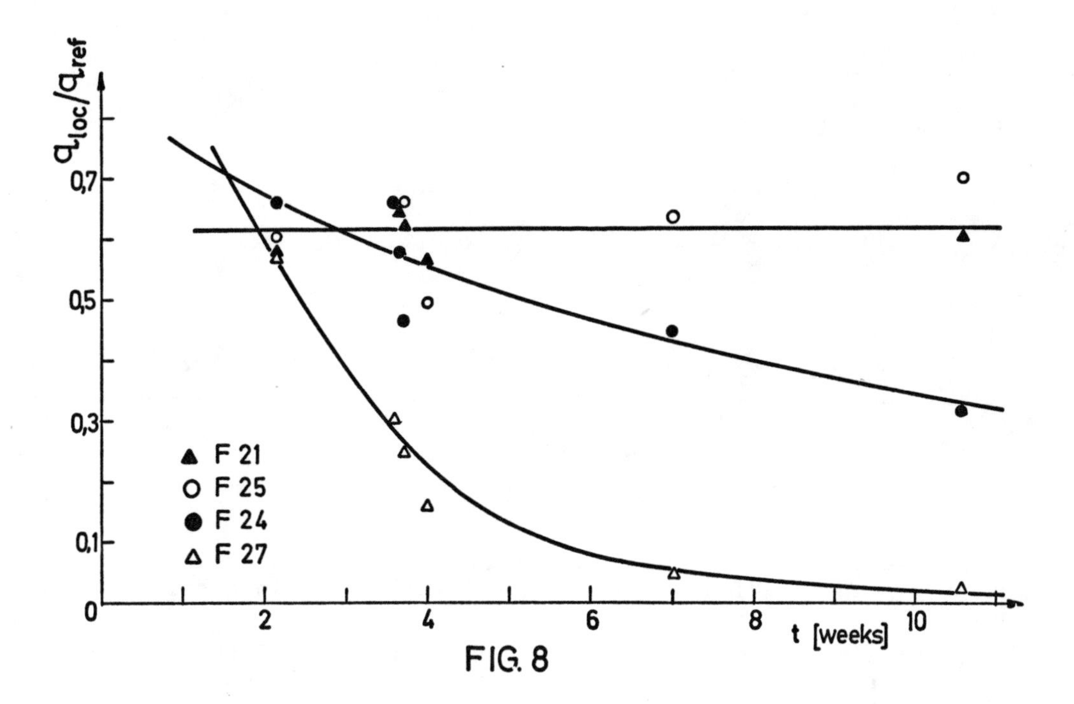

FIG. 8

4.2 Nonstationary measurement

In order to verify the nonstationary measurement of the fluxmeter, several typical nonstationary regimes have been investigated. In Figure 9 is given the response signal from one fluxmeter during the change of the flame termperature by switching off one coal mill and switching on another mill. The period of time for this operation is approximately 20 min. From the diagram it can be seen that the fluxmeter, after 2 min (137 sec) from the start of intervention by the mill, begins to register the change in a configuration of the flame. Essentially, this operation with the mills consists of the following steps: partly decreasing the capacity of one mill, turning on the second mill at full power, and switching off the first mill. In Fig. 9 these operations are identified. In the period a-b occurs the effect of decreasing capacity of the first mill. During this period the values of the heat flux decrease due to the change of the flame configureation. In period b-c, predominate is the effect of the mill which is starting operation, which in turn affects the increase of the heat flux. During period c-d the first mill ceases operation, so that the value of the fluxmeter reading is adjusted to the new stationary state. The characteristic frequency for the period a-b is $f_1 = 0.0048\ H_2$; for c-d it is $f_2 = 0.0042\ H_2$. When the limiting frequency of the fluxmeter is taken into account, it is obvious that the fluxmeter does not affect the measured heat flux in the transient investigation of the boiler screen tube.

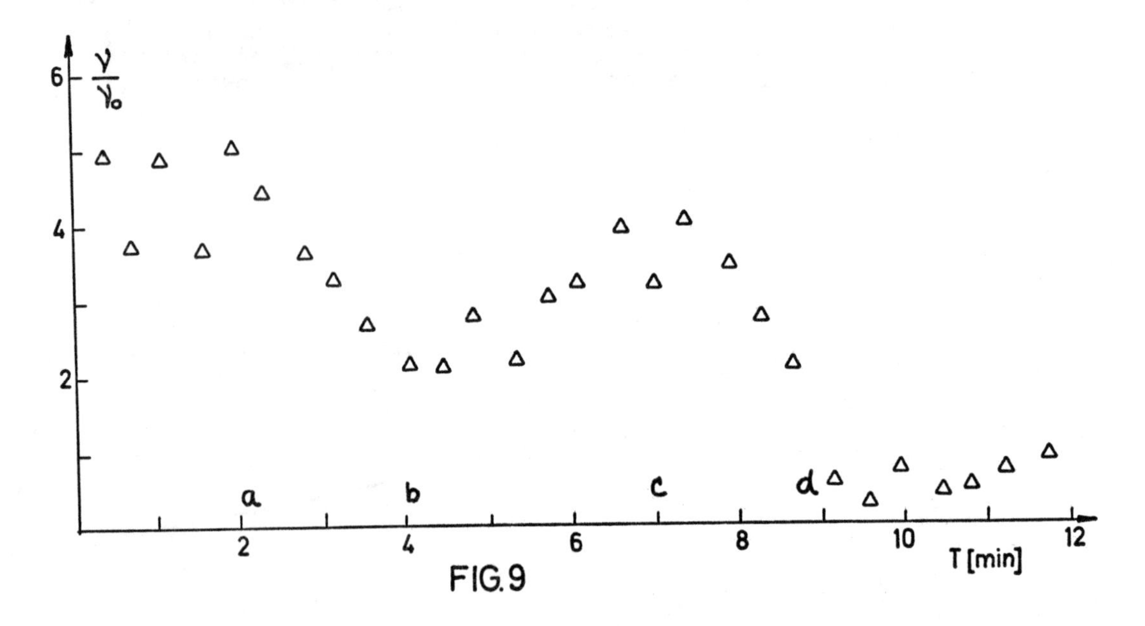

FIG.9

5. Conclusion

1. The developed resistant type radiant heat fluxmeter is a good instrument for studying the heat flux distribution in a boiler furnace.

2. Calibration methods used for the determination of the static and dynamic characteristics of the fluxmeter have shown reproducible results wiht sufficient accuracy.

3. The measurements performed in the boiler furnace of the power station have shown that the applied method of heat flux measurement gives repreducible results which can be used in analysis of temperature and heat flux distribution on boiler furnace surfaces.

4. It is shown that by comparing the fixed type fluxmeter measurements with measurements of a fluxmeter of known absoptivity, the effect of ash deposits on the heat transfer surface in the power atation boiler can be studied.

Nomenclature

θ_o	- temperature of the main body of the fluxmeter /^oC/
θ (r,z)	- temperature of the disc /^oC/
θ_m	- max. temperature of the disc
δ	- thickness of the receiving disc /cm/
λ	- thermal conductivity /$\frac{W}{cm^oC}$ /
λ_m	- mean value of thermal conductivity
H	- heat transfer coefficient /$\frac{W}{cm^2\ ^oC}$ /
q	- absorbed heat flux /$\frac{W}{cm^2}$ /
Jn(B)	- Bessel function with index n and argument B
R	- radius of the disc /cm/
$\nu_m=\theta_m-\theta_o$	- max. temperature difference
s	- complex variable
K_o	- sensitivity
k	- thermal diffusivity /$\frac{cm^2}{s}$ /

References

/1/ Sander H. Messungen von Feurerraum, "Heizflächenbelastungen an einem Flossenwandkessel," Energie, 1966, 18, 8.

/2/ Krasina, E.S. and others, "Primenenie Termozonda dlja issledovanija teploobmena v topkah parovyh kotlev." Teploenergetika, 1965, 2.

/3/ Meljnikov, V.K., "Isledovanje psiborov dlja izmerenija lučistyh teplovyh potokov v topvah," Teploenergetika, 1963, 7.

/4/ Semenovker, I.E., Gendelev, V.G., "Radiometričeskaja vstavka v ekrannye truby," Teploenergetika, 1970, 4.

/5/ Wall, T.F., "The Measurement and Prediction of Thermal Radiation from the Flame Zone of Industrial Tangentially Fired Pulverised Coal Furnaces," Ph.D. Thesis, Department of Chemical Engineering, University of Newcastle, Newcastle, Australia, 1971.

/6/ Vnukov, A.K., "Električeskij radiometr. - pribor dlja opredelenija teplovogo potoko lučistoj energii. - Teploenergetika, 1958, 8.

/7/ N. Afgan, Lj. Jovanović, P. Pavlović, V. Pisler, V. Jović, L. Jović, M. Studović, B. Arsić, S. Zarić, "Investigation of the Ash Deposits in Power Station Boilers," internal report (in Serbian).

/8/ Ševelev, V.M., Markov, B.L., "Izmerenie lučistyh karakteristik plameni radiometrom bez kondensirujuščih sredstv," IFŽ, 1965, 1.

/9/ Pikašov, V.S. and others, "Uzkougol´nyj radiometr-zond polnogo izlučenija dlja opedeljenija lučistyh karakteristik plameni," Teplofizika vyosokih temperatur, 1969, 2.

/10/ Mitor, V.V. "Teploobmen v topkah porovyh kotlov," MAŠGIZ M-L, 1963.

/11/ Gardon, R.A. "Transducer for the Measurement of Intense Thermal Radiation," Rev. Sci. Instr., 1953, 24,5, 366.

/12/ Moumoui, A., "Analyse des particularites du fouctionnement des radiometeres thermoelectriques a disque recepteur absorbant en regime permanent," Int. J. Heat Mass Transfer, 1967, 10.

/13/ Pavlović, P., "Fluxmetar za merenje lokalnog toplotnog optere-ćenja," XVI Jug. Conf. ETAN, Velenje, 1972.
/14/ Carslaw, H.S., Jaeger, J.C., "Conduction of Heat in Solids," Russian translation, Nauka, Moscow, 1964.
/15/ Hogan, C.L., Sawyer, R.B., "Thermal Conductivity of Metals at High Temperatures," J. Appl. Physics, 1952, 2.

Chapter 22

TEMPERATURE FIELD MEASUREMENT IN FLAMES BY EXTERNAL MEANS

N. N. Kondic

Westinghouse Electric Corporation, Tampa, Florida, U.S.A.

Abstract

A method is developed to measure the density- and temperature field in fluids or suspensions along a light path, bent in compliance with Fermat's Principle. With the light source and detector(s) located external to the system, the method is particularly advantageous in aggressive, hot gases and flames, where the ultimate goal is often to find the temperature field, which is interrelated with the fluid's density and refraction index.

NOMENCLATURE

a, b : coefficients defined below Eq. (15)
B, B_T: constants from Gladstone's law, Eq. (12): B in (m^3/kg), B_T in (deg K)
C : integration constant, defined by Eq. (8), dimensionless
D : distinction coefficient, defined by Eq. (16), dimensionless
F, f : general symbols for functions
k : temperature distribution coefficient, defined by Eq. (13), dimensionless
L : (= y_{max}), maximal width of the system, in (m)
m : refraction index of the fluid in the system, dimensionless
n : normale, the axis perpendicular to axis s
o : subscript pertaining to values and properties at coordinate y = 0 = r
p : fluid pressure, in (Bars)
q : (=z/L), dimensionless ordinate of the light path
R : (=ds/dn), curvature radius of the light path, in (m)
r : (=y/L), dimensionless abscissa of the light path
$\vec{r}$: position vector; intensity in (m)
s : arc, curvilinear coordinate along the light path
T : absolute temperature, in (deg K)
x : (=z'=dz/dy = tg θ) abbreviated notation for the derivative
y : abscissa of the light path, in (m)
z : ordinate of the light path, in (m)

Greek letters

ρ : fluid density, in (kg/m^3)
θ : (arc tg x), angle of the tangent of the light path
θ_o : (arc tg x_o), the angle shown in Fig. 1

1. Problem statement

Numerous systems encountered in Mechanical, Nuclear, Chemical Engineering etc. contain fluids or suspensions characterized by high temperatures, aggressive and poisoneous behavior and/or high velocities. Nevertheless, one is often concerned with finding local values of temperature or density in such systems. The fields of those two properties are especially important in regions close to system boundaries where steep gradients often occur, affecting in a substantial way the transport processes. Many systems of interest have at least one isobaric plane, or a plane with pressure distribution determined by an independent procedure. This means that if one can measure the temperature distribution in such a plane, these data can be easily converted into a density field, and vice versa. The conversion is done using the constitutive equation for the fluid contained within the boundaries.

Due to the hostile environment within the fluid (suspension) considered, an insertion of temperature- or other sensors of any kind is usually to be avoided. This eliminates also the taking of samples. Even if it may be possible, the samples could be destorted, yielding false informations on the properties of the fluid. Therefore, for the purpose of internal temperature- or density measurement, only remotely acting methods and devices should be considered: in other words, both the detectors and sources of any useful impulses must be externally located.

Among numerous "contactless" methods (a selection of which is outlined in the following Section) we are proposing a particular one which utilizes the following: a conventional light source or laser (but the latter is not compulsory), the standard photographic equipment, and an additional mathematical analysis on the engineering level, as outlined in Section 3.

2. Literature survey

Contactless measurements of density can be performed in several ways, out of which the following tentative systematization can be made:

1. Nuclear and atomic particles: neutrons, betas and alphas; the latter two being applicable only for short paths in liquid and solid absorbing matter.
2. Electromagnetic radiation (waves) :
   - 2.1. Visible light, including lasers and holography.
   - 2.2. X-rays, by their nature and utilization very close to 2.3.
   - 2.3. Gamma-rays, somethimes classified under 1., due to their origin.
3. Acoustic waves, including ultra-sound.
4. Tagging (labelling) of fluids using tracers: radioactive, chemical, colored etc.

The comprehensive study /1/ gives a closer insight into methods from above defined groups 1. and 2., but with practically no clues how to obtain local values of densities (concentration). Holography /2/ can give a"total picture" of the density field if there is an optical discontinuity present, for example in dispersions, bubble fields, etc. As a method it requires expensive equipment and special processing techniques.

Acoustic waves are a useful tool /3/, but with two inherrent disadvantages: similarly to 1., 2.2., and 2.3., they also report line-average properties, resulting from an integration along the direction of wave propagation. The second disadvantage of the sound is that its waves cannot be as finely collimated into a beam as it can be done in the application of methods listed above under 1. and 2.

Methods involving tagging techniques /4,5/ contaminate, in fact, the original system and introduce the danger of extended (radioactive or chemical) contamination in the case of leakage or explosions. In addition, they require expensive Health Physics procedures, personnel, monitoring instrumentation, etc.

Certain more advanced methods, which can utilize neutrons, gammas or X-rays, and which can report local values of densities,are only recently reported /6,7/.

Since light sources are less expensive, less dangerous and do not lose strength as the time progresses, a way was seeked to achieve the goal of temperature (density) field determination using the general laws of light propagation.

3. Analysis

According to /8/, Fermat's Principle rules that the light beam propagates along the path requiring the shortest time to reach the final point. In vectorial form, Fermat's Principle reads:

$$\frac{d}{ds}\left(m\ \frac{d\vec{r}}{ds}\right) = \text{grad } m \tag{1}$$

$d\vec{r}/ds = \vec{s}$, i.e. equal to the tangential unit vector (along the arc). The two-dimensional scalar counterparts of Eq. (1) are :

$$(dm/ds)\sin\theta + (m/R)\cos\theta = \partial m/\partial z \tag{2}$$

$$(dm/ds)\cos\theta - (m/R)\sin\theta = \partial m/\partial y \tag{3}$$

According to Fig. 1, z is the vertical axis and y the horizontal in the graph used to represent the light path in a plane. The real spatial orientation of these axes is discussed more closely in Section 5. In any case, y is the axis along which the variations of the field considered is mostly pronounced, and therefore investigated. Although the method is generally applicable to three-dimensional light paths, for the sake of simplicity of derivations and demonstration in practical use, we shall consider from now on only the z-y plane, assuming the.following:

$$\partial m/\partial z = 0 \tag{4}$$

Now, Eq. (2) yields:

$$dm/m = d(\ln m) = -(ds/R)\ \text{ctg}\ \theta = -(ds/R)(dy/dz) \tag{5}$$

Here, $R = \left[(1 + x^2)^{1.5}\right]/x'$, where $x = dz/dy$ and $x' = d^2z/dy^2 = dx/dy$, and $ds = dy\sqrt{1 + x^2}$; after all these substitutions and rearrangements, one obtains finally from Eq. (5) :

$$d(\ln m)/dx = -\left[x\ (1 + x^2)\right]^{-1} \tag{6}$$

The solution of the above differential equation is:

$$x = dz/dy = \left[\pm\sqrt{(m/C)^2 - 1}\right]^{-1} \tag{7}$$

C is the integration constant, which depends on the conditions at the system boundary. These conditions include both the fluid properties (such as m) and the angle of the light beam at the boundary. Application of Eq. (7) to the light entrance plane demonstrates these two effects on the value of the constant, C:

At $z = 0$, $x = x_o = \text{th}\ \theta_o$ and $m = m_o$. Consequently :

$$C = m_o (x_o^{-2} + 1)^{-0.5} = m_o \sin \theta_o \qquad (8)$$

Eq. (7) represents the semi-explicit analytical solution of the problem. The inverse equation can also be useful:

$$m(y) = C \sqrt{(dy/dz)^2 + 1} = C \sqrt{(x)^{-2} + 1} \qquad (9)$$

Since it generally applies that $m = F_o(\rho)$ or $\rho = F_1(m)$ and $T = F_2(m)$ for fluids of known composition — including suspensions — Eq. (9) correlates in fact the density or temperature with position. In other words, if we find out from the light path the field of the variable m, we know also the density- and temperature field, i.e. one gets the dependences:

$$\rho = F_1\left[m(y)\right] = F_\rho(y) \quad \text{or} \qquad (10)$$

$$T = F_2\left[m(y)\right] = F_T(y) = F(y) \qquad (11)$$

The only experimental input in the above equations consists of the data which the light path yields: one should only acquire the curve $y(z)$ or $z(y)$, where from one can obtain directly the distribution of the derivative $(dy/dz) = (1/x)$, which variable shows up in Eq. (9) and tells about the distribution of the refraction index, $m(y)$ or $m(z)$.

There are two general ways how to find out the necessary data on the distribution $T = F(y)$, as outlined below. In either way the fluid is known and its pressure completely defined (or measured independently) in the y-z plane, meaning that both functions $\rho(m)$ and $T(m)$ are known (usually from the standard literature).

A. The light path can be photographed. The path's possible fluctuations in time are discussed more closely in Section 5. Once the representative light path is found, its differentiation has to be performed. Then, one can use Eq. (9) in a "continuous" way as input into Eq. (11). On the other hand, instead of forming and operating on an "average" light path, i.e. z(y) curve, on can go more directly and form — on the base of photographs — an average $x(y)$ curve.

B. It can be proceeded as follows: (1) select the type of an analytical function $T = F(y)$, such as an algebraic- or exponential polynomial, Fourier series, etc. (2) Decide on the number of constant parameters (coefficients), i.e. on the order, j, of the function selected, $F = F_j$. In this way, if the function is an algebraic polynomial of order j, there are (j+1) coefficients to be determined: $T = a_o + a_1 y + a_2 y^2 + \ldots\ldots + a_j y^j$. (3) Select the most convenient experimental data to be measured. The total number of the independent data registered has to be (j+2), since the integration constant, C — from equations (7) and (8) — is also unknown.

The suggested data to be measured if the procedure B will be pursued are: two wall temperatures, which, being equal with the fluid wall-film temperatures, define easily the values $m(0) = m_o$ and $m(L)$, where L is the maximal width of the system along the y-axis. Further on, one should measure the entrance and exit slopes of the light beam, i.e. the values of $x(0) = x_o$ and $x(L)$. Finally, as the fifth helpful data, relatively easy to be measured, we propose a slope of the light path somewhere in the region $0 < y < L$. With five accessible data, $j_{max} = 3$. Since the approach B does not supply the experimenter with photographs, this, fifth information, could only be used in a special way, which is conditional in the following sense: the light path must have the exit from the fluid at the same side (plane) where it entered the system. This means that the total reflection occurs somewhere in the depth of the system. Then, the distance along the light path ordinate axis, z_{ie}, between the inlet and exit, defines the ordinate, z_r, where it holds that $(dz/dy)_r = \infty$. And just that ordinate is the fifth needed information. It can be easily shown that in systems where $\partial T/\partial z = 0$, $z_r = z_{ie}/2$, and these are the systems where Eq. (4) holds.

In systems without the total reflection the first four conditions, i.e. measurements to be performed, remain the same as described above; the fifth measurement could be done in a few ways. A relatively simple one would go along the following procedure (which is applicable <u>also</u> to systems with total reflection): one assumes that $y(z)$ is a polynomial of the order $(j+1) = 4$, rendering $m(z)$ as a polynomial of order $j = 3$, what follows from Eq. (9). Then, $T(z)$, being a function of m, is also a z-dependent function of the order j. The fifth independent measured information consists in simple recording of the difference between the light exit and inlet ordinates, $z(L) - z(0)$. Only, now, six conditions are necessary because there are six unknowns: five coefficients in the assumed function $y(z)$ of order 4, plus the unknown constant, C. The sixth condition (not requiring measurements) can be selected as: for y=0, z=0. The above listed six independent conditions, which include five experimental data, are sufficient to determine the $y(z)$ polynomial, which, in turn, via Eq. (9) and Eq. (11),defines the function $T(z)$ in terms of 4 coefficients.

The selection between the procedures A and B depends on financial factors, available time and instrumentation, as well as on the accuracy expected. In relation to the last item, a logical question arises: is the order j=3 and the assumed type of the $T(y)$ function sufficiently precise and reliable to enable the experimenter to determine as accuratelly as intended the temperature distribution along the y-axis? In special cases of dense suspensions, where the light beam becomes fully attenuated before leaving the system, only the procedure A is applicable for all practical purposes.

4. Numerical feasibility test

To demonstrate how a light path will deflect in a concrete practical system, an example is selected, the light beam trajectory calculated and represented graphically in Fig. 1.

According to /9/, Gladstone's law is applicable to gases such as CO_2 , which is our choice for the feasibility test. This law reads:

$$m = 1 + B\rho = 1 + (B_T/T) \tag{12}$$

We select now a hyperbolic temperature distribution, i.e.:

$$T = T_o \left[1 + (k'L)(y/L)\right]^{-1} = T_o\ (1 + k\ r)^{-1} \tag{13}$$

Substituting Eq. (13) in Eq. (12), the latter equation becomes:

$$m = 1 + (B_T/T_o)(1+kr) = \left(1+\frac{B_T}{T_o}\right) + \left(\frac{B_T}{T_o}k\right)r = C(a + br) \tag{14}$$

Introducing Eq. (14) into Eq. (7), performing the integration, utilizing the boundary condition z=0 at y=0 (i.e. q=0 at r=0) we have finally obtained:

$$q = (1/b)\ \ln\left\{\frac{\sqrt{(a+br)^2-1} + br + a}{\sqrt{a^2-1} + a}\right\} \tag{15}$$

Since $a = (1 + B_T/T_o)/C$ and $b = (B_T k)/(T_o C)$, it is obvious that the law $q(r)$ depends on the gas data as well as on the boundary condition, into which C is built in, according to Eq. (8).

With the values of the constant $B = 2.294 \times 10^{-4}\ m^3/kg$, as given in /9/, for CO_2 at $p = 1$ Bar (close to atmospheric pressure), $B_T = B\ p/R_{CO_2} = 0.123$ deg K.

For the temperature distribution, Eq. (13), two cases are selected:

Case 1: Decreasing temperature along the light beam (i.e. along y-axis)

$T(r=0)=T_o= 1273$ deg K (1832 F), $k = 10$, $tg\ \theta(r=0) = tg\ \theta_o=(dz/dy)_o= x_o=20$

Case 2: Increasing temperature along the light beam

$T(0) = T_o= 318$ deg K (113 F), $k = -0.8$, $tg\ \theta(0) = tg\ \theta_o =(dz/dy)_o= x_o=30$

All the above listed values define completely the coefficients a, b and C. The numerical interpretation of Eq. (15) yields the light paths which are shown in Fig. 1, together with the corresponding temperature distributions.

The practical applicability of the method outlined in this report depends, in fact, on the deviation of the light path curve from the initial light direction, i.e. from the "light entrance tangent". For a given temperature field, this deviation depends on the constant, C, i.e. on the entrance angle itself, $\theta_o=\text{arc tg}\ x_o = \text{arc sin}[C/m(0)]$, according to equations (7) and (8). Using the above described deviation, one convenient way to express the light path method sensitivity on C (i.e. on θ_o or x_o) could be through a "distinction coefficient", D, defined as follows:

$$D = \frac{z(L) - (dz/dy)_o L}{(dz/dy)_o L} = \frac{z(L)/L}{(dq/dr)_o} - 1 = \frac{q(1)}{x_o} - 1 \tag{16}$$

Normally, for any given temperature distribution, $q(1)$ also depends on x_o. Eq. (16) defines D as the relative difference of the light path exit ordinate and the ordinate of the "light entrance tangent" extended to the exit of the light path, i.e. to the abscissa $y = L$ (where $r = 1$).

The dependence $D(x_o)$ for both cases of temperature distribution is represented in Fig. 2, indicating a definite advantage of experiments with a large entrance angle. The practical limitation for this angle are imposed by concrete experimental conditions.

5. Experiments and data processing

Any convenient commercial light source — including, but not requiring the laser — can be used. If necessary, the entrance and exit windows for the light can be made out of special glass and/or externally cooled; being a strictly local action, such cooling would not destort the internal temperature field, especially if the fluid contained is moving.

The simplest practical situation is when one seeks the one-dimensional temperature field which varies in the y-direction. y-axis is perpendicular to the plane of the wall, as indicated in Fig. 1. Such a field may extend between two walls of a furnace or a combustion chamber. The plane of the light path introduced is defined with an axis which lies in the wall plane (z-axis) and the already defined y-axis, where these two axes are perpendiculat to each other. The orientation of the z-axis is to be selected according to the experimental requirements and it has to fulfill as much as possible the condition of Eq. (4).

If the data processing procedure B (Section 3.) is selected, the light path angle at the entrance is simply measured at the light source itself, while the exit angle can be found either by observing the extended free exit beam into an external isothermal fluid (e.g. air with dust), or by maximizing the signal from an attached standard light-detector while adjusting its angle. Inlet and exit light beam coordinates, as well as the boundary temperatures are to be measured using the standard laboratory instrumentation.

If procedure A is utilized, one has either to make a movie-film or to perform a random, stroboscopic photographing using a photo-camera. In either case, only the standard commercial equipment is needed. For the experiment itself, the optimal position of any photographic equipment involved is in the region perpendicular to the y-z- plane, properly centered as much as possible in order to minimize the optical destortion of the picture.

If a 3-dimensional temperature field is measured, two cameras have to be simultaneously used, and the stereo-photograph (or a hologram) has to be examined, utilizing equations (2), (3) plus another, similar one, pertaining to the third axis, which is perpendicular to y and z. Fig. 3 shows experimental arrangements.

The next step, the analysis of the light path, is to be done by determining the dependence $(dz/dy)_{ave} = \overline{x}(y)$ from the experimentally found curve, z(y). When Gladstone's law is applicable, a combination of equations (9) and (12) yields the final "operating" equation, a special case of Eq (11):

$$\overline{T}(y) = \frac{B_T}{C\sqrt{\left[\overline{x}(y)\right]^{-2} + 1} - 1} \tag{17}$$

The continuous filming or statistical sampling of the light path characteristics (based either on stroboscopic photographing or on an analog registration technique in procedure B) is necessary because of the following. Practically in all fluids of technical interest and at usual temperatures (meaning far above the absolute zero), there are persistent local fluctuations of density and temperature although the fluid may be in steady state as a whole. These fluctuations, /10/, are either caused by molecular phenomena or by the turbulent transport processes or both, where the latter cause is of decreasing importance in the wall layer of the fluid. The time-average values, $\overline{x}$, to be used in Eq. (17) can be determined either by a thorough statistical analysis, analog to those applied to turbulent flows, or one can use the approximate postulate: a linear averaging of the registered magnitude (in our case, this is basically z, where from x emerges) yields with a certain accuracy the real time-average value of the magnitude one seeks (which is T in our case). This "rule" is very often being accepted without too many reservations - except recently - in radiation attenuation methods for measurement of void fraction in two-phase systems. An improvement of the quoted postulate consists of the probabilistic, frequency-ruled ("weighted") averaging instead of the "linear" one, as it is in the above quoted "rule". Movie-frames or stroboscopic pictures of the light path enable the more accurate, probabilistic data processing. The final decision on the matter how precisely one has to elaborate on the data depends on the accuracy demands, finances, available time etc.

The only limitation for the method's application to transient temperature- and density fields lies in the described fluctuations. Therefore, slower or moderately fast transients, especially if they can be repeated in experiments, allow the measurements of transient local temperatures and densities utilizing the proposed new method.

6. Conclusions

The method described in the paper is characterized by following:

1. It yields local values of fluid temperature, translatable into density and other fluid properties (and vice versa) without internal sensors.
2. It uses only standard laboratory equipment and commercial instrumentation; a laser is preferred, but it is not a necessity; also, no interferometry is involved.
3. The method is inherrently precise, since only optical instruments and length measurements are involved; there are no simplifications in the mathematical analysis utilized or in the physical postulates.
4. As it results from the theory, the sensitivity of the method depends greatly on the entrance angle of the light beam; thus, it can be enhanced as much as local hardware conditions and the necessary space for the instrumentation permit.
5. There are no additional temperature limitations (superimposed to those ruled by the containment walls), because the temperature of the wall-windows for the penetration of the light beam can be maintained at a sufficiently low level, utilizing external local cooling.
6. The light attenuation by the fluid, or by a dense suspension (smoke, particles, fog, etc.) present in the chamber under investigation does not represent a serious limitation, since the light wavelength may be accordingly selected or the intensity may be increased. On the other hand, close to the wall, where the changes in property fields are important and mostly pronounced, evan a part of the light path can help to acquire the necessary informations on the local properties along the registrable trace of the light beam.
7. When measuring 2-dimensional fields, which extend in directions u and y (and u-axis is perpendicular to y and z), the method has a unique simplifying feature: The shape of the light path in y-z plane is fully independent of the temperature (density) field distribution along u-axis. This fact enables the experimental determination of such 2-dimensional fields by a sequence of independent light paths measurements, where the location of the y-z plane is varied along the u-axis. Each light path has to be analyzed separately, as outlined in previous chapters.

   Once the temperature field is 3-dimensional, i.e. one is confronted with T(u,y,z), the same procedure described in the above paragraph can be performed, except that the light path in each of the y-z planes has to be interpreted by utilizing simultaneously eqs. (2) and (3).

REFERENCES

/1/ "Two-Phase Flow Instrumentation", 8 papers, printing from the 11th National ASME/AIChE Heat Transfer Conference, Aug. 3-6, 1969, Minneapolis, Minn. USA.

/2/ Duffy, D.E., "Practical Application of Holographic Interferometry, "PLUS-186 (or R68ELS-93), General Electric Publications (Aug. 1969).

/3/ Solve'ev, A. A, "Acoustic Methods to Study the Inhomogenous Media State," Heat Transfer-Soviet Research, Vol. 4 No. 6 (Nov.-Dec. 1972).

/4/ Poletavkin, R. G., et al., "Tagged Atom Method of Investigating Water & Steam Content During Boiling," AEC-tr-4206, p. 16-20; IANSSR, 4-12 (April 1957).

/5/ Kondic, N. et al., "The Radioactive Particle Concentration Field Measurement in the Tube of Finite Cross-Section," Conf. on Dispersed Flowing Systems, Odessa, USSR, Sept. 1967.

/6/ Kondic, N. and Hahn, O. J., "Theory and Application of Parallel & Diverging Radiation Beam Method in 2-Phase Systems," Paper MT 1.5, Proc. of the 4th Internat. Heat Transfer Conference, Versailles, France, Sept. 1970.

/7/ McAdams, V. D., Kondic, N. & Baldwin, M. N., "Determination of Local Density by Gamma Scattering Using External Means," American Nuclear Society Meeting, Chicago, Illinois, USA, June 1973.

/8/ Born, M & Wolf, A., "Principles of Optics," p.121-7, Pergamon Press, N.Y.,1959.

/9/ Flügge, W., "Handbuch der Physik," Vo.24.,p.53, McGraw Hill, N. York, 1956.

/10/ Durao, D. et al., "Optical Measurements in a Pulsating Flame", Journal of Heat Transfer, May 1973, p. 227.

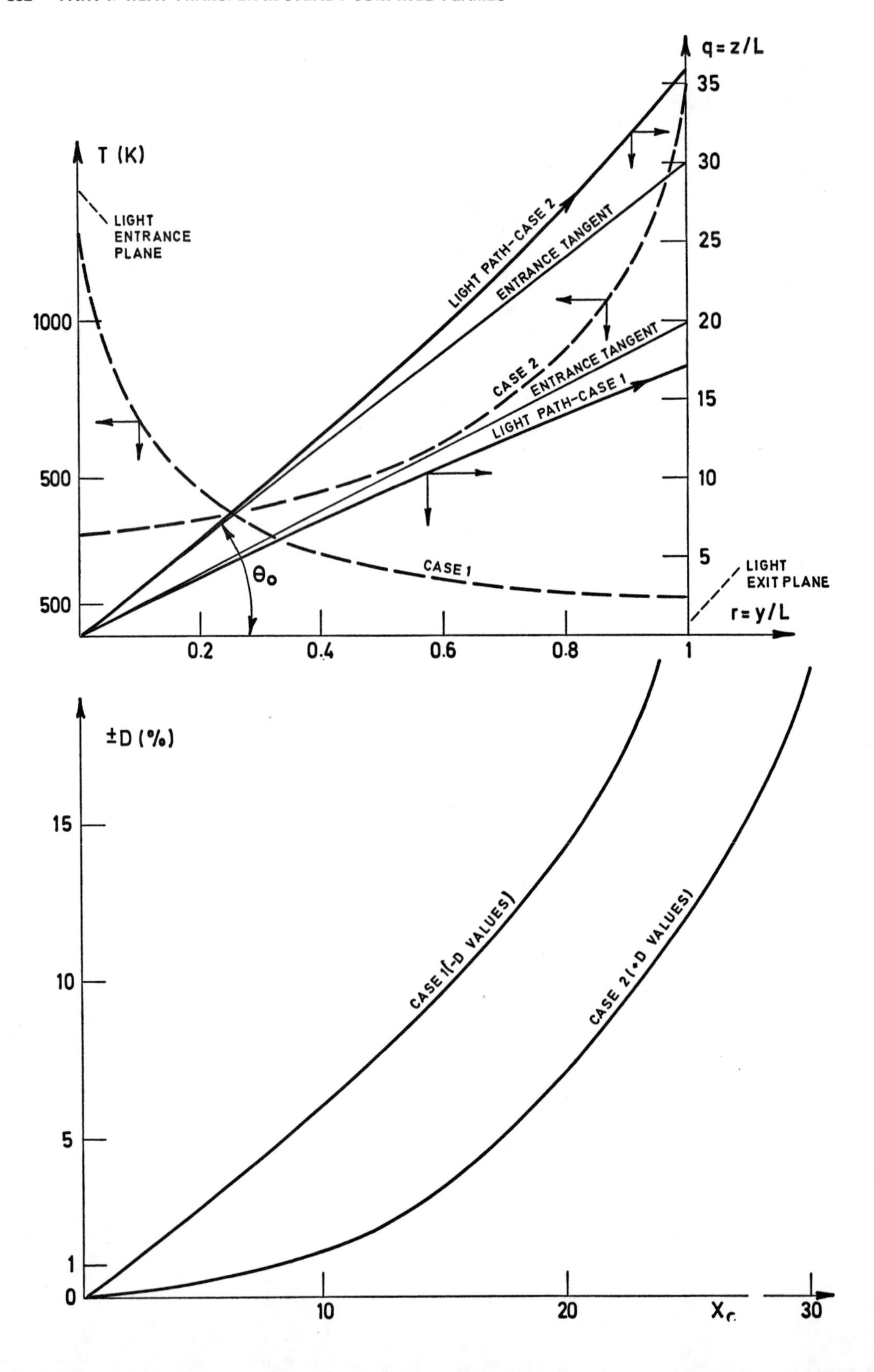

T (K)
q = z/L
LIGHT ENTRANCE PLANE
LIGHT PATH-CASE 2
ENTRANCE TANGENT
CASE 2
ENTRANCE TANGENT
LIGHT PATH-CASE 1
CASE 1
θ_o
LIGHT EXIT PLANE
r = y/L
1000
500
500
0.2
0.4
0.6
0.8
1
35
30
25
20
15
10
5
±D (%)
CASE 1(-D VALUES)
CASE 2(+D VALUES)
15
10
5
1
0
10
20
30
X_c

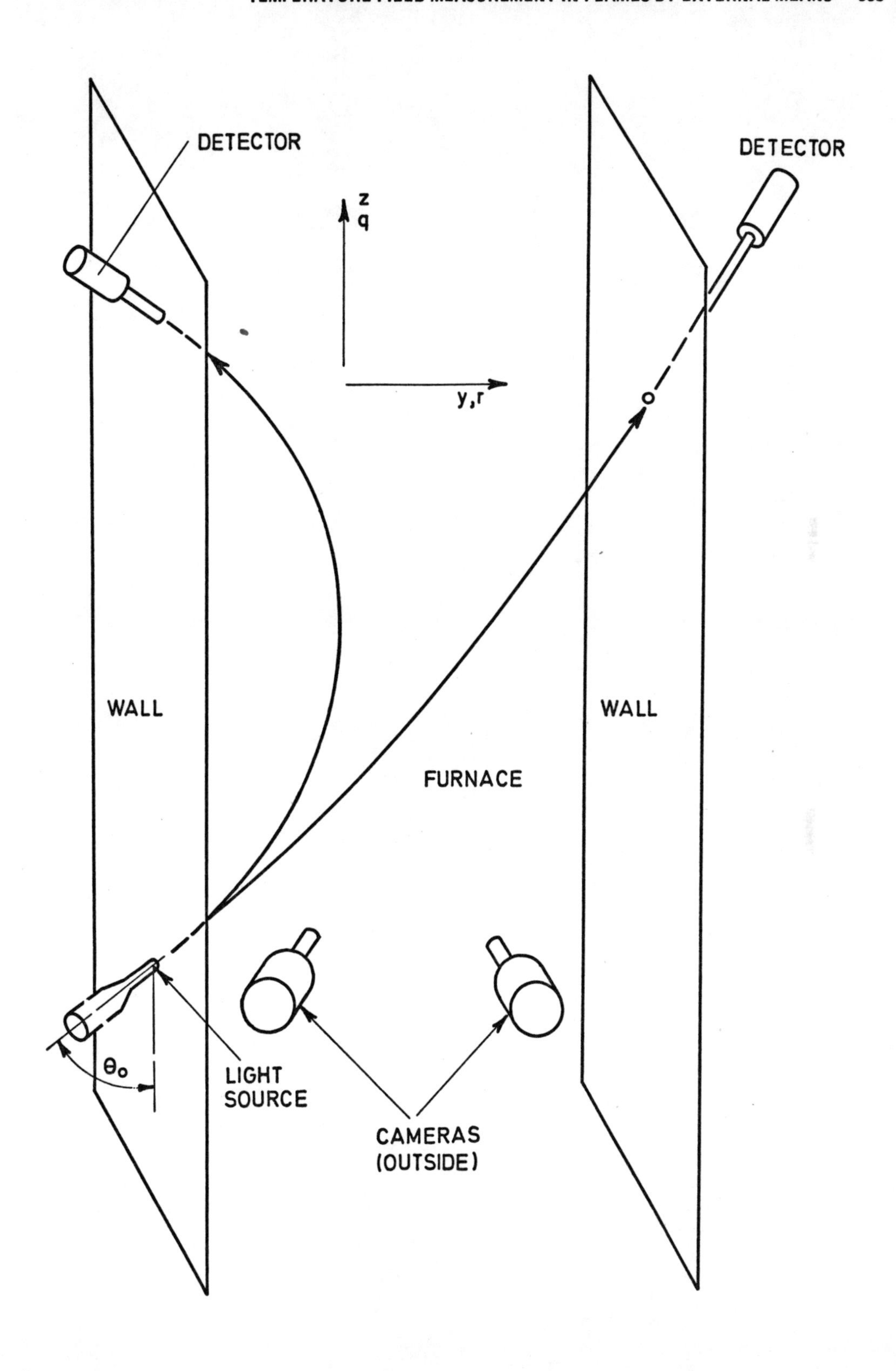
DETECTOR
DETECTOR
z
q
y,r
WALL
WALL
FURNACE
θo
LIGHT
SOURCE
CAMERAS
(OUTSIDE)

Chapter 23

AN EXPERIMENTAL AND ANALYTICAL DETERMINATION OF HEAT AND MASS TRANSFER IN A DIFFUSION FLAME

S. Abdel-Khalik, T. Tamaru and M. M. El-Wakil

University of Wisconsin, Madison, USA

ABSTRACT

The distributions as well as integrated convective and radiation heat transfer, and mass-transfer fluxes in a diffusion flame surrounding "cylindrical" burning drops of n-heptane are presented. These were calculated using experimentally obtained temperature and composition profiles surrounding the drops. These profiles were obtained by interferometry and gas chromatography. The results show that the radiation component of the heat transfer, often ignored, is comparable to the convective component.

NOMENCLATURE

B : transfer number $\left(= \frac{H_c}{Q}\frac{(Y_{O_2})_\infty}{j} + c_p\frac{(T_\infty - T_w)}{Q}\right)$

c : molar density (g-mole/cc)

c_p : specific heat (cal/g_m°K)

d : diameter of soot particle (micron)

D : cylinder diameter (mm)

D_{ij} : binary diffusion coefficient (cm²/sec)

f_v : volume fraction of soot

H_c : heat of combustion of fuel (cal/g_m)

H_v : latent heat of vaporization of fuel (cal/g_m)

j : mass of oxygen required for complete combustion of a unit mass of fuel

k : thermal conductivity (cal/cm sec°K)

L : cylinder length (mm)

$\dot{m}$: mass flow rate (g_m/sec)

$\dot{m}_f''$: mass flux of fuel vapor (g_m/cm²sec)

$\overline{\dot{m}_f''}$: fuel burning rate per unit area (g_m/cm²sec)

M_f : molecular weight of fuel (g_m/g-mole)

n : number of species in the mixture

$\underline{N}_i$: molar-flux vector of specie i

p : pressure (Kgf/cm²)

q : heat transfer rate (cal/sec)

q'' : heat flux (cal/cm²sec)

Q : heat reaching the fuel surface per gram of fuel burned (cal/g_m)

Re_∞ : Reynolds number ($=VD/\nu_\infty$)

R_o : universal gas constant (cal/g-mole°K)

T : absolute temperature (°K)

V : air velocity (cm/sec)

X : mole fraction

y : radial position relative to cylinder surface (mm)

Y : mass fraction

θ : polar angle from forward stagnation line

λ : wave length (micron)

ν : kinematic viscosity (cm²/sec)

SUBSCRIPTS

c : convection		r : radiation	
cw : cooling water		s : sensible	
f : fuel		w : cylinder wall	
in : inlet to cylinder		λ : latent	
out : outlet from cylinder		∞ : free stream	

1. INTRODUCTION

A considerable amount of research has been done on the problem of single fuel droplet combustion because of its fundamental importance for the understanding of the much complicated mechanism of combustion of fuel sprays. A relatively small number of studies, however, have been concerned with the detailed structure of the flame surrounding the droplet [1,2,3]. In sprays both detached and envelope flames take place. This study concerns itself with diffusion flames surrounding single drops.

A knowledge of the flame structure is essential for the determination of the heat and mass-transfer rates. The temperature and composition profiles in the flame surrounding a simulated burning droplet, obtained in a separate study, were used to evaluate the heat and mass-flux distributions around the drop and to make a complete energy balance for the system. The composition profiles were used to obtain the mass flux of the fuel vapor, and to evaluate the necessary optical constants to interpret the interferometer data into temperature data. The heat transfer convective and radiative components were then obtained.

2. THE EXPERIMENTAL MODEL

Liquid fuel-wetted cylinders were used to simulate the droplets. The cylindrical geometry was chosen because of the two-dimensional requirement of the interferometer. The cylinders were made of porous bronze and were constructed in the manner shown in Fig. 1. Two different size porous cylinders were used. The first was 12.8 mm in diameter and 25.4 mm long, while the other was 6.35 mm in diameter and 16 mm long.

A gravity feeding system was used to supply liquid n-heptane to the porous cylinders at controlled rates. The fuel flow rates entering and leaving the cylinder were adjusted so that a thin film of liquid fuel was always maintained around the porous cylinder surface while burning. Cooling water was allowed to flow through cooling jackets surrounding the fuel inlet and exit tubes in order to prevent the boiling of fuel inside the cylinder.

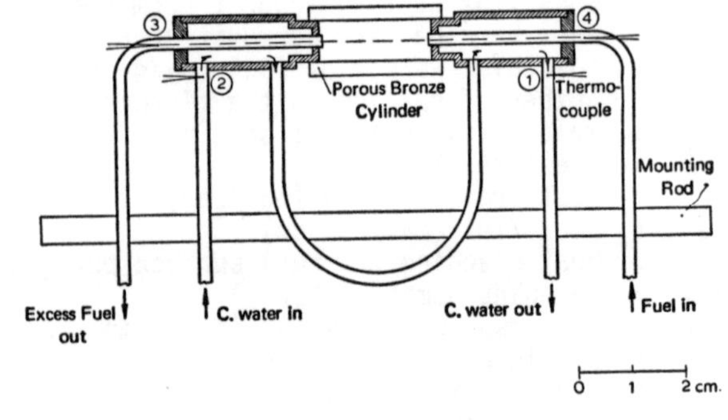

FIG. 1 The drop experimental model including fuel and cooling water circuits.

Four copper-constantan thermocouples were installed in the cooling water and fuel streams entering and leaving the large porous cylinder. These were used to measure the cooling water and fuel temperatures necessary for energy balance calculations.

The fuel inlet flow rate was measured by means of a calibrated capillary and manometer. The mass flow rates of the cooling water and excess fuel were collected and weighed over specified periods of time.

The wetted cylinder was allowed to burn steadily at atmospheric pressure in a uniform air flow field. The air flow was supplied by a convergent nozzle assembly that yielded a steady, uniform velocity, low turbulence air jet at various velocities. In all the cases examined in this work, the air velocity was kept below the extinction value so that an envelope flame was always established around the cylinder. Further details regarding the design and operation of the fuel and air flow systems may be found in [4].

3. EXPERIMENTAL APPARATUS, TECHNIQUES AND RESULTS

3.1 Composition profiles

The composition profiles in the flame surrounding the wetted cylinder were obtained by means of gas chromatography. Samples were withdrawn from the flame at points along the 0°, 45°, 90°, 135°, and 180° radial lines (measured from the forward stagnation line) by means of quartz microprobes. The probes were made from 5 mm diameter quartz tubes tapered down to 30-50μ orifices. The probes were mounted on a compound micrometer stand that could be indexed to measure probe tip positions along any of the sampled lines.

Care was taken to prevent the condensation of fuel vapor or any other high boiling point species that might be present in the samples by heating the different components of the sampling system. The samples withdrawn from the flame were dried and were then pressurized by mixing them with pure helium (sample: helium = 1:3). The sample-helium mixtures were analyzed by means of a three-column gas chromatograph with a thermal conductivity detector.
The sample size was 2 ml at 75°C and 1 atm. Constructional and operational details of the system are given in reference [4].

A sample plot of the obtained composition profiles in the flame is shown in Fig. 2. Concentrations of CO_2, CO, O_2, N_2, CH_4, C_2H_2, C_2H_4, and C_7H_{16} in dry samples were obtained. These profiles were used to evaluate the mass-flux distributions of the fuel vapor around the cylinder surface as will be explained in Sec. 4. They were also used to evaluate the molar refractivity distributions along the sampled radial lines, necessary for the determination of the temperature profiles by means of the interferometer.

The composition profiles were obtained along the five sampled radial lines for the 12.8 mm diameter cylinder while burning at air velocities of 12, 40 and 64 cm/sec. The small cylinder was only examined at 12 cm/sec, since at higher velocities the flame was visibly disturbed by the presence of the sampling probe.

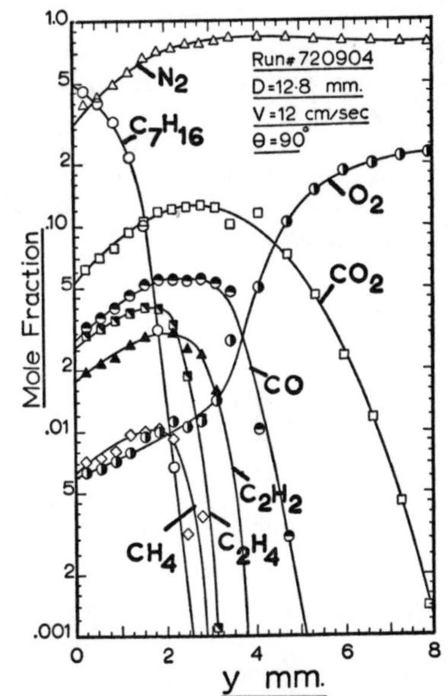

FIG. 2 Sample composition profiles along the 90° radial line of a burning n-heptane wetted cylinder.

3.2 Temperature profiles

The temperature profiles along the same sampled radial lines in the flame were determined by means of optical interferometry. The burning cylinder was mounted in the test section of a 7.5 cm diameter Mach-Zehnder interferometer. Interferograms for the burning cylinder were taken and analyzed to obtain the fringe shift distributions along the five radial lines. Because of the severe index of refraction gradients in the radial directions, refraction corrections were applied to the fringe shift data. This data was then used along with the molar refractivity profiles to obtain the temperature distributions in the flame. The interferometric equations used are given by Ross and El-Wakil [1]. In calculating the molar refractivity distributions, corrections were made to take into account the water vapor concentration present in the samples. The H_2O concentrations were assumed to be a constant ratio of those for CO_2. That ratio was taken to be equal to that in the equilibrium products of combustion of stoichiometric n-heptane-air mixture. This was used after checks of available experimental data.

A sample plot of the temperature distributions along the different radial lines is given in Fig. 3. These were used to evaluate the convective and radiative heat flux distributions around the cylinder as will be explained in Sections 5 and 7, below.

4. MASS-TRANSFER RESULTS

4.1 Mass-flux distributions

The mass flux distributions of the fuel vapor around the surface of the cylinder were obtained using the following assumptions:

1. The gas mixture at the surface is ideal.
2. Thermal diffusion is negligible.
3. The tangential component of the mass-flux vector at the cylinder surface is negligible.
4. There is no net flow of any of the gas components of the mixture (except the fuel vapor) through the surface of the cylinder.

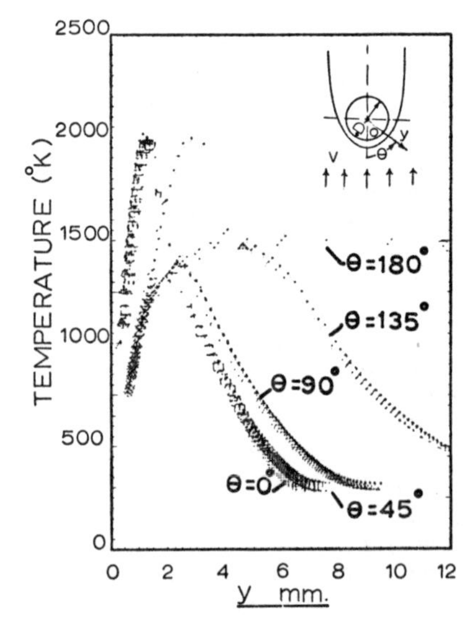

FIG. 3 Temperature profiles around a n-heptane wetted cylinder (D=12.8 mm, V=12.0 cm/sec)

Applying these assumptions to the Stefan-Maxwell equations [5],

$$\nabla X_i = \sum_{j=1}^{n} \frac{1}{cD_{ij}}(X_i N_j - X_j N_i)$$

the following equation is obtained.

$$[\dot{m}''_f]_w = -\frac{M_f}{R_o}\left[\frac{p}{T}\frac{\partial X_f}{\partial y} \Big/ \sum_{\substack{j=1 \\ j\neq f}}^{n} \frac{X_j}{D_{fj}}\right]_w \quad (1)$$

The concentration gradients of the fuel vapor and the concentrations of the different species at the cylinder surface were obtained from the composition profiles.

Corrections were applied to these quantities to take into account the water vapor concentrations in the mixtures. The surface temperature was taken to be equal to the saturation temperature of the fuel vapor corresponding to its partial pressure at that point. The binary diffusivities at the local surface temperature were evaluated using the Slattery-Bird equations [5]. The obtained mass-flux distributions are shown in Fig. 4.

4.2 Fuel burning rates

The fuel burning rates were evaluated by integrating the mass-flux distributions of the fuel vapor over the surface area of the cylinder. The burning rate per unit area is given by

$$\overline{\dot{m}''_f} = \frac{1}{2\pi}\int_0^{2\pi} [\dot{m}''_f]_w \, d\theta \tag{2}$$

The integration was performed using the trapezoidal rule. The obtained results along with the corresponding transfer numbers B and free stream Reynolds numbers Re_∞ are given in Table 1.

TABLE 1

Fuel Burning Rates

D (mm)	V (cm/sec)	Re_∞	B	$\overline{\dot{m}''_f}$ (mg_m/sec cm^2)
12.8	12.0	102	0.84	0.653
12.8	40.0	340	0.97	1.098
12.8	64.0	544	1.05	1.238
6.35	12.0	50.6	-	0.563

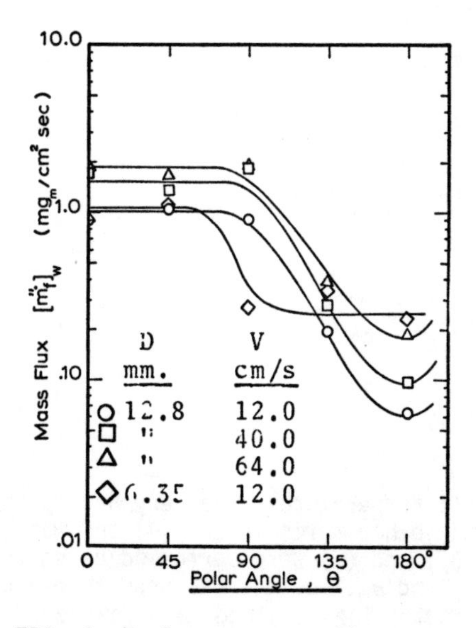

FIG. 4 Fuel vapor mass flux distributions.

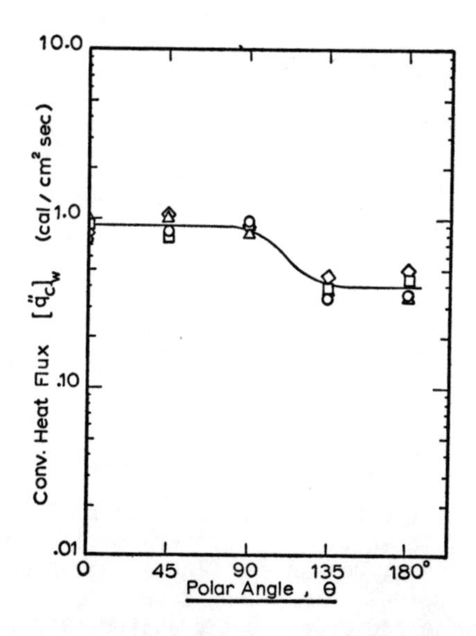

FIG. 5 Convective heat-flux distributions (symbols given in Fig. 4).

5. CONVECTIVE HEAT-FLUX DISTRIBUTIONS

The convective heat flux distributions around the cylinder were evaluated using Fourier's law. The tangential component of the heat flux vector was neglected because the wall temperature changed very slowly with θ. Thus

$$[q''_c]_w = -\left[k\,\frac{\partial T}{\partial y}\right]_w \tag{3}$$

The thermal conductivity of the mixture at the surface was evaluated using the equations given in reference [5] for perfect-gas mixtures. The radial temperature gradient at the wall was evaluated by fitting a straight line to the known surface temperature and the nearest six data points to the surface in the temperature profiles obtained from interferometry. The obtained convective heat-flux distributions are shown in Fig. 5. They were also integrated over the drop surface to obtain q_c.

6. ENERGY BALANCE

Energy balances were made for the 12.8 mm diameter porous cylinder while burning at different air velocities. These were used to evaluate the radiative heat transfer rates to the cylinder, which in turn, were used to evaluate the soot concentrations, and hence the soot contribution to the radiative heat-flux distributions as described in Section 7. This approach was necessitated by the lack of authoritative soot concentration distribution data in drop flames.

$$q_{cw} + q_{s_{\mathrm{out}}} + q_s + q_\lambda = q_c + q_r \tag{4}$$

where

$$q_{cw} = \dot{m}_{cw}\, c_{p_{cw}} (T_{cw_{\mathrm{out}}} - T_{cw_{\mathrm{in}}}) \tag{5}$$

$$q_{s_{\mathrm{out}}} = \dot{m}_{f_{\mathrm{out}}}\, c_{p_f} (T_{f_{\mathrm{out}}} - T_{f_{\mathrm{in}}}) \tag{6}$$

$$q_s = \frac{DL}{2}\int_0^{2\pi} [\dot{m}''_f]_w\, c_{p_f} (T_w - T_{f_{\mathrm{in}}})\, d\theta \tag{7}$$

$$q_\lambda = \frac{DL}{2}\int_0^{2\pi} [\dot{m}''_f]_w\, [H_v]_w\, d\theta \tag{8}$$

$$q_c = \frac{DL}{2}\int_0^{2\pi} [q''_c]_w\, d\theta \tag{9}$$

In all the cases examined, the fuel inlet and exit temperatures were very nearly the same, so that $q_{s_{out}}$ was very small compared to the other terms in Eq. (4) and was, therefore, neglected. The integrations in (7),(8), and (9) were performed using the trapezoidal rule. Upon knowledge of q_{cw}, q_s, q_λ, and q_c, Eq. (4) was used to evaluate the radiation heat transfer rate to the cylinder. The obtained results are given in Table 2.

TABLE 2

Energy Balance Results

D = 12.8 mm

V cm/sec	q_{cw}	q_s	q_λ	q_c	q_r
	cal/sec				
12.0	10.35	.198	.521	6.822	4.247
40.0	11.08	.362	.886	6.925	5.403
64.0	10.82	.416	.998	7.221	5.013

7. RADIATIVE HEAT-FLUX DISTRIBUTIONS

Radiative energy transfer from the flame to the cylinder was estimated using the experimental data of temperature, composition and the energy balance of the burning cylinder.

The major assumptions used are:

1. The diameter of the soot particles in the flame are within the applicable range of the Mie Theory, i.e., $(\pi d/\lambda) < 5$.
2. Light scattering is negligible.
3. Emissivity of the fuel surface is unity.
4. Contributions of dissociation, electron transitions and ionization to the absorption coefficient are negligible.
5. Unidimensional geometry of the flame at each particular angular position.

The radiative heat flux is composed of radiation from gases and soot. These were calculated by dividing the flame into finite isothermal layers. Interaction between gas and soot radiation was neglected.

Gas radiation was calculated using the Curtis-Godson approximation [6]. In this method, the transmittance of a given path through a nonisothermal gas is related to the transmittance through an equivalent isothermal gas. The relation between the nonisothermal and the isothermal gas is carried out by assigning an equivalent amount of isothermal absorbing material to act in place of the nonisothermal gas. That amount is based on a scaling temperature and a mean density that is obtained in the analysis. The heat flux calculation is then carried out using the isothermal gas method with the spectral band model.

It was found that the major gas radiation was contributed by the 4.3 and 2.7μ bands of CO_2 and 6.3 and 2.7μ bands of H_2O. Profiles of their radiative flux are shown in Fig. 6. The contributions of other emission bands (15 and 9.4μ bands of CO_2 and 1.87 and 1.38μ bands of H_2O) are negligible compared to the sum of radiative fluxes of the four bands above. It is interesting to note in the results that the gas radiation in the case of smaller cylinder is only slightly lower than that of the large cylinder and that the smaller cylinder shows the largest heat flux at $\theta = 180°$.

In the case of soot radiation, property values and spectral absorptance of a cloud of soot particles were given by Dalzell and Sarofim [7]. The radiative heat flux from soot depends upon the volume fraction of soot f_v. The value of f_v was assumed to be constant along each radial line, changing linearly with the polar angle θ from a value of zero along the forward stagnation line to a maximum value $f_{v,\max}$ along the 180° radial line. However, this maximum value is not known and is evaluated from the energy balance computations.

The distribution of the radiative heat flux from soot was calculated for different values of $f_{v,\max}$. Different emissivities were used for the flame layers depending on their temperature since they are dependent on wavelength [7].

The soot and gas radiation fluxes were integrated around the cylinder to obtain the total radiation heat transfer rate. These rates were then plotted against $f_{v,\max}$ as shown in Fig. 7. The radiative heat transfer rates obtained from the energy balance calculation were then used in conjunction with Fig. 7 to determine the value of $f_{v,\max}$ for each of the cases considered. These values of $f_{v,\max}$ were then used to calculate the corresponding heat flux distributions contributed by soot radiation as shown in Fig. 8.

Details of these computations are given in Ref. [8].

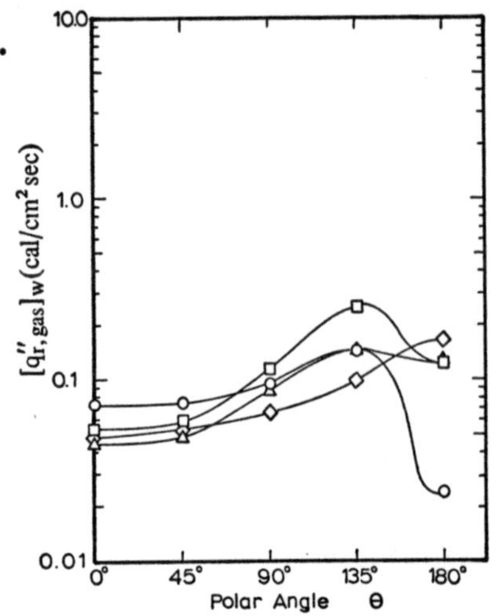

FIG. 6 Gas radiation heat-flux distributions (symbols given in Fig. 4).

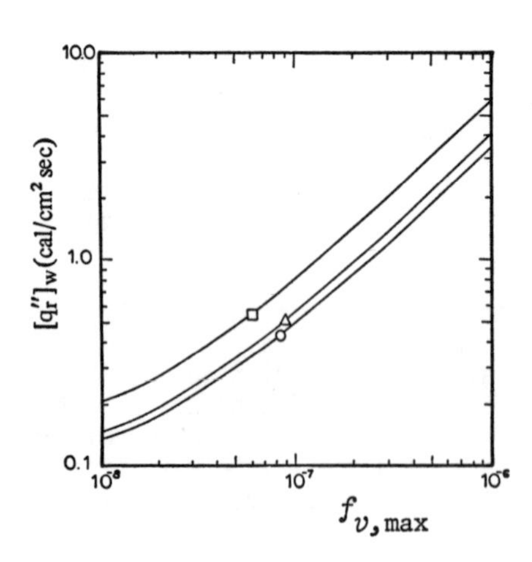

FIG. 7 Variation of calculated total radiative heat transfer rate with maximum soot concentration.

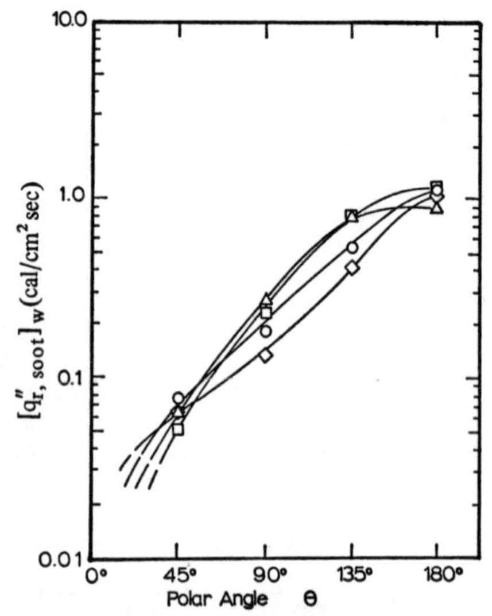

FIG. 8 Soot radiation heat flux distributions (symbols given in Fig. 4).

8. DISCUSSION

Examination of Fig. 4 reveals that there is a considerable difference between the values of the fuel vapor mass-flux in the leading and trailing halves of the flame. The mass-flux increases with the air velocity, which means that the burning rate will increase with the air velocity, a phenomena that has been experimentally observed by many investigators for both suspended droplets and wetted spheres. A comparison was made between our burning rates and those of Sami and Ogasawara [9].

A remarkable agreement was obtained between our results and their empirical formula when corrected to include the variation in the transfer number B. The correction was made by dividing the burning rate by the factor $B^{0.6}$ as suggested by Spalding's empirical formula [10].

Examination of the convective heat flux distributions given in Fig. 5 show that the air velocity does not seem to affect the convective heat flux to the same extent as it affects the mass flux of the fuel vapor. The convective heat transfer rate to the cylinder increases slightly with the air velocity as shown in Table 2.

The energy balance determined the magnitude of total radiation flux to the cylinder and consequently gave the only unknown parameter $f_{v,\max}$ for the assumed soot concentration profile around the cylinder.

Kunugi and Jinno [11] measured the soot concentration in a diffusion flame of mixed hydrocarbons (C_4H_8, C_4H_{10}, etc.) by a light scattering technique. Their maximum value of f_v was 3×10^{-7} at the top of the flame.

Considering the bigger H/C ratio of n-heptane compared to their fuel, the results obtained in present analysis ($0.6\text{-}0.9\times10^{-7}$) seem reasonable. On the other hand, Gollahalli and Brzustowski [3] measured soot concentration in the wake flame of n-pentane droplet, by the attenuation of laser beam. Their maximum value of f_v was the order of 10^{-6} which is considerably larger than either of the two above investigations.

9. CONCLUSIONS

1. It has been shown that heat and mass flux distributions around burning droplet can be experimentally obtained.
2. The air velocity has a much larger effect on the mass flux distribution than it does on the heat flux.
3. Radiation heat transfer to the cylinder is by no means negligible, being about 40% of the total heat transferred to the cylinder.
4. Radiation from the gases (CO_2 and H_2O) is 20% of the total radiation, the rest being soot radiation.

10. REFERENCES

[1] Ross, P.A. and El-Wakil, M.M., "A Two-Wavelength Interferometric Technique for the Study of Vaporization and Combustion of Fuels," *Liquid Rockets and Propellants*, Academic Press, New York, pp. 265, 1960.

[2] Aldred, J.W., Patel, J.C. and Williams, A., "The Mechanism of Combustion of Droplets and Spheres of Liquid n-Heptane," *Combustion and Flame, 17*, pp. 139, 1971.

[3] Gollahalli, S.R. and Brzustowski, T.A., "Experimental Studies on the Flame Structure in the Wake of a Burning Droplet," Fourteenth Symposium (Int.) on Combustion Proceedings. To be published.

[4] Abdel-khalik, S.I., "An Investigation of the Diffusion Flame Surrounding a Simulated Liquid Fuel Droplet," PhD Thesis, University of Wisconsin, 1973.

[5] Bird, R.B., Stewart, W.E. and Lightfoot, E.N., *Transport Phenomena*, John Wiley and Sons, Inc., New York, 1960.

[6] Siegel, R. and Howell, J.R., *Thermal Radiation Heat Transfer*, McGraw-Hill, New York, 1972.

[7] Dalzell, W.E. and Sarofim, A.F., "Optical Constants of Soot and their Application to Heat-Flux Calculations," *Journal of Heat Transfer, 91*, 1, pp. 100, 1969.

[8] Tamaru, T., "Radiative Heat Transfer in a Diffusion Flame," MS Thesis, University of Wisconsin, 1973.
[9] Sami, H. and Ogasawara, M., "Study on the Burning of a Fuel Drop in Heated and Pressurized Air Stream," Bulletin JSME, 13, 57, pp. 405, 1970.
[10] Spalding, D.B., "The Combustion of Liquid Fuels," Fourth Symposium (Int.) on Combustion, Williams and Wilkins, pp. 847, 1953.
[11] Kunigi, M. and Jinno, H., "Determination of Size and Concentration of Soot Particles in Diffusion Flames by a Light-Scattering Technique," Eleventh Symposium (Int.) on Combustion, pp. 257, 1967.

PART II: HEAT TRANSFER IN UNSTEADY CONFINED FLAMES

Chapter 24

HEAT TRANSFER FROM FLAMES IN INTERNAL-COMBUSTION ENGINES

W. J. D. Annad

Simon Engineering Laboratories, University of Manchester, England

Abstract

Available information on the heat transfer from the working fluid to the walls of reciprocating internal combustion engines, during the combustion period, is reviewed. Convection, gas radiation, and solid-body radiation all require consideration. In all three modes of transfer, problems requiring further investigation are identified.

NOMENCLATURE

C	:	soot concentration
k	:	absorption coefficient
k_a	:	equivalent grey-body absorption coefficient
k_1	:	k_a/p
L	:	flame thickness
p	:	gas pressure
S	:	surface area of flame
q_λ	:	monochromatic black-body emissive power at λ
V	:	volume of flame
ϵ_a	:	apparent grey-body emissivity
ϵ_λ	:	monochromatic emissivity at λ
λ	:	wavelength

1. Introduction

An ability to predict the magnitude of heat exchange between the working fluid in a reciprocating engine and the walls of the combustion chamber is of obvious importance to the engine designer. Estimation of the effect of heat loss on performance and efficiency requires, perhaps surprisingly, only a fairly rough approximation; complete elimination of heat loss would, typically, increase efficiency by only 5%. More accuracy is needed for an attempt to assess the probable bulk and cost associated with disposal of the heat rejected to the coolant. Most important, however, and requiring the most detailed information, is the need to estimate the thermal stresses imposed on critical components. It is thermal stress, and not breathing or combustion, that sets the limit on power obtainable. In making a choice between projected designs of different types, the assessment of thermal stress limits is crucial.

The first fundamental investigations of the problem, made by Nusselt /1/ and by Eichelberg /2/, were published just fifty years ago. Since that time, many papers on the topic have appeared. A full bibliography, confined to in-cylinder heat exchange, would now contain well over fifty items. A good general understanding of the nature of the processes involved has been reached, but many problems of practical application remain unsolved.

The present discussion is limited to that part of the engine cycle during which combustion takes place, and is confined to an examination of the effects of the flame. Combustion in reciprocating engines may be either homogeneous or diffusive. In the majority of spark-ignition engines, pre-mixed fuel and air are ignited by a spark, and combustion takes place by turbulent propagation of a flame through the mixture. In compression-ignition engines, localised fuel sprays are ignited by homogeneous explosion of a small amount of mixture formed from the spray, and thereafter burn diffusively. In both cases, some heat reaches the walls by convection. With homogeneous combustion, gas-phase radiation is present, and with diffusive combustion, solid-particle radiation occurs. These three modes of heat transfer will now be discussed separately.

2. Convective heat transfer in the reciprocating engine

2.1 General

Convective heat transfer takes place in reciprocating engines under conditions of rapidly varying gas pressure and temperature, and with local fluid velocities which may also vary more or less rapidly, depending upon the configuration of the combustion chamber. Heat flux into the containing walls varies continuously throughout the engine cycle, from a small negative value to a positive value reaching several MegaWatts per square metre. The flux variation lags behind the change of gas temperature. An exact theoretical analysis of heat transfer under these conditions is not possible for the existing turbulent régime, and recourse must be had to experiment. The only available method of measurement of the heat flux variation lies in the observation of the time variation of wall surface temperature, using thermocouples or resistance thermometers constructed as very thin films on the surface. The heat flux is deduced by differentiation of a Fourrier series fitted to the temperature observations. The precision attainable is only moderate.

At least a dozen studies of this nature have been published. The lag between heat flux and driving temperature difference is clearly perceptible, but the precision suffices for only a very rough determination of its magnitude. In general, investigators have concluded that, within the attainable precision, it is best to treat the observations as quasi-steady, and correlate them in terms of conventional Nusselt and Reynolds numbers. Arguments rage as to the representative lengths and velocities which should best be used in evaluating these dimensionless groups. Fortunately, these arguments need not concern us here.

2.2 Effects of confined combustion on the convective heat transfer

From the point of view of this Symposium, the important question to be answered is: does the existence of a flame change the convective heat transfer, directly or indirectly, apart from obvious effects of the associated changes of temperature and of gas properties?

It is unlikely that there is any direct effect. At one time, it was widely held that combustion would in itself produce a major increase of intensity of turbulence, but recent studies have shown conclusively that any increase of turbulence within the flame is small and transient.

An indirect effect is produced by the local velocity changes caused by expansion of the hot burned products. In the spark-ignition engine, expansion of the charge on passing through the propagating flame produces a clearly defined difference between the rate of flame movement and the speed of propagation relative to the unburned gas. Calculation for typical cases shows that the translational velocity represented by that difference can be of the order of 10 m/s. That is quite large compared with the fluid velocities normally found in this type of engine before the passage of the flame. The result can be seen particularly clearly in some experiments made at Manchester University by Summers /3/, using a single-cycle engine simulator. In this device, figure 2.2-1, a charge of fuel and air is induced through a conventional inlet valve, is compressed, and is then held at constant volume. Ignition by a spark plug may follow after any desired interval. The loss of heat from the compressed charge can be computed from observations of the rate of change of pressure, after fixture of the piston at inner dead centre.

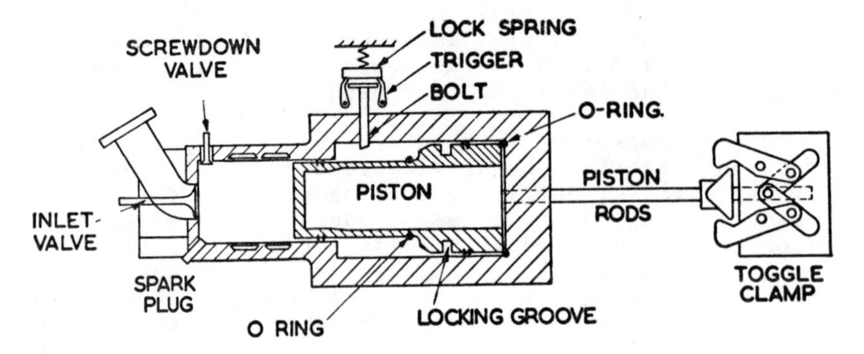

Fig. 2.2-1 Diagram of operation of the Manchester single-cycle engine simulator.

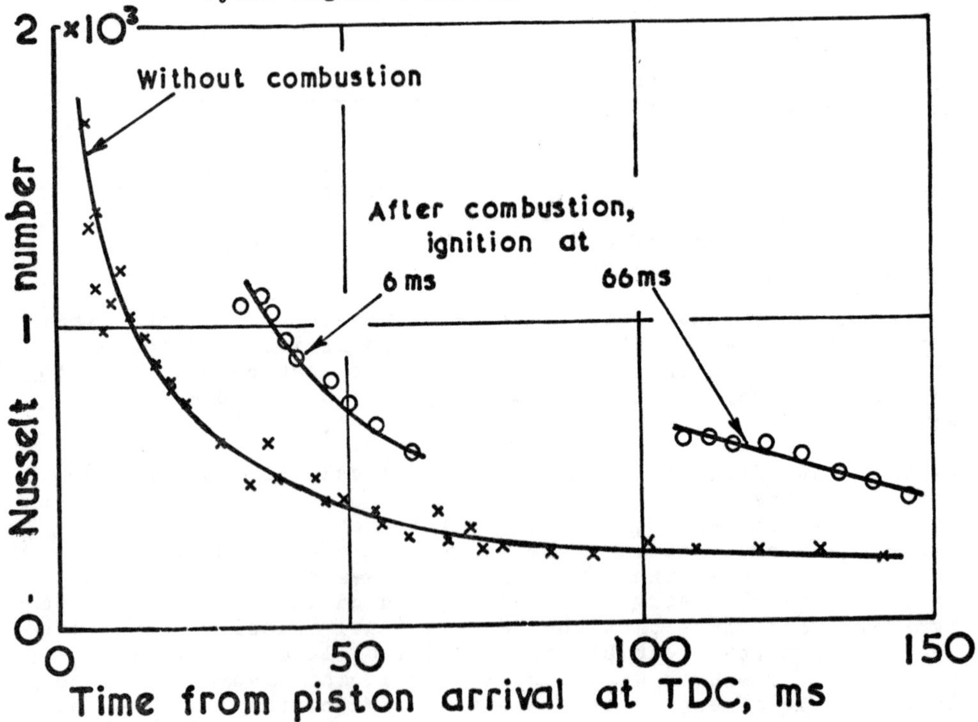

Fig. 2.2-2 Nusselt number variation with time after piston fixture at top dead centre, in the single-cycle simulator.

In experiments in which the charge was not fired, the Nusselt number diminished with time as shown in figure 2.2-2. Probable reasons are the increase of thickness with time of the thermal boundary layer on the walls, and decay of the initial turbulence and gross fluid motion. If the charge is fired, and observation resumed after combustion was complete, the Nusselt number was found to have increased markedly, as the figure shows. This is opposite to the effect to be expected, with significantly large initial Reynolds number, purely from the increase of fluid viscosity consequent on temperature change. It is clear, and is confirmed by the near equality of the increases observed at the two ignition times shown, that the cause is indeed an increase of fluid velocity.

Similar effects arise in compression-ignition engines. In indirect-injection engines, in particular, the air motion produced by expansion is deliberately exaggerated and channelled, to intensify mixing of the fuel and air. Measurements to confirm the effect on instantaneous heat flux are lacking for both swirl-chamber and ante-chamber types. Henein /4/ presents some observations made on an engine with an energy-cell combustion system, with analysis indicating that the high values of peak heat transfer observed are associated with charge velocities of the order of 500 m/s produced by outflow from the energy cell. In direct-injection engines, charge motion produced by combustion is much smaller and much less well defined, and the resulting velocities may well be relatively unimportant in super-position on those induced before combustion to ensure mixing. Woschni /5/ proposed that, in this case, the combustion-induced charge motion could be related empirically to the magnitude of the pressure rise produced, taken as a measure of the energy release. Woschni was, in fact, the first to embody an allowance for the combustion-induced motion in a predictive procedure for heat exchange in engines. His suggestion received less acceptance from other workers than it deserved, because it was originally bound up with a view, upon which debate centred, that radiation was not important. Now that the debate is settled (as will be seen in section 4, below), and the importance of radiation is firmly established, Woshni's basic idea merits reconsideration.

3. Radiant heat transfer in the spark-ignition engine

Under the conditions of homogeneous deflagration existing in the spark-ignition engine, no solid intermediate products are formed, and only gas-phase radiation has to be considered. In the combustion products, the only important emitters are carbon dioxide and water vapour.

It has been usual to assert, on the basis of some early assessments, that the maximum effective emissivity of a typical thickness of products is only about 0.1, and that the maximum radiant flux at peak temperature, being of the order of half a MegaWatt per square metre, is only about one-tenth of the maximum total flux typically observed. Since radiation diminishes much more rapidly than convection as temperature falls during expansion, the contribution of radiation to the total flux has been held to be neglegible.

Whilst this view may well be correct, no direct measurements have ever been made to substantiate it, and there are several grounds for question:

(a) No account has been taken of radiation from the flame itself, which in the prevailing turbulent conditions takes the form of a fairly thick, rather ill-defined, slightly luminous region.
(b) The pressure in the combustion chamber at the time of peak radiation is such that the partial pressures of carbon dioxide and of water vapour may each approach ten bar; existing measurements, recently summarised by Hottel and Sarofim /6/, relate to much lower pressure, and extrapolation appears doubtful.
(c) The temperature of the products is not uniform at the end of combustion. Material which is burned at the beginning of flame travel expands and is then recompressed, reaching at the end of the process a temperature several hundred degrees higher than does that which is burned last. Typically, this recompression is accompanied by production of visible yellow radiation, which is held to come mainly from carbon dioxide /7/.
(d) The temperature varies very rapidly. The equilibrating reactions are known /8/ to be fast enough at this time to maintain equilibrium very closely, but the question remains whether equilibrium radiation data can properly be applied.

Altogether, there is scope for a more penetrating analysis than has previously been published, or for the application of the experimental techniques that have so far only been deployed in investigation of the compression-ignition case.

4. Radiant heat transfer in the compression-ignition engine

4.1 Experimental evidence

As recently as 1967, debate continued as to the importance of the radiant contribution to total heat transfer in the compression-ignition engine. There was, indeed, published evidence well prior to that time, clearly indicating that radiation could account for an appreciable fraction of the total. Belinkii /9/ had published his pioneering measurements in 1955, and Ebersole et al /10/ had also reported some preliminary measurements in 1963, suggesting that the radiant contribution could reach 40% of the total. Nevertheless, many workers remained unconvinced, particularly as Belinkii's work was almost unknown outside the U.S.S.R. At the present date, we may confidently say that no doubt can remain. The publications of Oguri and Ho /11/, of Flynn et al /12/, and of Sitkei and Ramaniah /13/ agree in showing measured radiant heat transfer amounting to between 20% and 30% of the total, averaged over the complete engine cycle.

It is worth remarking here that conditions in small high-speed engines are very different from those found in boiler furnaces, so that direct application of furnace flame data is not possible. The pressure is of the order of 100 bar, and the intensity of energy release is very high, reaching the region of 10,000 MW/m^3 for a period of a millisecond or two in each cycle.

Belinkii made his measurements on a small (140 mm bore) open-chamber engine, using a technique which required the comparison of photographically-recorded spectra from the flame, from a beam of light passed through the flame by a tungsten-filament lamp, and from the lamp alone. The method yielded direct estimates of monochromatic emissivity of the flame, at several wavelengths between 0.4 and 0.6μm. Within this rather narrow range (limited by the sensitivity of the film), the emissivity appeared to be independent of the wavelength, and Belinkii concluded that the flame radiated as a grey body. His observations of the variation of emissivity with crank angle are shown in figure 4.1-1. He made also measurements of the apparent radiation temperature of the flame, using a two-colour method; these observations are briefly commented upon below, in comparing them with those of Flynn et al.

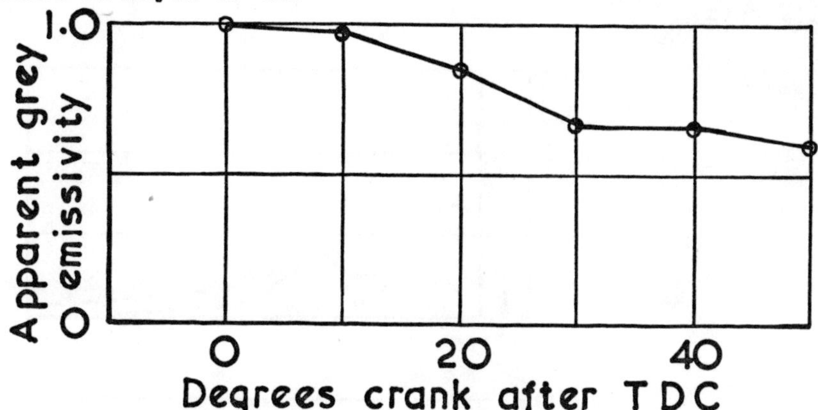

Fig. 4.1-1 Flame emissivity observations of Belinkii 2000 rev/min. Fuel:air ratio not stated.

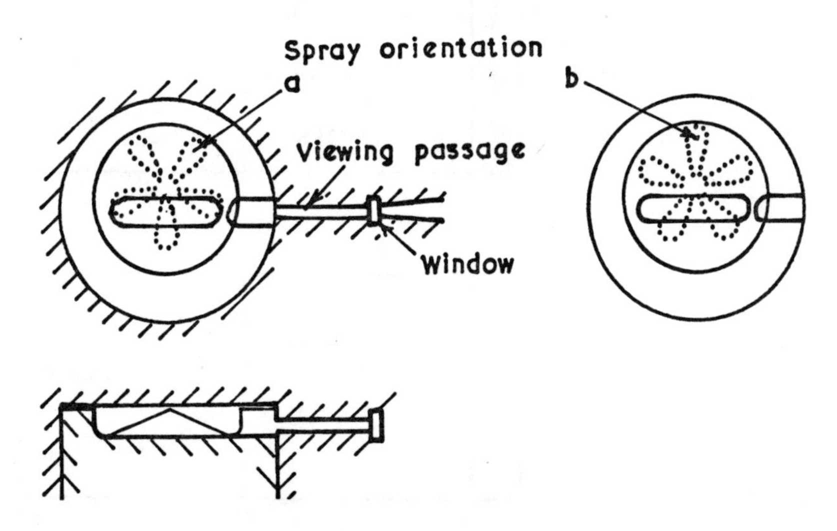

Fig. 4.1-2 Geometry of the combustion space in the engine used by Flynn et al.

Flynn et al have made the most detailed observations yet published. They used a monochromator to analyse the radiation from combustion in a small, direct-injection engine, with the configuration seen in figure 4.1-2. The viewing path, about 80 mm long, cut through the central cone of the toroidal combustion chamber. A five-hole nozzle could be arranged in either of the orientations shown. Some air rotation was provided. The intensity of radiation was measured at each of seven wavelengths, at one-degree intervals of crank angle; at each crank angle, readings were averaged over fifty cycles. At any given crank angle, the distribution of energy over the seven wave-lengths was used to reconstruct the entire energy spectrum, and to calculate the apparent radiation temperature and the apparent optical thickness kL.

The range of wavelength covered was much larger than in Belinkii's work.

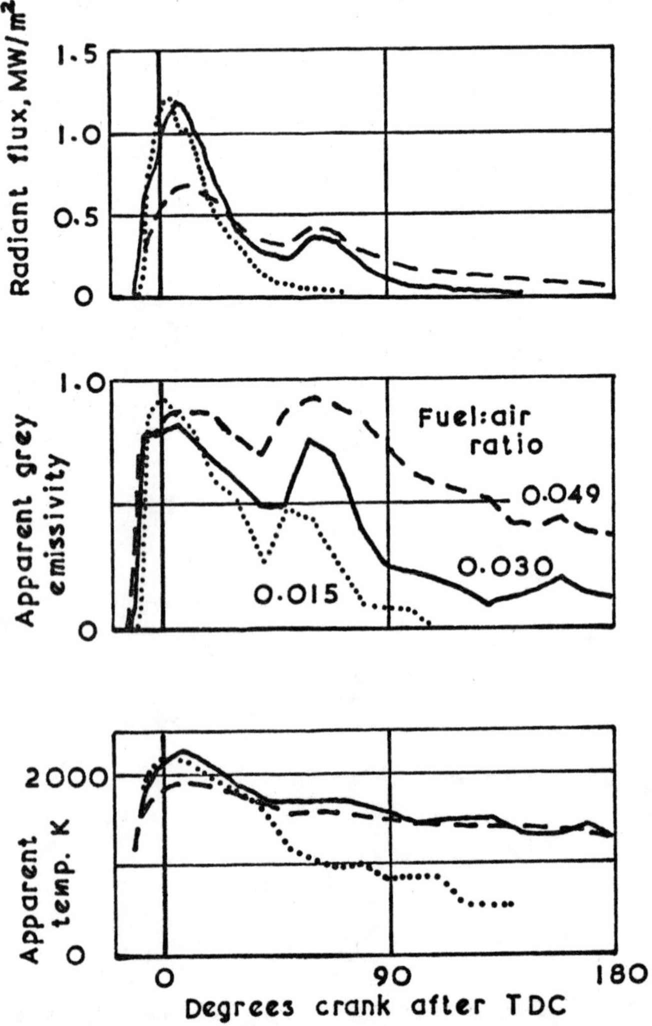

Fig. 4.1-3 Typical observations from ref /12/. 2000 rev/min.

The energy distribution was found to be somewhat non-grey, and fitted well to the equation

$$\epsilon_\lambda = 1 - \exp(-kL/\lambda^{0.95}) \tag{1}$$

suggested by Liebert and Hibbard /14/ for emission from clouds of small particles. However, for simplicity of presentation, results were converted to equivalent grey-body emissivity by using the equation:

$$\epsilon_a = \int_{0.5}^{10} \epsilon_\lambda q_\lambda \, d\lambda \Big/ \int_{0.5}^{10} q_\lambda \, d\lambda \tag{2}$$

Figure 4.1-3 shows typical results obtained at three fuel: air ratios.

During the period of maximum radiation, reached shortly after the end of the ignition delay, the apparent grey emissivity lies in the region 0.8 to 0.9. If an equivalent grey-body absorptivity k_a is defined by

$$\epsilon_a = 1 - \exp(-k_a L) \tag{3}$$

the value of $(k_a L)$ lies between 1.6 and 2.3. Some order-of-magnitude calculations on this basis lead to interesting conclusions. On the assumption that flame fills the piston bowl at this time, L is about 73 mm. Taking $(k_a L) = 2.0$ as average, the value of k_a must be about 27 m^{-1}. Field et al /15/ suggest that k_a is related to the soot cloud density C kg/m^3 by

$$k_a = 1200\, C\, m^{-1} \tag{4}$$

from this, C would be about 0.022 kg/m^3. Such a value is high compared to those predicted by the theory of Khan et al /16/ for a naturally-aspirated engine of similar size, as displayed in figure 4.1-4, but not impossibly so, particularly having regard to the uncertainty of the value assigned to L.

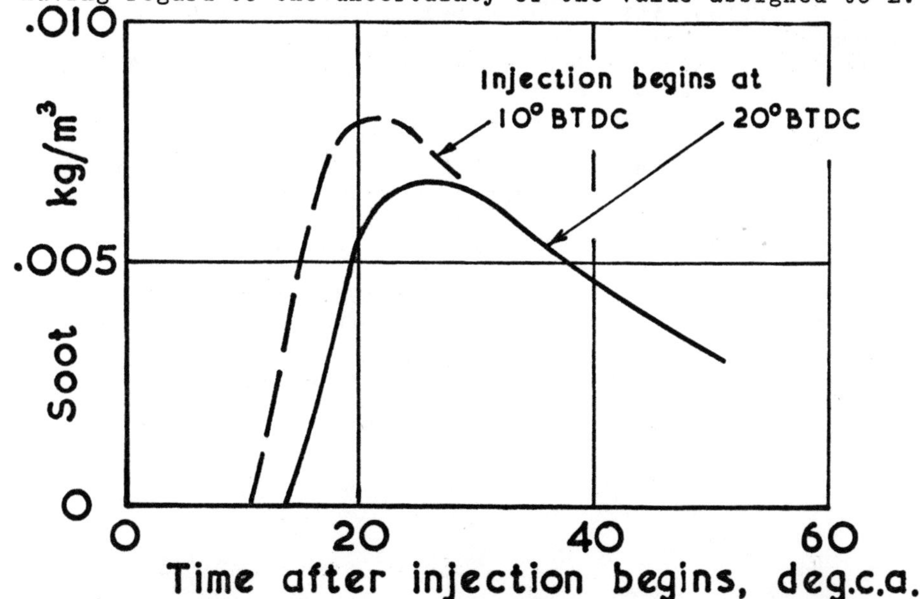

Fig. 4.1-4 Soot concentration values calculated by Khan et al for a small, naturally aspirated, direct-injection engine.

Sitkei and Ramaniah /13/ have proposed that k_a should be replaced by $k_1 p$, where p is the pressure in the combustion space, and that k_1 would then be a unique function of overall fuel:air ratio and of crank angle. Cylinder pressure values for the tests of reference /12/ are not available, but the pressure at the maximum fuelling rate would probably reach about 120 bar at the time of peak radiation. From the curves given in reference /13/, $k_1 p$ would be about 0.5 x 120 = 60 m^{-1} which is rather high. Although k is dependent upon p for gas radiation, it is not clear why proportionality should apply to soot radiation. The experiments of Flynn et al do not show any large or consistent variation of $k_a L$ with variation of initial pressure from 1 bar to 2.5 bar, at fixed overall fuel:air ratio.

After the first peak of radiation, figure 4.1-3 shows a second, lower peak associated with a temporary increase of emissivity. This was seen in nearly all runs reported with nozzle orientation (a), but was not present in a comparable run with orientation (b) of figure 4.1-2. The most probable explanation advanced was the late burning of fuel deposited in the actual viewing passage. No similar second peak is visible in the observations of Sitkei and Ramaniah, which were made in a flat-sided swirl chamber without any cavities.

Apart from this, probably spurious, peak, the apparent optical thickness diminishes approximately in inverse proportion to the increase of cylinder volume. This accords with the kinetic study of Khan et al, in suggesting that the number of radiating particles remains substantially constant during the later stages of combustion.

Broadly speaking, the emissivity and apparent optical thickness observed in reference /12/ at the peak radiation time showed very little variation with engine speed or fuel:air ratio, although the duration of high radiation of course varied. Fuel Cetane number had some effect, associated with changes of ignition delay, but fuel composition at fixed Cetane number (excluding special additives) had none.

The apparent radiation temperature likewise showed very little effect of any of the running parameters, except those affecting ignition delay. The reduction in peak temperature at high fuel:air ratio seen in figure 4.1-3 is an exception, which might indicate an absorption effect by hydrocarbon fragments. The observations of Belinkii /9/, in contrast, show a consistent increase of temperature with increase of fuel:air ratio, but the effect is not large. The maximum radiation temperature shown in figure 4.1-3, approximately 2300 K, is rather lower than Belinkii's values, which range from 2500 to 2560 K. The discrepancy could result from

Belinkii's assumption of grey-body radiation, in interpreting his two-colour intensity measurements.

4.2 A route to prediction of : radiative transfer

The published material summarised in the preceding section opens up a route to the calculation of the heat transfer by radiation, for the compression-ignition engine. There are still gaps and obstacles in this route, but its line can now be seen.

The first obstacle lies right at the beginning, in the need for a satisfactory model of the combustion process, that would enable both the rate of burning and the development of the flame shape to be predicted. Several such models have been proposed, applicable to particular combustion systems, but as yet they have been insufficiently tested to give confidence in selecting one, rather than another. In time, this obstacle will be surmounted. Then, the kinetics of soot development must be embodied, building on studies such as that of Khan and his co-workers. If the modelling of flame development is adequate to determine instantaneous volume V and surface area S, the flame thickness L may be taken as 3.4 V/S, the value given in /15/ for a particle cloud of arbitrary shape. Otherwise, one might follow Sitkei and Ramaniah in substituting the instantaneous volume and surface area of the combustion space for those of the flame - amounting to the assumption that the flame fills the space. The absorption coefficient k_a can in principle be derived from the soot concentration, leading to an estimate of emissivity. Verification is clearly needed, for the level of soot concentration found in engine flames. Here we come to a gap, in that the effective radiation temperature is needed and is not readily calculated. Perhaps this gap must be bridged by using empirical estimates, as has been done by Sitkei and Ramaniah. Once that is done, the radiant flux from the flame follows.

It is unlikely that we should ever be able to model the combustion process in sufficient detail to make it necessary to consider form factors for the radiant transfer from flame to wall. We will probably have to remain content to treat the flame as a grey body in a black box; but for most purposes this will be adequate.

Proposals of this kind often meet with the objection that there is great variability from cycle to cycle in the detail of flame initiation and development, and in the motion of radiating volumes and of obscuring chilled soot-clouds, but in fact the modelling of individual cycles is almost never desired. What is needed is the time-average thermal loading of the engine surfaces, so that a representation of average behaviour is sought. It does appear a perfectly practical aim, to model that average development successfully.

REFERENCES

/1/ Nusselt, W. : "Die Wärmeübergang in den Verbrennungskraftmaschinen". Z.D.V.I., Vol.67, p.692 (1923).
/2/ Eichelberg, G.: "Temperaturverlauf und Wärmespanning in Verbrennungsmotoren". Forsch.Ing.Wes., No.263 (1923).
/3/ Summers, I.G.S.: "Convective heat transfer in a rapid compression machine simulating a spark-ignition engine". M.Sc. Thesis, University of Manchester (1970).
/4/ Henein, N.A.: "Instantaneous heat transfer rates and coefficients between the gas and combustion chamber of a diesel engine". Trans S.A.E. (1965).
/5/ Woschni,G.: "A universally applicable equation for the instantaneous heat transfer coefficient in the internal combustion engine". Trans. S.A.E., Vol.76 (1967).
/6/ Hottel, H.C., and A.F.Sarofim: "Radiative transfer". McGraw Hill, New York (1967).
/7/ Withrow,L., and G.M.Rassweiler: "Spectroscopic studies of engine combustion". Ind.Eng.Chem., Vol.23 (1931).
/8/ Newhall,H.K.: "Kinetics of engine generated nitrogen oxides and carbon monoxide". Twelfth Symposium (International) on Combustion, Poitiers (1968).
/9/ Belinkii,L.M.: "Thermal radiation in the combustion chamber of a high-speed compression-ignition engine". Trudi N.I.L.D. No.1, Moscow (1955).
/10/ Ebersole, G.D., P.S. Myers and O.A. Uyehara: "The radiant and convective components of diesel engine heat transfer". S.A.E. Summer Meeting, paper 701C (1963).
/11/ Oguri,T., and Ho Hsi-Tang: "Radiant heat transfer in the cylinder of diesel engine". Japan Auto.Res.Inst.Tech. Memo. No.2, p.263 (1971).
/12/ Flynn,P., M.Mizusawa, O.A.Uyehara and P.S.Myers: "An experimental determination of the instantaneous potential radiant heat transfer within an operating diesel engine". Trans.S.A.E., Vol.81 (1972).
/13/ Sitkei,G., and G.V.Ramaniah: "A rational approach for calculation of heat transfer in diesel engines". Trans. S.A.E., Vol.81 (1972).
/14/ Liebert, C.H., and R.R.Hibbard: "Spectral emittance of soot". N.A.S.A. Tech.Note No.D-5647 (1970).
/15/ Field,M.A., D.W.Gill, B.B.Morgan and P.G.W. Hawksley: "Combustion of pulverised coal". B.C.U.R.A., Leatherhead, England (1967).
/16/ Khan,L.M., G.Greeves and D.M.Probert: "Prediction of soot and nitric oxide concentrations in diesel engine exhaust". I.Mech.E. Conference on air pollution control in vehicle engines, Solihull (1971).

Chapter 25

A METHOD FOR CALCULATING THE FORMATION AND COMBUSTION OF SOOT IN DIESEL ENGINES

I. M. Khan and G. Greeves

CAV Limited, London, England

Abstract

The calculation of radiative heat transfer in hydrocarbon flames requires knowledge of the temperature and soot concentration. There has been a lack of knowledge particularly on methods of calculating the soot concentration. A semi-empirical equation for the local rate of soot formation is proposed and this is applied to a jet mixing model of the direct injection diesel combustion process. The model developed predicts the experimentally observed variation of exhaust soot with engine fuelling, injection timing, injection rate, engine speed, and air swirl ratio. The model is then extended to include the calculation of soot particle coagulation and soot particle combustion. Some of the test conditions are recalculated with the extended model and although a substantial fraction of the soot formed is burnt, this fraction remains approximately constant so that the net soot in exhaust is controlled largely by the soot formation process.

Although the method was developed for calculating soot emissions from diesel engines, the technique is applicable to the calculation of thermal radiation in diesel and similar combustion systems.

NOMENCLATURE

A	:	Mass of air entrained, g.
A_t	:	Total mass of air in cylinder, g.
a	:	Mass of air consumed, g.
B	:	Coagulation constant, $cm^3 s^{-1}$.
C_c	:	Rate constant for soot combustion, g $^oK^{\frac{1}{2}}$/at s cm^2.
C_s	:	Soot formation rate coefficient, mg/Nm s.
D	:	Diffusivity constant, cm^{-1}.
D_p	:	Diameter of particle or coagulate, cm.
D_n	:	Nozzle hole diameter, cm.
E_c	:	Activation energy for soot combustion, cal/mole.
E_r	:	Air entrainment ratio.
E_s	:	Activation energy for soot formation, cal/mole.
H	:	Lower calorific value of fuel, cal/g.
M_a	:	Mass of micromixed air, g.
M_f	:	Mass of micromixed fuel, g.
M_p	:	Mass of particle or coagulate, g.
N	:	Coagulate number density after time Δt, cm^{-3}.
N_o	:	Initial coagulate number density, cm^{-3}.
n	:	Equivalence ratio exponent for soot formation.
P	:	Cylinder pressure, atm.

2.2 Micromixing and heat release

The macromixed quantities of fuel and air within the jet boundaries are assumed to micromix by turbulent diffusion according to the following equations (2).

$$\dot{M}_a = D \cdot V_f \ (A - Ma) \tag{6}$$

$$\dot{M}_f = D \cdot V_f \ (X - M_f) \tag{7}$$

Heat release is computed from the micromixed quantities of fuel and air and account is taken of chemical kinetics during the first phase of combustion. For the first phase of combustion, experimental values for ignition delay are used for the engine being modelled, and the rate of heat release from the fuel and air premixed during ignition delay is assumed to follow a triangular rate law with a base length of 6 degrees of crank angle. The instantaneous curves of fuel injected, air entrained, micromixed air, micromixed fuel, and heat release are shown in fig. 2 for a typical engine condition where the numbers 1, 2, 3, 4, 5 on the curves represent the following quantities.

1. Air entrained $= A(H/15A_t)$
2. Fuel injected $= X(H/A_t)$
3. Micromixed air $= M_a(H/15A_t)$
4. Micromixed fuel $= M_f(H/A_t)$
5. Rate of heat release

The diffusivity constant D controls the rate of micromixing and is chosen to give best match of the curves for calculated and experimental heat release. To account for the effect of air motion it has been found that the value of D must vary progressively with air swirl ratio and engine speed (1).

The thermodynamic mean temperature T is calculated from the heat release, the thermodynamic properties of the cylinder gases, and by deducting the heat loss to the walls. Variations of heat loss with changes in engine speed, injection timing, and fuelling are taken into account by using appropriate experimental data.

3. Soot Formation and Soot Combustion

Ideally the detailed nature of temperature and concentration history produced by the fuel/air mixing process should be taken into account in calculating soot formation and soot combustion. A much simpler approach is desirable at this stage, especially since detailed experimental information on these aspects is not yet available.

The approach adopted is to divide the entrained zone 'e' shown in figure 1 into three zones, namely the fuel rich zone 'c' where soot formation takes place, the products zone 'd' where intense reaction takes place and an outer zone which is mainly air. Also outside the entrained region there is a zone 'a' and it is assumed that the temperature of the air in this zone is given by

$$T_a = T_i \left(\frac{P}{P_i}\right)^{\frac{\gamma - 1}{\gamma}} \tag{8}$$

Equation 8 together with the values for heat release and heat loss enables the mean jet temperature to be calculated.

3.1 Model for soot formation

A review of the literature showed that little is known in quantitative terms on the mechanism of soot formation in flames but all indications are that under high temperature and short reaction time conditions encountered in a diesel combustion chamber, the overall mechanism of soot formation may be characterised by an Arrhenius type equation. This as well as the experimental observations on a number of diesel engines led to the following equation for the local rate of soot formation (1).

$$\frac{dS}{dt} = C_s \frac{V_u}{V_{ntp}} \phi_u^n P_u e^{-E_s/RT_u} \tag{9}$$

Engine experiments were carried out to determine a value of E_s = 40000 cal/mole for the fuel used by varying cycle temperature whilst holding other conditions constant. Similarly, experimental data which gave large changes in the ϕ_u history were used to determine a value of n = 3 for the equivalence ratio exponent. Thus according to equation 9 the rate of soot formation depends on the partial pressure P_u, the temperature T_u and the equivalence ratio ϕ_u history of the unburnt fuel. These variables must be defined in the soot formation zone. In the absence of knowledge on heat transfer into the fuel-rich zone, the temperature T_u in the soot formation zone is taken to be the thermodynamic mean temperature of the whole jet. The local equivalence ratio ϕ_u in the fuel rich zone is defined by

$$\phi_u(\text{mean}) = 15\ (X - x)/(A - a) \tag{10}$$

$$\phi_u = \frac{X}{x}\ \phi_u(\text{mean}) \tag{11}$$

The factor X/x is (fuel injected)/(fuel mixed) and represents the stratification of fuel within the jet.

3.2 Model for soot coagulation and soot combustion

After formation the soot is assumed to exist in the form of particles of 250Å diameter which then undergo coagulation to larger particle sizes and/or combustion. The equation for reduction in particle number density by coagulation is based on a

p_i : Cylinder pressure at ignition, atm.
p_j : Injection pressure, atm.
p_{O_2} : Partial pressure of oxygen, atm.
p_u : Partial pressure of unburnt fuel, kN/m^2
R : Universal gas constant, cal/g mole degK.
r : Radius of wall jet front from axis, cm.
S : Soot formation per unit volume, g/m^3 at NTP.
T : Thermodynamic mean temperature in cylinder, degK.
T_a : Temperature of air zone, degK.
T_c : Gas temperature for soot combustion, degK.
T_i : Temperature at ignition, degK.
T_u : Local temperature for soot formation, degK.
t : Time since beginning of injection, s.
U : Free jet front velocity, cm/s.
V_f : Jet front velocity, cm/s (= U or W).
V_{NTP} : Volume of cylinder contents at NTP, m^3.
V_u : Volume of soot formation zone, m^3.
W : Jet front velocity on the wall, cm/s.
W_o : Initial velocity on the wall, cm/s.
X : Fuel injected, g.
x : Fuel prepared for heat release neglecting chemical kinetics, g.
Y : Zone width.
y : Dimension across zone.
Z : Ratio of gas density relative to air at NTP.
γ : Isentropic index for air zone.
δ : Jet thickness on wall, cm.
δ_o : Initial jet thickness on wall, cm.
θ : Cone half angle.
ϕ_u : Local unburnt equivalence ratio for soot formation.
ϕ_u(mean) : Mean unburnt equivalence ratio in jet.
ϕ : Local original fuel/air equivalence ratio.

1. Introduction

The emission of soot and other pollutants in the exhaust of diesel engines is of particular interest to the authors' company. A model was developed to calculate the emission of soot and nitric oxide in the exhaust of a direct injection diesel engine (1) so that the effect of a number of combustion system parameters could be investigated. The nature of pollutant formation made it essential to calculate the fuel/air mixing, heat release, gas temperature, to take some account of the temperature and composition profiles within the combustion chamber, and to model the soot concentration history. This model has been developed to include soot combustion and it is the work on soot formation and soot combustion that is reported here and also the implications of this work in modelling heat transfer in the diesel engine.

The ability to calculate soot concentration is of direct interest to the calculation of thermal radiation in diesel and other combustion systems, and the technique used to model the stratified nature of temperatures in the diesel combustion chamber is of interest both for convective and radiative heat transfer calculations.

2. Model for Fuel/Air Mixing and Heat Release

The model is based on the fuel spray jet model of Grigg and Syed (2). This model predicts the rate of heat release mainly by calculating the rate of air entrainment, or macromixing, and the rate of micromixing of fuel and air within the airborne jet. This model has been extended to include the impingement of the jet on the wall and subsequent air entrainment by the wall jet (1) and is illustrated in figure 1. The model may be summarised as follows.

2.1 Air entrainment

The velocity of the front of the airborne jet is determined by a penetration equation based on the data of Schweitzer for penetration of a diesel fuel spray in a quiescent charge.

$$U = \left(\frac{0.260 \times 10^6 \; D_n \; P_j^{\frac{1}{2}}}{Zt}\right)^{\frac{1}{2}} \qquad (1)$$

Application of the momentum equation along the jet axis gives the cone angle and hence the jet width.

$$\tan^2 \theta = 1.09 \times 10^{-3} \; Z \qquad (2)$$

For transition of the jet front from a free jet to a wall jet, the loss in kinetic energy during transition is neglected and the equations of reference (3) are applied to describe the velocity and jet thickness on the wall.

$$W = W_o \left(\frac{r}{r_o}\right)^{-1.06} \qquad (3)$$

$$\delta = \delta_o \left(\frac{r}{r_o}\right)^{1.006} \qquad (4)$$

Equations 1, 2, 3, 4 have been solved to give the boundaries of the jet and the volume of air entrainment at any time from the beginning of fuel injection. After termination of fuel injection, a back edge of the jet forms and the jet boundaries and volume entrained are modified accordingly. No account has yet been taken of air motion, so a factor E_r is introduced to account for the relative variations of air entrainment which occur with changes in air swirl ratio and engine speed. The factor E_r is defined as:

$$E_r = \frac{\text{Actual air entrained}}{\text{Air entrained under stagnant conditions}} \qquad (5)$$

There is no knowledge as to the absolute value of E_r so a value of $E_r = 1.0$ is assigned for the engine conditions of figure 2. Also it has been found that if E_r is varied progressively with air swirl ratio and engine speed then the calculations of the model give a satisfactory agreement with experiment (1).

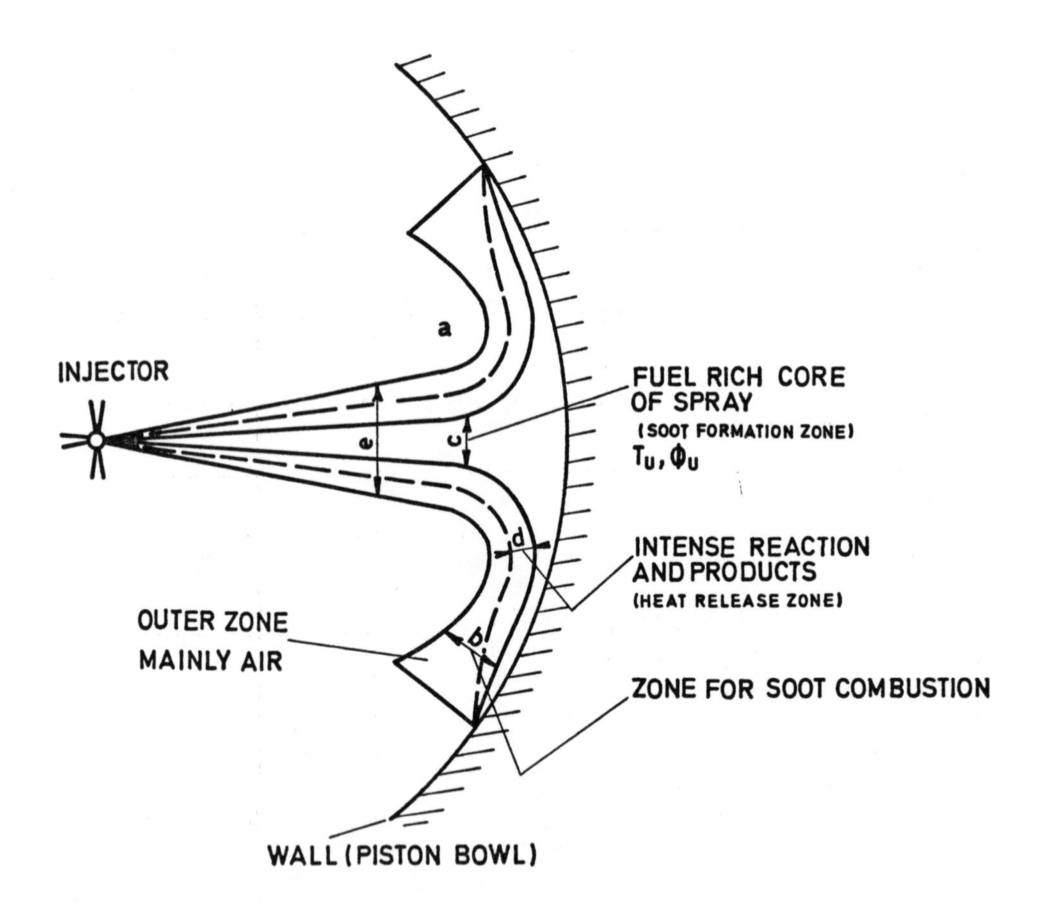

Fig. 1 Free jet and wall jet model for direct-injection combustion system showing zones of soot formation and soot combustion

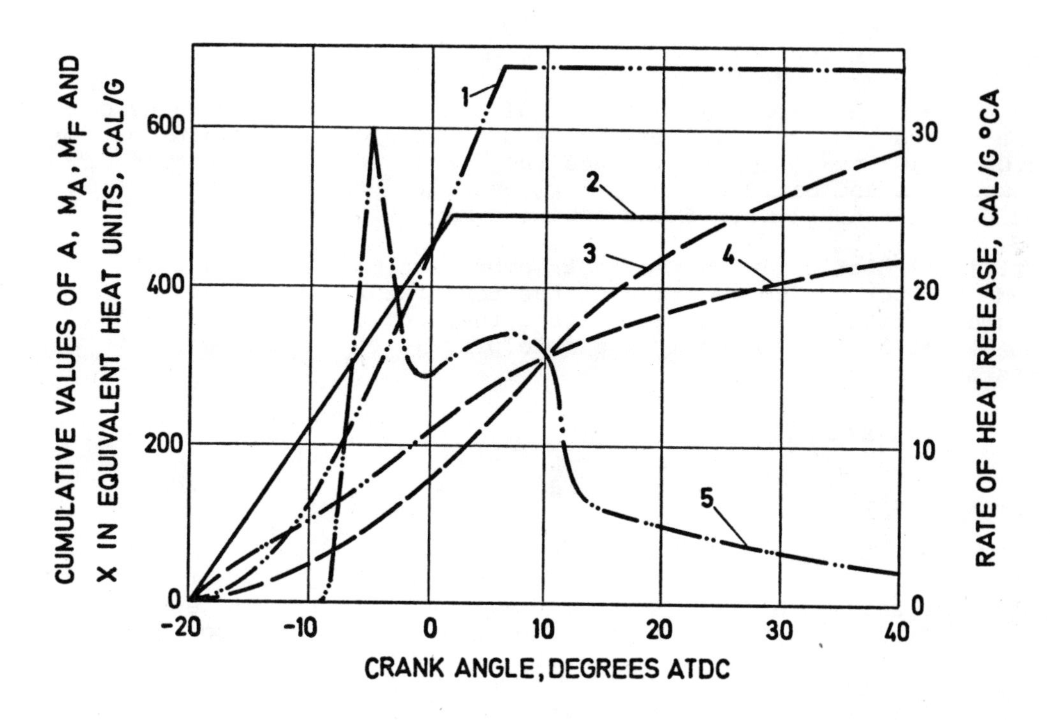

Fig. 2 Calculated instantaneous values for air entrainment micro-mixing, and rate of heat release versus crank angle; Engine speed = 2000 rev/min, Fuelling = 60 mm^3/stroke, Injection timeing = 20°btdc.

Whytlaw and Gray (4) type equation and is:

$$\frac{1}{N} - \frac{1}{N_o} = B\,\Delta t \qquad (12)$$

The constant B is chosen so that the final coagulate diameter in the exhaust agrees with the observations reported in reference (5) for one test condition. The mean particle size after coagulation at each step is calculated and the reduction in mass of soot by combustion is calculated by using an equation similar to Lee (6).

$$\frac{dM_p}{dt} = -C_c \pi\, D_p^2\, P_{O_2}\, T_c^{-\frac{1}{2}}\, e^{-E_c/RT_c} \qquad (13)$$

An activation energy E_c of 39300 cal/mole is used and a value of the constant C_c of 7.5 x 10^4 is used instead of Lee's value of 1.085 x 10^4 gm$^\circ$K$^{\frac{1}{2}}$/cm^2 at , since the former is for combustion of coagulates and based on the experimentally determined consumption of diesel exhaust soot in an engine compression cycle (5).

Trial calculations with the soot combustion model showed that some account had to be taken of the temperature and oxygen concentration stratification within the soot combustion zone 'b' in figure 1. Consequently a simple profile of the following form was assumed.

$$\phi = \frac{\text{constant}}{y} \qquad (14)$$

where $o < y > Y$

and $\phi = \phi_u$ at $y = 0$

Thus the assumed equivalence ratio within the zone 'b' decays from the equivalence ratio ϕ_u of the unburnt fuel zone 'c' to that of pure air at the outer boundary, the characteristic zone width being Y. The profile is divided into a number of elements and the temperature T_c for soot combustion is defined as follows. The adiabatic equilibrium temperature at the stoichiometric condition ϕ = 1.0 is calculated, account being taken of dissociation, and the temperatures in the elements of leaner mixture calculated from the stoichiometric temperature by assuming adiabatic mixing with air of temperature T_a (as given by equation 8) in the appropriate proportion.

At each step of the computer calculation the soot formed in the unburnt fuel zone is assumed to be transferred to the zone for soot combustion and to be uniformly distributed within the profile for subsequent coagulation and combustion calculations within each element. The contributions of each element to soot coagulation and soot combustion are summed to give the total remaining soot at any instant and the corresponding mean coagulate diameter.

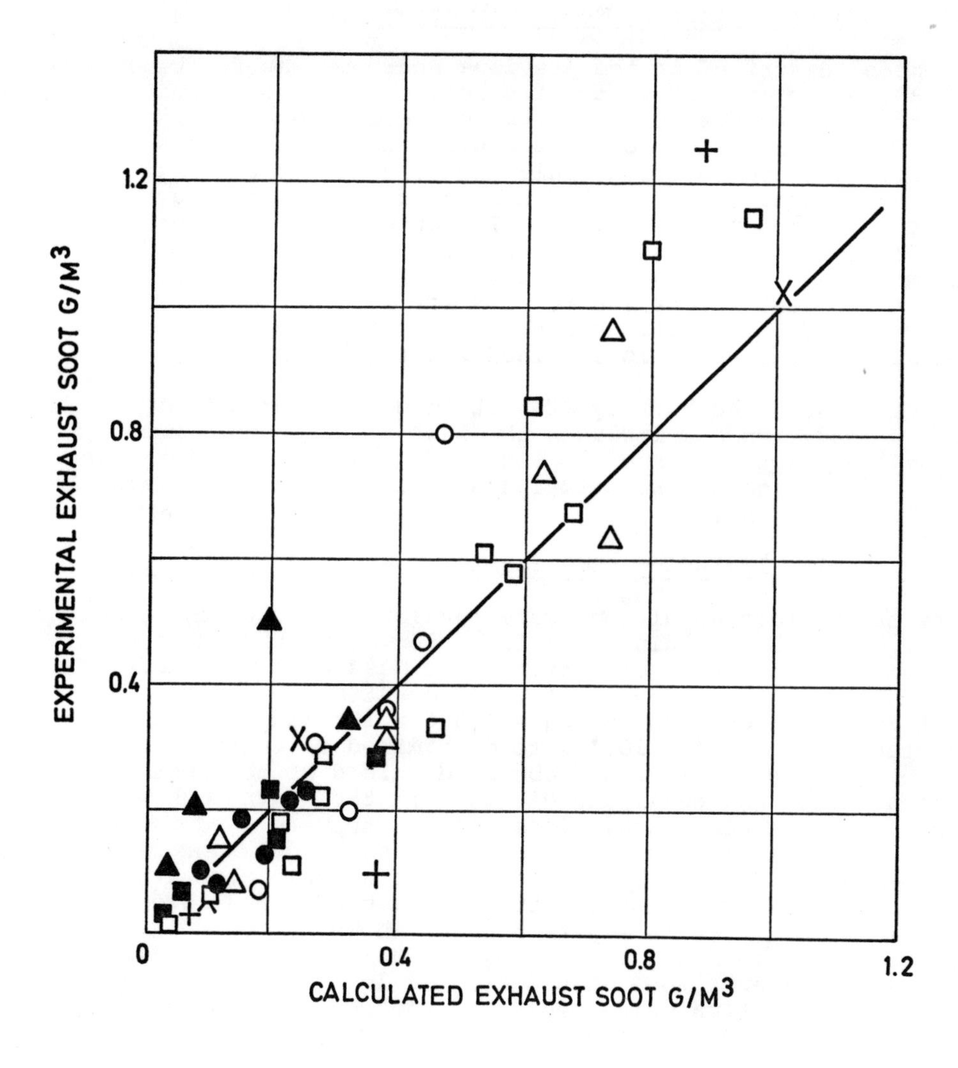

Fig. 3 Experimental exhaust soot versus calculated values for the model which neglects soot combustion; plotted points include conditions for two rates of injection, three levels of air swirl,
engine speed = 1100, 2000, 2700 rev/min,
injection timing = 35 to 5° BTDC, and
fuelling = 30 to 75 mm^3/stroke

4. Calculated Results and Comparison with Experiment

4.1 Results neglecting soot combustion

The model described in the previous sections was first used to calculate the exhaust soot while neglecting soot combustion, that is, the soot exhausted was entirely controlled by equation 9 for soot formation. The model was applied to three engines but to simplify the presentation, only the results for the engine A are presented here. Engine A had a bore of 98 mm , stroke of 127 mm and compression ratio of 16:1. The experimental values of exhaust soot are plotted versus the calculated vaues in figure 3 for a value of the soot formation constant of $C_S = 4.68 \times 10^5$ which was chosen for best fit of the results for 2000 rev/min. The agreement of calculation and experiment is considered to be good and similar agreement is obtained for engines B and C.

Figure 4 shows the history of soot formation for a typical engine condition and includes the instantaneous values of equivalence ratio ϕu and temperature T_u in the soot formation zone, and the soot concentration S in the cylinder referred to N.T.P. conditions.

4.2 Results including soot formation and soot combustion

Some of the test conditions were recalculated with the complete model, that is including soot formation, soot particle coagulation and soot combustion. The mass of soot exhausted is now determined by the net effect of equations 9 and 13. A soot formation constant of $C_S = 1.376 \times 10^6$ and coagulation constant of $B = 1.4 \times 10^{-4}$ were determined so that there was agreement of calculated and observed values of the soot concentration and coagulate diameter in the engine exhaust for a typical engine condition. These constants were used for all subsequent calculations. According to the model approximately 60% of the soot formed is burnt and in general this percentage remained almost constant for the engine conditions investigated. This is illustrated in figure 5 which shows the calculations for the effect of changing the timing of fuel injection. Also the calculated exhaust soot values are little different to the original prediction which neglected soot combustion and used a lower formation constant. Figure 6 shows the history of soot formed, soot released (that is, soot formed less soot burnt), and the mean coagulate diameter.

5. Validity of the Model

The results presented here and elsewhere (1) (7) indicate that a reasonable degree of success has been achieved in modelling the processes of fuel/air mixing, heat release and soot formation. The model, however, is a very simple approximation of the actual processes which occur.

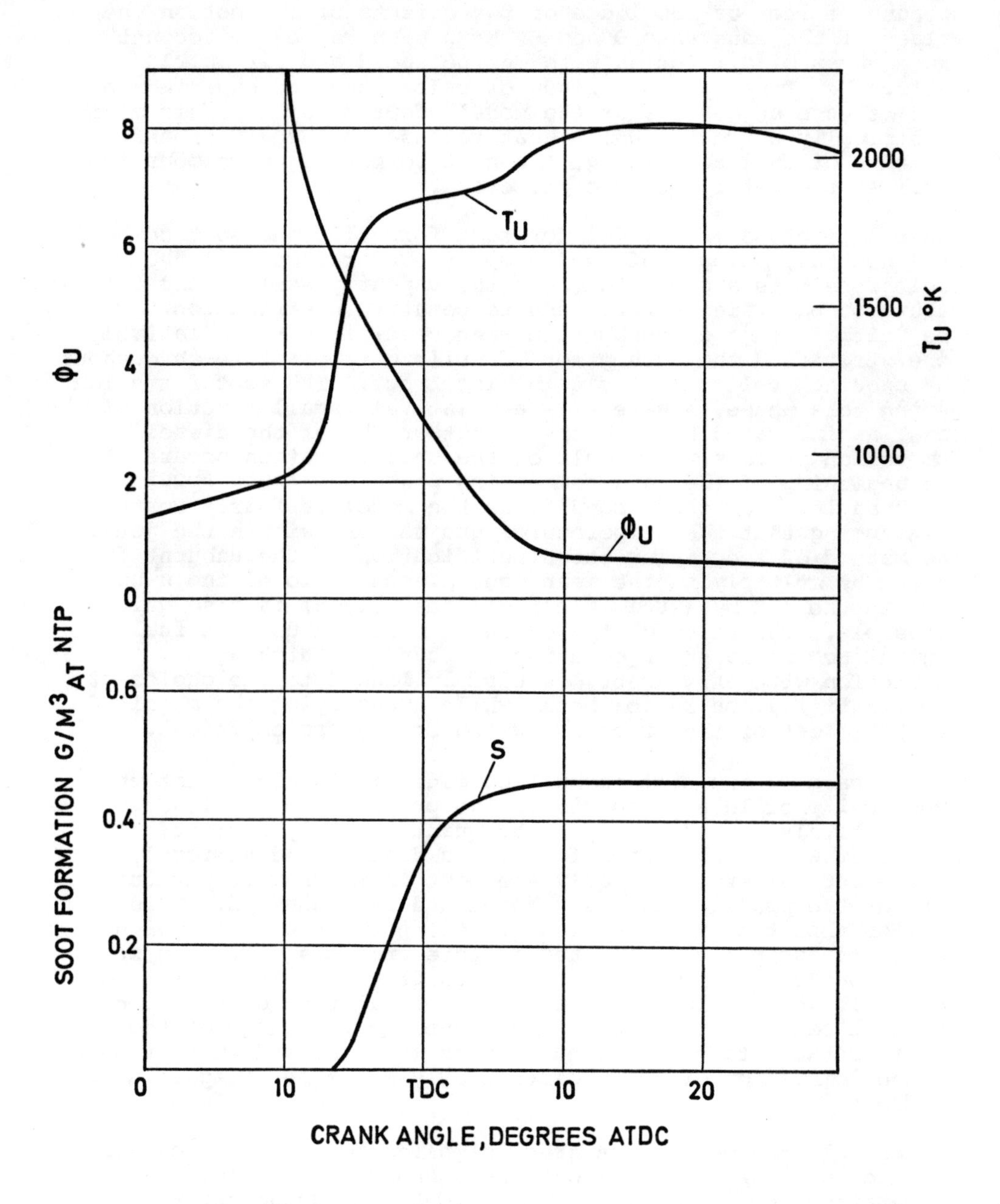

Fig. 4 Instantaneous values of ϕ_u, T_u and cululative soot formation; engine speed = 20s0 rev/min; fuelling = 60 mm^3/stroke; injection timing = 20° BTDC.

Firstly the basic direct-injection fuel/air mixing process of a fuel spray injected into a cross stream air swirl and bounded by a piston bowl wall has been greatly simplified. In particular, because of lack of knowledge of the effects of air motion the values of the constants D and E_r have been varied to account for the changes of air motion with engine speed and air swirl ratio. Nevertheless comparison of calculated and experimental heat release suggests that the model adopted for fuel/air mixing is basically correct. While heat release is largely dependent on the rate of fuel air mixing, the soot formation and combustion involves the details of the processes.

The relatively simple model for soot formation and soot combustion does not take account of fuel droplet evaporation (all the unburnt fuel present is assumed to be in the vapour phase) or the detailed nature of equivalence ratio and temperature distribution. During the initial stage of combustion when conditions are relatively more stratified the assumption of uniform temperature throughout the unburnt fuel is likely to be incorrect. The soot formation during this phase, however, is a relatively small fraction of the total as indicated by high speed photography of the diesel combustion process. The bulk of the soot formation occurs at the beginning of the main combustion phase and lasts about 10° crank angle. For these conditions the model is fairly correct in assuming that the temperatures are uniform within the jets. The method of specifying the stratification of the unburnt fuel charge by multiplying the mean equivalence ratio of the unburnt fuel in the jet by (fuel injected)/(fuel mixed) is also quite plausible. The value of the exponent n of the unburnt fuel equivalence ratio ϕ_u in equation 9, found by matching prediction with experiment, is closely linked to the choice of the stratification ratio; both, while quantifying the magnitude of the effect of the processes which occur, are empirical.

In the case of the soot combustion model it is clear that the empirical profile assumed within the products zone, although qualitatively correct, may not be quantitatively correct. It was assumed that the soot once formed was distributed uniformly throughout the profile. Only the soot in the low temperature part of the profile remains unburnt and is exhausted. This implies that the exhausted soot is the portion of soot which forms and escapes the high temperature products oxidation zone in the process of turbulent diffusion from the rich to the lean zone. It should also be noted that the validity of Lee's equation for soot combustion for conditions outside the range of the experimental data has been questioned (8), (9) and further data on the combustion of soot, particularly at higher temperatures, are needed.

In general, the model is a great simplication of the processes which occur. Experimental data are needed on the time histories of temperature and concentration distributions throughout the combustion chamber before a more detailed model of the diesel combustion process can be developed and verified.

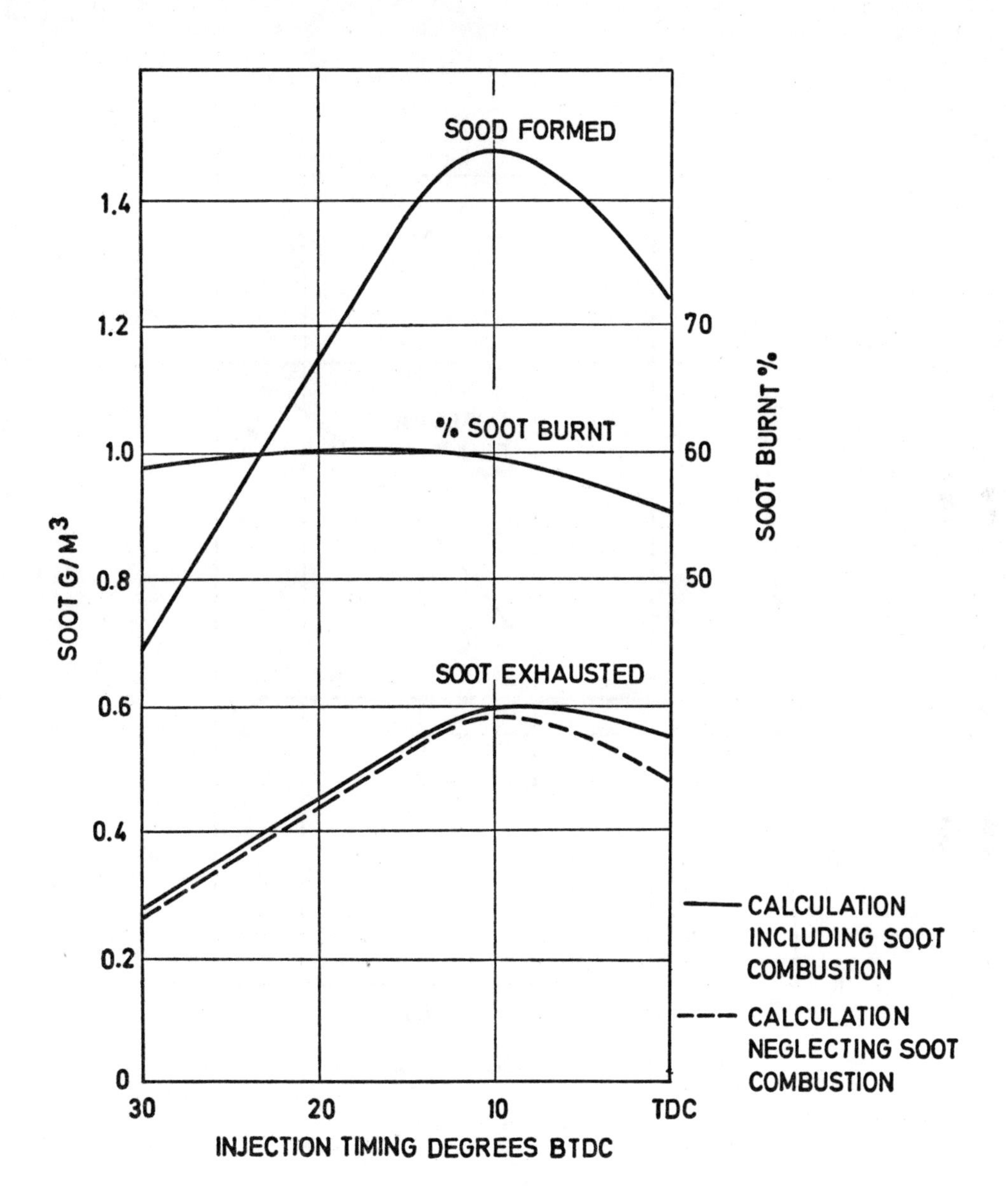

Fig. 5 Calculation of the effect of injection timing on exhaust soot using the model which includes soot combustion; engine speed= 2000 rev/min; fuelling = 60 mm^3/stroke

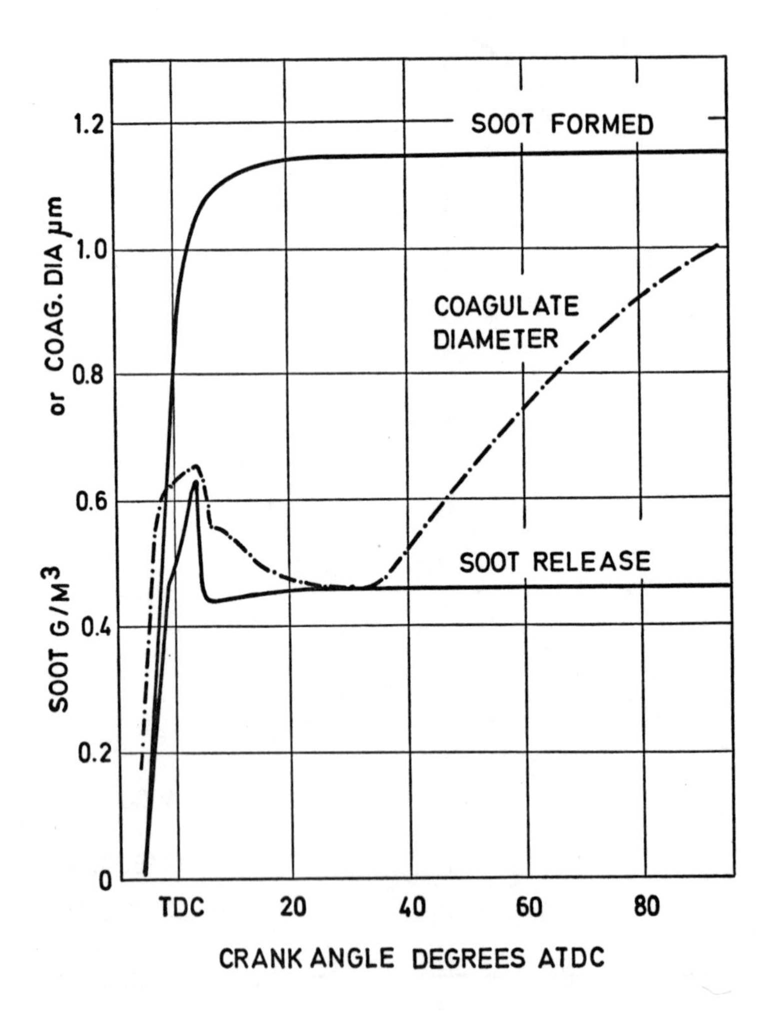

Fig. 6 Calculated instantanous values of the soot formed, the soot released, and the mean coagulated diameter of the soot released for the model which includes combustion; engine speed= 2000 rev/min, fuelling = 60 mm^3/stroke, injection timing = 20° BTDC.

6. Application of the Model to Studies of Heat Transfer

Previously the modelling of gas to wall heat transfer in diesel engines has assumed a simple formulation for the convective and radiative heat transfer, for example the equation by Annand (10). Annand's equation has been applied through the combustion part of the engine cycle (11) with the assumption of a mean gas velocity, a thermodynamic mean temperature for the whole of the cylinder contents, and in addition the form of the equation implies a radiating source of constant emissivity. Such assumptions take no account of the stratified nature of the diesel combustion process.

The model reported here gives data on the gas velocities of the mixing fuel jet, the temperature within the jet and also the history of soot concentration. This provides information for study of both convective and radiative heat transfer. The emissivity of the soot clouds, which are the main source of radiation in the diesel engine, can be calculated from the soot concentration history and the optical properties of the soot. Unfortunately time did not permit the carrying out of such an investigation here.

7. Conclusions

A relatively simple fuel jet mixing model has been developed to describe the fuel/air mixing and heat release process in a direct injection diesel engine. This model together with the proposed equation for soot formation has been used to successfully predict the effects of injection timing, injection rate, fuelling, air swirl ratio, and engine speed on exhaust soot concentration.

The model confirms that soot formation for a given fuel is controlled by the amount of insufficiently mixed fuel, its equivalence ratio and temperature. Also it appears that even though a substantial fraction of the soot formed may be burnt the net soot is determined largely by the characteristics of the soot formation process rather than by the soot combustion process. It is probably that the proposed soot formation equation will be applicable to other combustion systems such as the indirect injection engine and the gas turbine, provided adequate models representing the mixing processes are formulated though soot combustion may be more important for these systems.

The model, although relatively simple, provides more detailed information than previously available for the study of heat transfer in diesel engines. In particular a more realistic treatment of the radiative component of heat transfer is now possible.

ACKNOWLEDGEMENTS

The authors wish to thank the Directors of CAV Ltd. and Lucas Group Chief Engineer Mr Ewen McEwen for permission to publish this paper. In particular they are grateful to Dr A E W Austen, Technical Director and Mr H C Grigg, Chief Research Engineer, for their support and discussions.

REFERENCES

/ 1/ Khan, I.M., G. Greeves and D.M. Probert: "Predictions of Soot and Nitric Oxide Concentrations in Diesel Engine Exhaust". Symposium on Pollution Control in Transport Engines, I.Mech.E., London, November 1971.

/ 2/ Grigg, H.C., M.H. Syed: "The Problem of Predicting Rate of Heat Release in Diesel Engines". Symposium Diesel Combustion, Proc. I.Mech.E., London, 1969-70, Vol. 184 (pt. 3J), p.192.

/ 3/ Glauert, M.B.: "The Wall Jet". J. Fluid Mech., December 1956.

/ 4/ Green, H.L. and W.R. Lane: "Particulate Clouds, Dusts, Smokes and Mists". 2nd edition, E and F.N. Spon: London 1964.

/ 5/ Khan, I.M., C.H.T. Wang and B.E. Langridge: "Coagulation and Combustion of Soot Particles in Diesel Engines". Combustion and Flame, Vol. 17, p.409-419, 1971.

/ 6/ Lee, K.B., M.W. Thring and J.M. Beer: Combustion and Flame, Vol. 6, p.137, 1962.

/ 7/ Khan, I.M., G. Greeves and C.H.T. Wang: "Factors Affecting Smoke and Gaseous Emissions from Direct Injection Engines and a Method of Calculation". S.A.E. paper no. 730169.

/ 8/ Magnussen, B.F.: Thirteenth Symposium on Combustion 1971, p. 869.

/ 9/ Parker, K.H. and O. Guillon: Thirteenth Symposium on Combustion 1971, p.667.

/10/ Annand, W.J.D.: "Heat Transfer in the Cylinders of Reciprocating Internal Combustion Engines", Proc. I.Mech.E., London 1963, Vol. 177, p. 973.

/11/ Whitehouse N.D., W.J.D. Annand and T.H. Ma: "Heat Transfer in Compression Ignition Engines". Proc. I.Mech.E. London 1971, Vol. 185.

Chapter 26

FLAME RADIATION IN HIGH SPEED DIESEL ENGINES

G. Sitkei

Technical University of Budapest, Hungary

Abstract

In a direct injection engine experiments were made to determine flame emissivities at different operating conditions. Measurements have shown that the absorption factor of the flame varies similarly to the mass burning velocity of the cylinder charge. Also investigated is the effect of fuel additives, which decrease not only the smoke level in the exhaust but also the thermal radiation of the flame.

NOMENCLATURE

F : instantaneous value of the cylinder surface, m^2
k : absorption factor of the flame, 1/m atm
ℓ : mean-path-length, m
p : cylinder pressure, atm
T_b : black body temperature of flame, ^oK
T_f : true fame temperature, ^oK
V : instantaneous value of the cylinder volume, m^3
: flame emissivity

1. Introduction

Experiments conducted in the last few years by the author /1-3/ and other researchers /4,5/ have shown that the heat transfer due to flame radiation comprises a considerable portion of total heat transfer. Based on experimental data, new calculation methods were proposed to complete the existing ones /2,4/.

Radiation in Diesel engines is due to carbon particles in combustion gases. Unfortunately, very little experimentation has been made regarding the carbon formation and combustion during the combustion process. Only in the works of ROTAR /6/ and KHAN /7/ was an attempt made to investigate the process of carbon formation and to determine the amount of soot exiting through the exhaust. The work done by ROTAR is interesting because he has measured local carbon concentration during the combustion process. His experimental results are seen in Fig. 1.

KHAN /7/ examined the effect of directly controllable and fundamental variables on smoke formation. He stated that soot release is largely dependent on the diffusion burning of flame. An increase in duration of diffusion burning gives rise to more soot in the exhaust.

Our experiments formerly conducted have shown that the variation of the absorption factor of flame in prechambered engines is similar to the mass burning velocity of the cylinder charge. In direct injection

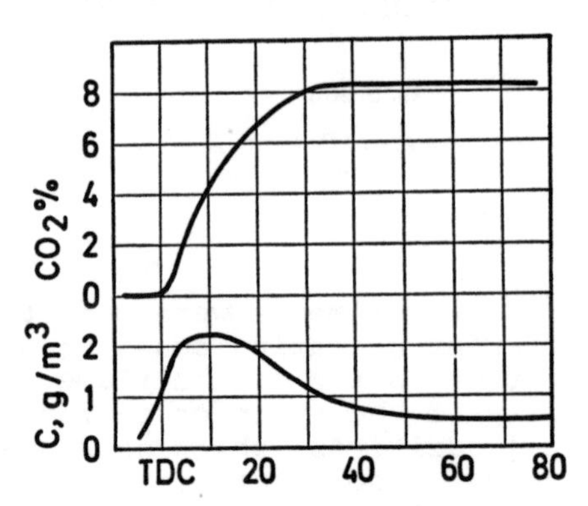

Fig. 1. Carbon and carbon dioxide concentration in a direct injection engine /6/.

engines, however, the heat release rate has a different course than in chambered engines. Therefore, it may be assumed that the flame emissivities in direct injection engines have also another history in the function of the crank angle. To answer this question some experiments were performed with a direct injection engine.

At the present time different fuel additives are used to reduce the net soot in the exhaust of Diesel engines. A decrease in carbon concentration will reduce the emissivity of the flame, and consequently the heat transferred by radiation. On the other hand, barium-ashes arising during combustion give additional particles, increasing the radiation capability of the flame. This combined effect will result in the estimated emissivity of the flame.

In his paper KHAN /7/ has pointed out that in a Diesel engine the premixed part of the fuel is rapidly burnt as a non-luminous flame. The rest of the fuel burns as a diffusion flame with its characteristic continuum radiation due to the presence of carbon particles. If this statement is true, the carbon concentration and the emissivity of the flame in the first period of combustion (kinetic flame) should be much lower than in the diffusion period of the flame. This opinion regarding the carbon concentration is expressed in Fig. 3.7 of KHAN's paper /7/.

A comparison of the expected total carbon concentration variation with the measured one (see Fig. 1) shows definite disagreement. As will be seen later from our experiments, the highest values of the flame emissivities are obtained in the first period of the combustion process. To explain this phenomenon, further information is needed. Therefore, investigations concerning the formation and combustion of soot during the combustion period are of great importance.

2. Test set-up for radiation measurements

The engine used for conducting our experiments is a two-cylinder two-stroke cycle direct injection Diesel engine (bore 84 mm, stroke 100 mm, compression ratio 15, engine speed 1000-1500 rpm) with a conical combustion chamber located in the cylinder head. Radiation measurements were taken through a quartz window of 14 mm diameter. The measurement technique and evaluation of results obtained were the same, as described in /1-3/.

For calculation of flame emissivities, the following equation was used:

$$\epsilon = \left(\frac{T_b}{T_f} \right)^4 \qquad (1)$$

From known emissivities the k absorption factor was determined. The k factor is related to other engine and operating parameters and emissivities in the following form:

$$\epsilon = \epsilon_o \quad (1 - e^{-kpl}) \qquad (2)$$

where ϵ_o - emissivity of an infinitely thick radiating agent, taken to be 0.8 in calculation, and the mean-path-length is defined by the equation

$$l = 3.6 \frac{V}{F}$$

For the engine under test the values of p can be read from the p - V diagram. The true flame temperature was taken from BELINSKIY's results /8/ with minor modification in the vicinity of tdc (Fig. 2).

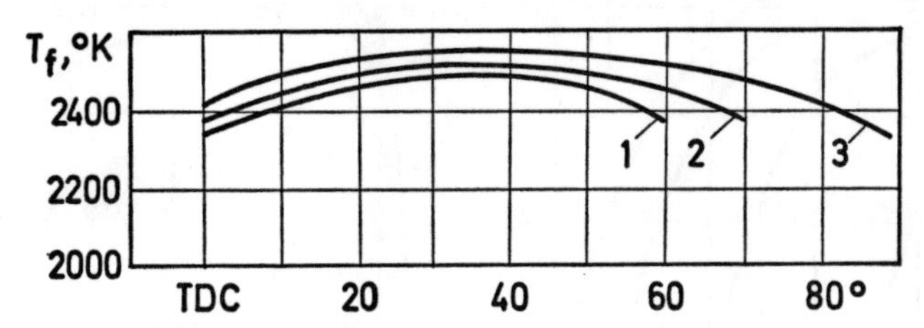

Fig. 2. True flame temperature versus crank angle /8/. 1 - no load, 2 - half load, 3 - full load.

The maximum value of the mean effective pressure in a two-stroke cycle engine is highly dependent upon the quality of scavenging. From earlier experiments the scavenging curves were known and thus the excess air factor could have been calculated as a function of mean effective pressure. This relationship is seen in Fig. 3.

The smoke in exhause was determined by a BOSCH-Smokemeter and the gravimetric values (g/m^3) of the smoke were calculated by means of correlation curves /9/. For direct estimation of variations in thermal loading temperature, measurements were taken in the inner surface of the combustion chamber wall.

3. Evaluation of results obtained

The variation of smoke concentration with and without fuel additives is given in Fig. 3. It can be observed that 1.0 per cent fuel additive gives a high reduction in the smoke level, especially at full load. The relatively high per cent value of the additive used in our experiments is to obtain as great a variation in emissivities as possible, because the cyclic variation of emissivities does not permit an accurate estimation if emissivities vary only slightly.

The emissivity values versus crank angle for different excess air factors are presented in Fig. 4. The emissivity attains a maximum value of 0.4 at full load. The maximum values of the emissivity are located at 10 deg. atdc practically for all loading conditions. This observation in direct injection engine is in contrast to those in chambered engines giving maximum values of about 30 deg. atdc.

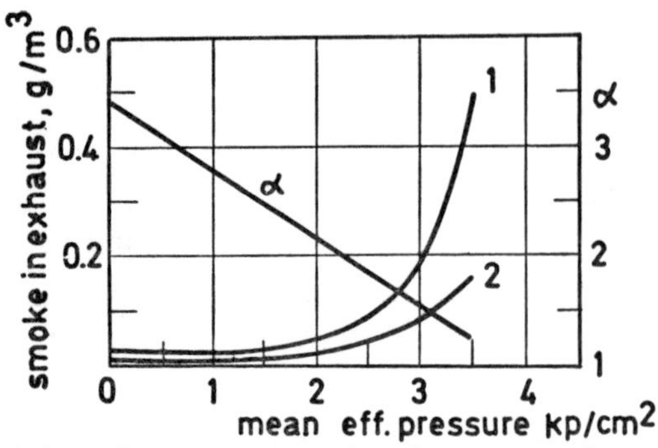

Fig. 3. Excess air factor and smoke concentration versus m.e.p. 1 - Diesel oil, 2 - Diesel oil + 1.0 per cent additive

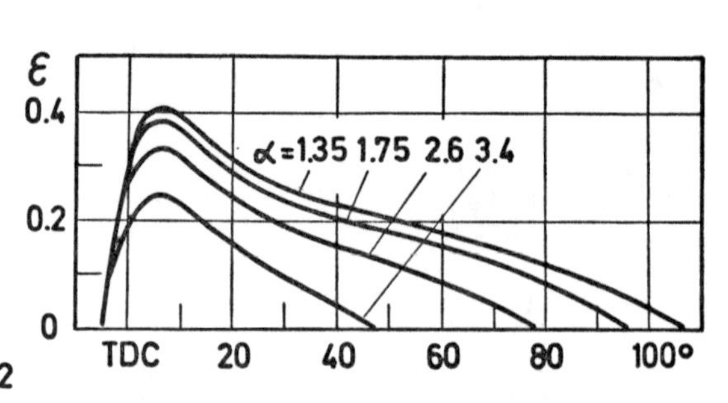

Fig. 4. Flame emissivities as a function of crank angle

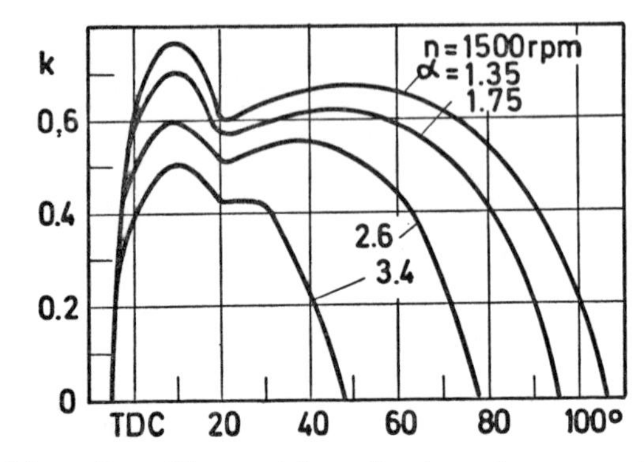

Fig. 5. Absorption factor k versus crank angle

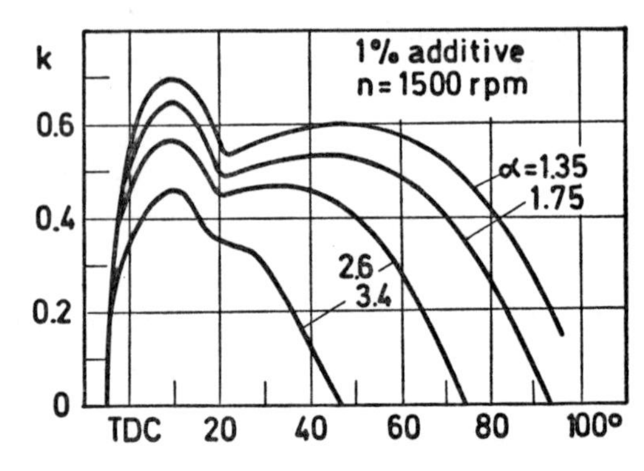

Fig. 6. Absorption factor k versus crank angle with fuel additive

Fig. 5 shows the relationship between the absorption factor k and crank angle for different excess air factors and without fuel additive. The course of the curves is characteristic for direct injection engines and is similar to the mass burning velocity of the cylinder charge. From these results an important conclusion can be drawn that the variation of emissivities in direct injection engines might be quite different from those of chambered engines, although their magnetudes are the same.

The variation of the absorption factor k as a function of the crank angle for different excess air factors and with a fuel additive is given in Fig. 6. The course of the curves is the same as in Fig. 5, but the k-values are lower. It can be also observed that the duration of combustion when using a fuel additive is shorter than without an additive. This gives a further reduction in heat transfer due to flame radiation.

The variation of heat transfer due to flame radiation with and without a fuel additive is seen in Fig. 7 for the engine tested. The integrated value of the heat transferred is 1950 kcal/h for Diesel oil and 1600 kcal/h using 1.0 per cent fuel additive. The reduction in flame radiation in this case is 18 per cent, which causes already a sensible decrease in heat load.

A direct estimation of the reduction of thermal loading can be seen in Fig. 8, where the cylinder head temperature is plotted against the mean effective pressure with and without a fuel additive. The reduction of temperature in the inner surface of the combustion chamber has a value of about 30°C at full load. This temperature decrease can be important if the initial thermal loading is too high.

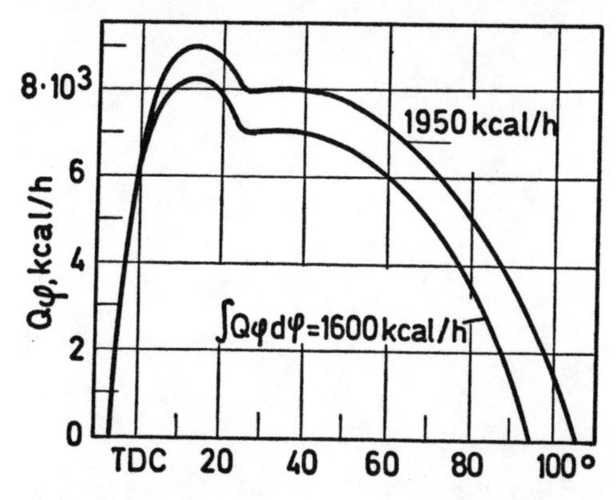

Fig. 7. Heat transfer due to flame radiation with and without fuel additive

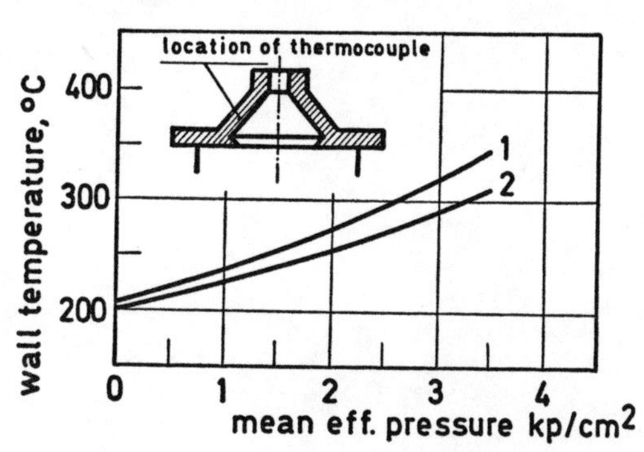

Fig. 8. Wall temperature versus mean effective pressure.
1 - Diesel oil, 2 - Diesel oil + 1.0 per cent additive

From these experiments it is evident that the use of fuel additive decreases not only the smoke level in the exhaust but also the thermal loading of the engine to a certain extent.

Another observation is that flame radiation in direct injection engines has its maximum in the first period of the combustion. This observation seems to be not in agreement without theoretical considerations concerning the mechanism of combustion in direct injection engines. Therefore, it may be assumed that the kinetic flame either does not appear in a clear form or causes carbon formation on its boundary surface. Further study of this phenomenon is highly desirable.

REFERENCES

/1/ Sitkei, G. und G. Ramanaiah: "Das Strahlungsvermögen der dieselmotorischen Flamme." Periodica Politechnica Budapest, No. 2. 1972.

/2/ Sitkei, G. and G. Ramanaiah: "A Rational Approach for Calculation of Heat Transfer in Diesel Engines." SAE paper 720027, Detroit, 1972.

/3/ Sitkei, G. and G. Ramanaiah: "Insight into the Mechanism of Luminous Flame Radiation in an Operating CI Engine." Archiwum Procesów Spalania, No. 4. 1971 (Poland).

/4/ Flynn, P. et al: "An Experimental Determination of the Instantaneous Potential Radiant Heat Transfer Within an Operating Diesel Engine." SAE Paper 720022, Detroit, 1972.

/5/ Oguri, T. and S. Inaba: "Radiant Heat Transfer in Diesel Engines." SAE Paper 720023, Detroit, 1972.

/6/ Rotar, I.: "Mixture Formation and Combustion in High Speed Direct Injection Engines." Trudy NILD, No. 1., 1955, Moscow.

/7/ Khan, I.: "Formation and Combustion of Carbon in a Diesel Engine." Paper presented at the Diesel Engine Combustion Symposium in London, April, 1970.

/8/ Belinskiy, L.: "Heat Radiation in the Combustion Chamber of a CI Engine." Trudy NILD, No. 1, 1955, Moscow.

/9/ Dodd, A. and Z. Holubecki: "The Measurement of Diesel Exhaust Smoke." MIRA Report No. 10, April, 1965.

PART III: OPEN FLAME HEAT TRANSFER

Chapter 27

RADIATION FROM POOL FLAMES

D. Burgess and M. Hertzberg

U.S. Department of the Interior, Bureau of Mines, Pittsburg Mining and Safety Research Center, Pittsburgh, Pa., U.S.A.

Abstract

Some radiation data from pool flames is summarized, and our understanding of the problem is reviewed. Spectral data yields a 1500°K temperature for hydrocarbon pool fires, which is consistent with the 40 percent maximum in the fraction of combustion energy radiated and with limit flame temperatures. A revised correlation of mass burning rate with $\Delta H_c/\Delta H_v$ is presented and derived fundamentally.

1. Introduction

The main interest in the radiations from fires concerns the need for estimating the heat flux to humans or structures nearby. The available data have been used, for example, to establish minimum distances between adjacent fuel storage tanks and minimum distances from hazardous operations to property lines /1/. We should like to review some old Bureau of Mines data that were gathered for this purpose, add some second thoughts relative to their interpretation, and present some more recent measurements that seem to demand a better rationalization.

2. Burning Rates - heat transfer from the flame to its fuel

A logical starting point is the presentation by Blinov and Khudiakov of the burning rates of pool fires as function of pool diameter. Curves for gasoline and for diesel oil are given in figure 1. As suggested by Hottel /2/, the heat flux to the liquid surface that provides steady-state evaporation to feed the flame is

$$\frac{q}{\pi d^2/4} = \frac{k_1\ (T_F - T_B)}{d} + U\ (T_F - T_B) + \sigma F\ (T_F{}^4 - T_B{}^4)(1 - e^{-\kappa d}). \qquad (1)$$

The first term on the right, conduction, is an edge effect that is important only at the smallest diameters. Since we could see little interest in mass fires of a few centimeter diameter, we merely persuaded ourselves that burning rate decreased linearly with increasing diameter (as in the conduction term of equation 1) and thereafter ignored conduction to the lip of the fuel container.

The second term on the right of equation 1, convection, pertains most clearly to the burning rate minimum at about 10 cm diameter. We observed that many flames sweep back and forth across the liquid surface at this diameter; they also assume odd shapes and show wide variations in burning rate /3/. Since these oddities tend to disappear in larger fires, we were overly happy to ignore convection and proceed to large fires in which the flame typically stood high above the fuel surface and burning rates were more nearly reproducible.

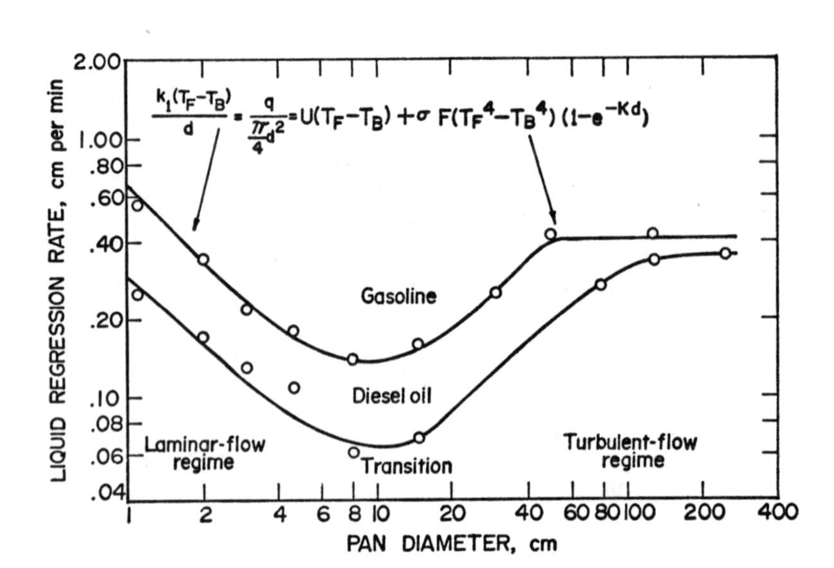

Figure 1. - Rates of diffusive burning of two liquid hydrocarbon blends.

Subsequent problems proved that we had been too hasty. Figure 1 shows that there is some difference between gasoline and diesel fuel that affects the burning rate at small diameter but not at large diameter. In ignoring this factor in small fires we erred badly in extrapolating to infinite diameter. Recently, Stark /4/ showed that some burning rates in small trays were inversely proportional to the fuel's boiling point. However, one of us has had some success in relating the convective contribution to the molecular diffusivity of the fuel /5/. Figure 2, for example, compares the sum of convective and radiative contributions with the data for gasoline. In the convective limit,

$$\dot{m}\ (\text{gm cm}^{-2}\text{min}^{-1}) \sim \left[D\,\alpha\ \frac{\Delta H_c}{\Delta H_v} \right]^{\frac{1}{2}}, \tag{2}$$

where m is the mass burning rate; D, the molecular diffusivity of the fuel; α, the average thermal diffusivity of flame gases, ΔH_c, the heat of combustion to CO_2 and water vapor, and ΔH_v, the sensible heat of vaporization. The minimum corresponds to the point at which the decreasing contribution from convection is about equal to the increasing contribution from radiation. In the convective range, the proportionality constant in equation 2 would be a mild function of pool radius r_o and the radius of convective eddies $\bar{r}_m$:

$$\dot{m} \sim \frac{1}{r_o^{\frac{1}{2}} \bar{r}_m^{\frac{1}{2}}} . \tag{3}$$

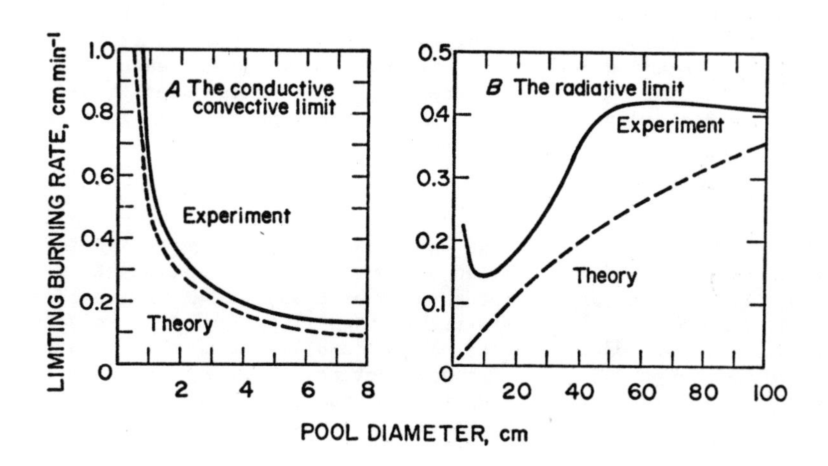

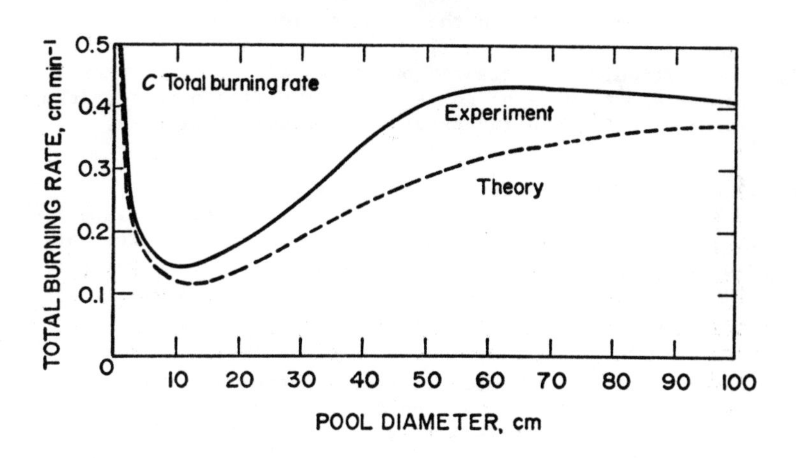

Figure 2. - The burning rate of pool flames of gasoline, theory and experiment.

The third term on the right of equation 1, radiation, whould apply primarily to large flames. We reasoned that if the flame shape factor, F, could be held constant, then by equation 1, the burning rate should vary simply as

$$v = v_\infty (1 - e^{-\varkappa d}), \tag{4}$$

d is the pool diameter, and κ is a constant related to emissivity and the ratio of pool diameter to flame thickness. Hoping to keep F and κ constant with varying pool diameter, we burned a variety of fuels under nearly windless conditions (out of doors at sundown on summer evenings). The results are shown in figure 3. The measurements represented by solid circles were used to evaluate κ, from which we drew the curves of figure 3 and calculated the burning rates, v_∞, at infinite diameter. Values of κ and v_∞ are given in table 1.

Finally the extrapolated burning rates, v_∞, were plotted as in figure 4. It was assumed from equation 5 that the burning rate would decrease with increasing ΔH_v; there are several easy ways to predict that heat transfer should increase with ΔH_c, so we plotted v_∞ against $\Delta H_c/\Delta H_v$. The line shown is given by

$$v_\infty \text{ (cm/min)} = 0.0076\ \Delta H_c/\Delta H_v \ . \qquad (5)$$

We were regrettably unaware that $(\Delta H_c/\Delta H_v)$ had already been designated as a diffusive transfer number /6/, called B. Our view of the straight line in figure 4 was that it was simply too good to be true; we had no justification, for example, for burning rate in cm/min rather than in $gms/cm^2 min$. We felt that heat transfers varied from mostly convective (methanol) to almost totally radiative (butane) and were unprepared for the argument that figure 4 suggests convective transfer to the liquid surface /7,8/ and for the continued interest in subdividing the heat flux into its convective and radiative components /9/.

Since liquid regression rates were measured, among other ways, with an array of thermocouples, it came to our attention that there was no temperature discontinuity in butane flames between the boiling liquid and the interconal gases. Therefore, butane is an example of a flame that can feed back heat to its fuel exclusively by radiation. For this reason, it is also a very hazardous fuel -- its cold vapors are about twice as dense as air and flow outward along the ground so that flame dimensions become independent of initial pool size.

There was a sharp temperature discontinuity at the fuel surface in methanol, benzene, and UDMH flames. This conforms to the absorption of flame radiations by these fuel vapors (Table 2). When a radiation meter was placed at the methanol surface, the radiant heat flux observed was about 25 percent of that required for fuel volatilization. For this reason it was stated previously that the methanol heat transfer was mostly convective.

Before the measurement was repeated with benzene, however, it occurred to us that this was not a pertinent observation at all. Some heat is surely transferred convectively from the hot benzene vapors to the relatively cold liquid; this might affect the vaporization somewhat like a change in flame emissivity or of shape factor, but this is not the "convection" of equation 1, which corresponds to a luminous zone sweeping across the liquid surface.

The subsequent measurements that have disturbed the simple empirical findings of figure 4 were burning tests of liquefied natural gas (LNG) conducted at Lake Charles, La. /10/. These were carried out in diked areas ranging from 2.4 to 38 m^2, in natural winds ranging upwards of 3 meters/sec. Twelve estimates of burning rate ranged from 0.80 to 1.23 cm/min; although all measurements were crude, the highest value, in particular, must be taken seriously since a 43 cm depth of liquid was consumed in 35 minutes. The extrapolated v_∞ from windless tests (table 1) was 0.60 cm/min.

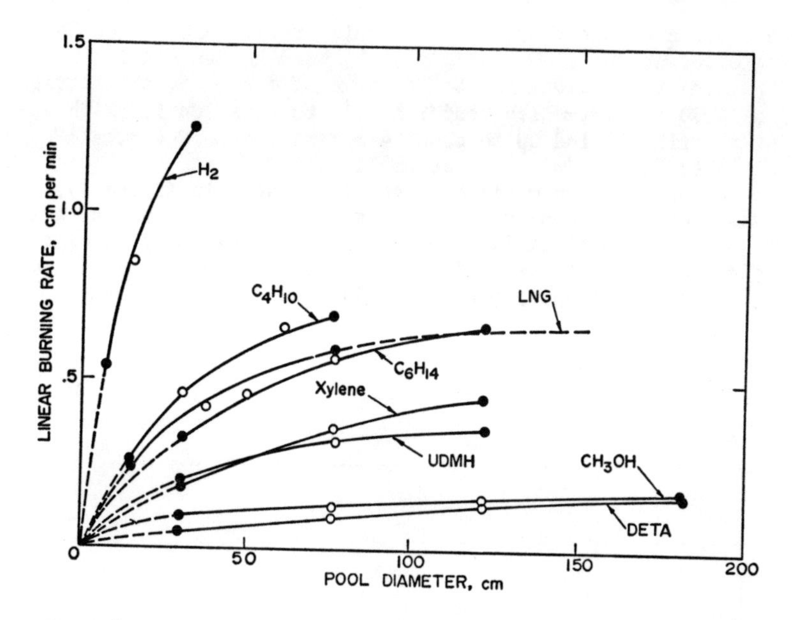

Figure 3. - Dependence of liquid burning rate on pool diameter.

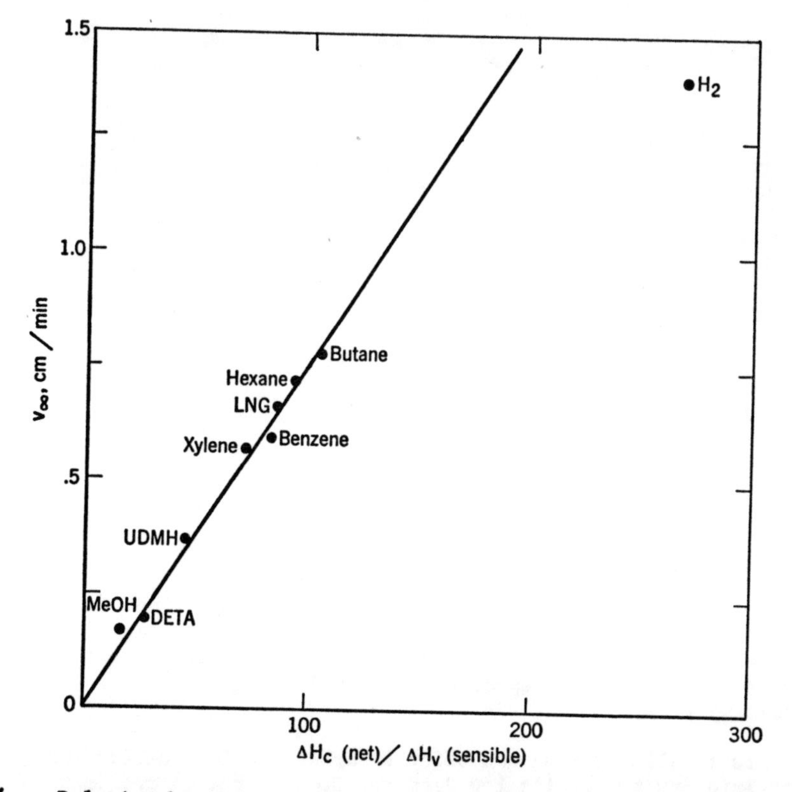

Figure 4. - Relationship between linear burning rates and thermochemical properties of fuels.

We do not believe that these higher burning rates were caused by the wind. The following observations seem pertinent. By burning benzene in neglible wind, we obtained burning rates along the solid curve of figure 5; the extrapolated value, v_∞, was 0.60 cm/min, which conforms well to equation 5. With any kind of natural or artificial wind up to about 4 meters/sec, however, all burning rates were close to 0.60 cm/min even at small pool diameter (with wind, benzene seems to have no burning rate minimum of the type shown in figure 1). Our explanation was that wind rumples the flame surface, increasing its emissivity, so that $e^{-\kappa d}$ goes to zero at smaller values of d. Significantly, we could not obtain a burning rate much above v_∞ in any kind of wind until the flame was torn, giving a bright, noisy (probably premixed) flame in the wake of the container's edge.

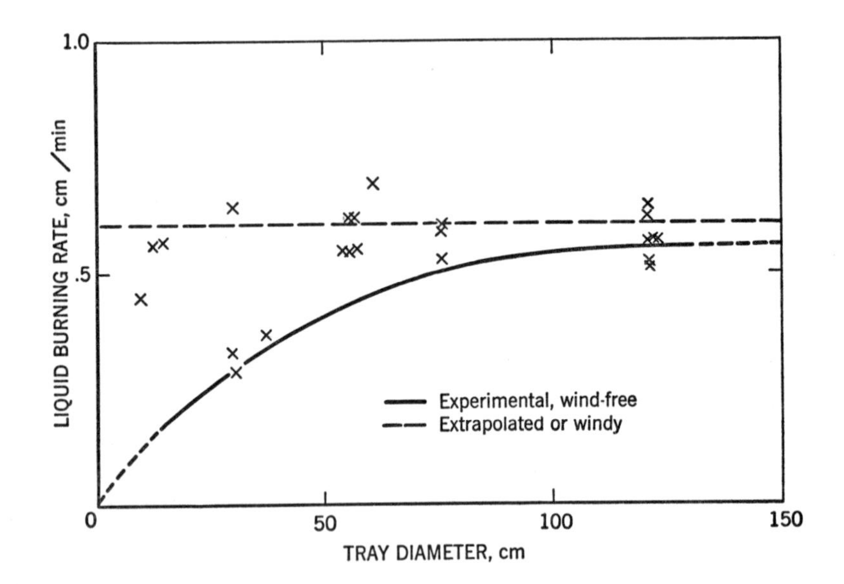

Figure 5. - Effect of wind on burning rate of benzene.

With hexane, we could make the burning rate approach but not exceed v_∞ in a 76 cm diameter tray with a 3 meter/sec wind. Further increases of wind speed merely spilled fuel out of the tray. With methanol we could lay the luminous zone flat across the liquid by imposing a wind and attained v = 0.14 cm/min (v_∞ = 0.17 cm/min) /11/.

The only rationalization we can offer to bring the LNG results into conformity is by invoking the fuel's low density, 0.42 gm/cc. A treatment of burning rate as affected by radiative heat transfer, given in the appendix, requires that a mass burning rate ($gm/cm^2 min$) be correlated with $\Delta H_c/\Delta H_v$.

All of the results of figure 4 are accordingly replotted on this basis in figure 6. The correlation is not impressive above xylene; however, the points for benzene and gasoline are within experimental scatter (compare figure 5). The butane value that was obtained by extrapolation from 76 cm may simply have been underestimated as was that for LNG. The point for liquid hydrogen, which was never to be taken seriously, is now omitted.

We think this situation arose through our ignoring the convective contribution at small pool diameters. This was relatively unimportant with the high molecular weight fuels, so that most of the v_∞ values of Table 1 are still acceptable. However, with LNG the burning rate curve of figure 3 was much too flat (crossing that of hexane) leading to the underestimate of v_∞. On this basis, the few data for liquid hydrogen give no estimate at all of a limiting rate of burning.

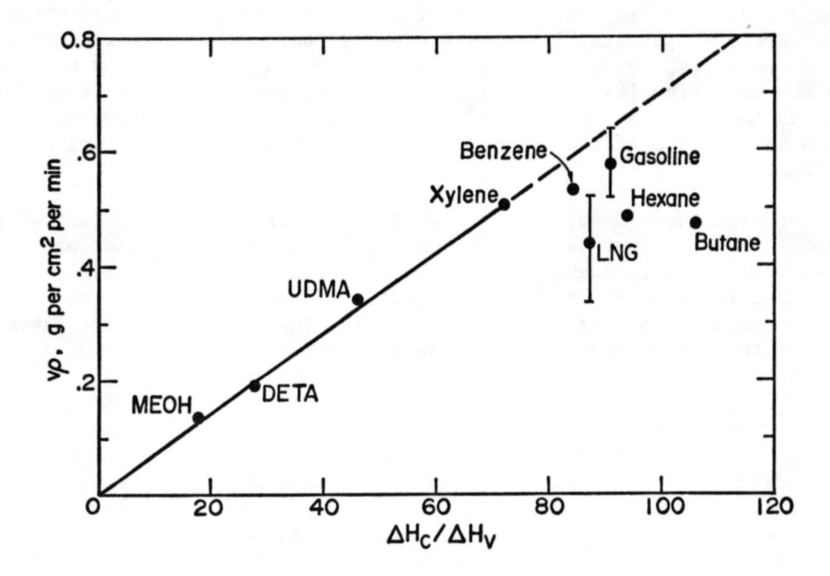

Figure 6. - Relationship between mass burning rates and thermochemical properties of fuels.

3. Radiation to the immediate surroundings

Total radiation was measured with three Eppley thermopiles using CaF_2 windows located in a horizontal plane typically one diameter above the level of the pool. The sensitivities of the thermopiles were such that they could be spaced from 5 to 20 diameters from the flame; their time constants ranged from 0.7 to 6 seconds. Gaseous diffusion flames were studied first, and some effort was made to show that the measurements were insensitive to flow rate of fuel. Thus the radiant power was figured as a percentage of total thermal power as listed in

Table 3. Assumptions were made that the flame's thermal power resulted from complete combustion to CO_2 and H_2O and that the measured fluxes were representative of the average flux on a spherical surface centered on the axis of fuel flow. These assumptions were not important to the initial objective of the work, that is, to estimate radiant flux at ground level from very large fires. The near-constancy of values for ethylene flames (Table 3) was surprising to us and led to an early conclusion that no size of air-supported diffusion flame was likely to radiate more than 35-40 percent of its thermal power to the immediate surroundings.

Subsequent measurements with pool fires were not much different. In this case, benzene provided the same high (35-40%) values of radiant power even at the smallest pool diameters (Table 4). Proving that aliphatic hydrocarbon fires would attain this level at some sufficiently large pool diameter was somewhat awkward. Various gasolines and hexane gave 30-40 percent of thermal power in a 122 cm diameter pool. Unfortunately LNG and butane could not conveniently be studied in bigger diameters than 76 cm. The Lake Charles tests provided data for large pools (bracketed in Table 4) but also with substantial natural winds. The radiation levels of gasoline fires were clearly lower in a wind; whether this was due to convective losses or merely to smoke obscuration was not determined. Not only the radiation levels but also the burning rates of LNG were hard to determine in the large wind-driven fires. The peak value of 34 percent radiation in Table 4 is accurate if the LNG burning rate stayed constant during changes of wind velocity.

Judging from these data, the Bureau of Mines has recommended that 40 percent of the total thermal power be taken as an upper limit on the radiation from accidental spill fires. There seems to be a consensus that this figure is about twice too high for pool fires in natural winds /12/. Fortunately, our overestimate compensates to some extent for the wind-induced tilt of the flame, which brings the flame-zone closer to ignitable structures.

4. Flame Temperatures

The radiation of some large fraction of a flame's thermal power is obviously achieved at the expense of the flame's temperature. Figure 7 shows computed flame temperatures for heptane-air with various percentages of loss of its heat of combustion. If one allows for a small fraction of incomplete combustion and for 40 percent radiation, the final flame temperature cannot much exceed 1500° K.

Table 5 gives some measurements of luminous zone temperatures in pool fires, all of approximately 30 cm diameter. All are of the order of 1500° K. Those of Kahrs are described in some detail since they provide a comparison of CO_2 and continuum temperatures that will be invoked later. Spectral scans (1-4.6 μ) were made of the luminous upper portion of the flame, both in absorption and in emission. The absorptivities appear as circles in the upper part of figure 8; a curve that seemed best to fit the variation of the emissivity of soot with wavelength is given by

$$\epsilon \sim \lambda^{-0.77}, \tag{6}$$

in which the exponent is not inconsistent with the findings of Siddall and McGrath /13/. The ϵ values were applied as corrections to the Planck radiation of a 1500° K black body to give curve a. Curve b is calculated from the smoothed relative radiance data of Kahrs by normalizing to the observed CO_2 peak emissivity at 4.4 microns from some recent unpublished work. It appears that the CO_2 temperature (4.4 microns) is almost exactly 1500° K while the continuum temperature would be taken as slightly above or slightly below 1500° K depending on the wavelength region.

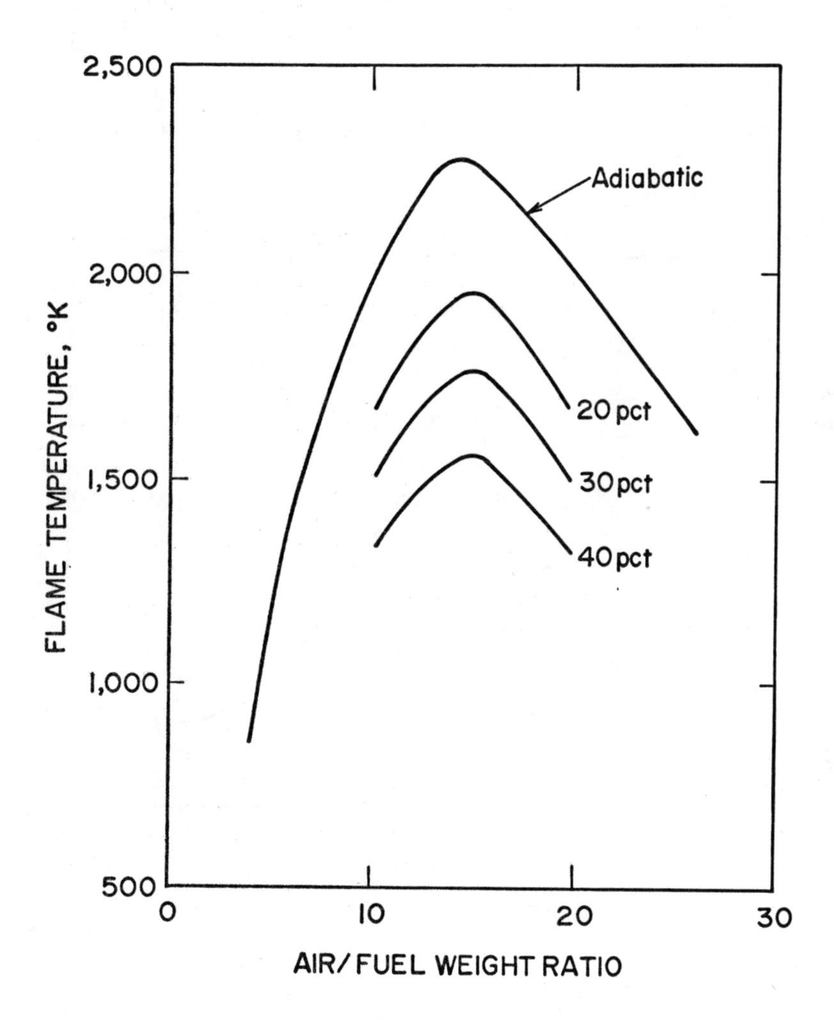

Flame 7. - Flame temperatures of Heptane-air with and without radiative loss.

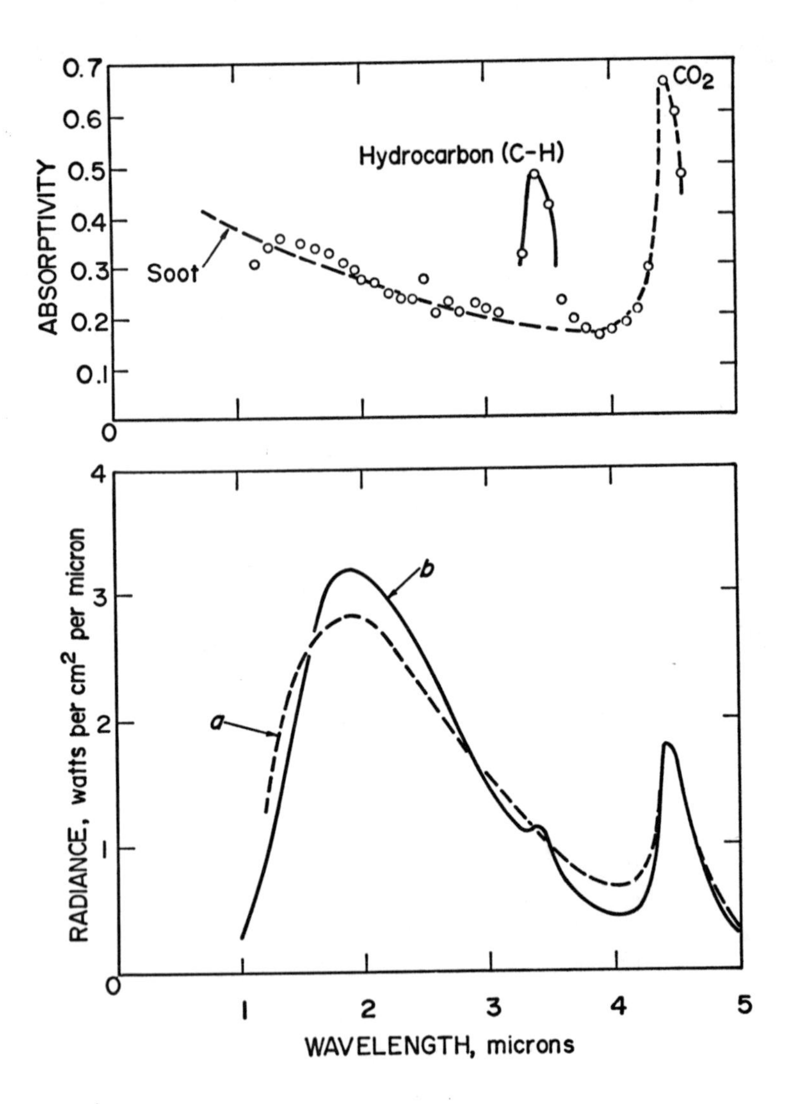

Figure 8. - Spectral absorbance of a pool fire. (top)
Spectral radiance of pool fires. (lower)

The striking thing about having a number of flames with luminous zone temperatures of 1500° K is the close approximation of this figure to limit flame temperatures in premixed systems. Table 6 lists the limit temperatures reported by Egerton /14/ as determined from lower flammable limits on a flat flame burner. These are somewhat lower than most reported values, probably because of the more careful technique of measuring lower limits. The final column of table 6 gives the calorific values of lean limit mixtures (lower limit concentration multiplied by standard heat of combustion) with concentrations measured for the most part in a consistent way with flammability tubes. The variability in $L\Delta H_c$ gives an idea of the variability of lean limit temperatures.

Since the fuel in a pool fire is not necessarily consumed in a lean mixture with air, it is pertinent to inquire if rich flame temperatures are not lower than 1500°K. A recent paper by Furno /15/ argues that the richest mixture that can burn in upward propagation is actually the limit mixture for downward propagation; that is, any richer mixture adjusts itself by diffusional demixing to have a higher oxygen concentration in order to sustain burning. Flame temperatures at the rich limits of downward propagation are, if anything, higher than lean limit temperatures.

Assuming that the approximation of luminous zone temperatures to limit temperatures is something more than coincidence, we are not sure what this implies. Are the upper reaches of a pool fire in a constant state of extinguishment and reignition? Or is combustion taking place in something like a "distributed reaction zone" with slowed chemical kinetics?

It is of interest to consider some recent data from a quite different flame that has about the same continuum temperature. Figure 9, curve a, shows the IR emission spectrum of an explosion involving 500 mg/liter of -14 μ coal dust in air. The radiation path length was about 30 cm atm. Clearly the continuum temperature is in the 1400-1500°K range although the CO_2 temperature is more like 1700°K. If the similarity in the surface temperatures of these relatively large coal particles and the submicron soot particles of a pool fire is something more than coincidence, then we should be thinking about the microscopics of the surface reactions by which these particles are being consumed.

Conclusions

Much of the paper is devoted to burning rates because this is the area of greatest uncertainty. An old correlation of linear regression rate with $\Delta H_c/\Delta H_v$ has not withstood the test of large-scale experiments. Many of the existing data can be correlated on the basis of mass burning rate; however, for this rationalization to be tenable we must admit that values for butane and hexane, in particular, have been grossly underestimated. As an indication of the magnitude of uncertainties, the present value of v_∞ for gasoline is nearly twice the burning rate given by Blinov and Khudiakov in 1958.

Some luminous zone temperatures are shown to be close to the limit flame temperatures for the same fuels premixed with air. These temperatures are also close to the maximum attainable if one assumes that 40 percent of the flame's thermal power is radiated during the combustion reactions. These observations are of some practical importance in that they permit the prediction of maximum radiation hazards.

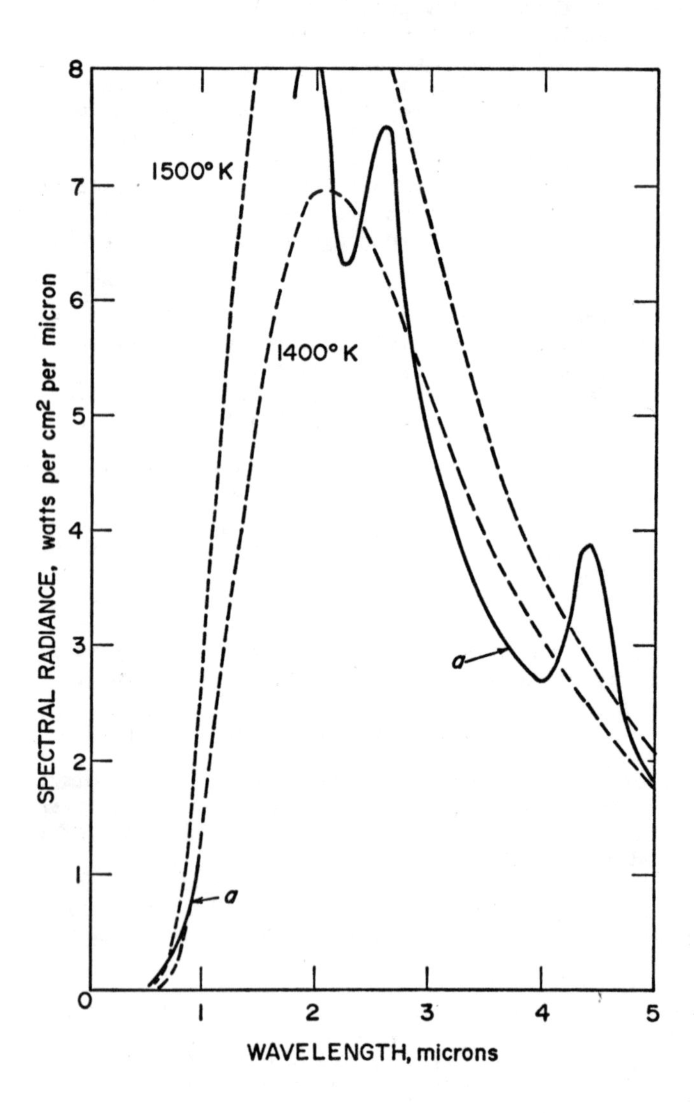

Figure 9. - Spectral radiance of coal dust explosion.

Table 1. - Summary of Computed Values Bearing on Radiative Hazards of Fires

Fuel	$\varkappa$, cm^{-1}	V_∞, cm/min	Thermal output per unit liquid surface, $kcal/cm^2$ min: Total	Radiated
Hexane	0.019	0.73	5.1	2.0
Butane	.027	.79	5.1	1.4
Benzene	.026	.60	5.1	1.8
Xylene	.012	.58	5.0	---
Methanol	.046	.17	.64	.11
UDMH	.025	.38	2.2	.60
Hydrogen	(0.07)	(1.4)	(2.8)	(0.7)
LNG	.030	.66	3.2	.74

Table 2. - Percentage of Absorption of Flame Radiations in Cells With CaF_2 Windows

Fuel	Absorbing medium, path length, temperature: Liquid fuel, 0.3 cm, 30°C	Fuel vapor, 8.9 cm, 100°C	Steam, 8.9 cm, 165°C
Methanol	100	27[1]	13
Hydrogen	--	0	33
UDMH	>98	43	18
Hexane	71	--	<6
Benzene	62	11	--

[1] 38 percent absorbed over 18.4 cm path.

Table 3. - Radiation by Gaseous Diffusion Flames

Burner Diameter (cm)	100 x Radiant Output/Thermal Output			
	Hydrogen	Methane	Butane	Ethylene
0.50	9.5	10.3	21.5	27.5
0.92	9.1	11.6	25.3	27.8
1.85	9.7	16.0	28.6	35.3
4.1	11.1	16.1	28.5	38.3
8.3	15.6	14.7	29.1	38.4
20.3	15.4	19.3*	28.0	31.9
40.7	16.9	23.3*	29.9	35.8

* Natural gas (95% CH_4)

Table 4. - Radiation by Liquid-Supported Diffusion Flames

Pool Diameter (cm)	100 x Radiant Output/Thermal Output				
	Methanol	Methane*	Butane	Gasoline	Benzene
7.6	16.2	--	--	----	35.0
15.2	16.5	--	--	----	--
30.5	--	21	19.9	----	--
45.7	--	--	20.5	----	34.5
76.	--	23	26.9	----	35.0
122.	17.0	--	--	30-40	36.0
153.	--	(15-24)	--	(16-27)	--
305.	--	(24-34)	--	(13-14)	--
610.	--	(20-27)	--	----	--

* Liquefied natural gas, ~ 95% CH_4.
** Bracketed values in natural winds above 3 m/sec.

Table 5. - Reported Flame Surface Temperatures of Pool Fires

Fuel	Temperature (°K)	Author	Method
Ethanol	1560	Rasbash et al	Hottel & Broughton
Benzene	1460	"	"
Petrol	1520	"	"
Kerosine	1480	"	"
Benzene	~ 1450	Kahrs et al	Scanning spectrometry
Mixed solvent	~ 1500	"	"

Table 6. - Limit Flame Temperatures as Calculated from Lower Flammable Limits in Air

Fuel	Limit Temperature (°K)	$L \cdot \Delta H_c$
Methane	1495	9.5
Ethane	--	10.1
Propane	1473	10.2
n-Butane	1506	--
n-Pentane	1431	10.8
n-Heptane	1495	11.2
Benzene	--	9.8
Ethanol	--	9.8
Average, 56 organic fuels	--	10.5

REFERENCES

/1/ Standard 59-A, "Liquefied Natural Gas", National Fire Protection Association, 60 Batterymarch St., Boston, Mass. 02110

/2/ Fire Research Abstracts & Reviews 1, 41-4 (1958) (Hottel, H. C.): "Review of Certain Laws Governing Diffusive Burning of Liquids".

/3/ Emmons, H.: "Observations on Pool Burning". The Use of Models in Fire Research, p. 50-67 Publication 786 NAS-NRC, Washington, D.C. (1961).

/4/ Stark, G. W. V.: "Liquid Spillage Fires". J. Chem. E. Symposium Series No. 33 (1972: Instu. Chem. Engrs., London) p. 71-78. See particularly Figure 9.

/5/ Hertzberg, M.: "The Theory of Free Ambient Fires......". Combustion and Flame, to be published.

/6/ Spalding, D. B.: "Some Fundamentals of Combustion" (1955) London: Batterworth.

/7/ Glassman, I., J. G. Hansel: "Some Thoughts and Experiments on Liquid Fuel Spreading, Steady Burning and Ignitibility in Quiescent Atmospheres". Fire Res. Abs. & Rev., 10, (1968), p. 217.

/8/ Spalding, D. B.: "The Burning Rate of Liquid Fuels from Open Trays By Natural Convection". Fire Res. Abs. & Rev. 4, (1962), p. 234-6.

/9/ Kelly, C. S.: "The Transfer of Radiation from a Flame to its Fuel". J. Fire and Flammability, Vol. 4, (1973), p. 56.

/10/ Bureau of Mines Report of Investigations 6099: "Fire and Explosion Hazards Associated with Liquified Natural Gas". D. S. Burgess and M. G. Zabetakis (1963).

/11/ Bureau of Mines Report of Investigations 5707: "Fire and Explosion Hazards Associated with the Production and Handling of Liquid Hydrogen". M. G. Zabetakis and D. S. Burgess (1961).

/12/ See for example, May, W. and W. McQueen: "Radiation from Large LNG Fires". Conference Proceedings on "LNG Importation and Terminal Safety", Boston, Mass., June 13-14, 1972. Available from Committee on Hazardous Materials, National Academy of Sciences, Washington, D.C.

/13/ Siddall, R. G. and I. A. McGrath: "The Emissivity of Luminous Flames". Ninth Symposium (International) on Combustion, p. 102-110, Academic Press (1963).

/14/ Egerton, Sir Alfred: Fourth Symposium (International) on Combustion (Williams & Wilkins) p. 4-13 (1952).

/15/ Furno, A. L., E. B. Cook, J. M. Kuchta, and D. S. Burgess: "Some Observations on Near-Limit Flames". Thirteenth Symposium (Int.) on Combustion, p. 593-9, The Combustion Institute, Pittsburgh, Pa. (1971).

Appendix A

The empirical relationship

$$v\ (\text{cm/min}) = 0.0076\ \frac{\Delta H_c}{\Delta H_v} \qquad \text{(A-1)}$$

has been treated from the viewpoint of convective heat transfer from the flame to the vaporizing liquid. However, we are not aware of any rationalization of equation A-1 on the basis of radiative heat transfer. The law may be derived on the basis of a mass burning rate rather than a linear burning in the following manner.

Idealize and laminarize the enthalpy flow balance for the burning pool flame. The liquid, initially at T_u, absorbs energy from the flame and vaporizes at a rate v(cm/min). The vapors mix with entrained air as they rise by buoyancy. They are further heated to ignition and react, generating sensible heat. Burned gases at T_f then radiate energy until they reach some low temperature T_e, at which point they merely convect heat to the surroundings, as they exit from the pool flame at velocity $\bar{v}_e$.

Now the flux of radiating enthalpy per unit area of flame-zone surface is $\bar{\epsilon}_f\ \sigma\ T_f^4$, where $\bar{\epsilon}_f$ is the flame emissivity (as viewed by the surroundings). The flux of convecting enthalpy is $\bar{c}_p\ \bar{\rho}_e\ \bar{v}_e\ (T_e-T_u)$. The fraction radiated is therefore the radiated part divided by the sum of radiation plus convection:

$$f = \frac{\epsilon_f\ \sigma\ T_f^4}{\epsilon_f\ \sigma\ T_f^4 + c_p \rho_e v_e\ (T_e - T_u)}\ . \qquad \text{(A-2)}$$

We now equate this sum with the source enthalpy of the combustion reaction. We assume a source strength just equal to the minimum flux required to sustain a propagating flame at a limiting rate. This is $(S_u)_a\ c\ \Delta H_c$, where $(S_u)_a$ is the limit burning velocity for natural convection, and c is the fuel concentration. Equating enthalpy source and sink rates per unit area of flame surface gives

$$(S_u)_a\ \ c\ \ \Delta H_c = \bar{\epsilon}_f \sigma\ \ T_f^4 + \bar{c}_p \bar{\rho}_e \bar{v}_e\ (T_e - T_u)\ . \qquad \text{(A-3)}$$

Now the limiting burning rate for a large diameter pool under radiative control is given as follows /5/:

$$v\ (\text{cm/min}) = \frac{\frac{1}{2}\ \epsilon_f\ (h)\ \ F_{f\rightarrow \ell}\ \sigma\ \ T_f^4}{\rho_\ell\ (C_A T_S - C_\ell T_u + \Delta H_v)} \qquad \text{(A-4)}$$

where $\epsilon_f(h)$ is the flame emissivity as viewed by the absorbing fuel reservoir, and $F_{f\rightarrow\ell}$ is a shape factor. Eliminating $\sigma\ T_f^4$ and the convective term from equations A-2,3,4, and setting $\Delta H_v \gg C_A T_S - C_\ell T_u$ (the usual condition) gives

$$v\ \ (\text{cm/min}) = \tfrac{1}{2}\ f\ \frac{\epsilon_f\ (h)}{\bar{\epsilon}_f}\ \ c\ \frac{F_{f\rightarrow\ell}}{\rho_\ell}\ (S_u)_a\ \frac{\Delta H_c}{\Delta H_v}\ , \qquad \text{(A-5)}$$

and

$$v\rho_\ell\left(\frac{gm}{cm^2 min}\right) = \beta \frac{\Delta H_c}{\Delta H_v} . \quad \text{(A-6)}$$

This is equivalent to equation A-1. Realistic values of the parameters for hydrocarbon flames are $\rho_\ell = 0.75$ gm cm^{-3}; $F_{f\rightarrow\ell} = 0.40$; $(S_u)_a = 6.0$ cm sec^{-1}; $f = 0.50$; $c = 0.96 \times 10^{-4}$ gm cm^{-3}; and $\epsilon_f(h)/\bar{\epsilon}_f = 1.5$.

These values give $\beta = 0.51 \times 10^{-2}$ gm cm^{-2} min^{-1} in fair agreement with the measured value in figure 6. It may be significant that the magnitude of the constant in equation A-1 is properly predicted by a source function that is controlled by the limit burning velocity for natural convection. This is consistent with the observation that the temperatures of the emitting flame zones are indeed nearly those of flames at their flammability limits ($T_f \sim 1500°$ K).

Chapter 28

HEAT TRANSFER BY RADIATION FROM FIRES OF LIQUID FUELS IN TANKS

P. G. Seeger

Forschungsstelle für Brandschutztechnik an der Universität Karlsruhe, Karlsruhe, Germany

Abstract

Luminous flames generally occurring in fires from liquid fuels may be considered as black or grey bodies if they have reached a certain flame thickness. The effect of the radiation intensity on the environmental surfaces may, therefore, be calculated by a relationship based on the Stefan-Boltzmann Law. Fire tests with model tanks have shown that the measured radiation intensities are nearly identical with those values established theoretically.

NOMENCLATURE

a : distance between tank wall and irradiated area element
a' : equivalent distance between flame and irradiated area element
b : equivalent breadth of flame
c : height above tank opening
c_1 : constant
d : diameter of tank
Fr : Froude number; $Fr = u^2/gd$
g : gravitational acceleration
h : height of tank
l : flame length
$\dot{q}$: radiation intensity
T : absolute temperature
u : mean velocity of fuel gas or -vapor
ϵ : emissivity
σ : Stefan-Boltzmann constant; $\sigma = 5.77 \times 10^{-12} W/cm^2 K$
φ : configuration factor

1. Introduction

One of the problems offering itself when tank farms are being installed by firms of the mineral oil- and chemical industry is the required distance between the single tanks in order to avoid heat transfer in case of a tank fire which would endanger other tanks in the vicinity. Knowledge of the radiation intensities produced by a fire is also essential for the arrangement of sprinkler installations for cooling of tank walls and upper surfaces of tanks.

Heat released from a burning tank of liquid fuels will be transferred by radiation as well as by convection of the combustion gases. Convevtive heat transfer occurs particularly when a strong wind forces fumes and combustion gases in the direction of adjacent tanks or in a case where tanks are placed in a common trap room and this room catsches fire due to spilt oil. These cases, however, will not be considered in the follpwing since the present study only deals with the heat transfer by radiation.

2. Fundamentals of Heat Transfer by Radiation

The thermal radiation of a flame is composed of the radiation of individual components of combustion gases, in particular of carbon dioxide and water vapor which absorb and emit in certain narrow wave length ranges - the so-called bands - and of the radiation of glowing soot particles disengaged by incomplete combustion which emit a continuous spectrum. The spectral intensity of gas radiation depends on temperature, thickness of flame layer and the partial pressure of the corresponding constituent of gas. The intensity of the radiation from soot particles also depends on temperature and flame thickness but, in addition, on the concentration of soot, its size and structure. Whereas the calculation of the heat transfer by gas radiation offers no serious difficulties if the above named values are known, the calculation of heat transfer by soot radiation is much more difficult since properties, size and distribution of soot particles within a flame depend on a great number of factors affecting them. There are luminous and non-luinous flames depending on whether soot formation and consequently soot radiation are present or not.

Due to heavy development of soot, generally occurring in fires of liquid fuels where combustion is incomplete, flames are of the luminous type. Flames of a thickness exceeding 1 m may be considered more or less as black or grey bodies respectively which have an emissivity between 0.9 to 1. This allows to apply the Stefan-Boltzmann Law thus omitting difficulties in determining the soot radiation. Based on this assumption, the radiation intensity $\dot{q}$ which may appear at any element of a surface in the vicinity of the burning object can be calculated by means of the following equation:

$$\dot{q} = \varphi \epsilon \sigma T^4 . \tag{1}$$

The configuration factor φ reflects the fractional part of the heat emitted by a luminous body falling upon the receiving area element in the surroundings. For technically meaningful cases the expressions required to determine the appropriate configuration factors may be taken from the literature [1,2,3]. Expressions required for subsequent calculations to evaluate the configuration factor are as shown under [1,2] :

1. Configuration factor for an area element placed in parallel with a rectangular surface and the rectangular surface where one corner of the rectangular surface strikes the perpendicular from the center of the area element:

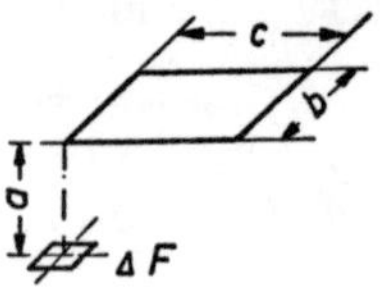

c b a ΔF

$$\varphi = \frac{1}{2\pi}\left(\frac{B}{\sqrt{1+B^2}}\tan^{-1}\frac{C}{\sqrt{1+B^2}} + \frac{C}{\sqrt{1+C^2}}\tan^{-1}\frac{B}{\sqrt{1+C^2}}\right) \qquad (2)$$

with B = b/a and C = c/a.

2. Configuration factor for an area element placed vertically to a rectangular surface with respect to the rectangular surface where one side of the rectangular surface lies flush with the area element and the normal of one corner of the rectangular surface intersects the center of the area element:

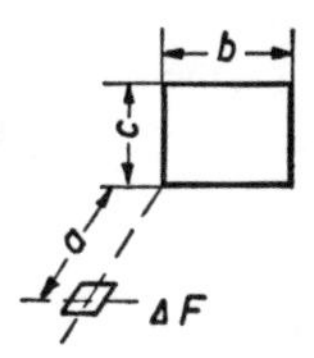

$$\varphi = \frac{1}{2\pi}\left(\tan^{-1}B - \frac{1}{\sqrt{1+C^2}}\tan^{-1}\frac{2}{\sqrt{1+C^2}}\right) \qquad (3)$$

with B = b/a and C = c/a.

In [1] configuration factors can be derived directly from diagrams which are corresponding to Equations (2) and (3).

If the geometrical relationship of two surfaces placed in parallel with or in quadrature to each other does not meet above requirements, this may be attained by splitting up one of the surfaces into portions or by enlarging it. For these diminished or enlarged rectangular areas the configuration factors can be determined and the wanted configuration factor may be found by addition or subtraction as appropriate.

3. Calculation of Radiation Intensities at an Arbitrary Point in the Environment of a Burning Tank

The heat transferred by radiation from a tank fire to the invironment is composed of the radiation emitted from the flame and from the tank wall. To calculate the configuration factor, the assumption is made that the flame ist precisely seated on the tank opening and has the shape of a cylinder with a diameter equal to the tank diameter an a height equal to the flame height. An effect of wind on the combustion process, shape of flame and configuration factor will be out of consideration.

Since the configuration factor only depends on the angle enclosing the radiating area but not on the shape of the area, the surface of the flame- or tank cylinder may be replaced by a rectangular surface at the distance a as shown in Fig. 1 allowing hereby an elementary calculation of the configuration factor according to Equation (2) or Equation (3). The error caused in this calculation due to the omission of area portions at the upper and lower boundary of the tank-or flame cylinder is so trifling that it may be neglected.

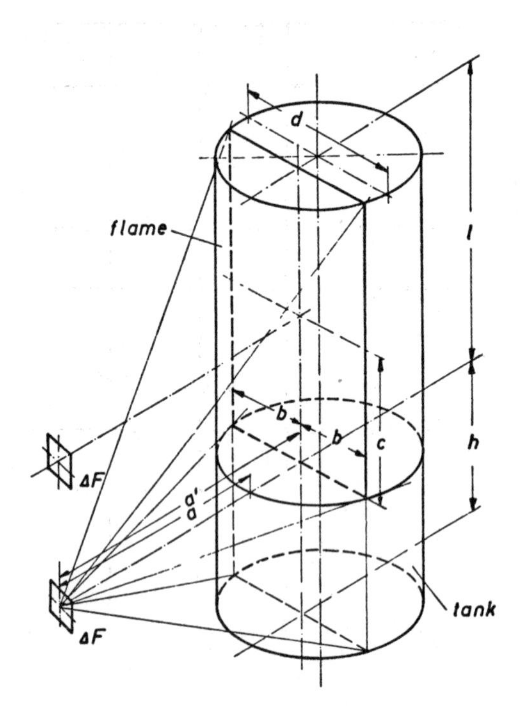

Fig. 1. Substitution of flame- and tank cylinder by a rectangular surface to determine the configuration factors for the flame or tank wall and an area element in parallel with the flame- or tank axis

The configuration factor governing the radiation intensity being much greater where two areas are in parallel with each other than in a case where the areas are at right angles with the other geometrical relations being equal, and considering that for safety reasons the higher value ist the preferable one, calculations contained herein have been based entirely in the configuration factor according to Equation (2).

Flame heights required to determine the configuration factor in cases where the problem relates to the flame radiation are factors yet unknown as far as tanks of excessive diameters are concerned. Previous measurments of the heights of flames originating from burning tanks have been restricted so far to tank diameters up to 22.9 m [4].

Flames occurring in tank fires are buoyant diffusion flames constituting the transition phase between laminar and purely turbulent flames where in most cases the momentum carried with the upward flow of fuel vapour - as compared with the momentum produced by buoyancy - may be neglected. For such flames the assumption, as stated by Thomas [5], can be made that the flame height related to the tank diameter depends only on the Froude number.

There are some theoretical and experimentical studies on hand [6-12] proving that the interrelation between the relative flame height and the Froude number may be described by the following exponential equation:

$$l/d = c_1 Fr^n \tag{4}$$

In this equation c_1 is a constant depending on the fuel used. In carrying out their tests with flames of gaseous and liquid fuels, the authors of above mentioned studies [6-12] nearly agreed in determining the value 0.2 as being the exponent n of the Froude number.

Since according to Blinov and Khudiakov [4,13] the burning rate for fuels being liquid in their normal state is about 4 mm/min if tank diameters exceed 1 m, corresponding thus to a velocity of

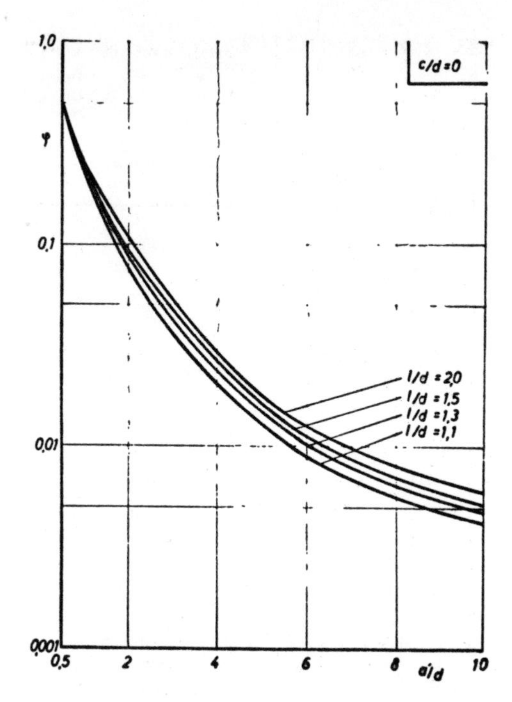

Fig. 2. Pattern of configuration factors for the flame and an area element in parallel with the flame axis at tank height as a function of the equivalent distance related to the tank diameter for various flame heights

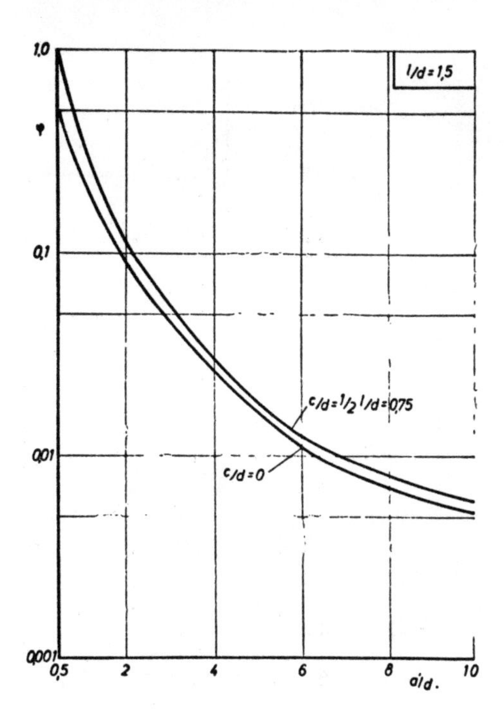

Fig. 3. Pattern of configuration factors for the flame and an area element in parallel with the flame axis as a function of the equivalent distance related to the tank diameter for two different positions with respect to the tank opening

about 10 mm/s of the upward flow of vapor within the tank, Froude numbers in real tank fires with tank diameters ranging from 10 to 100 m are in the order of 10^{-6} to 10^{-7}. For these burning rates or Froude numbers the flame heights related to tank diameters range between 1 and 2 as can be seen from [12,14,15].

However, the error caused by inaccurate determination of the flame height is unimportant as can be seen from Fig. 2. In this diagram the pattern of configuration factors for the flame and an area element placed in parallel with the flame axis at tank height has been plotted as a function of the equivalent distance a′ related to the tank diameter for various flame heights. The effect of the position of the irradiated area element on the configuration factor is illustrated in Fig. 3 where the pattern of configuration factors - in the case of the area element being placed at tank height as well as at the height of the flame center - has been drawn as a function of the equivalent distance related to the tank diameter.

The determination of the radiation intensity according to Equation (1) is done at that moment when the burning rate of the fire is

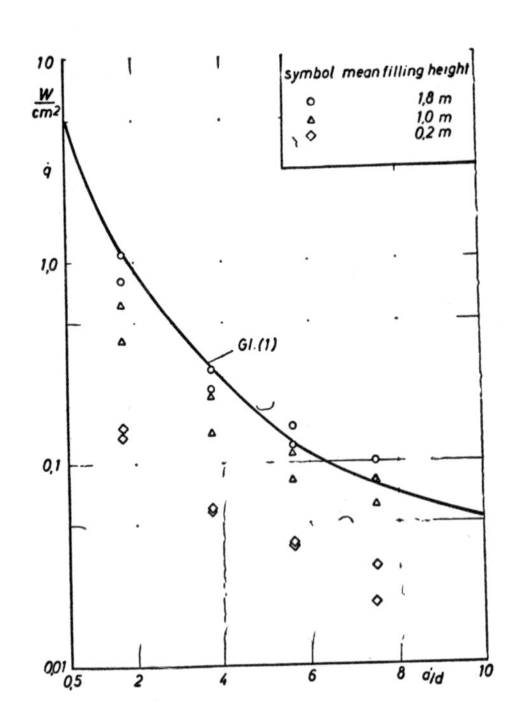

Fig. 4. Radiation intensities at tank height vertical to the equivalent distance related to the tank diameter with the tank wall cooled.
Diameter of tank 1.6 m

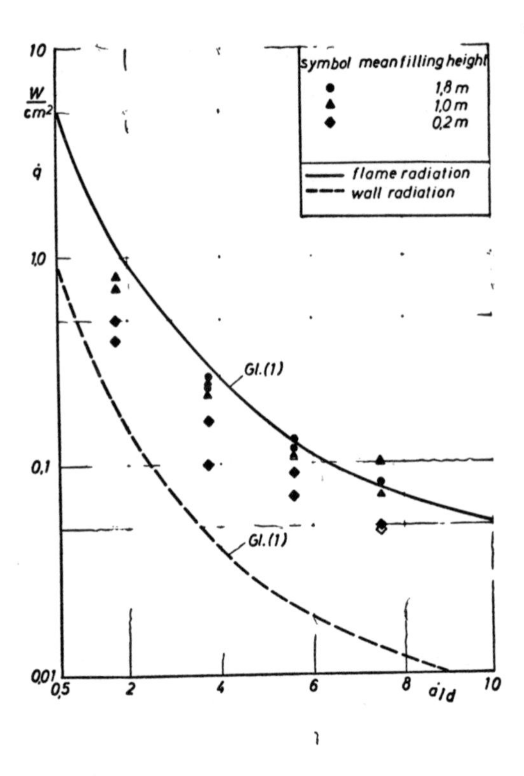

Fig. 5. Radiation intensities at tank height vertical to the tank axis as a function of the equivalent distance related to the tank diameter with the tank wall uncooled.
Diameter of the tank 1.6 m

nearly constant. It is assumed that during this period no absorption or dispersion of radiation emitted from the burning tank occurs in the atmosphere between the burning tank and the irradiated area element due to water vapor, carbon dioxide and dust. Another presupposition is that temperature and emissivity of the flame are the same whatever the position of the flame may be and that these factors do not change during the process of burning.

4. Comparison of Calculated Radiation Intensities with Test Results

To ascertain to what extent Equation (1) reflects the radiation intensities of real tank fires, the Forschungsstelle für Brandschutztechnik performed experimental test fires using 3 model tanks, each 2 m high, the diameters of which were 1.28, 1.6 m and 2 m [16]. In some tests the tank walls were cooled with water. The fuel used was gas oil with a boiling range of 167 to 367 °C and a residue of 1.5 Vol-% (gas oil III) which under conditions of a tank fire produces a flame of intensive luminosity. With three different mean filling heights - 1.8 m, 1.0 m and 0.2 m - for the fuel contained in the tanks measurements of the radiation intensities were made

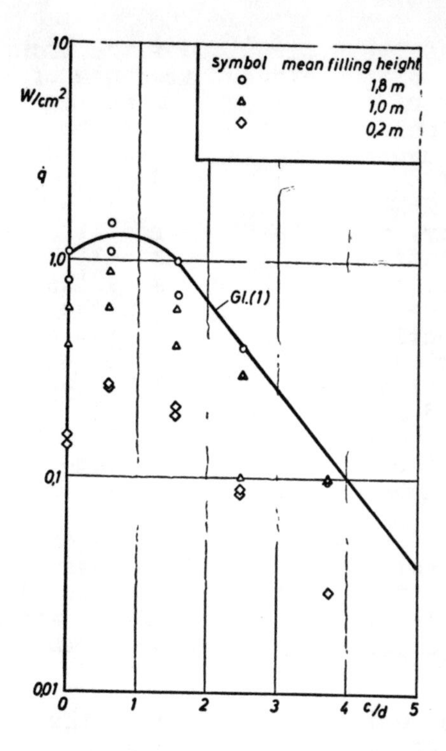

Fig. 6. Radiation intensities at a distance of 3 m from the tank axis as a function of the height above tank opening related to the tank diameter with the tank wall cooled.
Diameter of tank 1.6 m

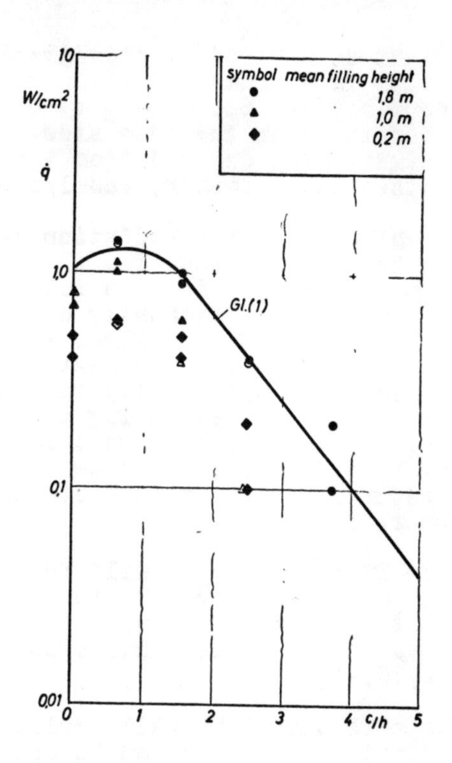

Fig. 7. Radiation intensities at a distance of 3 m from the tank axis as a function of the height above tank opening related to the tank diameter with the tank wall uncooled.
Diameter of tank 1.6 m

by means of a total radiation pyrometer at various points vertically to and in parallel with the tank axis. In addition, the flame temperature was measured with a total radiation pyrometer and the flame height determined by photographing.

In Figs. 4 and 5 the radiation intensities which during the test had been measured on the tank with a diameter of 1.6 m at tank height, vertically to the tank axis, are plotted as a function of the equivalent distance related to the tank diameter. Figs. 6 and 7 illustrate the radiation intensities measured in the same test at a distance of 3 m from the tank axis along the height above the tank opening related to the tank diameter. As proved by a comparison with the pattern of the radiation intensity computed according to Equation (1) which is shown on these diagrams as an unbroken curve, values measured on the completely filled tank fit well into the theoretical correlation. The calculation was based on a flame emissivity of 0.9 and a flame temperature of 900 °C. This value coincides with the results of measurements made during the tests described herein and also with data in the literature [14,16]. The evaluation of the configuration factor was based on the flame height determined by photographing. As the evaluation of the flame height is extremely difficult because of the cloudy structure of the flame and the momentary character of the photography and in order to be on the safe side, the calculation was based on the maximum value derived from the experiments. This value, related to the tank diameter, was $l/d = 1.5$.

In calculating the radiation intensity, only the flame radiation has been considered. The thermal radiation from the tank wall has been disregarded for the following reasons: On the one side the tank wall is sufficiently cooled by the fuel contained in the tank if completely filled, thus preventing any thermal radiation. On the other side, the heat flux emitted from the tank wall indeed increases proportionally to the drop of the liquid level but - with the emissivity being the same and a temperature not yet jeopardizing the tank wall stability - attains only a fraction of the heat flux emitted from the flame due to the fact that temperatures enter into the Stefan-Boltzmann Law raised to the 4th power.

From Figs. 4 to 7 it will further be noted that radiation intensities decrease according to the fall of the liquid surface which is due to the fact that with the liquid level dropping the flame withdraws into the tank inside as has been proved by adequate tests [12,16,17].

The effect of tank wall cooling on the emitted heat flux or the radiation intensity can be seen from Figs. 4 and 5 and Figs. 6 and 7. It proves hereby that the flame radiation which reduces according to the lowering of the liquid level is being compensated by the thermal radiation from the uncooled tank wall portion above the surface of the liquid. This is quite clearly demonstrated in Fig. 5 whwere also the pattern of the radiation intensity at tank height caused by the thermal radiation from the tank wall as computed by applying Equation (1) has been plotted as a dotted line as a function of the equivalent distance related to the tank diameter. The calculation was based on a emissivity of 0.9 [2] and a average temperature of the tank wall of 500 °C as found experimentally. With respect to the configuration factor, the maximum value derived unter conditions of heat being radiated from the total height of the tank wall has been introduced.

4. Summary

Luminous flames generally occurring in fires from liquid fuels may be considered as black or grey bodies from a certain flame thickness upward. The radiation intensity caused by the thermal radiation of the flame which is bearing upon an area element at any arbitrary point in the vicinity of a burning tank may, therefore, be calculated according to a relation based on the Stefan-Boltzmann Law. In order to find out to what extent this relation reflects the radiation intensities occurring in real tank fires, test fires on model tanks were undertaken at the Forschungsstelle für Brandschutztechnik during which the radiation intensities were measured at various points. As was shown by comparison, the measured radiation intensities essentially agreed with the values obtained by theoretical calculations. In addition, the proportion of thermal radiation from the tank wall, as referred to the total amount of heat transfer, has been ascertained clearly by adequate experiments in the course of which the tank walls were cooled with water.

REFERENCES

[1] "VDI-Wärmeatlas, Berechnungsblätter für den Wärmeübergang". VDI-Verlag, Düsseldorf (1963).

[2] Pepperhoff, W.: "Temperaturstrahlung". Dr. D. Steinkopff, Darmstadt (1956).

[3] McGuire, J.H.: "Heat Transfer by Radiation". Fire Research Special Report No. 2, H. M. Stationery Office, London (1953).

[4] Blinov, V. and Khudiakov, G.N.: "Certain Laws Governing Diffusive Burning of Liquids". Doklady Akademii Nauk SSSR, Vol. 113, No. 5 (1957), pp. 1094-1098.

[5] Thomas, P.H.: "Some Studies of Models in Fire Research". 1st International Fire Protection Seminar of the VFDB, VFDB-Zeitschrift, Vol. 9 (1960), Special Nr. 3, pp. 96-101.

[6] Thomas, P.H.: "The Size of Flames from Natural Fires". Ninth Symposium (International) on Combustion, Academic Press, New York - London (1963), pp. 844-859.

[7] Sunavala, P.D.: "Dynamics of the Buoyant Diffusion Flame". Journal of the Institute of Fuel, Vol. 40 (1967), pp. 492-497.

[8] Steward, F.R.: "Linear Flame Heights for Various Fuels". Combustion and Flame, Vol. 8 (1964), pp. 171-178.

[9] Hess, K.: "Flammenlänge und Flammenstabilität". Dr.-Ing.-Thesis, Technische Hochschule Karlsruhe (1964).

[10] Kremer, H.: "Die Ausbreitung inhomogener turbulenter Freistrahlen und turbulenter Diffusionsflammen". Dr.-Ing.-Thesis, Technische Hochschule Karlsruhe (1964).

[11] Putnam, A.A. and Speich, C.F.: "A Model Study of the Interaction of Multiple Turbulent Diffusion Flames". Ninth Symposium (International) on Combustion, Academic Press, New York - London (1963), pp. 867-877.

[12] Seeger, P.G. and Werthenbach, H.G.: "Diffusionsflammen mit extrem niedriger Strömungsgeschwindigkeit". Chemie-Ingenieur-Technik, Vol. 42 (1970), No. 5, pp. 282-286.

[13] Hottel, H.C.: "Certain Laws Governing Diffusive Burning of Liquids". Fire Research Abstracts and Reviews, Vol. 1 (1958/59), No. 2, pp. 41-44.

[14] Atallah, S. and Allan D.S.: "Safe Separation Distances from Liquid Fires". Fire Technology, Vol. 7 (1971), No. 1, pp. 47-56.

[15] Werthenbach, H.G.: "Flammenlänge bei Bränden von Flüssigkeitsbehältern". Verfahrenstechnik, Vol. 5 (1971), No. 3, pp. 115-118.

[16] Werthenbach, H.G.: "Brände von Erdölprodukten in Tanks - Versuche und Rechenmodell". Research Report Nr. 21, Arbeitsgemeinschaft Feuerschutz (AGF), (1971).

[17] Werthenbach, H.G.: "Brände von Erdölprodukten in Tanks - Versuche und Rechenmodell". VDI-Zeitschrift, Vol. 115 (1973), No. 5, pp. 383-388.

Chapter 29

FLAME RADIATION AS A MECHANISM OF FIRE SPREAD IN FORESTS

H. P. Telisin

Far East Forestry Research Institute, Khabarovsk, USSR

Abstract

Flame radiation within and above the fuel bed is assumed to be the dominating ignition process. On the basis of this assumption equations for the rate of fire spread are derived. Interdependences of other fire behaviour characteristics with flame radiation parameters are also discussed. Agreement of observed and theoretical values of the rate of fire spread and the length of the burning zone is shown.

NOMENCLATURE

c : specific heat of fuel
D : length of burning zone
D_0 : length of burning zone for backfires
h : height of fuel bed
h : mean free path of radiation within the fuel bed
I_0 : radiative heat flux from the flame zone within the fuel bed
I : radiative heat flux from flames above the fuel bed
k : attenuation coefficient of radiation in flames
L : flame length
M : moisture content of fuel
q_0 : energy required to ignite the unit volume of fuel
q : energy required to ignite the square unit of fuel bed
R_0 : rate of flame downspread
R : rate of fire spread
T_f : absolute flame temperature
T : absolute fuel temperature
t : fuel temperature
t_i : ignition temperature of fuel
α : flame angle to fuel bed
ϵ : emissivity of fuel particles
ϵ_0, ϵ_1 : flame emissivity
ϵ' : emissivity of the flame-fuel system
ρ_b : bulk density of fuel bed
ρ_f : density of fuel
σ : surface-volume ratio of fuel
φ : configuration factor of heat exchange by radiation

I. Introduction

The rate of fire spread has an important effect on every aspect of fire behaviour in forests. And the spread rate itself is greatly influenced by some characteristics of fire behaviour. The mechanism of this complex interdependence has long been the subject of many investigations.

In the earliest analyses /I,2/ fire spread process was interpreted as heat transfer to the unburnt fuel ahead of the fire by convection of flames. Fuel particles were assumed to be ignited by continuous flame envelopment of the surface fuel for some distance of the leading edge of the flame zone plus random flame contacts at greater distances. We belive that this mechanism can possibly take place only at unstable and strong winds.

Radiative heat transfer from the burning zone through fuel bed /3,4/ and radiation from flames above the fuel bed /5,6/ and both flame and fuel bed radiation /7,8/ were also discussed as mechanisms of fire spread. But the equations, derived for the rate of fire spread, in conditions when flame radiation above the fuel bed contributed greatly to the fire spread process, seemed to be valid only for fuel beds of a certain, rather small thickness.

Measurements of forest fire characteristics at various fuel and weather conditions were carried out /9,IO,II,I2/ thus permitting to verify different fire spread theories.

This paper describes a version of the theory of fire spread by flame radiation. The main difference of this version from the previous ones /5,6,7/ is the assumption that radiation from flames above the fuel bed preheats only a surface layer of fuel bed. The thickness of this layer is equal to the mean free path of radiation within the fuel bed. If the fuel bed is thicker than this layer, the lower part of it is ignited by fuel bed radiation only. And when thickness of the fuel bed is less than this mean free path, a certain part of radiative heat would be lost from the fuel preheating process.

2. Assumptions

The supposition that flame radiation above and within the fuel bed is the dominating mechanism of fire spread demands to introduce some assumptions on fuel bed properties and burning conditions.

Fuel particles are assumed to be thin enough to prevent large temperature differences through their thickness and secure the uniform heating. This simplifies the discussion as permits to disregard the thermal conductivity of fuels. As it was shown /I3/, foliage, grass and other forest fuels with thickness up to $2 \cdot 10^{-3}$m can be considered "thin".

Fuel bed is supposed to be loose to secure favourable oxygen conditions so that they would not limit the rate of burning. In this case heat exchange conditions only govern the rate of fire spread.

Radiative heat transfer in a horizontal direction can be the dominating ignition more likely in still air or stable winds and moderate slopes. At gusty or very high winds and steep slopes other mechanisms than flame radiation can probably control the fire spread. Discussion of these mechanisms is out of the scope of our paper.

3. Energy required to ignite forest fuels

The heat q_o required to ignite the unit volume of the fuel is given by

$$q_o = \rho_b \left[c(t_i - t) + 2{,}6 \cdot 10^4 M \right] \quad J/m^3 \qquad (I)$$

Fuel temperatures t higher than $50°C$ can hardly be expected in forests, and it is considerably less the ignition point $t_i = 300°C$. So natural fluctuations of fuel temperature have negligible effect on q_o and the ignition temperature rise can be considered constant. Assuming $t_i - t = 280°C$ and $c = I,4 \cdot IO^3$ J/kg·°C /I,7/ we get

$$q_o = 2,6 \cdot IO^4 \rho_b (I5 + M) \quad J/m^3 \qquad (2)$$

The heat required to ignite the unit area of the fuel bed is described by

$$q = 2,6 \cdot IO^4 \rho_b h(I5 + M) \quad J/m^2 \qquad (3)$$

4. Fuel bed radiation as a mechanism of flame downspread

This is the simplest type of flame spread which can be observed in forests, for example, when burning goes deep into fuel bed from its surface, or when fire is spreading against a very strong wind. A scetch of the flame spreading down the vertical fuel layer is shown in Figure I. Flame downspread here, as in the case of burning liquids in basins /I4/, is controlled by radiative heat flux from the flame zone onto the burning surface. This flux, expressed by Stephan Boltzmann law, can be written as

$$I_o = \frac{5,67 \cdot IO^{-8}(T_f^4 - T^4)}{\frac{1}{\epsilon} + \frac{1}{\epsilon_0} - 1} \quad W/m^2 \qquad (4)$$

Fuel surface can be considered to have an emissivity of unity, and the fuel temperature is very small comparatively to that of the flame, so the above equation can be simplified to

$$I_o = 5,67 \cdot IO^{-8} \epsilon_0 T_f^4 \quad W/m^2 \qquad (5)$$

Dividing (5) on (2) we get a simplified formula for the rate of flame downspread and also for the rate of fire spread by fuel bed radiation

$$R_o = 2,2 \cdot IO^{-I2} \frac{\epsilon_0 T_f^4}{\rho_b (I5 + M)} \quad m/sec \qquad (6)$$

To calculate the flame zone emissivity by means of equation (6), observations of the rate of flame downspread along vertical beds of oak leaves were carried out. Flame length, bulk density and moisture content of fuel were also measured. The results are given in Table I together with M.J.Woolliscroft data on fuel bed radiation /IO,II,I2/.

Table I. Flame length and flame zone emissivity

Flame length, m	0,3	0,6	I,0*	I,5*	2,5*
Radiative heat flux through the fuel bed, $W/m^2 \cdot IO^4$	I,0	I,6	2,2*	3,0*	5,7*
Flame temperature, °C	850	850	940*	I050*	I080*
Flame zone emissivity	0,II	0,I7	0,I8	0,I8	0,30

* Woolliscroft data

The flame zone emissivity depends on the flame thickness in the direction of the flux and an effective attenuation coefficient, accordingly to the well known exponential law /7,I5/. In the case of fuel bed radiation the flame thickness is the flame length, so

$$\epsilon_0 = 1 - e^{-kL} \qquad (7)$$

where e is the base of Naperian logarithm.

The application of this law to Table I leads to

$$\epsilon_0 = 1 - e^{-0,I6(L + 0,375)} \qquad (8)$$

where 0,375 accounts for radiation from glowing embers in the fuel bed. It means that radiative heat flux from glowing embers is equal to that of flame 0,375 m long.

The attenuation coefficient 0,I6 here is the mean value for all experiments incorporated in Table I. However, observations revealed a trend of the coefficient to increase with the increasing of moisture content of fuel. For example, when moisture content increases from I2 to 34%, the attenuation coefficient increases from 0,I5 to 0,I7. The obvious explanation of the fact is the higher vapour pressure within the flame zone of more moist fuels.

5. Flame radiation above the fuel bed

Radiative heat flux from flames above the fuel bed onto the surface of the fuel ahead of the burning zone, accordingly to /I5/, is given by

$$I = 5,67 \cdot IO^{-8} \epsilon' (T_f^4 - T^4) \varphi L \quad W/m \qquad (9)$$

Heat losses by fuel bed cooling should also be added to this equation but, as it was shown by /I6/, they can be considered negligible in approximate analyses. As emissivity of fuel is near unity, the emissivity of flame-fuel system is near to that of the flame. Assuming this and disregarding T^4 as very small in comparison with T_f^4, we have the simplified equation for I

$$I = 5,67 \cdot IO^{-8} \epsilon_1 T_f^4 \varphi L \quad W/m \qquad (IO)$$

The configuration factor φ in this equation shows what part of radiation from the flame front is directed onto the fuel bed surface. When the flame front is straight and infinitely wide, the factor depends on the flame angle to the fuel bed surface and, as it was shown /4,5,6/, can be given by

$$\varphi = 0,5(1 + \cos\alpha) \qquad (II)$$

For flames of not infinite width (as in crib fires) the configuration factor can be estimated with the help of Table 2.

Table 2. Configuration factor for flames of finite width (flame is normal to the fuel layer)

Ratio of flame length to the fuel bed width	0,I	0,5	I,0	I,5	2,0	5,0
Configuration factor	0,40	0,30	0,24	0,20	0,I7	0,IO

The flame emissivity for radiation above the fuel bed depends on the length of the burning zone and, similarly to (8), is given by

$$\epsilon_1 = 1 - e^{-0,I6D} \qquad (I2)$$

6. Flame radiation as a mechanism of fire spread in forests

The flame downspread process along the vertical fuel bed has been discussed above. It has been shown that the propagation mechanism of backfires at high winds is governed by the same laws. Now the propagation of head fires in horizontal and slightly inclined fuel beds is to be considered. Here, in addition to fuel bed radiation, radiative heat is also transmitted to the fuel bed surface from the flame front above the fuel. This additional heat flux dries and preheats surface particles of fuel before they become exposed to fuel bed radiation. So they require less energy to ignite by fuel bed radiation than the lower ones when they come in contact with the flame. The lower particles can be ignited by fuel bed radiation a little later as the upper particles are not transparent and the thermal conductivity of contacts between adjacent particles is negligible /I7/. So the burning spreads faster along the surface of the fuel bed than along its depth, and the burning front within the fuel bed becomes inclined (Fig.2).

The conclusion is that to estimate the rate of fire spread we have to deal with the energy required to ignite only the surface layer of the fuel bed but not the fuel bed as a whole. This energy is given by

$$q = 2{,}6\cdot 10^4 h_1 \rho_b (15 + M) \qquad J/m^2 \tag{I3}$$

Here the depth of the surface layer h_1, preheated by flame radiation ahead of the burning zone, is equal to the mean free path of radiation within the fuel bed, which is described by /3/ as

$$h_1 = 4\frac{\rho_f}{\rho_b \sigma} \quad m \tag{I4}$$

It follows from this equation that the measure of $h_1 \rho_b$ depends on the fuel density and the surface-volume ratio only

$$h_1 \rho_b = 4\frac{\rho_f}{\sigma} \tag{I5}$$

and so can be considered constant for a particular type of fuel. It is in fact the weight of fuel in the unit area of the surface layer. Estimations on data of forest fire observations show that for most of forest fuels $h_1 \rho_b = 0{,}5\text{–}0{,}8$ kg/m^2. Only for grass this characteristic is 0,44 kg/m^2 and it is slightly more than equation (I5) gives. The possible explanation of the fact is that thin particles of grass are transparent for heat radiative flux. Special investigation shows that infra-red radiation penetrates into cellulosic materials up to $1{,}6\cdot 10^{-4}$ m /I8/, and grass leaves are just of that thickness.

If fuel loading is less than $h_1 \rho_b$, a certain part of radiative flux would go through voids in the fuel bed and so fall out the ignition process. When fuel height is more than h_1 and so fuel loading is more than $h_1 \rho_b$, the lower part of fuel bed would be ignited within the flame zone by fuel bed radiation only. Here the energy required for the ignition of fuel bed unit area would depend on the surface layer thickness, and the fuel bed thickness would influence the size of flame thus controlling the heat flux to the unburnt fuel.

On the base of equations (6), (I0) and (I3) the rate of fire

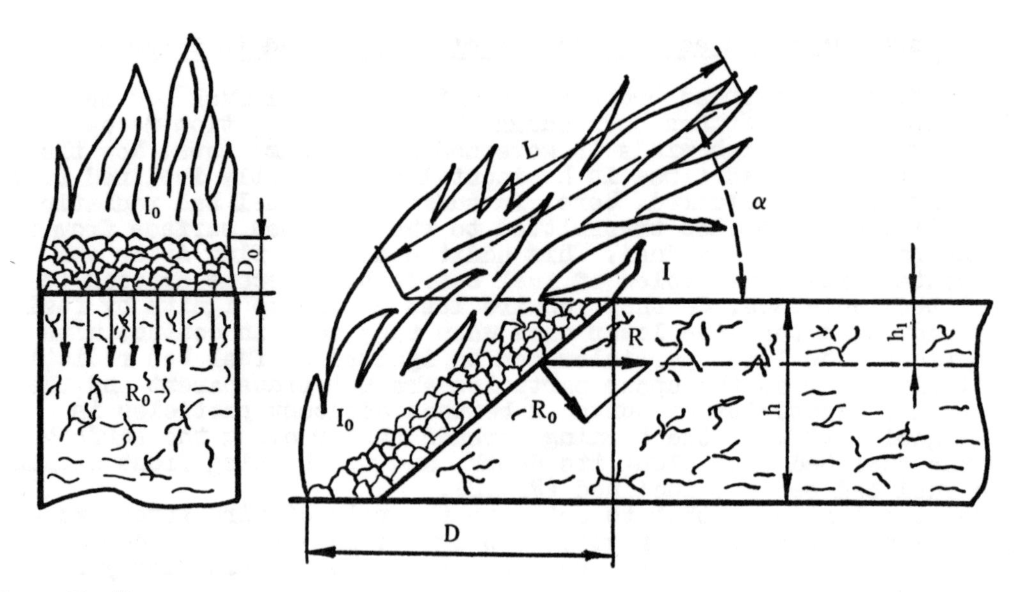

Fig. I Flame downspread process

Fig. 2 Process of fire spread along the thick fuel bed composed from thin particles of fuel

spread by the joint action of fuel bed radiation and flame radiation can be written as

$$R = \frac{2{,}2\cdot 10^{-12}\epsilon_0 T_f^4}{\rho_b(15 + M)} + \frac{2{,}2\cdot 10^{-12}\epsilon_1 T_f^4 \varphi L}{h_1 \rho_b(15 + M)} \quad \text{m/sec} \qquad (\text{I6})$$

and modified to

$$R = \frac{2{,}2\cdot 10^{-12}\epsilon_0 T_f^4}{\rho_b(15 + M)}\left(1 + \frac{\epsilon_1 \varphi L}{h_1 \epsilon_0}\right) \quad \text{m/sec} \qquad (\text{I7})$$

Taking into an account equation (6) we have

$$R = R_0\left(1 + \frac{\epsilon_1 \varphi L}{h_1 \epsilon_0}\right) \quad \text{m/sec} \qquad (\text{I8})$$

where R_0 is the rate of fire spread by fuel bed radiation only, and $\frac{\epsilon_1 \varphi L}{h_1 \epsilon_0}$ is the ratio of the contribution of flame radiation to fire spread to that of fuel bed radiation.

Data of more than 50 observations on the rate of forest fire spread, including those published /8,9,IO,II,I2/ were used for verifying the final equation. The comparison of observed and estimated rates of spread revealed satisfactory correspondence at various weather and fuel conditions except a few cases shown in Table 3. Thus the stated theory is shown to be valid at least for grass and heather fires.

Table 3 shows estimated rates of spread and those observed by Woolliscroft /IO,II,I2/ together with other characteristics of fire behaviour. Some discrepancy should be noted only for three fires where gorse was the dominant fuel. Perhaps, gorse as fuel has some peculiarity influencing fire spread but is not taken into account by this theory.

Table 3. Observed (Woolliscroft) and estimated values of forest fire characteristics

(a) Observed

Fire number	Fuel type	M (%)	ρ_b (kg/m^3)	h (m)	L (m)	T_f (^{o}K)	α (deg)	D (m)	R (m/sec)
I	Bracken	40	I0,6	0,30	0,9	I328	46	0,6	0,0I3
2	Heather	4I	7,3	0,50	I,6	I323	49	I,I	0,086
3	Gorse	60	2,9	0,40	2,2	I353	40	I,4	0,300
4	Heather, grass (5%)	50	7,2	0,50	I,0	I2I3	85	0,8	0,0I5
5	Heather	30	2,4	0,50	2,0	I323	45	I,0	0,088
6	Heather, grass (I4%)	27,5	2,7	0,40	I,3	I343	40	I,8	0,2I0
7	Gorse, heather, grass	I7	7,2	0,35	I,7	I283	50	I,I	0,028
8	Gorse	35	3,I	I,50	5,0	I353	60	3,0	0,I50
9	Grass*	I8	0,38	0,40	I,3	I343	40	I,8	0,2I0
I0	Gorse, heather, grass	40	5,8	0,35	2,0	I253	55	I,4	0,I46
II	Grass*	25	0,73	0,35	2,0	I253	55	I,4	0,I46

* It is fire Nr 6, assuming only grass burnt
** It is fire Nr I0, assuming only grass burnt

(b) Estimated

Fire number	ϵ_o	ϵ_1	ψ	$\frac{\epsilon_1 \varphi L}{\epsilon_o}$	h (m)	R_o (m/sec)	D (m)	R (m/sec)
I	0,I9	0,I0	0,84	0,40	0,06	0,002	0,5	0,0I5
2	0,27	0,I6	0,83	0,8	0,06	0,005	0,9	0,069
3	0,34	0,20	0,88	I,2	0,2I	0,020	I,4	0,I32
4	0,20	0,I2	0,54	0,3	0,07	0,0025	0,4	0,0I3
5	0,32	0,I5	0,85	0,5	0,I8	0,028	0,7	0,I04
6	0,24	0,25	0,88	I,2	0,I6	0,0I5	I,4	0,I25
7	0,28	0,I6	0,82	0,8	0,09	0,007	0,9	0,07I
8	0,58	0,38	0,75	2,I	0,26	0,03I	2,4	0,284
9	0,23	0,25	0,88	I,2	I,I3	0,I28	2,3	0,263
I0	0,32	0,20	0,79	I,0	0,I0	0,0055	I,I	0,060
II	0,32	0,20	0,79	I,0	0,59	0,0565	I,6	0,I53

It can be concluded from equation (I8) that there is the least limit of the rate of fire spread. It is equal to the rate of flame downspread along vertical fuel beds, where only fuel bed radiation controls the propagation of fire. The rate of backfire spread is very near to this "slow" limit. Head fires with their long and thick flames propagate much faster, and the ratio of head fire spread to the slow spread limit discussed first by /7/, accordingly to equation (I8), is given by

$$\frac{R}{R_o} = 1 + \frac{\epsilon_1 \varphi L}{\epsilon_o h_1} \qquad (19)$$

As the rate of spread multiplied by the residence time gives the length of the burning zone /2/, and the residence time is constant for any rate of fire spread provided the fuel bed properties do not vary, the above equation can be transformed into

$$\frac{D}{D_o} = 1 + \frac{\epsilon_1 \varphi L}{\epsilon_o h_1} \qquad (20)$$

where D_o is the length of the burning zone when the spread is "slow" (flame downspread or backfire). Hence

$$D_o = \frac{D}{1 + \frac{\epsilon_1 \varphi L}{\epsilon_o h_1}} \quad m \qquad (21)$$

Estimations of D_o by means of equation (21) on data of observations of forest fire behaviour, those of Table 3 including, showed that

$$D_o = h_1 \qquad (22)$$

and this is quite explicable for thin fuels with their short time of burnout. Inserting (22) into (21) we get

$$D = h_1 + \frac{\epsilon_1 \varphi L}{\epsilon_o} \quad m \qquad (23)$$

The last equation is in satisfactory agreement with the observed values of the length of the burning zone (see Table 3). Closer agreement could hardly be expected, taking into account variable winds and ununiform fuels in forests.

7. Conclusion

By justifying the concept of forest fire spread by radiation from flames this report evidences the necessity for a more rigorous investigation of the problem. A satisfactory forest fire spread theory must be capable of predicting fire behaviour characteristics completely in terms of properties of fuel and weather conditions, rather than partly in terms of flame size and flame angle, which themselves are necessary to predict. This paper is hoped to be a step to the development of such a theory.

8. Acknowledgements

The author wishes to thank Mr. P.H.Thomas for his considerable and constant assistance.Thanks are also due to Mr. C.E.Van Wagner, N.P.Kurbatsky and M.A.Sophronov for very useful consults.

REFERENCES

/1/ Fons, L.W.: "Analysis of fire spread in light forest fuels". Jour. Agr. Res., 72, 95-121 (1946)

/2/ Byram, G.M., H.B. Clements, E.R. Elliot and P.M. George: "An experimental study of model fires". Technical Report No.3, Forest Service, U.S. Dept. of Agr. Southeast. Forest Expt. Sta., 36 (1964)

/3/ Thomas, P.H.: "On the rate of spread of fire in cities and forests". Fire Res. Sta., Fire Res. Note 649, 9, (1967)

/4/ Emmons, H.W.: "Fire in the forest". Fire Res. Abstr. and Rev., 5, (2), I63-I78 (I964)
/5/ Van Wagner, C.E.: "Calculation on forest fire spread by flame radiation". Canada Dept. of Forestry and Rural Development, Pub. II85, 28 (I967)
/6/ Telisin, H.P.: "Dependability of the rate of forest fire spread on weather conditions". Far East Forestry Res. Inst., Sbornick trudov, 7, Khabarovsk, 39I-405 (I965)
/7/ Thomas, P.H.: "The contribution of flame radiation to fire spread in forests". Fire Res. Sta., Fire Res. Note 594, I9 (I965)
/8/ Thomas, P.H. "Rates of spread of some wind-driven fires". Forestry, 44, 2, I55-I75 (I97I)
/9/ Van Wagner, C.E.: "Fire behaviour mechanisms in a red pine plantation: field and laboratory evidence". Canada, Forestry branch, Dept. Pub. I229, 30 (I968)
/I0/ Woolliscroft, M.J.: "A report on forest fire fieldwork (New Forest, Mar. I968)". Joint Fire Research Organization Fire Research Note No. 744 (I969)
/II/ Woolliscroft, M.J.: "A report on forest fire fieldwork (New Forest, Mar. I966)". Joint Fire Research Organization Fire Research Note No. 693 (I968)
/I2/ Woolliscroft, M.J.: "Notes on forest fire fieldwork (New Forest, I967)". Joint Fire Research Organization Fire Research Note No. 740 (I969)
/I3/ Gundar, S.V. and H.P. Telisin: "Flammable characteristics of forest fuel particles in radiative heat flux". Combustion and fires in forests. USSR Acad. of Sci., Siberian Branch, Krasnoiarsk, I46-I52 (I973)
/I4/ Blinov, V.I. and G.N. Khudiakov: "Diffusive burning of liquids". USSR Acad. of Sci., Moscow, II3 (I96I)
/I5/ Blokh, A.G.: "Principles of heat exchange by radiation". Leningrad, 33I (I962)
/I6/ Fons, W.L.: "Project fire model. Summary Progress Report - II". U.S. Forest Serv., Dept. Agri., Southeastern Forest Expt. Sta., 55 (I962)
/I7/ Kantorovitch, B.V.: "Principles of the theory of burning and pyrolisis of solid fuels". USSR Acad. of Sci., Moscow, 43I-465 (I95I)
/I8/ Alexandrov, V.V. and G.S. Aravin: "Wood transparence influence on ignition by radiation and methods of its determination". Combustion and fires in forests. USSR Acad. of Sci., Siberian Branch, Krasnoiarsk, I26-I45 (I973)

Chapter 30

FABRIC IGNITION AND THE BURN INJURY HAZARD*

Wolfgang Wulff and Pandeli Durbetaki

Georgia Institute of Technology, Atlanta, Georgia, U.S.A.

Abstract

Burn-injury probability is proposed as a quantitative measure of the burn injury hazard from apparel fires, and is related conceptually to laboratory test methods. Ignition probabilities were measured under laboratory conditions. The measurements yielded the mean ignition time. The experimental results show, that fabrics respond deterministically under laboratory conditions.

NOMENCLATURE

C	:	ratio of heated fabric height over thickness of gas boundary layer
c, c_p	:	specific heat and specific heat at constant pressure
$\underline{E}$	:	activation energy
$\bar{F}$	:	average sample view factor with respect to its environment
h_c	:	convection heat transfer coefficient
$\bar{h}_c$	:	average (front and back face average or time average) conv. heat transf. coeff.
Δi	:	reaction enthalpy
k	:	thermal conductivity
k_d, k_e, k_g	:	frequency factors of desorption, oxidation and pyrolysis respectively
n	:	reaction order
N_{Fo}	:	Fourier number, Eq.5
$P(i/j)$	:	probability of event i for given event j
q^*	:	nondimensional heat flux
R	:	universal gas constant
s	:	estimate for standard deviation σ
T	:	absolute temperature
W_o	:	radiative power flux
z	:	general variable
α	:	absortance
δ	:	fabric thickness
ε	:	decomposable mass fraction
θ	:	nondimensional temperature, Table 2
λ	:	fraction of decomposable mass fraction
ρ	:	fabric density
$\tilde{\rho}$	:	reflectance
σ	:	standard deviation
$\tau, \tau^*, \langle\tau\rangle$	:	time, nondimensional and median
χ^*	:	standard normal variable, Eq. 4

*The work presented here resulted from ongoing research which has been funded by the National Science Foundation, first as Grant No. GK-27189 and now under the RANN Program as Grant No. GI-31882.

Subscripts

b	:	boundary
c	:	convection
d	:	desorption
e	:	exposure or exothermic
g	:	gas (pyrolysate)
i	:	ignition
ℓ	:	laboratory
m	:	melting
r	:	reference
rad	:	radiative
∞, o	:	environment or initial

1. Introduction

The general health hazard from fires involves the generation of toxic gases, smoke and vapors, and the infliction of burn injuries. Apparel fires produce primarily burn injuries the severity of which is assessed in terms of the body area covered by the burn and by the depth of tissue destruction.

There are three courses of action required to reduce losses from clothing fires: Firstly, educational campains to arouse the awareness of the public toward the fire hazard and to increase the safety consciousness, secondly, the modification of combustible materials to reduce ignition probability and, thirdly, the modification of frequently involved ignition sources to cause fewer accidental ignitions than is presently the case. The two latter approaches require, at least presently, legal standards for their implementation. Standards must be rational, reasonable and effective. Whether a standard is reasonable and effective, is primarily a political and economical judgement which ought to be supported by combined utility and probability studies. However, the question of rationality should be decided on physical arguments involving the description of material responses to fire, and on relevant statistical evidence describing the stochastic behavior of people. The fabric response to heating is to be determined by laboratory experiments.

A large number of flammability tests have evolved and specific material response to heating is being measured without clear conceptual connection to any measure of injury hazard. Representative examples regarding fabric-related burn injury hazards are undoubtedly the flammability tests which constitute essentially the measurement of flame propagation speed. Slow burning has been considered desirable and associated with relatively high safety. However, there appears to be no convincing argument that low fabric flame propagation rates ought to be desirable in general for reducing the burn injury potential unless the overall heat transfer process, including the propagation rates, are properly related to the victim's defensive response and to the fabric's fire extinguishment characteristics.

2. The Fabric-Related Burn Injury Hazard and the Burn Injury Probability

Acceptable fabric flammability standards must rationally relate laboratory test results to burn injury hazard. They must take into account, firstly, the stochastic events which lead from retail distribution of the fabric to any one of many, conceivable accident-prone situations and, secondly, the combination of stochastic human behavior in a fire with the predominantly deterministic processes of ignition, combustion and thermal tissue decomposition. In particular, flammability standards as well as their conceptual connection with laboratory tests are ineffective without a quantitative measure for the potential burn injury hazard.

Myron Tribus [1] introduced in 1970 the probability P(B) that a fabric might contribute toward burn injury, as the quantitative measure of the fabric flammability hazard. This so-called burn injury probability P(B) is a function of the partial probabilities which are associated with the occurrences of all possible events and circumstances leading from fabric retail distribution to burn injury, such as the tayloring of a fabric into a garment of a particular style, the laundering or drycleaning of the garment and its accidental exposure to an ignition source, the processes leading to ignition, the extinguishment of the garment fire and the denaturalization of tissue. The so-called decision tree [1] facilitates the systematic graphical representation of all possible events.

These events fall into two classes, namely the selection processes for which the associated probabilities are computed from relative frequencies, and the physicochemical processes the probabilities of which depend on characteristic times.

Selection processes constitute consumer choices and supplier responses to these choices. The associated probabilities equal the fractions of the total sample which are available prior to a particular choice. The stochastic nature of human preferences requires that relative frequencies be derived by statistical means from observation, or, in the case of new materials, from the best available experience with similar materials.

The physicochemical processes, starting with fabric exposure to an ignition source and leading via fabric ignition to combustion, flame spread and tissue destruction, are all transient in nature and therefore depend on characteristic times. This was recognized first by Evans, Wulff and Zuber [2] who postulated that the probability of ignition for given exposure P(I/E) is primarily a function of the ratio of exposure time τ_e over ignition time τ_i

$$P(I/E) = f(\tau_e/\tau_i) \tag{1}$$

This principle applies quite generally in that the probability of a physicochemical process to succeed depends on the ratio of the time during which such process is allowed to proceed, over the time which the process requires for its completion.

Consequently, the quantitative prediction of the burn injury hazard for a given fabric is achieved by (a) enumeration of all relevant events which could lead from fabric certification to burn injury, (b) assessment of the partial probability for each event, (c) computation of the overall probabilities for every possible succession of events which result in burn injury, and (d) summation of the overall probabilities. The resulting sum or, better yet, its negative logarithm represents a <u>rational</u> measure of burn injury hazard. The overall probabilities for a succession of independent events are the products of partial probabilities for events promoting and of complementary probabilities for those preventing the burning. For example

$$P(B) = P(E)\cdot P(I/E)\cdot[1 - P(Ex/I)]\cdot P(S/I)\cdot P(B/S) \tag{2}$$

expresses the probability of burn injury as the product of the probability of exposure P(E), times the probability of ignition for given exposure, times the complementary probability of extinguishment after ignition [1 - P(Ex/I)], times the probability of tissue destruction for given flame spread.

Little effort has been spent on the study of selection processes, as it lies in the nature of material standards to reflect material properties which are exhibited most strongly in physicochemical process. The first and most crucial process is ignition, in which the fabric plays an extremely deterministic role. Ignition characteristics of a fabric or garment are assessed through its probability of ignition after exposure. This is discussed next as a typical example of treating physicochemical processes in the context of the burn injury hazard.

3. Probability of Ignition After Exposure

Equation 1 expresses the probability of ignition after given exposure as a function of the ratio of exposure time over ignition time. Whereas the exposure time τ_e reflects the stochastic behavior of human activity and even the human response to fire, the ignition time τ_i depends on fabric properties and exposure conditions and is thus the central dependent material characteristic of ignition probability.

For operational expediency probability of ignition after given exposure, P(I/E), is derived from the probability of ignition under laboratory exposure conditions, $P_\ell(I/E)$ and the probability that laboratory exposure conditions exist in real life [3]. The former probability will be discussed further, the latter one is presently being investigated.

Experimentally obtained accumulative ignition frequency under fixed exposure conditions exhibit Gaussian character. Therefore, the probability of ignition under laboratory conditions is formally

$$P_\ell(I/E) = \int_{-\infty}^{\chi^*} \exp(-z^2/2)dz/\sqrt{2\pi} \tag{3}$$

where $$\chi^* = (\tau_e/\langle\tau_i\rangle - 1)/\sigma \tag{4}$$

In Eq. 4 the symbols τ_e, $\langle\tau_i\rangle$, and σ represent the exposure time, the median ignition time and the standard deviation of ignition time from the median ignition time, respectively. Both the median ignition time $\langle\tau_i\rangle$ and the standard deviation σ depend on heating intensity and fabric condition. These two parameters are sufficient to describe $P_\ell(I/E)$ and are obtained through probit analysis [4], performed on data from ignition frequency measurements.

Such measurements can be performed only under a selected number of experimental conditions. For generally applicable flammability standards the ignition frequency measurements must be supported and supplemented by a sufficiently well verified modeling analysis for the prediction of ignition time from fundamental fabric properties and exposure conditions.

Thermophysical fabric properties which are relevant to the description of preignition processes have been reported earlier [3, 4].

3.1. Measurement of Ignition Time

The ignition time τ_i in Eq. 4 for the ignition probability was measured under radiative and gas flame heating modes. Single measurements of τ_i as a function of heating intensity were first used as an estimate of the median ignition time $\langle\tau_i\rangle$. Ignition time frequency was then measured under controlled radiative heating intensity and exposure time to yield both $\langle\tau_i\rangle$ and the standard deviation σ in Eq. 4.

The ignition time τ_i of a non-melting fabric is defined here as the time span between the sudden exposure of the fabric to a time-invarient, uniformly distributed source of heating and the occurrence of a flame. For fabrics which melt prior to ignition, the time τ_m of melt- through is used instead of the ignition time because the events leading to melt-through are similar to those prior to ignition. Experiments as well as analytical modeling can thus be generalized to the study of fabric destruction.

Based on this definition of fabric destruction times a Radiative Ignition Time Apparatus (RITA) was constructed with the objective of achieving [3]:

a. uniform, time-invariant radiative fabric heating on 25mm-diameter samples at power fluxes between 0.25 and 15 W/cm^2,
b. sample exposure transients below two milliseconds at both the start and the end of the preselected exposure interval,
c. automatic exposure time control between one and thirty seconds,
d. remote infrared detection of flames and melt-through conditions,

e. environmental humidity control between 5 and 95% r.h.,
f. environmental temperature control between 5 and 35°C.

To exhibit the common correlation of fabric destruction time as a function of heating intensity, the destruction time data were normalized on the basis of a one-dimensional ignition model for an inert, semi-transparent fabric. In Figure 1 are shown the nondimensional destruction times or Fourier numbers

$$(N_{Fo})_{i,m} = (k/\delta)\tau_{i,m}/(\rho c \delta) \tag{5}$$

versus the nondimensional radiative power flux (equivalent Biot Number)

$$(q^*)_{rad} = (1-\tilde{\rho})W_o/[(k/\delta)(T_{i,m} - T_o)] \tag{6}$$

where ρ, k, c and δ stand for fabric density, conductivity, specific heat and thickness, respectively, $T_{i,m}$ and T_o for destruction and initial temperatures, respectively and $\tilde{\rho}$ and W_o represent, respectively the optical reflectance and the radiative power flux.

It should be observed that for each fabric a single experimental curve is obtained representing with relatively little scatter the ignition time as a function of heating intensity. The variation from fabric to fabric must be attributed to differences in optical thickness and effects due to thermal decomposition prior to destruction.

The straight line (a) in Fig. 1 shows the analytical result obtained from numerical integration for inert heating of GIRCFF Fabric No. 10 without convective losses. Curve (b) represents the numerical solution to a lumped-parameter ignition model which includes convective cooling effects, evaluated for GIRCFF Fabric No.10.

Gas flame ignition times were measured by suddenly exposing the stationary fabric samples to the gas flame and also by moving the sample through the flame at a prescribed velocity. The Convective Ignition Time Apparatus (CITA) used in carrying out gas flame ignition time measurements was designed to meet the following conditions[3]:

a. Specimen heating by a gaseous fuel-air flame provided by a variable fuel rate and variable air rate burner.
b. Specimen exposure to the flame, for the static tests, through a shutter providing full exposure of the sample in five-hundredths of a second or less.
c. Specimen side of shutter not to exceed 60°C, to avoid preheating of the sample.
d. Specimen transport for the dynamic tests at speeds in the range of 2 cm/s to 25 cm/s.

The convective film coefficient was derived from the temperature history of a 200-mesh stainless steel screen, exposed to the flame in the same geometrical configuration as the fabric. The convective film coefficients, averaged over projected front and back faces of the screen, ranged in magnitude from $2h_c = 0.00436$ to $0.01013\ W/cm^2K$, and were used to infer the heating intensity form measured gas flame temperatures.

The results of the static tests are summarized in Figure 2. Curve (a) drawn on this figure represents the solution to the inert heating analysis and the two broken lines show the range of radiative ignition data from Figure 1 for comparison. The results of the dynamic tests are presented in Figure 3 for the igniting fabric. All the tests were conducted with the fabric placed at the lowest setting above the top of the burner and at the highest heating intensity. The data points at zero speed were obtained from the stationary tests.

Table 1 is the summary of the probit analysis performed on ignition frequency data. The sixth column indicates that simple ignition time measurements, obtained for cotton under single shutter operation with RITA, agree well with the median ignition time as derived from ignition statistics. The last column exhibits the surprisingly low value of s which is the experimental extimate for the standard deviation σ of ignition time in Eq. 4. It is for cellulose at 30% r.h. only eight

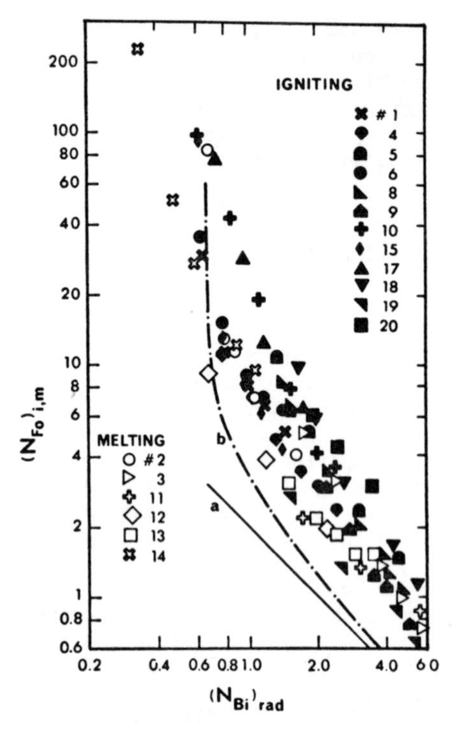

Figure 1. Normalized Ignition or Melting vs Normalized Radiative Heating Intensity.

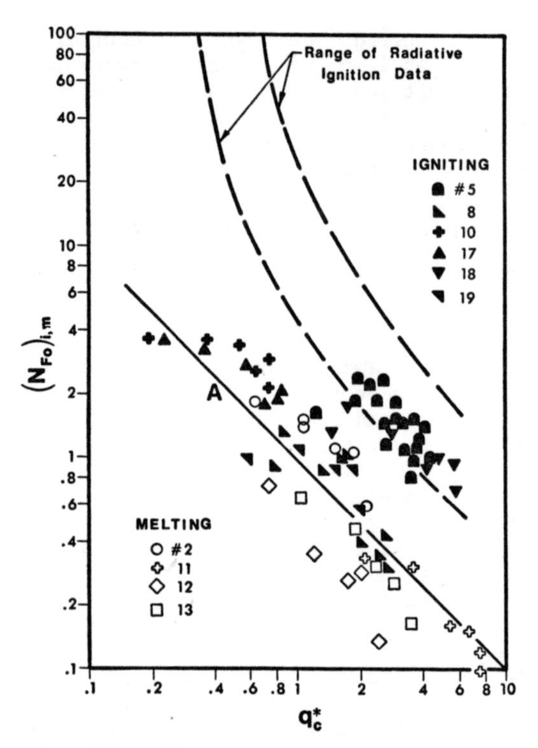

Figure 2. Normalized Destruction Time vs Normalized Convective Heat Flux.

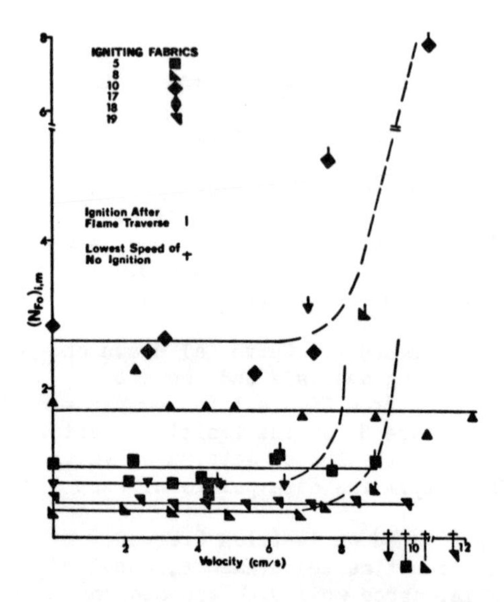

Figure 3. Normalized Destruction Time vs Sample Transport System Velocity for the Igniting Fabrics.

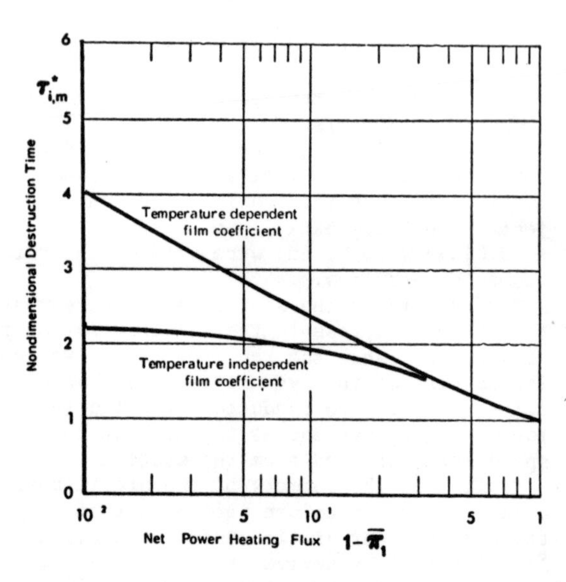

Figure 4. Effect of Free Convection on Ignition Time.

Table 1. Ignition Time Statistics

Fabric	W_o	Rel. Hum.	No. of Tests	Median Ignition Time $<\tau_i>$	Relative Mean $<\tau_i>/\tau_i$	Standard Deviation of Ignition Mean Time s_μ	Standard Deviation of Ignition Time
	W/cm^2	%		s			s
Cotton	6.50	30	77	31.93	1.174	0.38	2.55
	9.52	30	25	13.08	1.022	0.14	0.353
	13.80	30	18	7.11	1.030	-	0.0781
	6.50	90	17	35.49	-	-	0.175
	9.52	90	43	15.75	-	0.57	2.62
	13.80	90	32	7.91	-	0.07	0.246
Nylon	6.50	30	86	12.61	1.68	0.32	2.25
	9.52	30	42	6.15	1.40	0.15	0.70
	13.80	30	57	3.10	1.11	0.028	0.13
	6.50	90	66	11.63	-	0.19	1.11
	9.52	90	59	5.87	-	0.062	0.35
	13.80	90	79	3.23	-	0.24	1.62

percent at the low heating intensity and decreases to one percent at the high intensity. The column before last lists the expected standard deviation of the median ignition time and shows the reliability of the ignition time statistics.

3.2. Analytical Prediction of Ignition Time

A distributed parameter analysis has been developed for the inert, one-dimensional radiative heating of a semi-transparent fabric with in-depth absorption. The results of the numerical solution showed that thermal difusion is unimportant in all but two rather opaque fabrics, namely cotton denim ($296g/m^2$) and wool ($199g/m^2$) because in general the fabric temperature rises uniformly throughout the fabric during most of the preignition process. Consequently integral methods are justified for the prediction of ignition time.

A lumped-parameter model was developed to describe thermal energy storage, moisture desorption and pyrolysis in the fabric, that is, in the condensed phase, oxidation in the boundary layer and convective cooling of the fabric. The attainment of a prescribed mean fabric temperature is employed as ignition criterion. As shown in Reference [3], the three governing equations describe energy conservation in the fabric and three material decompositions on the basis of Arrhenius reactions laws, namely one for desorption, one for pyrolysis and one for oxidation. For the case of radiative heating, these equations are, in terms of the non-dimensional groups defined in Table 2:

$$d\theta/d\tau^* = 1 - \bar{\pi}_1(\theta+1)^{5/4} - \pi_2[(1+\pi_o\theta)^4-(1-\pi_o)^4] - \pi_3(1-\lambda_d)^{n_d}\exp[-\pi_4/(1+\pi_o\theta)]$$
$$-\pi_6(1-\lambda_g)^{n_g}\exp[-\pi_7/(1+\pi_o\theta)] + \pi_9/(1-\lambda_g)^{n_g}\exp[-\pi_{10}/(1+\pi_o\theta)] \quad (7)$$

$$d\lambda_d/d\tau^* = \pi_5(1-\lambda_d)^{n_g}\exp[-\pi_4/(1+\pi_o\theta)] \quad (8)$$

$$d\lambda_g/d\tau^* = \pi_8(1-\lambda_g)^{n_g}\exp[-\pi_7/(1+\pi_o\theta)] \quad (9)$$

The equations for gas flame ignition differ only slightly in form and definition of groups [3] from the above equations. The non-dimensional ignition time for radiative heating is given in terms of thirteen scaling parameters, i.e.

TABLE 2. DEFINITION OF RADIATIVE IGNITION PARAMETERS

Parameter Symbol	Definition	Interpretation
π_o	$1-T_\infty/T_{i,m}$	initial condition
$\bar{\pi}_1$	$2(h_c)_r(T_{i,m}-T_\infty)/(\alpha W_o)[(T_{i,m}-T_\infty)/(T_r-T_\infty)]^{1/4}$	convective cooling
π_2	$2\sigma\bar{F}_{s\infty}T_{i,m}^4/W_o$	radiative cooling
π_3	$\varepsilon_d(\rho\delta)_o(\Delta i)_d k_d/(\alpha W_o)$	power requirement for desorption
π_4	$E_d/(RT_{i,m})$	desorption rate dependence on temperature
π_5	$(\rho\delta)_o c(T_{i,m}-T_\infty)k_d/(\alpha W_o)$	characteristic desorption rate
π_6	$\varepsilon_g(\rho\delta)_o(\Delta i)_g k_g/(\alpha W_o)$	power requirement for gasification
π_7	$E_g/(RT_{i,m})$	gasification rate dependence on temperature
π_8	$\pi_5\, kg/k_d$	characteristic desorption rate
π_9	$(-\Delta i)_e/(c_v + Cc_p)_b \cdot h_c/(\alpha W_o) \cdot (\rho\delta)_b/[\varepsilon_g(\rho\delta)_o]\, k_e/k_g$	thermal interaction between gas and solid phases
π_{10}	$(E_e - E_g)/(RT_{i,m})$	temperature dependence of thermal feedback from boundary layer
θ	$(T - T_{i,m})/(T_{i,m} - T_\infty)$	fabric temperature
τ^*	$\alpha W_o\,\tau / [\rho\delta c(T_{i,m} - T_o)]$	time
n_d, n_g		reaction orders

$\tau^*_{i,m} = f(\pi_o, \bar{\pi}_1, \pi_2, \ldots, \pi_{10}; n_d, n_g)$; gas flame ignition requires only eleven such parameters.

Partial modeling rules can be developed from Eqs. 7, 8 and 9. Firstly, inert heating without convective losses leads simply to $\tau^*_{i,m} = 1$ or $\tau_{i,m} = \rho c \delta (T_{i,m} - T_o) / (\alpha W_o)$, but this solution does not agree in general with experimental data. Secondly, the effect of convective losses during radiative heating is characterized by $\bar{\pi}_1$ and assessed by omitting radiative cooling ($\pi_2 = 0$) and chemical reactions ($\pi_3 = \pi_6 = \pi_9 = 0$). The corresponding closed-form integration leads to Fig. 4 which reveals that convective cooling can effectively prolong ignition time and even prevent ignition at low heating intensities. The effect is pronounced when the temperature variation of the convective film coefficient is considered.

When strong endothermal reactions, such as desorption or melting, force the fabric temperature to remain constant at T_j during such reactions, closed form integrations yield for radiative heating with n endothermal reactions (and constant convective film coefficient)

$$\tau^*_{i,m} = -1/\pi_1 \ln(1 - \pi_1) + \sum_{j=1}^{n} (\pi_{20})_j / [1 - \pi_1 (\pi_{21})_j] \tag{10}$$

where $\pi_1 = 2\bar{h}_c (T_{i,m} - T_\infty)/(\alpha W_o)$ (11)

and $(\pi_{20})_j = (\varepsilon \Delta i)_j / [c(T_{i,m} - T_\infty)]$, $(\pi_{21})_j = (T_j - T_\infty)/(T_{i,m} - T_\infty)$ (12)

Other closed-form solutions exist [3] for the case of zero-order reactions, but they have no practical significance.

Finally, the influence on ignition time of parameters which describe general chemical reactions and radiative cooling in accordance with Eq. 7, must be investigated through numerical integration of Eqs. 7, 8 and 9. A typical result of such an integration by Runge-Kutta procedure is presented for radiative heating in Figures 5, 6 and 7. The nondimensional temperature rise $\theta(\tau^*)$ from room to ignition temperatures is shown in Figure 5 along with the histories of moisture desorption $\lambda_d(\tau^*)$ and pyrolysis $\lambda_g(\tau^*)$. The respective contributions to the ignition time due to the energy requirements to compensate for convective losses, radiative losses and desorption are shown in Figure 6 as functions of the convective cooling parameter $\bar{\pi}_1$. The time delay due to pyrolysis, $\Delta\tau^*_i$, is presented in Figure 7 as function of the gasification parameter π_8, along with the degree of gasification at the instant of ignition.

However, the comparison of computed with measured ignition times (Fig. 8) indicates that the present ignition model is still inadequate to predict ignition times at very low and very high heating intensities. It is suspected at this time that outgassing reduces convective cooling sufficiently at high heating intensities where there is not enough time for diffusion and bouyancy to remove the pyrolysates. Thus, outgassing could be responsible for the greater ignition delay at high heating intensities, while the low pyrolsis rates at low heating rates reduce convective cooling only insignificantly, realtive to the cooling on the inert metal screen with which the convective film coefficient had been measured.

4. Conclusions

Burn injury probability is an operationally feasable, quantitative measure of the burn injury hazard from garment fires. The partial probability of fabric ignition for given exposure characterize that fabric contribution to the burn injury probability which is most strongly related to the thermophysical fabric properties. Representing a transient physicochemical process, the probability of ignition after exposure depends on characteristic times, more specifically, on the ratio of exopsure over ignition times. The central dependent variable in the description of ignition probabiliry is the ignition time.

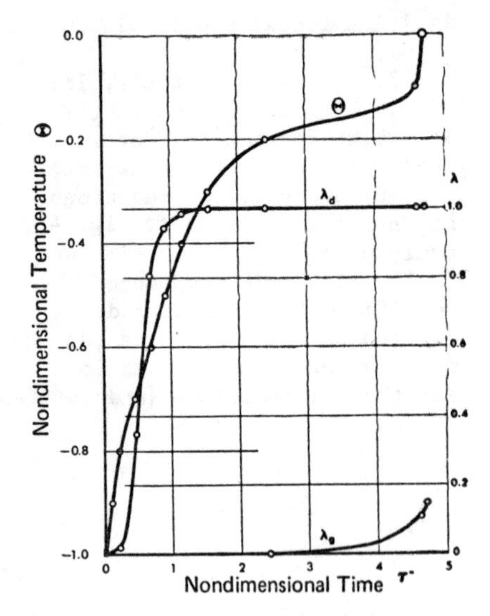

Figure 5. Computed Temperature Rise Θ, Desorption λ_d And Gasification λ_g During Ignition Process.

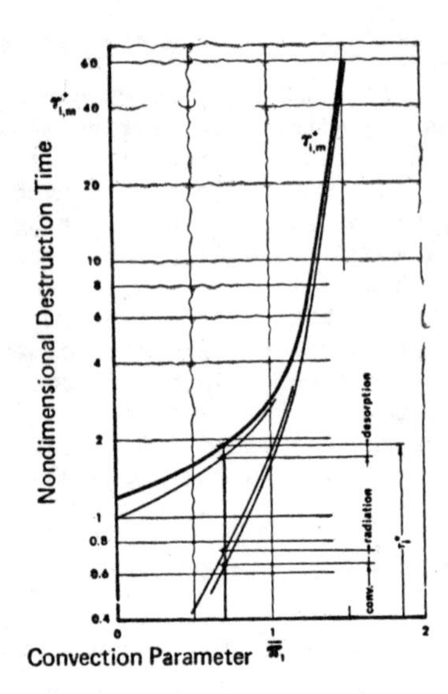

Figure 6. Convection and Moisture Effects on Destruction Time (Computed).

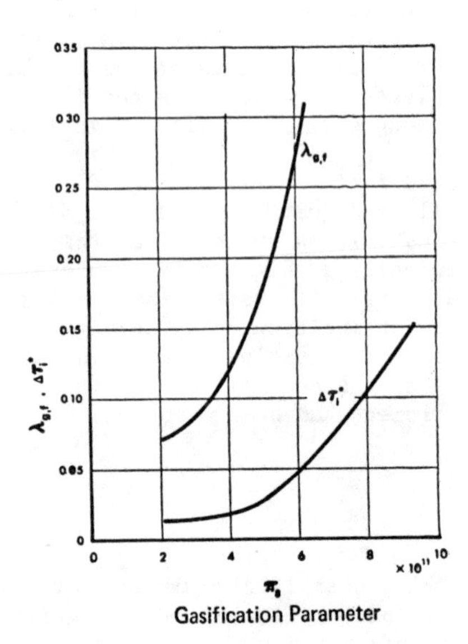

Figure 7. Endothermic Gasification Effects: Ignition Time Delay $\Delta\tau_i^*$ and Fabric Decomposition λ_g.

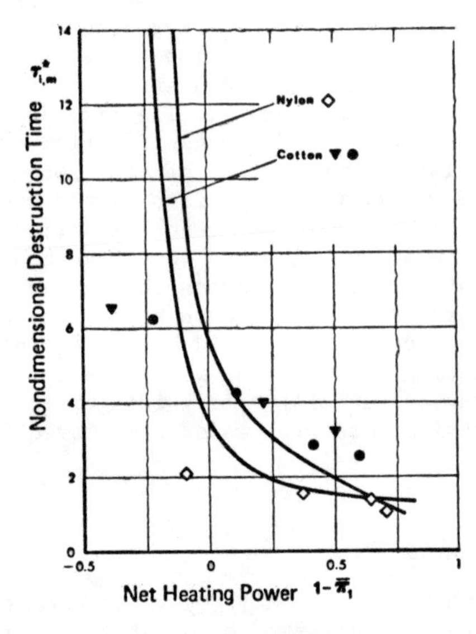

Figure 8. Comparison of Experimental And Analytical Results of Ignition Time Analysis.

Fabric response to heating under laboratory conditions is highly deterministic: The typical standard deviation from the median ignition temperature was found to be only five percent of the ignition time itself.

Convective cooling affects significantly the ignition process under radiative heating. The evolution of pyrolysates exerts a considerable influence on the convective heat transfer from or to the fabric.

Chemical kinetics plays a large role in the preignition processes. Moisture desorption can require up to 1/5 of the sensible heat supplied to the fabric during the preignition process. The energy requirements by pyrolysis increases with decreasing heating intensity.

Finally, ignition criteria should be developed which involve gas phase reactions rather than condensed phase temperature.

REFERENCES

/1/ Tribus, M.: "Decision Analysis Approach to Satisfying the Requirements of the Flammable Fabrics Act". An Address delivered by assistant Secretary of Commerce for Science and Technology at the Textile and Needle Trades Division, American Society of Quality Control at Greensboro, N. C. (February 12, 1970)

/2/ Evans, R. B., W. Wulff and N. Zuber: "The Study of Hazards from Burning Apparel and the Relation of Hazards to Test Methods". Research Proposal submitted by the School of Mechanical Engineering, Georgia Institute of Technology, to the Government-Industry Research Committee on Fabric Flammability (1970)

/3/ Durbetaki, P., W. Wulff, et al.: "Study of Hazards from Burning Apparel and the Relation of Hazards to Test Methods". Second Final Report, School of Mechanical Engineering, Georgia Institute of Technology, Atlanta, Georgia 30332, funded by the National Science Foundation under the RANN Program, Grant No. GI-31882, submitted to the Government-Industry Research Committee on Fabric Flammability, Office of Flammable Fabircs, National Bureau of Standards, Washington, D. C. 20234 (1972)

/4/ Wulff, W., N. Zuber, A. Alkidas and R. W. Hess: "Ignition of Fabrics under Radiative Heating". Combustion Science and Technology, Vol. 6, pp. 321-334 (1973)

Chapter 31

HEAT TRANSFER FROM TURBULENT FREE-JET FLAMES TO PLANE SURFACES

H. Kremer, E. Buhr[1] and R. Haupt

University of Trier-Kaiserslautern, Kaiserslautern and Gaswärme-Institut e.V., Essen, Germany

Abstract

The results on theoretical and experimental investigations of convective heat transfer are presented. Measured heat flux density distributions, obtained with chemically reacting gas jets (turbulent free jet flames) impinging at plane surfaces at different angles of impact, are discussed. The influence of surface temperature on heat flux densities and heat transfer coefficients is shown.

NOMENCLATURE

c_p	:	specific heat at constant pressure
c_s	:	specific heat of copper
d	:	burner tube diamter
h	:	enthalpy
h_{chem}	:	chemical enthalpy
K	:	correction factor in Eq. (2)
K_v	:	correction factor in Eq. (4)
Le	:	Lewis number
n	:	aeration coefficient (ratio of actual and stoichiometric volumetric air flow rates)
$\dot{q}$	:	heat flux density (based on unit area)
Pr	:	Prandtl number
Re_d	:	Reynolds number referring to burner tube diameter
T	:	absolute Temperature (K)
V	:	velocity on jet flame axis
U_x	:	radial velocity, parallel to wall
x	:	radial coordinate
y	:	axial coordinate on burner axis
α	:	heat transfer coefficient
δ	:	angle of impact between jet axis and plate surface
λ	:	thermal conductivity
ρ_s	:	density of copper plate material
μ	:	dynamical viscosity
τ	:	time

Subscripts

chem	:	chemical
e	:	condition at the edge of the boundary layer
f	:	frozen state
K	:	copper
w	:	condition at the plate surface
R	:	condition at the heat flux probe surface
M	:	condition at the plate surface, close to the heat flux probe

[1] Now with BASF AG, Ludwigshafen, formerly with Gaswärme-Institut e.V., Essen, Germany

1. Introduction

Heat transfer in conventional industrial furnaces occurs in general by thermal radiation, while convective heat transfer contributes only a miner portion. In modern "rapid heating" furnaces flames and hot combustion products impinge directly at the surface of the solid charge and heat it mainly by convection. In such furnaces heat flux densities, much higher than in "radiation furnaces", can be achieved.

In spite of many advantages of "convection furnaces" (smaller size, shorter heating-up periods) there is no simple method so far to determine convective heat transfer in these furnaces. The pheno-menomenological hehaviour of hot gas flows is known in principle, but the complex interaction of aerodynamics, chemical kinetics, thermodynamics, and various geometrical variables cannot be expres-sed in simple mathematical terms. Also the experimental verification of the different influences is extremely difficult to realize and therefore leads mostly to approximations for simplified theoretical models. The authors performed during the last years systematic inve-stigations of convective heat transfer from chemically reacting gases (turbulent jet flames) to solid surfaces. In the first part of the investigations a flat plate, as the most simple geometrical confi-guration, was used in order to minimize the number of variables. Besides the steady state heat flux at constant plate temperatures the unsteady heating process with transient surface temperatures was investigated as well.

2. Theoretical considerations

The analytical investigation starts from the conservation equations for momentum, mass and energy of the individual atomic species.
A general solution of this system of partial differential equations is not known. For some well-known geometries this system can be re-duced to a system of ordinary differential equations by means of suitable integral transformations (for some other configurations certain series substitutions lead to solutions) the solution of which is possible under certain suppositions (e.g. the thermophysical properties of the fluid must be known) by applying adequate numerical methods.
Such calculations have been performed by various authors [1, 2, 3].

The correlation of the numerical calculations by Fay and Ridell [3] led to the well-known equation for stagnation point heat transfer of laminar dissociated air streams

$$\dot{q} = 0.76\,Pr^{-0.6}\left(\frac{\rho_w\,\mu_w}{\rho_e\,\mu_e}\right)^{0.1}(\rho_e\,\mu_e)^{0.5}\left(\frac{\partial U_e}{\partial x}\right)^{0.5}(h_e - h_w)\left[1 + (Le^{0.52} - 1)\,\frac{h_{e,chem}}{h_e}\right] \qquad (1)$$

Since the present investigation is dealing with turbulent free jets it had to be scrutinized whether the above equation which was derived for lamiar flows is still valid in the case of turbulent impinging jets.
According to *Prandtl* [4] a turbulent boundary layer can be subdivided into several zones with different physical properties. The laminar sublayer is dominated by molecular transport processes, and therefore

contributes most to the heat transfer resistance, as compared to the upper mainly turbulent fluid lavers. As a first approximation therefore the calculation of convective heat transfer from turbulent jets can be based on the laws for laminar flow conditions.

Besides this basic simplification the influence of molecular diffusion, appearing in the second term of Eq. (1), on convective heat transfer in the case of burning technical methane-air mixtures for aeration coefficients $n > 1$ was studied. Therefore the chemical equilibrium compositions were determined for adiabatic flame temperatures as a function of the aeration coefficient and used to calculate Lewis numbers Le and chemical enthalpies Δh_{chem}.

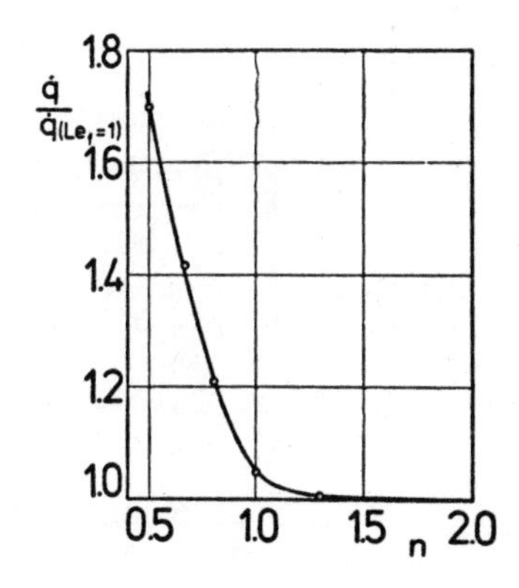

Fig. 1

Dependence of stagnation point heat flux on aeration coefficient and enthalpy ratio ($\dot{q}$ takes into account the variation of Le with n)

Fig. 1 shows that with technical heating processes molecular diffusion plays a minor role for values of the aeration coefficient above one due to fairly low adiabatic flame temperatures which lead to low radical concentrations. Thus, as a consequence of strong temperature gradients, only heat conduction in the fluid boundary layer close to the surface has to be considered.

For the tangential velocity gradient $(\partial U_e/\partial x)_e$ at the edge of the boundary layer an expression was derived [8] which is based on the equation by *Schlichting* [5] for the spatial velocity distribution in the vicinity of the stagnation point, the equation for the velocity distribution in turbulent jet diffusion according to *Kremer* [6], and the boundary layer thickness for viscous flow after *Homann* [7]

$$\dot{q}_w = 0.195\ K\ Pr^{-0.6}\ \rho_e\ v_{St}\ \Delta h \qquad (2)$$

The correction factor K takes care of actual flow conditions in free jets as well as of the temperature dependence of different thermophysical fluid properties in comparison to an air jet. K was determined from the experimental results.

By means of Eq. (2) the stagnation point heat flux density of free jet flames can be calculated if the temperature distributions at the edge of the boundary layer, the thermophysical properties of the fluid components corresponding to these temperatures $(Pr, \rho, \mu, \ldots) = f(T_e)$ and the surface temperature of the plate are known.

3. Experimental investigations

3.1 Measurement of heat flux densities

The experimental investigations concerned the measurement of heat flux density distributions (steady and unsteady) as well as the measurement of temperature, static pressure and concentration distributions close to the plate surface. The probes wihich were used to measure steady and unsteady heat flux densities are shown in Fig. 2 and Fig. 3 inserted into the flat plates.

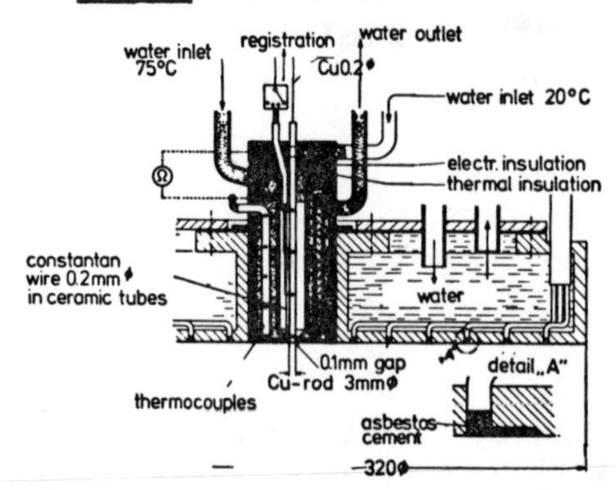

Fig. 2 Probe to measure steady-state heat flux

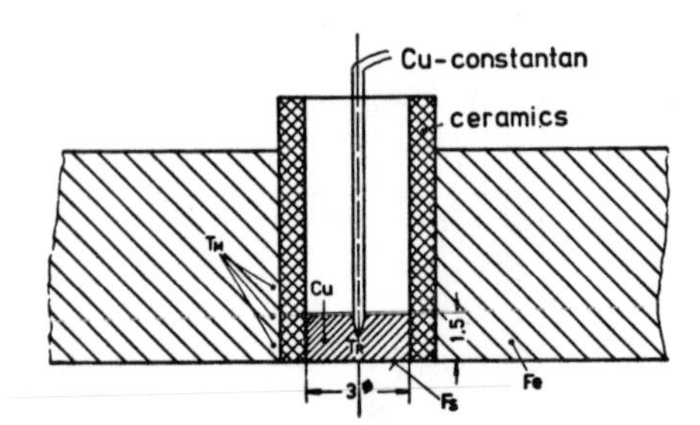

Fig. 3 Probe to measure unsteady heat flux

The steady-state heat flux densities were measured by means of a copper rod being inserted with one end into the flat plate flush with the plate surface which was exposed to the hot combustion gases. The copper rod was surrounded by an annular air-filled space and this again was inclosed in a cylindrical opening which was cooled by water at medium rod temperature. The other end of the rod was cooled by cold water. At the tip of the rod in the flat plate a 0.1 mm gap, controlled by electrical resistance measurement, provided thermal insulation between cooled housing and rod. This set-up caused negligible heat losses from the copper rod to its surroundings. The temperature gradient along the copper rod was used to determine the heat flux density by means of the expression

$$\dot{q} = \frac{\lambda(T)}{S} \Delta T \tag{3}$$

The unsteady heat fluxes are measured by a probe according to Fig. 3. A small, flat cylinder of pure copper was inserted tight into the opening of a ceramic tube. The temperature of the copper cylinder was registered during the time of heating by means of a UV recorder. The heat flux density can be evaluated as a function of the corresponding surface temperature T_e according to the equation

$$\dot{q}_W (T_e) = K_v (T_e, T_M) \rho_K s_K c_s \frac{dT}{d\tau} \tag{4}$$

Eq. (4) contains a factor K_v which corrects for erros caused by thermal conduction and radiation of heat from the copper cylinder due to imperfect insulation. Preliminary tests lead to calibration curves $K_v = f(T_R, T_M)$.

Heat losses of the copper cylinder to the surroundings limited for certain temperature differences between T_R (copper cylinder), T_M (plate surface in the vicinity of the probe) and T_o (ambient temperature) the maximum value of T_R due to the balance between supplied heat $\dot{q}_{zu}$ and heat losses $\dot{q}_{ab}$ [9].

3.2 Temperature measurements

The determination of enthalpies at the edge of the stagnation point boundary layer required measuring the temperature of the reacting gases as precisely as possible.
A miniature thermocouple (Pt-PtRh wire, diameter 0.1 mm) was used in oder to avoid flow disturbances and to obtain high spatial resolution. The thermocouple bead was coated with catalytically inactive Al_2O_3 in order to prevent radical recombinations at its surface.
Radiation and conduction heat losses of the thermocouple were compensated by electric heating, according to a method by *Berkebusch* [10].

4. Results

4.1 Steady-state heat flux densities

Radial and axial distributions of steady-state heat flux densities were measured in the stagnation point region of turbulent, partly premixed methane-air jet flames which impinged at hydraulically smooth, cooled plates. Fig. 4 contains some typical measured axial distributions for various nozzle Reynolds numbers.

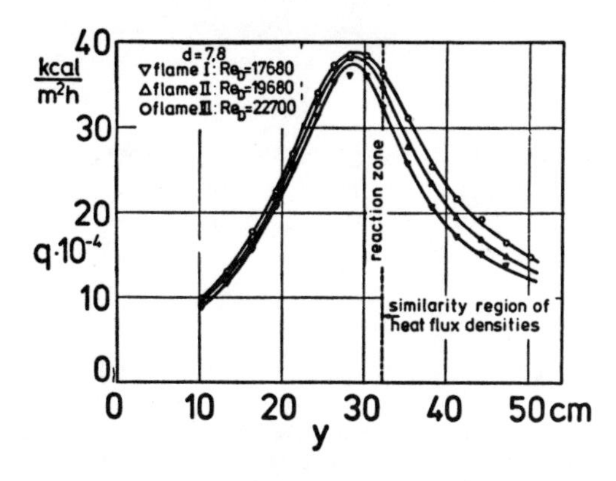

Fig. 4
Axial heat flux density distributions for various Reynolds numbers ($\delta = 90^o$)

The heat flux density increases with growing burner distance and reaches a maximum close to the axial location of the reaction zone, beyond this region it drops hyperbolically in all investigated cases because of decreasing jet temperature and flow velocity. Due to the assumptions made, concerning the distributions of enthalpy and momentum, Eq. (2) is only valid in the regions of decreasing heat flux densities. With partly pre-mixed turbulent flames this "similarity" region is mainly determined by the location of the reaction zone and starts where the methane conversion just has been completed (see dashed line in Fig. 4).
Typical radial heat flux density distributions are show in Fig. 5.

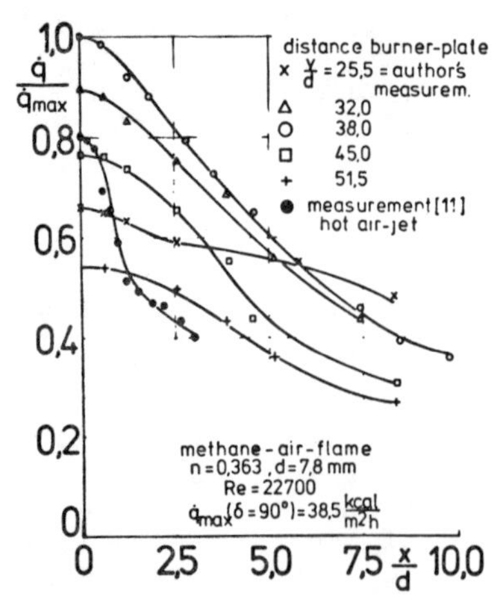

Fig. 5

Radial heat flux density distributions for various distances from the burner nozzle (δ = 90°)

Also the radial distributions can be presented in a "similarity plot" by means of an appropriate transformation of coordinates, in the present case for nozzle distances beyond y/d = 30.
At oblique angles of impact along the section of the plate surface covered by the flow, a wall jet starts to form, to which the simplifying conditions of rotational symmetry do no longer apply. Due to the conditions for the development of the boundary layers also the axial distributions of heat flux densities are influenced by the radial distributions in the measuring plane. Fig. 6 shows axial distributions for partly pre-mixed methane-air flames.

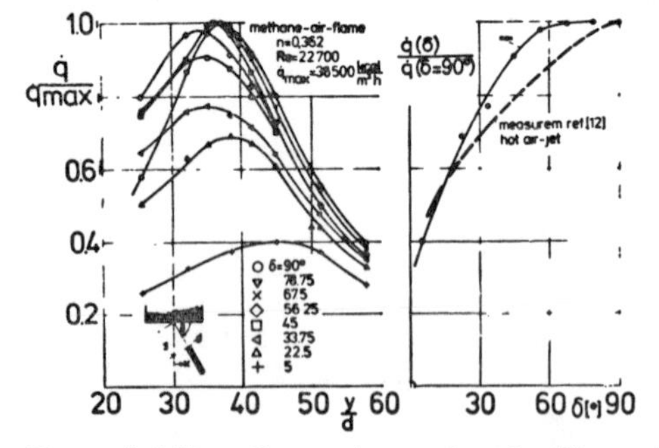

Fig. 6

Heat flux density distributions at the burner axis for various angles of impact ($\delta \neq$ const.)

The shift of maximum heat flux densities with smaller angles of impact to shorter distances between burner and plate and with larger angles of impact to longer distances results from the unsymmetrical deformation of the flame reaction zone close to the plate. In the right part of Fig. 6 the maximum values of heat flux density are plotted as a function of the angle of impact. For reasons of comparison data obtained by *Perry* [12] for a hot air jet are also reproduced. It can be observed that with the flames the decrease of the plotted ratio of heat flux densities occurs only for angles δ > 60° while with a hot air jet the decrease already starts with small deviations from the right angle.
The characteristic behaviour of radial distributions for several angles of impact with a constant distance between burner nozzle and plate can be found in Fig. 7. The degressively falling maxima with decreasing angles can be explained by the altered conditions of fluid flow and combustion reactions. An inclination of the burner axis towards the plate causes the location of the reaction zone to be shifted in flow direction corresponding to the pattern of the stream lines.

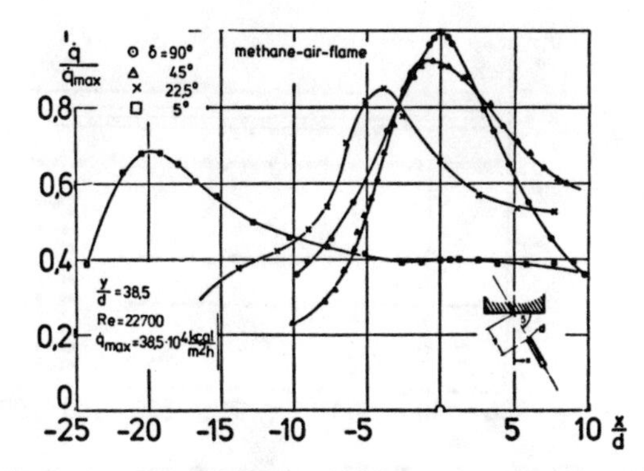

Fig. 7 Heat flux density distributions in the measuring plane at various angles of impact ($\delta \neq$ const.)

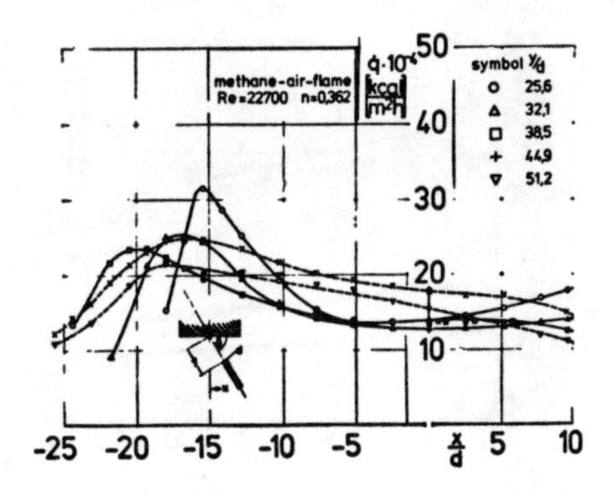

Fig. 8 Heat flux density distributions in the measuring plane at various axial distances y/d (δ = 5°)

The influence of unsymmetrical distributions can be inferred from Fig. 8. These results were obtained at a constant angle $\delta = 5^{\circ}$ for various distances between burner nozzle and plate. The heat flux density distributions, indicated by dashed lines, refer to those distances at which the flame fronts have already reached the center line of the jet. The solid lines apply to those cases where the (deformed) zones are to be found at a certain distance from the burner axis. Beyond the maxima again a similar behaviour of all distributions can be detected, i.e. they practically coincide after suitable transformation of coordinates.

4.2 Unsteady heat flux densities

Heat flux densities were measured with constant temperature of the cooled plate and also during the heatingperiod of the transient heat flux probe as a function of temperature (Fig. 9). The measured heat fluxes drop in a linear manner with rising probe surface temperatures or with decreasing temperature differences between flame and probe surface respectively. Also with surface temperatures $T_R > 600$ °C, conditions under which one could expect more favourable ignition conditions for unburned mixture components and an improvement of the conditions for the recombination of free radicals due to smaller temperature gradients between edge of the boundary layer and plate surface, no additional heat flux was recorded. From these results can be concluded that the transport of chemical energy by radicals across the boundary layer, as a consequence of fairly low flame temperatures and the thereby caused low radical concentrations, has practically no importance for the transfer of heat from technical flames with aeration coefficients $n > 1$. The curves shown in Fig. 9 which are valid for a nozzle Reynolds number Re_d = 22,700 are qualitatively the same for other Reynolds numbers. The fairly large scattering of measured values, occurring mainly in the lower temperature region can be traced back to the graphical differentiation of the curves $T = f(\tau)$ which is very sensitive to even small errors at the determination of slopes, especially with large gradients.
Using as a reference temperature the maximum (measured) flame temperature T_F and the surface temperature T_R a heat transfer coefficient α can be defined

$$\alpha = \frac{\dot{q}}{T_F - T_R}. \tag{5}$$

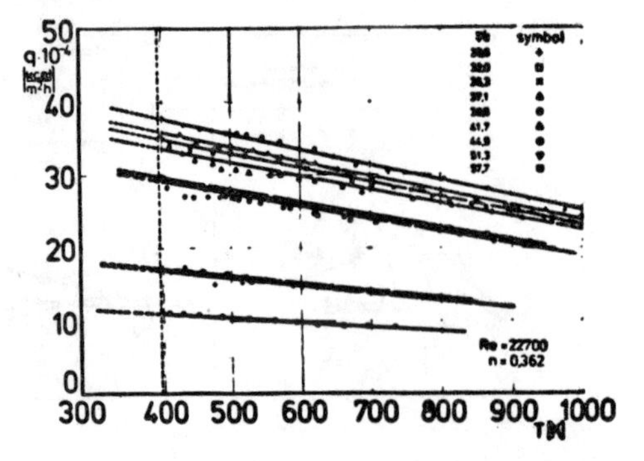

Fig. 9 Influence of probe surface temperatures on measured heat fluxes for various axial distances between plate and burner ($\delta = 90^{\circ}$)

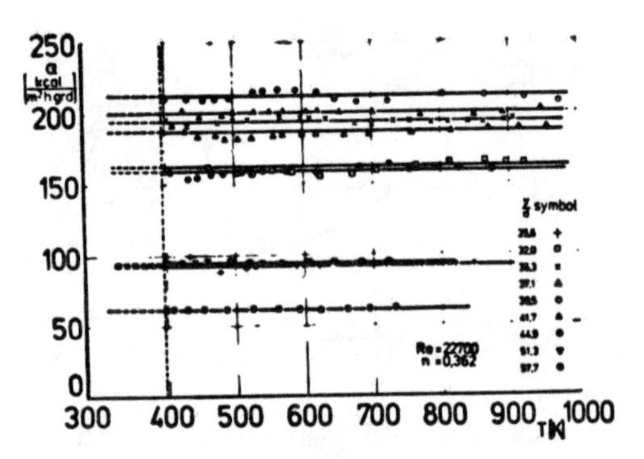

Fig. 10 Influence of probe surface temperatures on the heat transfer coefficient for various axial distances between plate and burner ($\delta = 90^{\circ}$)

The evaluation of Fig. 9 according to Eq. (5) lead to a distribution of the heat transfer coefficient α which was found to be independent of surface temperature (Fig. 10). Fig. 10 shows that the heat transfer coefficient is merely a function of the distributions of fluid velocity and enthalpy within the flame region. A plot of the heat transfer coefficient versus the axial distance between burner nozzle and heat flux probe shows that α follows a curve similar to that of the heat flux density $\dot{q}$ (Fig. 11).

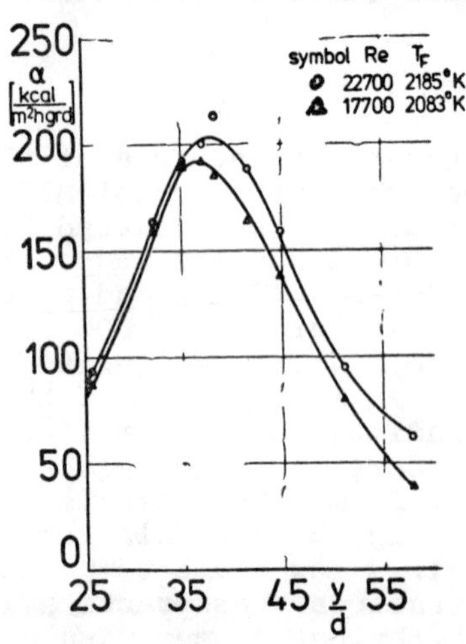

Fig. 11

Stagnation point heat transfer coefficients of turbulent, partly pre-mixed methane-air jet flames

4.3 Comparison between theoretical and experimental results

The measured temperatures in the boundary layers, corrected by means of the compensating method mentioned earlier, were used to calculate thermophysical fluid properties necessary to evaluate Eq. (2). The correction factor K which was obtained by comparing measured heat fluxes with those calculated by Eq. (2) has a mean value of K = 0.190.

Fig. 12 demonstrates that the introduction of this correction factor leads to a good agreement between calculated and measured stagnation point heat flux distributions.

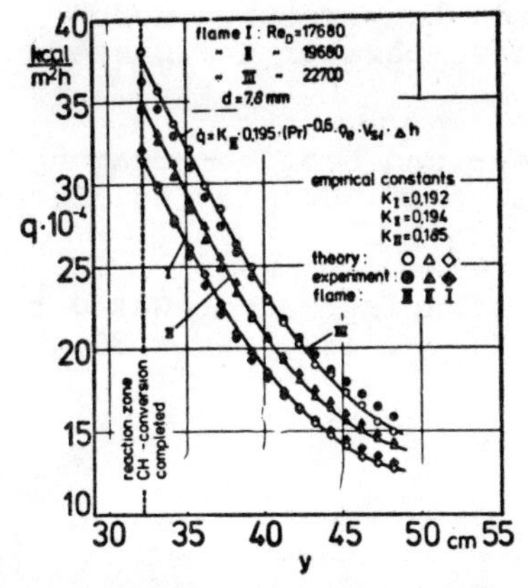

Fig. 12

Calculated and measured heat flux density distributions at the stagnation point of turbulent, partly pre-mixed methane-air flames

The maximum deviation observed was about 9 per cent.
Taking into account the complex conditions of aerodynamics, chemical kinetics, and thermodynamics, as well as the geometrical variables at the stagnation points of turbulent free jet flames, the observed agreement can be regarded as quite satisfactory.

REFERENCES

[1] Rosner, D.E.: "Effects of Diffusion and Chemical Reaction on Convective Heat Transfer". ARS Journal, Jan. 1960, p. 114-115

[2] Sibulkin, M.: "Heat Transfer Near the Forward Stagnation Point of a Body of Revolution". J. Aeron. Sci., Aug. 1952, p. 570-571

[3] Fay, J.A., F.R. Ridell: "Theory of Stagnation Point Heat Transfer in Dissociated Air". J. Aeron. Sci. 25, 1958, p. 73-86

[4] Prandtl, L.: "Zur Berechnung der Grenzschichten". ZAMM 18, 1938, p. 77-82

[5] Schlichting, H.: "Grenzschicht Theorie". Verlag G. Braun, Karlsruhe 1965

[6] Kremer, H.: "Zur Ausbreitung inhomogener turbulenter Freistrahlen und turbulenter Diffusionsflammen". Diss. TH Karlsruhe 1966

[7] Homann, F.: "Der Einfluß großer Zähigkeit bei der Strömung um den Zylinder und um die Kugel". ZAMM 16, 1936, p. 153-164

[8] Buhr, E.: "Über den Wärmefluß in Staupunkten von turbulenten Freistrahlflammen an gekühlte Platten". Diss. TH Aachen 1969

[9] Kremer, H., E. Buhr, R. Haupt: "Wärmeübergang bei der Verbrennung in Grenzschichten"; unpublished report

[10] Berkebusch, F.: "Zur Messung von Flammentemperaturen durch Thermoelemente, insbesondere über die Temperatur der Bunsenflamme". Wied. Ann. 67, 1899, p. 649

[11] Schlünder, E.U., V. Gnielinski: "Wärme- und Stoffübertragung zwischen Gut und aufprallendem Düsenstrahl". Chemie-Ing.-Techn. 39, issue 9/10, 1967, p. 578-584

[12] Perry, K.P.: "Heat Transfer by Convection from a Hot Gas Jet to a Plane Surface". Proc. Inst.Mech.Engrs. 168, issue 30, 1954, p. 775-780

Chapter 32

HEAT AND MASS TRANSFER CONSIDERATIONS IN SUPER-CRITICAL BIPROPELLANT DROPLET COMBUSTION

R. Natarajan

Indian Institute of Technology, Madras, India

Abstract

Although theoretical studies of droplet combustion have, as a rule, allowed for the possibility of attainment of the critical point by the burning drops, and have produced transient gas-phase theories of super-critical combustion, experiments have not verified the occurrence of this phenomenon. Consideration of the influence of mass transfer on heat transfer to burning drops is shown here to lead to the speculation that the critical point may never represent one of the operating conditions in droplet combustion.

NOMENCLATURE

B	:	transfer number, dimensionless
C_p, C_v	:	specific heats at constant pressure and constant volume, L^2/t^2T
d	:	droplet diameter, L
k	:	thermal conductivity, ML/t^3T
Nu, Pr, Re	:	Nusselt, Prandtl, Reynolds numbers; dimensionless
P	:	Pressure, M/t^2L
t_b	:	burning time, t
T	:	temperature, T
λ	:	evaporation constant, L^2/t
ρ	:	density, M/L^3

Subscripts

cr	:	critical condition
l	:	the liquid

1. Introduction

With the pressure levels of operation of the combustion chambers of contemporary diesel engines and liquid bipropellant rocket engines well above the critical pressure of the fuel and/or of the oxidizer (1), there have been several recent investigations of "super-critical bipropellant droplet combustion".

One of the first investigations on this topic was by Spalding (2) in 1959; although, not much attention and importance was given to this problem till much later. In fact, as recently as in 1964, in answer to a question regarding the subsequent history of a drop injected into a super-critical atmosphere, Lewis (3) chided the questioner for being "on one of his favorite hobby-horses". However, Lewis clarified the distinction between flash vaporization and super-critical vaporization in so far as during the latter process " there was no sudden change of density or other physical properties which there was with flash vaporization".

2. Super-critical bipropellant droplet combustion

At the outset, it is important to distinguish between droplet combustion at super-critical pressures and super-critical combustion.

Extrapolation to higher pressures of the commonly made assumption in low-pressure droplet combustion theories that the droplet surface temperature is equal to the boiling temperature leads to the conclusion that the droplet would attain the critical point when the ambient pressure equals the critical pressure. At the critical point, the transfer number tends to infinity and infinite burning rates are predicted, quite contrary to reality. There has thus been a need to explain and understand droplet combustion at and above the critical pressure of the fuel.

Even at low ambient pressures, it is now realized that the heat and mass transfer processes associated with the wet-bulb phenomenon limit the liquid surface temperature to the steady-state wet-bulb temperature, which is less than the boiling temperature. Thus, if the liquid drop attains the critical temperature, it should do so at a high super-critical pressure.

In the context of the above discussion, super-critical combustion would imply the attainment of the critical point by the burning drop before consumption, whereas combustion at all (or any) super-critical pressures may not involve super-critical combustion.

3. The thermodynamic critical point

3.1 Single-component fluids

The most general description of the critical point for a single-component system, which can be logically extended to multi-component systems, defines the critical point as the state at which the liquid and vapor phases become continuously identical and the intensive properties of each co-existing phase are indistinguishable from those of the other (4). Actually, the physical changes during this type of phase transition are not sudden, but gradual, over a range of temperature and pressure, taking place by a piecemeal transformation of small portions of the substance from one state to another. Giaque et al (5) have devised the word " Pernt" to fulfill the need to refer to the approximate temperature and pressure ranges of these gradual transitions.

Rice (6) presents theoretical and experimental evidence to show that the idea of a definite liquid-vapor critical point, with unambiguous critical temperature, pressure and volume is probably only an approximation; there actually being a critical region.

The values of the critical parameters and the exact behavior of a pure substance near the critical point are very sensitive to impurities, the details of past history, the rate of heating and the way it is handled. That the critical parameters for even CO_2, which is probably the most investigated pure substance as regards critical phenomena, are not unambiguously known is evident from the use of two alternative sets of critical values by Sengers (7).

3.2 Singularities (or anomalies) in thermodynamic and transport properties at the critical point

3.2.1 Thermodynamic properties

Since, by definition, the intensive properties of the liquid and vapor phases become identical at the critical point, the latent heat of vaporization, the surface tension and the refractive index difference should all vanish at the critical point.

For a "classical" or van der Waal's type of critical point, the free energy is everywhere an analytic function of density and temperature, and $(\partial P/\partial v)$ and $(\partial^2 P/\partial v^2)$ set equal to zero at the critical point, while the higher derivatives are non-zero. Thus the isotherm at Tcr undergoes a point of inflection at the critical pressure and volume on a P-v diagram.

Levelt Sengers (8) shows that a different type of critical behavior would be encountered, if not the first two, but the first four derivatives of P with respect to v are set equal to zero at the critical point. The case of only three zero derivatives would violate mechanical stability. In general, the nature of anomalies in several thermodynamic properties would depend on the number of derivatives $(\partial^n P/\partial v^n)_T$ set equal to zero at the critical point.

A classical description of the critical point leads to infinities in the following thermodynamic properties: (i) constant pressure specific heat; (ii) isothermal compressibility (iii) isobaric thermal expansion coefficient and (iv) Joule-Thomson coefficient.

The classical description does not require a singularity in C_v. In fact, C_v is always finite for a van der Waal's gas. However, Levelt Sengers (8) and Rowlinson (9) point out that several recent determinations reveal a tendency of this quantity to diverge near the critical point. It is shown that the classical description is inadequate and that the free energy is a non-analytic function of density and temperature. Rowlinson shows that the critical point is marked by a strong infinity in C_p and a much weaker infinity in C_v.

3.2.2 Transport Properties

Not only is the theory of the non-equilibrium transport properties not so well developed as that of the equilibrium thermodynamic properties, but also experimental measurements are extremely difficult in the critical region, as the infinities in compressibility and C_p make it almost impossible to establish stable gradients of velocity and temperature in the fluid.

It is generally believed that the viscosity exhibits no anomalous behavior near the critical point (9) although some workers have reported a small anomalous increase in this region (7).

As regards thermal conductivity, there flourishes a serious controversy in the literature whether it exhibits a pronounced anomaly in the critical region. Sengers (7) points out that papers reporting an anomalous increase are generally shown scant credibility because of the high probability of convection. He concludes that the existence of a pronounced anomaly has been verified by convection-free measurements. It remains to be shown whether the anomaly in k is related to that in the specific heats. Rowlinson (9) affirms that k exhibits a singularity at the critical point, but the order of its infinity, although not accurately known, is stated to be less than that for Cp. This means that the thermal diffusivity, $(k/\ell C_p)$ would tend to zero at the critical point.

Contrary to the above conclusions, however, Shiralkar and Griffith (10) have considered the viscosity and thermal conductivity of CO_2 to fall sharply at the critical temperature.

For a classical type of critical point, the diffusion coefficient is considered to approach zero at the critical point (11).

3.3 Multi-component fluids

Although many of the properties of the critical region of single component fluids are paralleled in multi-component systems, the latter exhibit a much more complicated phase behavior, which can not be explained as an extension of the behavior of pure substances, mainly as a result of variation of composition of the multi-component systems (12).

The critical state for a multi-component system is defined, as for a pure substance, as that state where all intensive properties of each of the two co-existing phases become continuously identical. But, unlike in the case of a pure substance, for which the state of identity of the two phases is at the maximum temperature and the maximum pressure at which the two phases coexist, for systems of more than one component, the critical state, in general, will not be at the maximum values of temperature and pressure at which two phases can coexist in equilibrium for a mixture of fixed composition. This maximum temperature and pressure have been called " Cricondentherm" or " maxcondentherm" and " Cricondenbar" or "maxcondenbar" respectively. The fact that the maxcondentherm and the maxcondenbar do not coincide with the critical point gives rise to retrograde condensation or vaporization in the vicinity of the critical point; these are discussed in detail in Ref.4.

4. Heat transfer studies in the critical region

The uncertainties in the thermodynamic and transport properties, and the strong dependence of these properties on temperature in the critical region pose tremendous difficulties in the correlation, prediction and evaluation of heat transfer performance in this region. There is also some doubt regarding the validity of the normal constitutive relations in this region.

Experiments have indicated unusual and peculiar behavior of the heat transfer coefficients between solid surfaces and fluids in the critical region. Shitsman (13) has reported an impairment of forced convection heat transfer to water at super-critical pressures while heating, and conversely an improvement in the heat transfer while cooling. Hall et al (14) and Wood and Smith (15) have reported a sharp increase in forced convection heat transfer coefficients for turbulent flow of CO_2 in the critical region.

It should, however, be noted that the concept of a heat transfer coefficient ceases to have the significance attached to it under normal conditions, since it is no longer even approximately independent of the heat flux. The unusual behavior of the heat transfer coefficient in the critical region has been explained in terms of its strong dependence on the heat flux (10).

Several investigators, however, have found that the usual free and forced convection correlations can be used even near the critical state, if the thermophysical properties are suitably evaluated.

The nature of the fluid investigated appears to have a pronounced effect on the heat transfer performance in the critical region. In forced convection, O2, H2, N2 and N2-O4 have been found to produce minimum heat transfer coefficients, and CO2 and Freon-12, maximum heat transfer coefficients. With water, some have reported an impairment and some an improvement in the heat transfer. Bramall and Daniels (16) caution that it is dangerous to assume that all fluids behave in the same manner in the critical region.

In summary, it appears that heat transfer processes in the critical region are not fully understood. The occurrence of the super-critical heating phenomenon (meaning the attainment of the critical point by the fluid during the heat transfer process) has not been verified experimentally. It is not clear how exactly the singularities in the thermophysical properties in the critical region affect the heat transfer processes. Substantial differences from sub-critical behavior have also not been discovered at super-critical pressures.

5. Studies on the vaporization and combustion of bipropellant drops at super-critical pressures

5.1 Theoretical studies

Spalding (2) has shown that the quasi-steady theory is not even a first-order approximation to the actual transient problem. He has developed a transient theory of super-critical combustion, assuming a priori that the burning droplet could attain the critical point, and replacing it by a point source of fuel vapor. As Brzustowski (17) has pointed out, Spalding's theory does not attempt to describe the details of the processes occurring in the immediate vicinity of the fuel drop, nor does it define the lower limit of pressure at which it is applicable.

Levine (18) seems to have been the first to question the validity of the a priori assumption that a burning drop could attain the critical point. He has speculated on the question of whether combustion of liquids at super-critical pressures would be different than at sub-critical pressures. He argued that since, in practice, the partial pressure at the drop surface would be less than the chamber pressure, no critical phenomena characteristic of a single component system could be expected to occur.

Wieber (19) has concluded on the basis of digital computations based on a low-pressure quasi-steady vaporization model that burning drops could attain the critical point at super-critical pressures.

Brzustowski (17) has attempted to predict the nature of the physical processes involved in super-critical droplet combustion. He has envisaged the change of phase at super-critical pressures to be accomplished by the passage of a transient temperature wave into the drop, all subsequent processes involving only gas-phase phenomena. The lower limit of pressure for this description of the combustion process to be applicable is that value at which the latent heat is zero. He has not, however, attempted to establish an explicit value of this pressure in terms of the critical pressure of the fuel. His predictions in regard to the sequence of events at super-critical pressures have not been verified experimentally.

Rosner (20) has developed a distributed-source model, in which the fuel was initially assumed to be uniformly distributed in a finite -size spherical region. He found substantial agreement between the results of his model with those of Spalding's point-source model (1). He has identified the quasi-steady envelope flame model and his model as asymptotic theories, useful under conditions far removed from the critical pressure of the fuel. He has also indicated that even at super-critical pressures, phase change might occur at sub-critical temperatures.

Manrique and Borman (21) have investigated the vaporization of liquid CO_2 drops in N_2 atmospheres at super-critical pressures and temperatures, employing a spherically symmetric quasi-steady droplet vaporization model. Their calculations included and showed the importance of the effects of thermodynamic non-idealities, solubility of gaseous N_2 in the CO_2 liquid, and the effect of total pressure on vapor pressure (Poynting effect). They found that the low pressure model was not properly posed for high pressure conditions in that it had more than one or no analytical solutions at high pressure levels, and predicted physically unrealistic steady state conditions under certain ambient conditions. Beyond stating that super-critical temperatures could only be reached by an unsteady process, they do not clarify whether there exists a possibility of attainment of the critical point (of the two-component system).

Lazar and Faeth (22) have also taken into consideration real gas effects, ambient gas solubility and variation of properties in the gas phase. They also found that the low pressure model was unsatisfactory. Their high pressure model predicted that super-critical burning would be achieved at a fuel reduced pressure of the order of 2 to 2.5. They clarified, however, that the drops would approach the critical mixing point due to ambient gas solubility, as opposed to the critical point for a pure liquid-vapor system. They claim that available experimental evidence showed that transport properties were not appreciably anomalous near a critical mixing point, unlike in the case of a pure liquid-vapor critical point. In any case, from the nature of the critical state, the latent heat, surface tension and refractive index difference should all vanish at the critical mixing point also. In agreement with Ref.21,Lazar and Faeth obtained a maximum limiting pressure where steady burning was possible; this was denoted as the "critical burning pressure", and was always greater than the critical pressure of the pure fuel. Their calculations showed that at pressures greater than the critical burning pressure, the drop temperature continued to rise until the critical temperature of the fuel was reached, when the drop was assumed to gasify.

In a subsequent investigation on similar lines by Canada and Faeth (23), critical burning was assumed to occur, with the low-pressure theory, when the liquid surface temperature was equal to the fuel critical temperature; and with the high-pressure theory, wher the liquid surface reached its critical mixing point for the conditions of the combustion process, as indicated by the equality of the liquid and gas phase composition at this state. It should, however, be pointed out that at the critical mixing point, not only the liquid and gas phase composition, but also all intensive properties of the two phases should become identical.

In summary, theoretical studies of droplet combustion have, as a rule, allowed for the possibility of attainment of the critical point by the burning drops before consumption. In any event, the low-pressure quasi-steady theories have been shown to be highly inadequate and recent studies have been directed toward the development of high pressure models and theories, taking into consideration effects of non-ideality, ambient gas solubility, multi-component critical phenomena etc. Transient gas-phase theories to describe the processes after the attainment of the critical point have also been developed.

5.2 Experimental studies

The difficulties encountered in experimental investigations of high pressure droplet combustion, which have severely limited the number of such investigations, have been summarized in Ref.24.

5.2.1 Suspended drop experiments

(a) Combustion studies

The high-speed motion pictures of Rabin et al (25), reproduced by Lambiris et al (26), obtained at a fuel reduced pressure of 1.36, showed nothing unusual, with the drops stilldiscernible and having discrete droplet-gas interfaces.

The interference effects of the supporting filament, prevalence of pyrolysis, and excessive soot formation of the aromatic fuel, aniline, in the high pressure experiments of Brzustowski and Natarajan (27) did not permit any definitive conclusions regarding super-critical combustion.

Faeth et al (28) have considered explicitly the problem of combustion of fuel drops at pressures " sufficiently high for the drops to reach their critical temperature during combustion" . Most of their conclusions regarding super-critical combustion were based on temperature records provided by the thermocouple used for supporting the drops, and their drop photographs were not of sufficiently good quality to enable any definitive conclusions. They believe that super-critical combustion was observed in their experiments, although " no unusual changes in the combustion zone were detected as the droplet passed through its critical temperature".

(b) Vaporization studies

Savery and Borman (29) have considered the attainment of the critical state in the gas phase at the drop interface to be indicated by the initiating condition for a completely unsteady temperature history, as measured by the supporting thermocouple. Their photographs of vaporizing drops, however, offered no definitive information regarding super-critical vaporization.

5.2.2 Porous sphere experiments

Canada and Faeth (23) claim that their porous sphere experiments at super-critical pressures, under natural convection conditions, indicated the approach of critical burning conditions at values of pressures, well-predicted by their low and high pressure analyses. No experimental criterion for the attainment of the critical burning condition has been indicated by them, however.

At pressures exceeding the critical combustion condition, they envisage the process to be similar to the porous sphere combustion of fuel gas. In this regime, no liquid interface would be observed, and a range of fuel flow rates, subject to blow-off and quenching limits, could be accommodated by the sphere at a given pressure, in contrast to the single fuel flow rate possible for liquid fuel combustion at pressures below the critical burning condition.

5.2.3 Falling droplet experiments

The only experimental investigation to date at high ambient pressures, using this technique, which simulates most closely the actual conditions in a combustion chamber, has been that of Natarajan and Brzustowski (30). Their high-quality photographs showed that even at the highest pressure investigated (corresponding to a fuel reduced pressure of 1.26), liquid drops were clearly discernible through the flame zones, with drops smaller than about 400 microns in diameter appearing very nearly spherical, indicating that the droplet temperature was considerably less than the critical temperature (at which the surface tension tends to zero). A reliable indication of the attainment of critical point by the burning drops is expected to be the vanishing of surface tension at the critical point, resulting in break-up and shattering of the drops.

5.2.4 Vaporization of liquid sprays at high pressures

Newman and Brzustowski (31) found, in their photographic investigation of the vaporization of liquid CO2 sprays in gaseous CO2-N2 atmospheres, that under some operating conditions, at both sub-critical and super-critical pressures, the liquid phase disappeared completely within the field of view. This, they rightly pointed out, meant that the liquid-gas interface, which from the photographic point of view was a source of reflected light, was no longer present. They, however, characterized this process as "vaporization" at sub-critical pressures, and "gasification" at super-critical pressures. Thus, even this excellent study of high pressure spray evaporation over a wide and important range of parameters relevant to super-critical vaporization, has not produced an unequivocal experimental criterion for this phenomenon.

5.2.5 Vaporization of drops on a hot surface at high pressures

While Adadevoh et al (32) assumed that the liquid drop would pass through its critical point, Temple-Pediani (33) found that even super-critical conditions of ambient pressure and surface temperture, the phase boundary between liquid and gas was always distinct during the lifetime, suggesting that all the liquid evaporated before it attained Tcr.

In summary, no satisfactory experimentally verifiable criterion for the attainment of the critical point by an evaporating or burning droplet has been put forward, so that this question still remains unsolved.

6. Effect of mass transfer on heat transfer in droplet combustion

A number of investigations on droplet combustion have considered the effect of mass transfer in reducing the heat transfer to evaporating and burning droplets, and have been reviewed in Ref.34. This has been referred to as " mass transfer shielding" or " blocking effect" or "blowing effect", and is considered to be a function only of the transfer number, B, which represents the ratio of the sensible heat requirements to the convective heat load in the boundary layer surrounding the burning drop. Hoffman and Ross (35) have theoretically verified the validity of the above concept. In droplet combustion, the transfer number is difficult to define unequivocally, in view of the intrinsically non-steady nature of the processes (36), In the present discussion it represents the steady-state value, in the absence of radiative transfer.

In the following discussion, 4 shielding functions are examined:
1. $f(B) = 1/(1+B)$. This function has been derived from laminar boundary layer theory by Eisenklam et al (37), and has been used fairly successfully in Refs. 30, 37 and 38 to correlate the experimental burning rates. The same function is shown in Ref.34 to result when superheating of the evolved vapor is considered.
2. $f(B) = 2/(2+B)$. This function was derived in Ref.37 from slow viscous flow theory.
3. $f(B) = \ln(1+B)/B$. This function has been derived in Ref.37 from stagnant film theory and is in concurrence with Marxman (39).
4. $f(B) = B/(e^B-1)$. This function has been derived in Ref.37 from an energy balance around a vaporizing drop and has been used, for example, in Refs.19 and 29. (It can be shown that their Z factor is in fact equal to B here).

Figure 1 shows a comparison of the four shielding functions. All of them tend to 1 as $B \to 0$ and tend to 0 as $B \to \infty$. $B/(e^B-1)$ appears to be the strongest shielding function, and $\ln(1+B)/B$ the weakest.

Denoting the Nusselt number in the absence of both convection and mass transfer by Nu_o^o, the Nusselt number in the absence of convection by Nu^o, the Nusselt number in the absence of mass transfer by Nu_o, and the Nusselt number in the presence of both convection and mass transfer by Nu, it can be seen that in the case of spheres, $Nu_o^o=2$, representing the case of a solid sphere in stagnant surroundings. With the above notation, the mass transfer shielding functions represent the ratios Nu/Nu_o; and Nu^o/Nu_o^o in the limit of zero convection. The relation between Nu_o^o and Nu_o is usually represented in the form

$$Nu_o = Nu_o^o + A\, Re^{\frac{1}{2}}\, Pr^{1/3} \qquad (1)$$

In order to investigate the effect of mass transfer on heat transfer, let us consider non-convective conditions and examine the shielding functions, Nu^o/Nu_o^o.

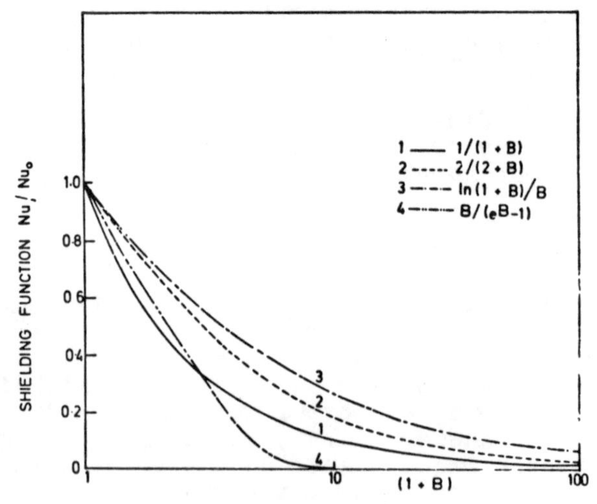

FIG.1 COMPARISON OF MASS TRANSFER SHIELDING FUNCTIONS

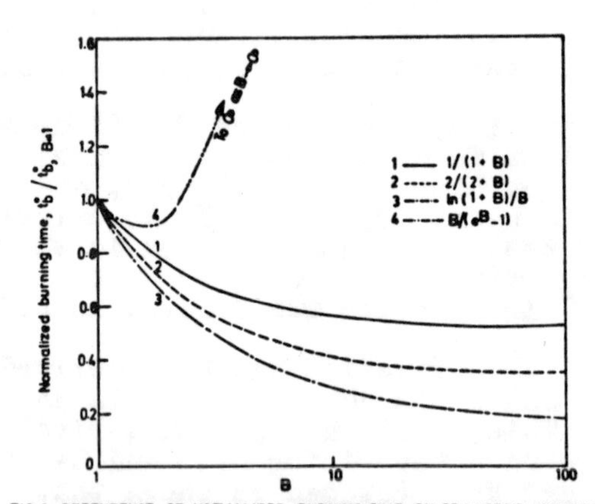

FIG.2 DEPENDENCE OF NORMALIZED BURNING TIME ON TRANSFER NUMBER (AND ON PRESSURE) IN THE ABSENCE OF CONVECTION, BUT INCLUDING MASS TRANSFER SHIELDING

In droplet combustion, the evaporation constant, λ is related to Nu for heat transfer as follows: $Nu = \ell_1 C_p \lambda / 4kB$ (2)

With $t_b = d^2/\lambda$ and $Nu_o^o = 2$, $t_b^o = \ell_1 C_p d^2 / 8kBf(B)$ (3)

Figure 2 shows the normalized burning time,

$$t_b^* = t_b^o / t_{b,B=1}^o = f(1)/Bf(B) \quad (4)$$

plotted against B for the four shielding functions. While this also represents the ratio of burning times under identical convection conditions, in order to examine the influence of mass transfer on heat transfer, convection-free conditions have been considered.

Table I summarizes a comparison of the normalized burning times, for each of the four shielding functions, in the limit as $B \to 0$, and as $B \to \infty$. All the functions predict infinite burning time as $B \to 0$, meaning that it corresponds to the case of no mass transfer. In the case of the exponential function, the normalized burning time decreases initially, attains a minimum of about 0.90 at $B \approx 1.45$, and then increases to ∞ as $B \to \infty$.

TABLE - I

COMPARISON OF NORMALIZED BURNING TIMES, $t^* = t_b^o / t_{b,B=1}^o$

Shielding Function	Limiting Value of t^* as $B \to 0$	Limiting value of t^* as $B \to \infty$	Remarks
1. $1/(1+B)$	∞	0.5	Does not predict zero burning time
2. $2/(2+B)$	∞	0.33	"
3. $\ln(1+B)/B$	∞	0	Weakest shielding: predicts zero burning time as $B \to \infty$
4. $B/(e^B - 1)$	∞	∞	Strongest shielding: predicts infinite burning time as $B \to \infty$

Thus the use of three of the four shielding functions above results in non-zero burning times at $B \to \infty$. Explicit consideration of the net heat transfer available to increase the liquid temperature of a burning drop (34) shows that as a result of the reduction in the convective contribution to negligible proportions, and because the liquid specific heat increases very rapidly to very high values near the critical point, the attainment of the critical point by a burning drop before consumption would seem to be precluded.

7. Concluding remarks

Neither theoretical nor experimental investigations, to date, have been able unequivocally to establish the possibility or verify the existence of super-critical bipropellant droplet combustion. The critical state is such a difficult state to attain (and maintain) that it seems very probable that it is not one of the operating conditions in droplet combustion, especially because of the very unfavorable conditions of heat transfer in the presence of mass transfer shielding in the critical region. In this connection, it must be pointed out that unless critical pressure and critical temperature are attained simultaneously, the critical state would not be encountered; attainment of the critical pressure at super-critical temperatures would result in superheated vapor states and attainment of the critical temperature at super-critical pressures would result in compressed liquid states.

Figure 3 represents the (non-equilibrium) vaporization processes that are generally agreed to result on the injection of a cold (liquid temperature less than the corresponding boiling temperature) liquid drop into a hot atmosphere, at a pressure $P_1 < Pcr$. The initial state of the liquid at injection is represented by C. The liquid is generally agreed to attain a steady-state wet-bulb temperature, T_D, at which the latent heat is supplied; $T_D < T_B$. E represents the state of vapor at the wet-bulb temperature.

Figure 3 also represents the vaporization processes that are expected to result on the injection of a cold (liquid temperature less than Tcr) liquid drop into a hot atmosphere, at a pressure $P_2 >$ Pcr. The initial state of liquid at injection is represented by C'. Brzustowski (17) postulated that the phase change should occur

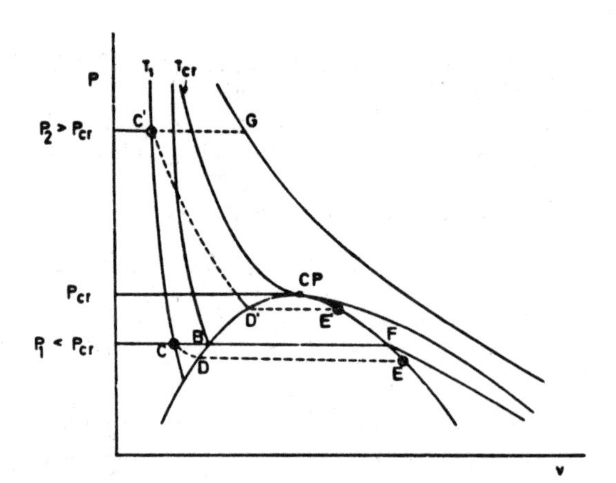

FIG. 3 DROPLET VAPORIZATION PROCESSES AT SUB-CRITICAL AND SUPER-CRITICAL PRESSURES

along C'G, with a transient temperature wave moving into the droplet. It can be seen that the critical point is not encountered in this process. His model, although unrealistic, was self-consistent at super-critical pressures. At and above the critical pressure, $(P_{f,A}/P)$ was equal to 1, and the liquid could be considered to be injected into its own gas. The drop would then be expected to pass through the critical point at the critical pressure. However, at sub-critical pressures, $(P_{f,A}/P)$ would be < 1, and it is no longer a single-component system. Noting that at the sub-critical pressure P_1, phase change is along CDE and not along CBF, it is postulated that with liquid injected at state C', it would undergo the process C'D'E' during vaporization, and that heat and mass transfer considerations impose a maximum limit on the liquid temperature, which would be less than Tcr.

High pressure droplet combustion might, then, be still describable in terms of vaporizing liquid droplets, but the quasi-steady theory would have to be replaced by a transient theory taking into consideration effects of thermodynamic non-idealities, ambient gas solubility, multi-component fluid mixtures, retrograde phenomena, liquid and gas phase pyrolysis of the fuel, production and combustion of soot and transport phenomena in the wakes of droplets, and other high pressure effects which are being recognized as important in high pressure droplet combustion.

REFERENCES

[1] Rosner, D.E., AIAA Journal, 5, 163 (1967)
[2] Spalding, D.B., ARS Journal, 29, 828 (1959)
[3] Lewis, J.D., J.Royal Aero.Soc., 68, 743 (1964)
[4] Sage, B.H. and Lacey, W.N., Volumetric and Phase Behavior of Hydrocarbons, Gulf Pub.Co., Houston (1949)
[5] Giaque, W.F., Brodale, G.E., Fisher, R.A. and Hornung, E.W., J.Chem.Phys., 42, 1 (1965)
[6] Rice, O.K., Thermodynamics and Physics of Matter, Princeton Univ.Press, 419 (1955)
[7] Sengers, J.V., Critical Phenomena, N.B.S.Misc.Pub.273, 165 (1965)
[8] Levelt Sengers, J.M.H., Ind.Eng.Chem.Fund., 9, 470 (1970)

[9] Rowlinson,J.S.,Symp.on Heat Transfer and Fluid Dynamics of Near Critical Fluids, Bristol (1968)
[10] Shiralkar,B.S.and Griffith,P.,ASME Paper 68-HT-39 (1968)
[11] Lorentzen,H.L.and Hansen,B.B.,Critical Phenomena,N.B.S. Misc. Pub.273, 213 (1965)
[12] Sage,B.H.,Thermodynamics of Multi-component Systems,Reinhold, New York (1965)
[13] Shitsman,M.E.,High Temperature, 1, 237 (1963)
[14] Hall,W.B., Jackson,J.D. and Khan,S.A.,3rd Annual Conference on Heat Transfer,I Mech.E. Section,257 (1966)
[15] Wood,R.D.and Smith,J.M.,A.I.Ch.E.J., 10, 180 (1964)
[16] Bramall,J.W.and Daniels,T.C.,Symp.on Heat Transfer and Fluid Dynamics of Near Critical Fluids,Bristol (1968)
[17] Brzustowski,T.A.,Can.J.Chem.Eng.,43, 30 (1965)
[18] Levine,R.S.,Combustion and Propulsion, 5th AGARD Colloquium, Pergamon Press, 170 (1963)
[19] Wieber,P.R.,AIAA J,1, 2764 (1964)
[20] Rosner,D.E.,Aero Chem Res.Labs.Inc.,TP-128 (1966)
[21] Manrique,J.A. and Borman,G.L.,Int.J.Heat Mass Transfer, 12, 1081, (1969)
[22] Lazar,R.S.and Faeth,G.M.,13th Symp.(Int) on Comb.,801 (1971)
[23] Canada,G.S. and Faeth,G.M.,14th Symp.(Int) on Combustion, Pennsylvania,Aug. (1972)
[24] Natarajan,R.,J.Phys.E:Sci.Inst., 5, 1189 (1972)
[25] Rabin,E., Schallenmuller,A.R. and Lawhead,R.B., AFOSR TR 60-75 (1960)
[26] Lambiris,S.,Combs,L.P. and Levine,R.S.,Combustion and Propulsion, 5th AGARD Colloquium, Pergamon Press, 569 (1963)
[27] Brzustowski,T.A., Natarajan,R., Can.J.Chem.Eng., 44, 194 (1966)
[28] Faeth,G.M.,Dominicis,D.P., Tulpinsky,J.F. and Olson, D.R., 12th Symp.(Int) on Combustion, 9 (1969)
[29] Savery,C.W. and Borman,G.L.,AIAA Paper No. 70-6 (1970)
[30] Natarajan,R., Brzustowski,TA,Comb.Sci.Tech., 2, 259 (1970)
[31] Newman,J.A.and Brzustowski,T.A.,AIAA Paper No.70-8 (1970)
[32] Adadevoh,JK,Uyehara,OA,and Myers,PS, SAE Paper 701 B (1963)
[33] Temple-Pediani,R.W., Proc.Inst.Mech.Engrs.,184, 677 (1969)
[34] Natarajan,R.,1st Nat. Heat Mass Transfer Conf,India (1971)
[35] Hoffman,TW,Ross,LL,Int.J.Heat Mass Transfer,15, 599 (1972)
[36] Natarajan,R.,Comb.and Flame (in Press)
[37] Eisenklam,P., Arunachalam,S.A. and Weston,J.A., 11th Symp.(Int) on Combustion, 715 (1967)
[38] Nuruzzaman,A.S.M., Hedley,A.B. and Beer,J.M., 13th Symp.(Int) on Combustion, 787 (1971)
[39] Marxman,G.A., 10th Symp.(Int) on Combustion, 1337, (1965).

Chapter 33

SOOT OXIDATION IN LAMINAR HYDROCARBON FLAMES

A. Feugier

Institut Francais du Pétrole, Rueil-Maison, France

Abstract

Soot particles are oxidized as a result of the supply of oxygen by diffusion along the streamlines of fuel-rich combustion gases. By means of an optical method, variations in soot particle concentration could be followed in time, and this enabled a kinetic expression to be deduced between 2000 and 2300 °K for the oxidation of these particles.

1. Introduction

In recent years the problem of soot in flames has been the subject of numerous research projects which have undeniably contributed to a better understanding of this phenomenon. A great deal of information, in particular concerning the formation of carbon, can be found in recent articles by Gaydon and Wolfhard (1) , Palmer and Cullis (2), Homann (3) , and Homann and Wagner (4). Conversely, only a few authors have undertaken measurements of soot oxidation rates, e.g., Lee, Thring and Beer (5) On the basis of their findings concerning laminar diffusion flames, Lee and al. have proposed a semi-empirical rate equation for the combustion of soot particles, with an activation energy of 39,300 cal/mole. The authors themselves, however, conclude that the validity of this equation still has to be checked by enlarging the range of the experimental variables.

The object of this work is to determine the combustion rate of soot particles at temperatures greater than 2000 °K. For this purpose we selected an experimental device identical to the one used by Fenimore and Jones (6), in which soot particles are made to oxidize by subjecting fuel-rich combustion gases to an oxidizing flow.

2. Experimental

The burner, shown in a schematic drawing in Fig. 1, is similar to the one described by Fenimore and Jones (6). Briefly, it consists of a 24 mm diameter porous bronze plate sintered to a stainless water-cooled coil. Surrounding this burner is a 27.5 cm^2 annulus also made of a cooled porous bronze plate. The central part is fed with an ethane-oxygen mixture burning with an equivalence ratio φ of 2.70 or 2.90 and at an unburnt-gas flow velocity of 10 cm sec^{-1} (related to room temperature). φ is the fuel/oxygen ratio in mixture to the fuel/oxygen ratio in stoechiometric mixture. The annular part is fed with a constant flow of an oxygen-nitrogen mixture whose ratio may vary between 0.4 and infinity. The burner can be moved both horizontally and vertically. It is enclosed in a vessel fitted with quartz windows.

When the annular mixture diffuses into the central part a secondary diffusion flame appears (Fig. 1). Only inside the cone are there still any carbon particles, thus making it luminous. The cone height h_o obviously depends on the composition of the annular mixture. For a given mixture it increases proportionately to an increase in the nitrogen-oxygen ratio.

A Monopesk-600 Hilger-Engis monochromator was used for measuring the emissions of light. This instrument is equipped with a 1200 lines/mm grating. The horizontal entrance and exit slits of the monochromator were respectively set at 200 μ (equivalence ratio of 2.70) and 25 μ (equivalence ratio of 2.90). A vertical entrance slit of 0.5 mm was used. In this way the height of the area under observation

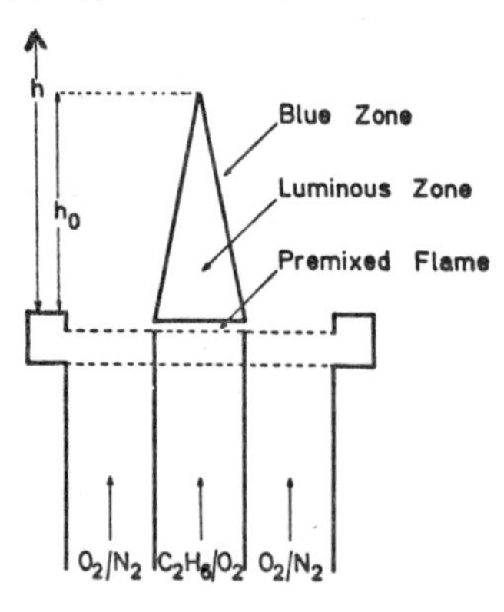

Fig. 1 - Schematic sketch of burner

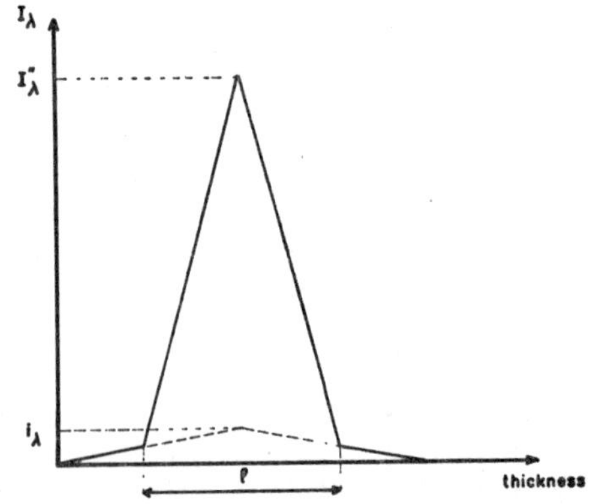

Fig.2 - Schematic intensity profile through a cross section of the flame

was 2 mm after focusing the image of the flame on the monochromator entrance slit. An RCA 1P28 photomultiplier detector was used. The signal thus produced was sent to an electronic integrator. In this way a mean luminous intensity value could be measured when done over a sufficient time period.

3. Carbon particle temperature and concentration measurements

The interaction of an incident radiation with a particle is a function of the following parameters : (1) the complex refractive index defined in terms of real refractive index n_1 and absorption index n_2 : $m = n_1 (1 - in_2)$, (2) the ratio of the characteristic particle size to the incident radiation wavelength, and (3) particle shape. For spherical particles, the general solution is due to Mie (7). When $\pi d/\lambda < 0.6/m$, in which d is particle diameter, the equations are simplified. Under these conditions K_d (scattering coefficient) and K_a (absorption coefficient) can be expressed as follows :

$$K_d = N \frac{\pi d^2}{4} \frac{8}{3} \left(\frac{\pi d}{\lambda}\right)^4 \left| \frac{m^2 - 1}{m^2 + 2} \right| \tag{1}$$

$$K_a = -N \frac{\pi d^2}{4} \, 4 \, \frac{\pi d}{\lambda} I_m \left(\frac{m^2 - 1}{m^2 + 2} \right) \tag{2}$$

where N = the number of particles per unit volume, and I_m = the imaginary part of the term in parentheses.

From the results obtained by Dalzell and Sarofim (8) with propane and acetylene flame soots, it may be assumed that the refractive index value of ethane flame soot is constant in the visible range and equal to $m = 1.56 - 0.48\,i$, while also assuming this value to be independent of temperature. Under the conditions described here and when the wavelength is varied between 4500 and 6100 Å, it may be seen that the use of the preceding relations implies that the diameter of the particles is less than approximately 500 Å and that, in addition, the influence of scattering is relatively slight so that it can be ignored.

With spherical particles having a volume fraction of $X_p = N\pi d^3/6$, the absorption coefficient can be given as : $K_a = 6\pi AX_p/\lambda$, with : (3)

$$A = \frac{6n_1^2 n_2}{\left[n_1^2(1 - n_2^2) + 2\right]^2 + 4 n_1^4 n_2^2}$$

After having experimentally verified that, under actual conditions, the absorption of a beam of monochromatic radiation through a flame with a thickness of 1 is relatively very low, the intensity I_λ received by the monochromator per solid-angle unit normal to the surface of area S is :

$$I_\lambda = 2\alpha Sc_1 \lambda^{-5} K_a \, 1 \exp\left(-\frac{c_2}{T_p\lambda}\right) \quad (4)$$

with α = constant which, for a given wavelength, depends only on the apparatus.
c_1 = 0.588 x 10^{-5} erg x cm^2 x sec^{-1}
c_2 = 1.438 cm x degree
T_p = particle temperature.

By replacing the burner with a calibrated tungsten ribbon lamp, the intensity I'_λ received by the monochromator through the same optical system is :

$$I'_\lambda = 2\alpha S' c_1 \lambda^{-5} e' \exp\left(-\frac{c_2}{\lambda T'}\right) \quad (5)$$

with S = S' if the image of the filament uniformly fills the monochromator entrance slit, and where e' is the emissivity of tungsten at the true temperature of filament T' and at wavelength λ (the variation of e' with wavelength was taken from data in De Vos (9)). The true temperature is determined from a calibration against a black body using red light.

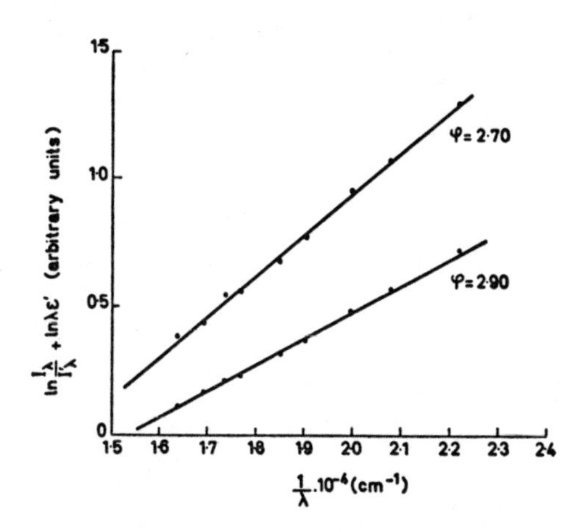

Fig. 3 - Linear relationship between $\ln(I_\lambda/I'_\lambda) + \ln \lambda e'$ and $1/\lambda$ for two equivalence ratios : φ = 2.70 and φ = 2.90. Slope = c_2 $(1/T' - 1/T_p)$ with c_2 = 1.438 cm x degree, T' = 1500°K and T_p : particle temperature.

The ratio of the emission I_λ of the soot cloud to the emission I'_λ from the tungsten lamp is :

$$\frac{I_\lambda}{I'_\lambda} = \frac{K_a 1}{e'} \exp\left[-\frac{c_2}{\lambda}\left(\frac{1}{T_p} - \frac{1}{T'}\right)\right] \quad (6)$$

By plotting In (I_λ / I'_λ) + In $\lambda e'$ against $1/\lambda$, as in Fig. 3, an excellent straight line is obtained. The slope of this line can be used to determine soot particle temperature. The intensity measurements were made on the center-line of flames burning without annular fluid, at wavelengths varying between 4500 and 6100 Å. The mean carbon particle temperatures obtained were 1790°K and 1680 °K for flames with equivalence ratios of 2.70 and 2.90 respectively. These temperatures were appreciably constant for several centimeters beyond the reaction zone.

Relation (6) also makes it possible to determine the mean soot volume fractions, X_p. These values were 0.85 x 10^{-9} and 5.30 x 10^{-9} respectively for flames with an equivalence ratio of 2.70 and 2.90.

4. Results and discussion

When the annular fluid contains oxygen, the latter, after diffusion in the central part, oxidizes the reducer species, i.e., mainly soot particles, carbon monoxide and hydrogen. At the same time, the latter two species also diffuse in the annular part. The particles are, however, assumed not to diffuse because of their relatively large size.The continuous spectrum emitted by the luminous gases is due essentially to radiation from carbon particles. We should not, however, entirely ignore the contribution of the continuous emission due to the process of association, i.e. $CO + O \rightarrow CO_2 + h\nu$, whose maximum intensity is 3500 - 4500 Å (10). The latter emission, which gives a bluish color to CO/O_2 flames, is probably the cause of the blue coloring observed in the diffusion flame surrounding the luminous cone (Fig. 1).

For a given analysis height if the burner is moved horizontally in relation to the monochromator entrance slit, an intensity profile like the one given schematically in Fig. 2 can be plotted. In this way it is possible to determine, with quite good accuracy, the thickness of the blue zone as well as that of the luminous zone 1. In order to determine the intensity on the center line emitted by the only luminous zone and by only carbon-particle radiation. value I_λ is subtracted from value i_λ. The latter value represents the intensity due to the CO+O continuum which can be found throughout the entire thickness of the flame. The only measurements taken into consideration are those in which the i_λ value is no greater than 10 % of the total intensity.

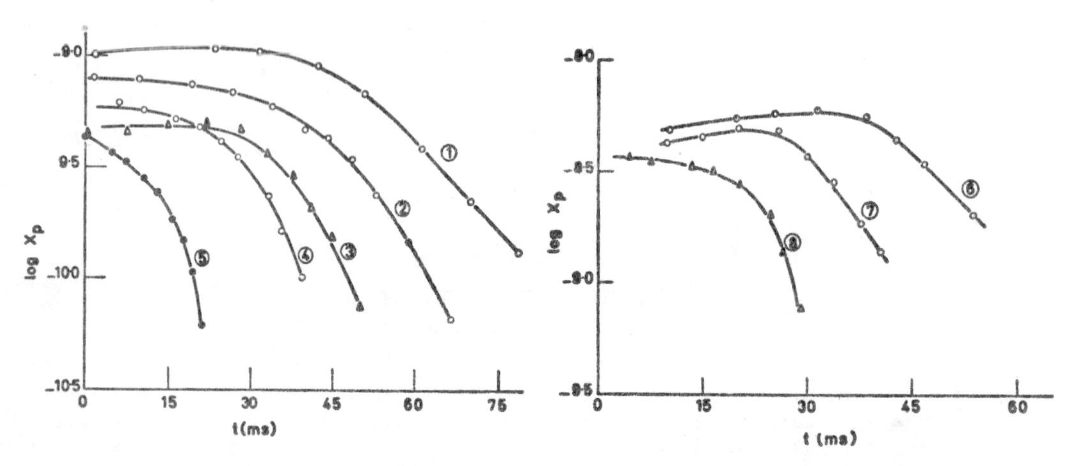

Fig. 4 and 5 - Log (soot volume fraction) versus time, for different values of N_2/O_2 ratio of the annular fluid.

Equivalence ratios of central mixture = 2.70 (fig. 4) and 2.90 (fig. 5).

For a given equivalence ratio for the central mixture and a given composition of the annular fluid, we determined the mean temperatures T_p and the mean volume fractions X_p of the particles over a thickness l in the only luminous area. The results given in Figs. 4 and 5 correspond to equivalence ratios of 2.70 and 2.90 respectively for the central mixture. Variations in log X_p were plotted versus time t. Each curve corresponds to a given composition of the annular fluid. The input conditions are listed in Table 1.

Table 1

Run No	N_2/O_2 ratio of the annular fluid rate (flow : 500 l/h)	Equivalence ratio of the central mixture
1	61/39	2.70 { C_2H_6 flow:71 l/h; O_2 flow:92 l/h
2	56/44	
3	47/53	
4	36/64	
5	0/100	
6	36/64	2.90 { C_2H_6 flow:74 l/h; O_2 flow:89,2 l/h
7	24.5/75.5	
8	10/90	

While the soot concentration varies only slightly during the first milliseconds of analysis (a time period which, for a given mixture strength, varies in inverse proportion to the oxygen concentration in the annular fluid), the temperature gradient is, on the contrary, relatively greater (Figs. 6 and 7)

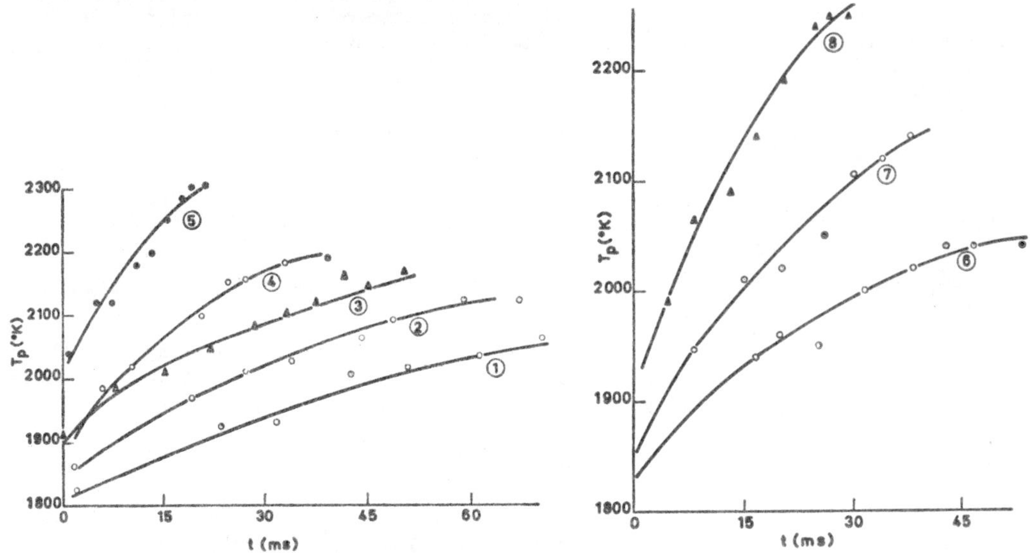

Figs.6 and 7 - Particle temperature versus time, for different values of N_2/O_2 ratio of the annular fluid. Equivalence ratios of the central mixture = 2.70 (Fig. 6) and 2.90 (Fig. 7).

This shows that the oxygen diffuses into the central part very early. There may be two simultaneous reasons for this : (1) pyrolysis of the hydrocarbons present in the combustion gases of the central part, accompanied by soot formation, (2) a very low soot oxidation rate due to the very low concentration of oxygen available for this particular reaction.

4.1. The steady-state condition for any species X_i requires that its concentration at any position in the flame be described by[1]:

$$D_i \nabla^2 X_i - \frac{d}{dh}(VX_i) + V_f - V_d = 0 \qquad (7)$$

where D_i = diffusion coefficient of species i present with a volume fraction X_i,

∇^2 = Laplacian operator,

V = linear velocity in the h direction (assuming a zero radial velocity component),

V_f and V_d = formation and disappearance rate of species i respectively.

If this equation is applied to soot particles, it is observed that the soot concentration decrease obeys this law only during the final milliseconds of the analysis. This implies that in this zone $V_f = 0$ and that oxygen diffusion no longer limits the different phenomena. This law may be written :

$$- K \frac{dT}{dh} - \frac{d \ln X_p}{dt} = k_p x_{O_2} \exp\left(- \frac{E}{RT_p}\right) \qquad (8)$$

in which $K = V_o/T_o$ with V_o being the ratio of total central-mixture flow to the central burner area, and T_o being room temperature. E represents the activation energy. Actually, the first term in Eq. (8) can be ignored. x_{O_2} is the amount of oxygen required per mole of initial fuel for the temperature profile T of the combustion gases, initially at unshielded flame temperature, to follow the experimental profile. This parameter is calculated by assuming the thermodynamic equilibrium established at all times as well as particle velocity and temperature to be equal to gas velocity and temperature (5, 11, 12). For example, let us consider the central mixture with an equivalence ratio of 2.70. In order to heat gases initially at a temperature of 1680 °K to a temperature equal to 2185 °K, 0.190 mole of oxygen per mole of ethane is required with an annular fluid made up of 70 % O_2 and 30 % N_2.

Figure 8 shows the good linear relation obtained by plotting the variations of $\ln(-\Delta \ln X_p/\Delta t) - \ln x_{O_2}$ as a function of $1/T = 1/T_p$.

Each point is numbered so that its origin can be identified in Table 1 and on the corresponding curves in Figs. 4 - 7. An average is given for several values taken in the end portions of these curves.

An apparent activation energy E is thus determined equal to 33 kcal/mole and a k_p value equal to 1.3×10^6 sec^{-1}.

4.2 - It is possible to interpret the results otherwise, by assuming that the reaction takes place at the external surface of particles. In this way, the specific rate of combustion of particles is as follows :

$$\frac{1}{N \pi d^2} \frac{dm}{dt} = \frac{\rho \, do}{6 X_{p_o}^{1/3} X_p^{1/3}} \frac{d X_p}{dt} = k'_p x_{O_2} \exp(-E'/RT)$$

where ρ = the particle density and m the mass. The subscript zero corresponds to the values without annular fluid.

After integration, and under a logarithmic form, the following expression can be obtained :

$$\ln\left[- \frac{\Delta X_p^{1/3}}{\Delta t}\right] - \ln x_{O_2} = \ln k''_p - \frac{E'}{RT}$$

with : $k''_p = 2 k'_p X_{p_o}^{1/3} / d_o \rho$.

Fig. 9 shows the good linear relation obtained by plotting the left hand side versus 1/T. It can be seen that, for each equivalence ratio, the slope is identical.

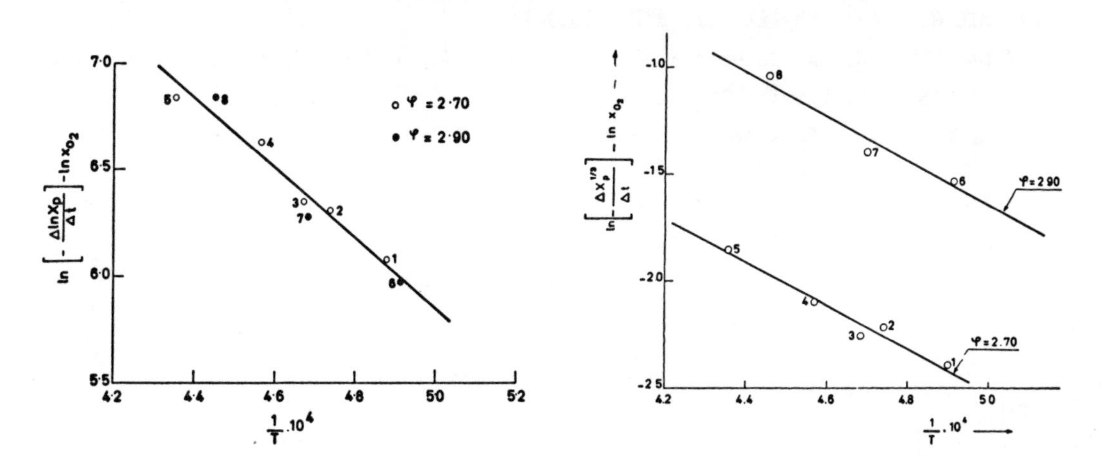

Figs. 8 and 9 - Slope E/R = 16,500 cal (fig. 8) and 10,000 cal (fig. 9). Each experimental point is numbered and thus can be identified in Table 1 and the corresponding curves in figs. 4 to 7.

The ordinates are different because the initial values of soot concentrations and particle diameters are different for each equivalence ratio. An activation energy, equal to about 20 kcal/mole, can be derived, and a k' value equal to 3.8 x 10^{-2} g.cm^{-2}. sec^{-1}, by assuming an initial particle diameter equal to 300 Å.

5. Conclusions

In this work, combustion gases from fuel-rich ethane-oxygen flames were subjected to an oxidizing flow made up of a nitrogen-oxygen mixture in varying proportions. By means of an optical method, variations in soot particle concentration could be followed in time. It was observed that the disapperance rate of particles obeys the following law : $V_d = 1.3 \times 10^6 \; x_{O_2} \; X_p \; \exp(-E/RT)$ sec^{-1}, with an apparent activation energy E equal to about 33,000 cal/mole.
If now it is assumed that the reaction takes place at the external surface of particles, an activation energy of about 20 kcal/mole can be derived for the specific rate of combustion, and compared with the value of 39,300 cal/mole obtained by Lee, Thring and Beer some years ago (5); but this latter value was deduced within a temperature range lower than the range used in this work. Our results are in better agreement with those of Park and Appleton (13), which indicate that the soot oxidation rate exhibits a dependence on the temperature which is not constant over the whole of the temperature range, between 1700 and 4000 °K. Particularly, over the temperature range of interest here ,a temperature coefficient near the value which is proposed in this work, can be estimated.

References

/1/ GAYDON A.G., and WOLFHARD H.G. : "Flames, their structure, radiation and temperature", 3 rd. edition, Chapman and Hall, London (1970), ch. VIII.

/2/ PALMER H.B. and CULLIS C.F. : "The formation of carbon from gases in chemistry and physics of carbon" (P.L. Walcker, Ed.) Vol. 1, Marcel Dekker, New-York, Arnold, London (1965), p. 261.

/3/ HOMANN, K.H. : Combustion and flame 11, 265 (1967).

/4/ HOMANN K.H. and WAGNER H.G. : Proc. Roy. Soc. (London) A 307, 141 (1968).

/5/ LEE K.B., THRING, M.W. and BEER J.M. : Combustion and flame 6, 137 (1962).

/6/ FENIMORE C.P. and JONES G.W. : Combustion and flame 13, 303 (1969).

/7/ MIE G. Ann. Physik, 25, 377 (1908).

/8/ DALZELL W.H. and SAROFIM A.F. : Trans. ASME, J. Heat Transfer 91, 100 (1969).

/9/ DE VOS J. ;. Physica 20, 690 (1954).

/10/ GAYDON A.G. : "The spectroscopy of flames " Chapman and Hall, London (1957) p. 91.

/11/ KUHN G. and TANKIN R.S. : J. Quant. Spectr. Rad. Transfer 8, 1281 (1968).

/12/MILIKAN R.C. : J. Opt. Soc. Amer., 51, 535 (1961).

/13/ PARK C. and APPLETON J.P. : Comb. and flame 20, 369, (1973).

Chapter 34

THE EXTINCTION OF SPHERICAL DIFFUSION FLAMES

G. I. Sivashinsky and C. Gutfinger

Faculty of Mechanical Engineering
Technion–Israel Institute of Technology, Haifa, Israel

Abstract

We examine the problem of existence limits of a diffusion flame in a stagnant gas stabilized at a diffusional fuel point source. This problem may serve as a quasi steady-state model for the description of combustion and extinction of fuel from evaporating drops.

The problem was solved under the assumption of a strong temperature dependence of the reaction rate. This condition enabled the treatment of the problem using an asymptotic approach with the dimensionless activation energy N as a large parameter.

As a result of this investigation it was found that for a sufficiently small intensity of fuel source the steady-state combustion problem has no solution leading, in practice, to the extinction of the flame. Furthermore, it was found that incomplete fuel consumption in the course of the reaction occurs even under combustion conditions.

Nomenclature

A : intensity of the surface source concentration
C^* : concentration
C : dimensionless concentration (C^*/C_o^*)
c_p : heat capacity
E : activation energy
G : heat release
K : dimensionless heat release
L : Lewis number
ℓ_T : thermal thickness of the flame defined by Eq. (4)
M : chemical symbol for a substance
N : dimensionless activation energy
Q : intensity of the fuel source
q : dimensionless intensity of the fuel source
R : gas constant
r^* : radial coordinate
r : radial coordinate, dimensionless (r^*/ℓ_T)
S : defined by Eq. (22)
T^* : temperature
T : dimensionless temperature (T^*/T_f^*)
W : chemical reaction rate
X : inner problem coordinate
Z : pre-exponential factor

Greek Symbols:

ε : dimensionless surroundings temperature (T_o^*/T_f^*)
ρ : density

λ : thermal conductivity coefficient
κ : dimensionless thermal conductivity coefficient (λ/λ_f)
Φ : defined by Eq. (8)
ν : defined by Eq. (19)
θ : defined by Eq. (22)

Subscripts:

f : value of quantity at the flame front
o : value of quantity at the surroundings

Superscript:

* : dimensional quantity

1. Introduction

A diffusion flame front is a surface located inside a chemical reaction zone in which the reacting components come into contact and react as they diffuse from different sides towards this surface [1,2,3,4]. In the present work we examine a diffusion flame of a very simple chemical reaction of the type

$$M_1 + M_2 \rightarrow \text{combustion products} \quad (1)$$

Substance M_1 will be taken as the combustible component of the gas and substance M_2 will be taken as the oxidizing agent. Substance M_1 diffuses towards the reaction zone from a certain concentrated source. Such a source might be, for example, a sublimating drop of fuel with dimensions small compared to the diameter of the flame surrounding it.

At a point rather removed from the flame, the oxidant concentration is equal to C_o^* and the temperature is low enough such that the reaction practically does not proceed at all. If the flame is by some means ignited, then in the case that it is stabilized, the temperature and reactant concentration distributions take the form schematically depicted in Figure 1.

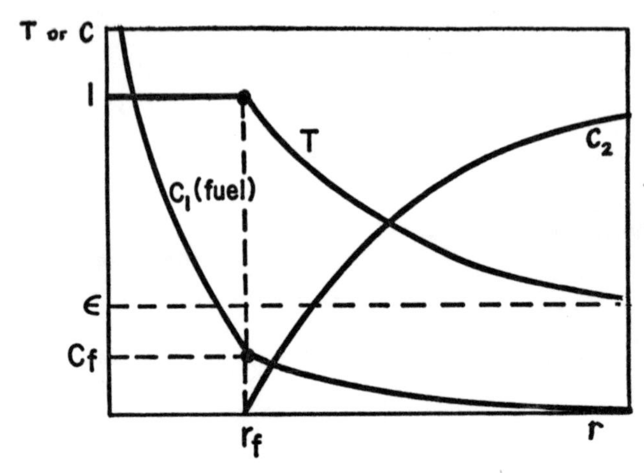

Figure 1 : Temperature and concentration profiles in a diffusion flame. (Outer expansion.)

As seen, the lines of C_1 and C_2 represent the respective dimensionless fuel and oxidant concentrations as functions of the dimensionless radial coordinate r. T is the dimensionless temperature in units of the flame front temperature T_f^*, while ε is the dimensionless temperature of the surrounding space.

Values for T_f^*, C_f, and r_f are initially unknown and must be determined in the course of a solution of the problem. We note that the combustible component M_1, generally speaking, does not vanish upon leaving the reaction zone but rather combusts only partially ($C_f \neq 0$). However, oxidant M_2 must fully combust in the zone. Otherwise, all the conditions exist for a continuation of the reaction at $r<r_f$, namely, high temperature and non-zero concentrations of both reacting components. However, at $r>r_f$, the existence of non-zero fuel and oxidant concentrations is not a sufficient condition for reaction due to a sharp drop in the temperature. Thus, the appearance of the combustible component of the gas past the reaction zone does not contradict equilibrium conditions.

Hence, as will be demonstrated below, incomplete combustion occurs not only due to extinction of the small fuel drops but also due to the 'slipping' of the combustible component through the reaction zone.

2. Fundamental Assumptions

We shall consider that the chemical reaction rate W is related to the temperature and reactant concentrations by the Arrhenius equation. For a reaction of Eq. (1), we have:

$$W = Z\rho^2 C_o^{*2} C_1 C_2 \exp(-E/RT_f^* T) \tag{2}$$

in which E is the activation energy, R is the universal gas constant, Z is the pre-exponential factor, and ρ is the density of the gas. The dependence of ρ on temperature is taken in the form:

$$\rho = \rho_f/T \tag{3}$$

in which ρ_f is the density at T_f^*.

We usually find a strong temperature dependence for the reaction rate, that is, $N = E/RT_f^* >> 1$.

It will be shown below that in order to satisfy the requirements of asymptotic matching [5] the magnitude of the flame front radius r_f is of the order $N\exp(N/2)$ in comparison to the flame radius at zero activation energy. With this condition in mind we define the thermal flame thickness ℓ_T as:

$$\ell_T = N\exp\frac{N}{2} \sqrt{\lambda_f \left(ZC_o c_p \rho_f^2\right)^{-1}} \tag{4}$$

3. Theoretical Analysis

The structure of a stationary diffusion flame is described by a system of equations for the diffusion of the reacting components and the equation for thermal conduction. For the dimensionless variables chosen above, these equations for the case of spherical symmetry have the form:

$$\frac{1}{r^2}\frac{d}{dr} r^2\kappa(T)\frac{dT}{dr} + K\Phi(T,C_1,C_2) = 0 \tag{5}$$

$$\frac{1}{L_1}\frac{1}{r^2}\frac{d}{dr} r^2\kappa(T)\frac{dC_1}{dr} - \Phi(T,C_1,C_2) = 0 \tag{6}$$

$$\frac{1}{L_2}\frac{1}{r^2}\frac{d}{dr} r^2\kappa(T)\frac{dC_2}{dr} - \Phi(T,C_1,C_2) = 0 \tag{7}$$

Here (T) is the dimensionless thermal conductivity in units of λ_f. L_1 and L_2 are the Lewis numbers corresponding to diffusion of the fuel and oxidant, respectively. $K = GC_o^*/T_f^*c_p$ in which G is the heat release of the reaction and Φ is the dimensionless chemical reaction rate:

$$\Phi = N^2 \exp N C_1 C_2 T^{-2} \exp(-N/T) \tag{8}$$

A solution for system (5)-(7) must satisfy the following boundary conditions:

$$T(+\infty) = \varepsilon \quad ; \quad T(0) = 1 \tag{9}$$

$$C_1(+\infty) = 0 \quad ; \quad -\kappa\frac{1}{L_1} r^2 \frac{dC_1}{dr} \to q \quad \text{as } r\to 0 \tag{10}$$

$$C_2(+\infty) = 1 \quad ; \quad C_2(0) = 0 \tag{11}$$

Here, $4\pi q$ is the dimensionless source intensity of the combustible component which is related to the source intensity Q by the formula

$$q = Qc_p/C_o^*\ell_T\lambda_f 4\pi \tag{12}$$

The system of Eqs. (5)-(7) along with boundary conditions (9)-(11) fully defines the mathematical framework of the problem.

The zone of the major heat evolution and material conversion is located in the vicinity of $r=r_f$. As will be confirmed by satisfying the principle of asymptotic matching, the width of this zone is of the order of 1/N relative to the thermal thickness of the flame. At a point far from the reaction when $|r-r_f|\gg(1/N)$, the reaction zone with accuracy to an error exponentially small compared to 1/N may be considered a source concentrated on the surface of a sphere of radius $r=r_f$, hence, the reaction rate in Eqs. (5)-(7) may be replaced by the corresponding surface δ-function

$$\Phi \to A\delta(r-r_f) \tag{13}$$

in which A is the intensity of the surface source, the definition of which requires examination of the structure of the reaction zone. In examining system (5)-(7) in conjunction with substitution (13) and boundary conditions (9)-(11), it is not difficult to find that at $r<r_f$

$$T = 1 \quad ; \quad C_2 = 0 \quad ; \quad \frac{dC_1}{dr} = -\frac{qL_1}{r^2\kappa} \tag{14}$$

while at $r>r_f$

$$\left.\begin{aligned} &\int_1^T \kappa(T)dT = \left(\frac{r_f}{r} - 1\right)\int_\varepsilon^1 \kappa(T)dt \\ &C_1 = C_f\frac{T-\varepsilon}{1-\varepsilon} \quad ; \quad C_2 = \frac{1-T}{1-\varepsilon} \end{aligned}\right\} \tag{15}$$

at $r=r_f$

$$\frac{1}{K}\left[\frac{dT}{dr}\right] + A = 0 \quad ; \quad \frac{1}{L_1}\left[\frac{dC_1}{dr}\right] - A = 0 \quad ; \quad \frac{1}{L_2}\left[\frac{dC_2}{dr}\right] - A = 0 \tag{16}$$

Hence,

$$\frac{1}{L_1}\left[\frac{dC_1}{dr}\right] = \frac{1}{L_2}\left[\frac{dC_2}{dr}\right] \quad ; \quad \frac{1}{K}\left[\frac{dT}{dr}\right] = -\frac{1}{L_2}\left[\frac{dC_2}{dr}\right] \tag{17}$$

Substituting expressions (14) and (15) into (17), we obtain

$$K = L_2(1-\varepsilon) \tag{18}$$

$$C_f = \frac{qL_1}{\nu r_f} - \frac{L_1}{L_2} \quad ; \quad \nu = \frac{1}{1-\varepsilon}\int_{\varepsilon}^{1} \kappa(T)dT \tag{19}$$

From Eq. (18) written for dimensional variables the value of T_f^* is obtained:

$$T_f^* = T_o^* + GC_o^*/c_pL_2 \tag{20}$$

Now we have completely determined the values of ℓ_T, N, q, K, and ε which depend on the front temperature T_f^*.

In order to find r_f and C_f, we turn to an examination of the inner problem related to the structure of the reaction zone. We introduce the corresponding inner variable

$$X = (r-r_f)/N \tag{21}$$

Temperature and concentrations in the region of the reaction zone are sought in the form of the following asymptotic expansions

$$T = 1 + \frac{1}{N}\Theta(X) + \ldots, \quad C_2 = \frac{1}{N}S_2(X) + \ldots, \quad C_1 = C_f + \frac{1}{N}S_1(N) + \ldots \tag{22}$$

Then, for the first approximation of the inner problem, system (5)-(7) together with (18) takes the form:

$$\frac{d^2\Theta}{dX^2} + L_2(1-\varepsilon)C_fS_2\exp\Theta = 0 \tag{23}$$

$$\frac{d^2S_2}{dX^2} - L_2C_fS_2\exp\Theta = 0 \tag{24}$$

$$\frac{d^2S_1}{dX^2} - L_1C_fS_2\exp\Theta = 0 \tag{25}$$

The last equation of system (23)-(25) is actually not significant because Θ and S_2 may be determined independently of S_1 from the first two equations of the system.

The conditions for matching with the outer solution (14) and (15) are the boundary conditions for (23) and (24).

$$S_2(-\infty) = \Theta(-\infty) = 0 \tag{26}$$

$$\frac{d\Theta}{dX}(+\infty) = \frac{dT}{dr}\bigg|_{r=r_f+0} = -\frac{\nu(1-\varepsilon)}{r_f} \tag{27}$$

$$\frac{dS_2}{dX}(+\infty) = \frac{dC_2}{dr}\bigg|_{r=r_f+0} = \frac{\nu}{r_f} \tag{28}$$

Eliminating the rate of chemical reaction from (23) and (24) and taking into account (26) we obtain

$$\Theta + (1-\varepsilon)S_2 = 0 \tag{29}$$

Hence, Eq. (23) takes the form:

$$\frac{d^2\Theta}{dX^2} - L_2C_f\Theta\exp\Theta = 0 \tag{30}$$

Taking (26) into account, we have

$$\left(\frac{d\Theta}{dX}\right)^2\bigg|_{X=+\infty} = 2L_2C_f \tag{31}$$

Considering Eqs. (31), (27), and (19), we obtain an equation for the determination of r_f

$$\frac{1}{r_f}^2 - 2\frac{qL_1L_2}{\nu^3(1-\varepsilon)^2}\left(\frac{1}{r_f}\right) + \frac{2L_1}{\nu^2(1-\varepsilon)^2} = 0 \tag{32}$$

Eq. (32) has two roots, but it is obvious that only the solution for which r_f increases with an increase in flow q has physical significance, thus:

$$\frac{1}{r_f} = \frac{qL_1L_2}{\nu^3(1-\varepsilon)^2} - \sqrt{\frac{q^2L_1^2L_2^2}{\nu^6(1-\varepsilon)^4} - \frac{2L_1}{\nu^2(1-\varepsilon)^2}} \tag{33}$$

From Eq. (33), it is easily seen that for a decrease in q, the front radius r_f diminishes. At a certain critical flow q_c and the correspondirg radius r_c, the problem ceases to have a stationary solution and the flame is extinguished (Figure 2).

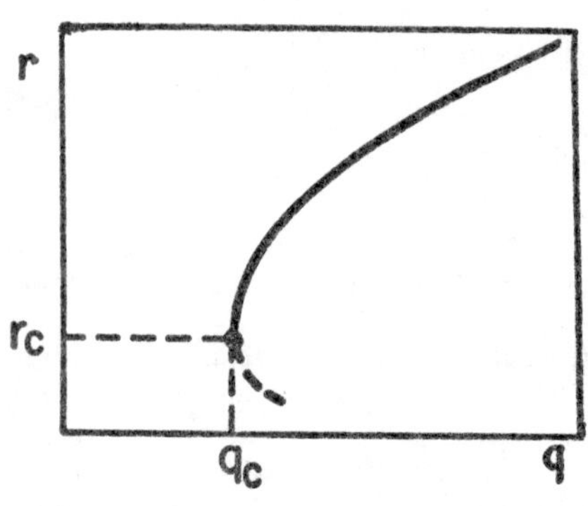

Figure 2 : Flame front radius as a function of the fuel source intensity.

From Eq. (33) we get:

$$r_c = \frac{\nu(1-\varepsilon)}{\sqrt{2L_1}} \qquad ; \qquad q_c = \frac{\nu^2(1-\varepsilon)}{L_2}\sqrt{2/L_1} \qquad (34)$$

The flow of the uncombusted portion of the fuel passing through the reaction zone is non-zero and equals $q-\nu r_f/L_2$. With an increase in q, this fuel loss vanishes completely and the relationship between q and r_f takes the simple form of

$$r_f \simeq L_2 q/\nu \qquad \text{as} \quad q\to\infty \qquad (35)$$

Conclusions

The problem presented may be considered as a quasi-stationary model for the description of the combustion of a drop. During the course of combustion, the drop diminishes, the rate of combustible gas evaporating from the surface of the drop decreases and at a certain finite value we arrive at conditions under which the continuation of combustion is impossible. Not all the combustible gas evaporating from the surface of the drop is consumed in the reaction zone. A portion of the combustible component passes through the flame front, does not react with the oxidant, and goes off into the surrounding space. For sufficiently large burning drops, the fuel loss is negligibly small but this loss increases with decreasing drop size.

References

1. B. Lewis and G. Von Elbe: *Combustion, Flames, and Explosions of Gases.* Academic Press, New York and London, 1961.

2. F. A. Williams: *Combustion Theory.* Addison Wesley, Massachusetts and London, 1965.

3. Ya. B. Zeldovich: On Combustion of Non-mixed Gases (O gorenii neperemeshannykh gazov). *Zurn. Techn. Fiz.*, 19, No. 6, (1949).

4. D. B. Spalding: A Theory of Extinction of Diffusion Flames. *Fuel*, 33, No. 3, 255-273, (1954).

5. M. Van Dyke: *Perturbation Methods in Fluid Mechanics.* Academic Press, New York, London, 1964.

INDEX